THE UNITED STATES OF AMERICA

ORIGINAL
TO BE GIVEN TO
THE PERSON NATURALIZED

No. 4763279

CERTIFICATE OF NATURALIZATION

Petition No. 47274

Personal description of holder as of date of naturalization Age 63 years; sex Female color White; complexion Fair color of eyes Hazel; color of hair Gray; height 5 feet 8½ inches; weight 130 pounds; visible distinctive marks none
Marital status Married former nationality France

I certify that the description above given is true, and that the photograph affixed hereto is a likeness of me.

Sign here

Consuelo Vanderbilt Balsan
(Complete and true signature of holder)

(SECURELY AND PERMANENTLY AFFIX PHOTOGRAPH HERE)

Consuelo Vanderbilt-Balsan

UNITED STATES OF AMERICA
DISTRICT OF NEW JERSEY } ss:

Be it known, that CONSUELO VANDERBILT BALSAN then residing at Lentana, Florida having petitioned to be admitted a citizen of the United States of America, and at a term of the District Court of The United States held pursuant to law at Newark on November 8th 1940 the court having found that the petitioner intends to reside permanently in the United States, had in all respects complied with the Naturalization Laws of the United States in such case applicable, and was entitled to be so admitted, the court thereupon ordered that the petitioner be admitted as a citizen of the United States of America.

In testimony whereof the seal of the court is hereunto affixed this 8th day of November in the year of our Lord nineteen hundred and Forty and of our Independence the one hundred and Sixty-Fifth.

Seal

Benjamin F. Havens,
Clerk of the U. S. District Court.
By [signature] Deputy Clerk.

DEPARTMENT OF LABOR

CONSUELO AND ALVA VANDERBILT

CONSUELO AND ALVA VANDERBILT

The Story of a Daughter and a Mother in the Gilded Age

Amanda Mackenzie Stuart

HarperCollins*Publishers*

HarperCollins books may be purchased for educational, business, or sales promotional use. For information, please write: Special Markets Department, HarperCollins Publishers, 10 East 53rd Street, New York, NY 10022.

Originally published in 2005 in Great Britain by HarperCollins Publishers.

FIRST EDITION

Printed on acid-free paper

Library of Congress Cataloging-in-Publication Data

Stuart, Amanda Mackenzie.
[Consuelo and Alva]
Consuelo and Alva Vanderbilt : the story of a daughter and a mother in the Gilded Age / Amanda Mackenzie Stuart.—1st ed.
p. cm.
Originally published: Consuelo and Alva. London : HarperCollins, 2005.
Includes bibliographical references and index.
ISBN-13: 978-0-06-621418-4
ISBN-10: 0-06-621418-1
1.Balsan, Consuelo Vanderbilt. 2. Belmont, Alva, 1853–1933. 3. Mothers and daughters—United States—Biography. 4. Rich people—United States—Biography. 5. Suffragists—United States—Biography. 6. United States—Biography. 7. Great Britain—Biography. I. Title.

CT3150.B34S83 2006
973.8092'2—dc22
[B]

2005054805

06 07 08 09 10 RPL 10 9 8 7 6 5 4 3 2 1

To my daughters, Daisy and Marianna

CONTENTS

PART FOUR

LIST OF ILLUSTRATIONS

FIRST PICTURE SECTION

SECOND PICTURE SECTION

The Gold Room at Marble House © *Richard Cheek for the Preservation Society of Newport County*

Consuelo's bedroom at Marble House © *Richard Cheek for the Preservation Society of Newport County*

A photograph of a portrait of Alva by Benjamin Charles Porter, *courtesy of the Preservation Society of Newport County*

Consuelo styled as an English duchess by Carolus-Duran © *Jarrold Publishing/Blenheim Palace, reproduced by kind permission of the Duke of Marlborough*

The 9th Duke and Duchess of Marlborough with Their Sons the Marquess of Blandford and Lord Ivor Spencer-Churchill by John Singer Sargent, 1905 © *Jarrold Publishing/Blenheim Palace, reproduced by kind permission of the Duke of Marlborough*

The Duchess of Marlborough and Her Son, Lord Ivor Spencer-Churchill by Giovanni Boldini, 1906, *courtesy of Corbis*

The water terraces at Blenheim designed by Achille Duchêne © *Corbis*

Consuelo at home at Casa Alva, Palm Beach, May 1950 © *Toni Frissell, Getty Images/LIFE*

Consuelo and Jacques Balsan on Fisher Island, Maine, 1952, *courtesy of the Duke of Marlborough*

THIRD PICTURE SECTION

Consuelo at a bazaar at Stratford Town Hall, *courtesy of Serena Balfour*

Menu card for the dinner given by the Playground Association of America, 31 March 1908, *courtesy of Serena Balfour*

Consuelo surrounded by knights at Blenheim House, 1904, *courtesy of Serena Balfour*

The Marquess of Blandford, Lord Ivor Churchill and a friend in Wengen, 1912, *courtesy of Serena Balfour*

'Golf Miss Gilliat Facet, 1911', *courtesy of Serena Balfour*

ENDPAPERS

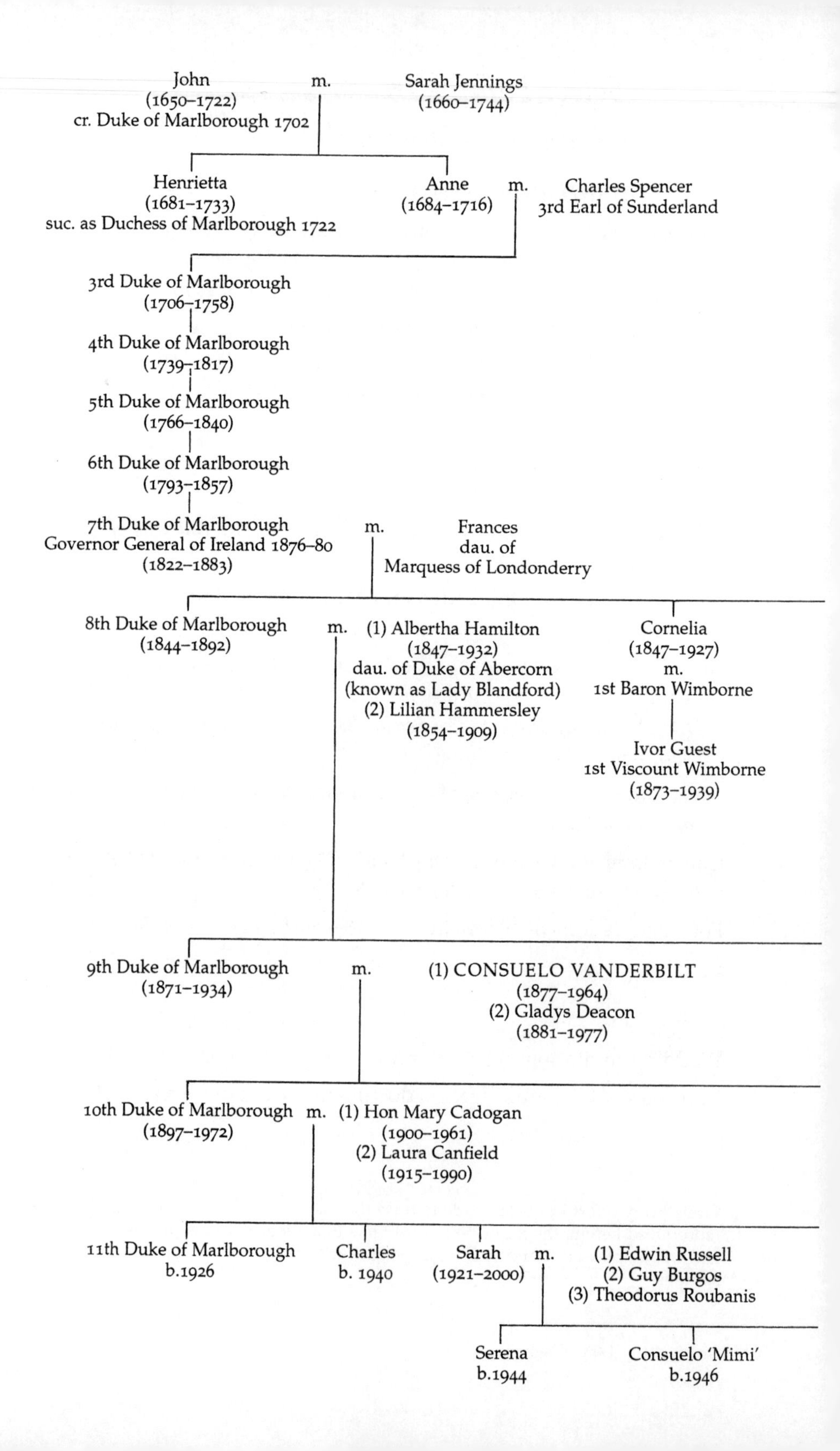
John
(1650–1722)
cr. Duke of Marlborough 1702
m.
Sarah Jennings
(1660–1744)
Henrietta
(1681–1733)
suc. as Duchess of Marlborough 1722
Anne
(1684–1716)
m.
Charles Spencer
3rd Earl of Sunderland
3rd Duke of Marlborough
(1706–1758)
4th Duke of Marlborough
(1739–1817)
5th Duke of Marlborough
(1766–1840)
6th Duke of Marlborough
(1793–1857)
7th Duke of Marlborough
Governor General of Ireland 1876–80
(1822–1883)
m.
Frances
dau. of
Marquess of Londonderry
8th Duke of Marlborough
(1844–1892)
m.
(1) Albertha Hamilton
(1847–1932)
dau. of Duke of Abercorn
(known as Lady Blandford)
(2) Lilian Hammersley
(1854–1909)
Cornelia
(1847–1927)
m.
1st Baron Wimborne
Ivor Guest
1st Viscount Wimborne
(1873–1939)
9th Duke of Marlborough
(1871–1934)
m.
(1) CONSUELO VANDERBILT
(1877–1964)
(2) Gladys Deacon
(1881–1977)
10th Duke of Marlborough
(1897–1972)
m.
(1) Hon Mary Cadogan
(1900–1961)
(2) Laura Canfield
(1915–1990)
11th Duke of Marlborough
b.1926
Charles
b. 1940
Sarah
(1921–2000)
m.
(1) Edwin Russell
(2) Guy Burgos
(3) Theodorus Roubanis
Serena
b.1944
Consuelo 'Mimi'
b.1946

Select Family Tree of the Spencer-Churchills mentioned in the text

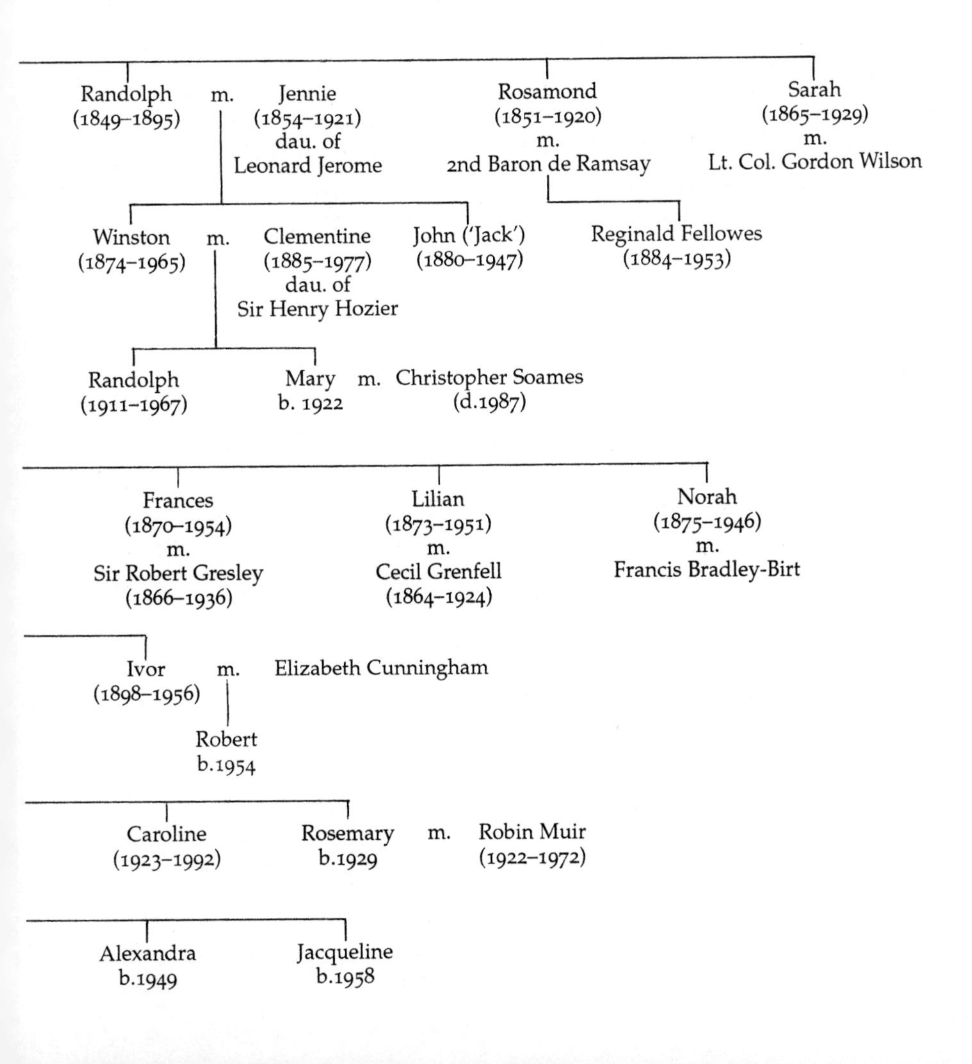

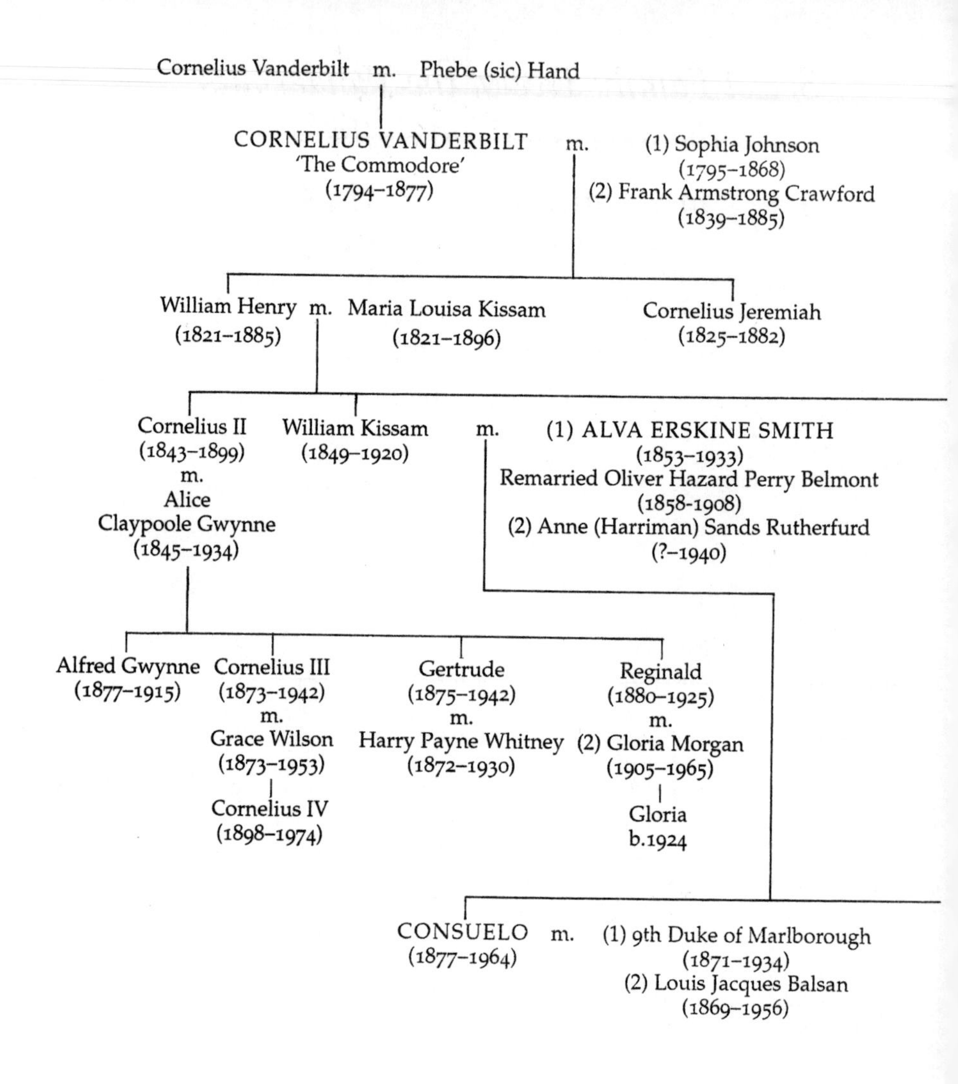
Cornelius Vanderbilt m. Phebe (sic) Hand
CORNELIUS VANDERBILT
'The Commodore'
(1794–1877)
m.
(1) Sophia Johnson
(1795–1868)
(2) Frank Armstrong Crawford
(1839–1885)
William Henry m. Maria Louisa Kissam
(1821–1885)
(1821–1896)
Cornelius Jeremiah
(1825–1882)
Cornelius II
(1843–1899)
m.
Alice
Claypoole Gwynne
(1845–1934)
William Kissam
(1849–1920)
m.
(1) ALVA ERSKINE SMITH
(1853–1933)
Remarried Oliver Hazard Perry Belmont
(1858-1908)
(2) Anne (Harriman) Sands Rutherfurd
(?–1940)
Alfred Gwynne
(1877–1915)
Cornelius III
(1873–1942)
m.
Grace Wilson
(1873–1953)
Cornelius IV
(1898–1974)
Gertrude
(1875–1942)
m.
Harry Payne Whitney
(1872–1930)
Reginald
(1880–1925)
m.
(2) Gloria Morgan
(1905–1965)
Gloria
b.1924
CONSUELO
(1877–1964)
m.
(1) 9th Duke of Marlborough
(1871–1934)
(2) Louis Jacques Balsan
(1869–1956)

Select Family Tree of the Vanderbilts mentioned in the text

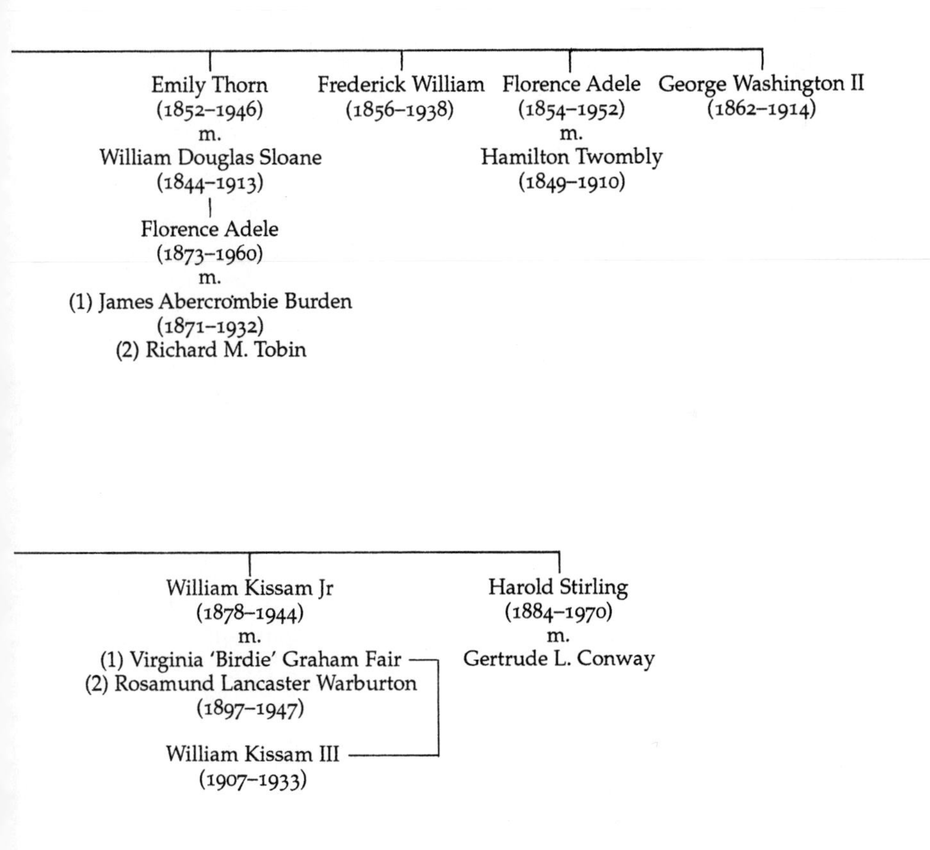

PREFACE

This book began with a story. Some time ago, I took my eighteen-year-old daughter and a young Australian friend to visit Blenheim Palace in Oxfordshire, in order that the young Australian could have a glimpse of English 'heritage' before she went home. The guides at Blenheim Palace are free to talk about its history as they please, but there is one tale which engages visitors powerfully – the story of Consuelo Vanderbilt, an American heiress said to have been compelled by her socially ambitious mother to marry the 9th Duke of Marlborough in 1895, bringing a generous Vanderbilt dowry to an English palace sorely in need. Anyone wandering through the state rooms of Blenheim soon encounters two very different portraits of Consuelo. The first, by Carolus-Duran, was painted when she was seventeen against a classical English landscape and suggests an enigmatic but dynamic young woman, as yet little more than a girl. The second portrait, far more uneasy and much more famous, was painted eleven years later by John Singer Sargent. Here, Consuelo and the 9th Duke have been placed in their historical context beneath a bust of John Churchill, 1st Duke of Marlborough, but at a distance from each other. Each inhabits a markedly private space linked only by their eldest son. Neither looks happy; but while the Duke gazes steadily outwards, Sargent has painted Consuelo glancing anxiously to one side with a striking air of melancholy.

It soon became apparent that our guide had little time for the 9th Duke. 'This is Sunny,' she said, gesturing at the Sargent portrait. 'Sunny by name but most certainly not Sunny by nature.' She glared severely at my daughter. 'Consuelo was your age when she came to Blenheim. You're probably still at school. But she got out in the end. Thank Heavens.'

Afterwards, the young Australian professed to be enthralled by English heritage, so we moved on to the nearby church of St Martin in Bladon, where Winston Churchill is buried in the churchyard with other members of the Spencer-Churchill family. Since his death, relatives buried alongside him have thoughtfully been re-defined so that visitors can understand the relationship with Churchill at a glance. In one grave in the corner of the plot, however, an inscription reads: 'Consuelo Vanderbilt wife of the ninth Duke.' On the other side, her headstone is inscribed: 'In loving memory of Consuelo Vanderbilt Balsan – mother of the tenth Duke of Marlborough – born 2nd March 1877 – died 6th December 1964.'

This was startling. Consuelo had clearly remarried. So why had she come back? Surely no-one had compelled her to burial in a Bladon churchyard, even if she had been forced to Blenheim Palace in life? This puzzle was soon replaced by another. Limited research revealed that there were writers who rejected the allegation that Consuelo had been forced to marry the Duke and there were those, even in her lifetime, who asserted the story was a flat lie. This was followed by a further conundrum. It emerged that Consuelo's mother, Alva, villainess-in-chief, eventually became a leader in the fight for women's suffrage in America. How could anyone square even rudimentary feminism with ordering her daughter to marry a duke? One writer suggested she might have undergone a conversion to suffragism as an act of penance, but even on superficial acquaintance, Alva Belmont (as she later became) did not seem the penitential type.

It soon became clear that an account of what had happened and why would have to explore Alva's life as well as Consuelo's but there were further complications. The story of Consuelo's first marriage had inspired others, notably Edith Wharton's last (unfinished) novel, *The Buccaneers*. There were obstacles in the way of non-fiction, however. Consuelo and Alva left few private papers, and surviving sources were far from impartial. Both made attempts at autobiography. Consuelo's memoir *The Glitter and the Gold* was published in 1952. Alva started her memoirs twice, once in 1917 and again after 1928, but neither version was completed. At the same time, the lives of both women were frequently the subject of comment in the press. Alva in particular encouraged this and intermittently attempted to influence and re-edit press narratives, including those relating to Consuelo. Mother and daughter spent a considerable part of their later lives thousands of miles apart, separated by the Atlantic, and both eventually preferred to be defined by events and activities beyond Consuelo's life as Duchess of Marlborough. In spite of these difficulties, however, their lives prove more illuminating side-by-side than taken singly. They continued to influence each other; their interests and tastes eventually converged; and they found themselves defined and bound together for ever by the story of Consuelo's first marriage.

CONSUELO AND ALVA VANDERBILT

Prologue

IN NOVEMBER 1895, shortly before the New York wedding of Consuelo Vanderbilt to the 9th Duke of Marlborough, her cousin Gertrude raged at her journal about the unhappy lot of heiresses. 'You don't know what the position of an heiress is! You can't imagine,' she wrote, nib scratching paper with anger. 'There is no one in all the world who loves her for herself. No one. She cannot do this, that and the other simply because she is known by sight and will be talked about . . . the world points at her and says "watch what she does, who she likes, who she sees, remember she is an heiress," and those who seem to forget this fact are those who really remember it most vividly.'[1]

Had she been in a position to read her cousin's journal, Miss Consuelo Vanderbilt would have agreed, for the world she knew was certainly watching and pointing. In the days leading up to 6 November 1895, preparations for her wedding dominated the front pages of New York's popular newspapers, relegating to second, third and fourth place the advancing popularity of bloomers as cycling dress, New York State elections and a war of independence in Cuba. One newspaper, the *New York World*, led the field in examining the bride-to-be, providing its readers with a helpful list of her most important characteristics:

Age: Eighteen years
Chin: Pointed, indicating vivacity
Color of hair: Black
Color of eyes: Dark brown
Eyebrows: Delicately arched
Nose: Rather slightly retroussé

Weight: One hundred and sixteen and one half pounds
Foot: Slender, with arched instep
Size of shoe: Number three. AA last
Length of foot: Eight and one half inches
Length of hand: Six inches
Waist measure: Twenty inches
Marriage settlement: $10,000,000
Ultimate fortune: $25,000,000 (estimated).[2]

As wedding preparations entered the final phase, a few enterprising passers-by near St Thomas Episcopalian Church on Fifth Avenue managed to peep in at the construction of the most spectacular wedding floral display ever assembled in a New York church: great flambeaux of pink and white roses on feathery palms at the end of pews; vaulting arches of asparagus fern, palm foliage and chrysanthemums; orchids suspended from the gallery; vines wound round the organ columns; floral gates constructed from small pink posies; and sweeping strands of lilies, ivy and holly swaggering from dome to floor, feats of festooning over 95-feet long. There was no shortage of other detail available for those with insufficient initiative – or interest – to seek it out for themselves. The press even provided lingering descriptions of the bridal underwear: 'It is delightful to know that the clasps of Miss Vanderbilt's stocking supporters are of gold, and that her corset-covers and chemises are embroidered with rosebuds in relief,' said the society magazine *Town Topics*. 'If the present methods of reporting the movements and details of the life and clothes of these young people are pursued until the day of the wedding, I look for some revelations that would startle even a Parisian café lounger.'[3]

On the morning of 6 November, it was soon apparent that uninhibited staring was the order of the day. The wedding was at noon. At 72nd Street and Madison, where the bride was dressing at her mother's house, the crowds started to gather before 9 a.m. and soon lined the entire route to St Thomas Church, twenty blocks away at 53rd Street and Fifth Avenue. At 72nd Street no-one was allowed within a hundred and fifty feet of Alva Vanderbilt's new home, but all kinds of tactics were deployed by certain individuals determined to subvert the rules. By 10.30 a.m. the crowd had grown to around two thousand people, and windows in every house in the neighbourhood were crammed with spectators enjoying small private parties of their own. Curiosity was not confined to the common herd. Gertrude Vanderbilt's outburst at the manner in which heiresses were watched was more than vindicated for lorgnettes were much in evidence and the *New York Times* noted that halfway down the block towards Park Avenue, women could be seen standing in bay windows peering out through opera glasses for glimpses of the bridal party.[4]

As the morning wore on, crowd control outside St Thomas Church became increasingly difficult. Reporters fell back on military metaphor. Acting Inspector Cartwright was in overall command of forces; Captain Strauss was deployed with his platoon at the home of the bride; further detachments were stationed by the church preventing incursions on the left flank; and all faced a difficult and unpredictable enemy. 'Picture to yourself a space 15-feet wide and 100-feet long,' said the *New York World*, 'tightly packed with young women, old women, pretty women, ugly women, fat women, thin women – all struggling and pushing and squeezing to break through police lines. Then imagine all those women quarrelling one with another, then struggling and pushing and squeezing and begging, imploring, threatening and coaxing the police to let them pass.'[5]

The most pressing problem for Inspector Cartwright was that some of the women came from good homes which put him in an invidious position and obliged him to give orders that clubs should not be used. 'Had you seen those policemen pressed to desperation turn and push that throng back inch by inch, half crushing an arm here, poking a waist or a neck there, collecting three women in an armful and hurling them back, the crowd would have reminded you of nothing so much as an obstreperous herd of cattle,' said one reporter.[6]

There were compelling reasons for such intense curiosity. In an age of international marriages, of trade between American money and European titles, this was the grandest. The *New York World* went straight to the point: 'Miss Consuelo Vanderbilt is one of the greatest heiresses in America. The Duke of Marlborough is probably the most eligible peer in Great Britain . . . From the standpoint of Fifth Avenue it will be the most desirable alliance ever made by an American heiress up to date.'[7] The engagement was presented by others as an all-American tale that held out hope for everyone, even the poorest. 'The world is actively engaged in making its fortune,' said Frank Lewis Ford in an article in *Munsey's Magazine*. 'Though sometimes it calls the task by another name and says that it is earning its living, acquiring a competency, building up a business, or what not. And there stand the Vanderbilts, with living earned, competency acquired, business built up, fortune made.'[8]

But to those in the know – and there were many people in the know thanks to the nature of the press in late-nineteenth-century New York – this was all too simple. The newspapers buzzed with rumours. Was it true that this wedding would break the heart of at least one eligible American bachelor? Was the English Duke simply marrying the eighteen-year-old bride for her money? And where were the Vanderbilts? Was it really possible that none of them had been invited? In America in the 1890s the lives of the very rich provided community entertainment supplanted later by cinema and television. This wedding alone was a soap opera with enough story lines to satisfy the most avid audience, laced with the faint but thrilling possibility of further drama at the church door. Might the eligible bachelor appear? Would rogue Vanderbilts attempt to force an entry? Could there be a misguided attempt to kidnap Consuelo to prevent her from marrying a blackguard, and stop her fortune from leaving the country? It was certainly worth waiting around to find out.

The wedding guests seemed just as determined to make the most of the occasion as the crowd. Some of them arrived as early as 10 a.m., a full two hours before the ceremony was due to start. This presented the embattled Inspector Cartwright with another headache because those in the crowd who had arrived early to obtain a good vantage point now refused to make way for the wedding guests, and had to be forced back inch by inch across the street. The guests, meanwhile, were required to dismount from their carriages and make their way to the church on foot. Here there were further difficulties, for the doors of the church were still tightly closed, creating a genteel but tense scrum as the *ton* barged each other out of the way. When St Thomas Church finally opened its doors, the gentlemen ushers, selected by Alva for their experience in seating guests at weddings, had to call for police reinforcements. The guests rushed past the sexton, whose task it was to match invitations to faces and expel interlopers. They ignored the efforts of the ushers to place them according to Mrs Vanderbilt's list. Some of them (women again) climbed over the floral decorations, deposited themselves in the central aisle reserved for the guests of honour and were only removed to less prestigious seating by the gentlemen ushers with the utmost difficulty. Others stood on their pews to peer at fellow guests over the foliage – the floral display was widely held to be a great success but it undeniably obscured the sight lines.

Inside the church, official entertainment began with an organ recital from Dr George Warren, as sixty members of Walter Damrosch's Symphony Orchestra moved into their desks. From 11 a.m., the orchestra played through a musical equivalent of the floral display in a programme at which even commercial classical radio stations might now baulk: the Prize Song from Wagner's *Die Meistersingers*; the Bridal Chorus from *Lohengrin*; Beethoven's Leonore Overture No. 3; Tchaikovsky's Andante Cantabile String Quartet; the Grand March from Wagner's *Tannhauser*; Fugue in G Minor by Bach; and the Overture to *A Mid-Summer Night's Dream* by Mendelssohn, all in the space of fifty-five minutes. Fifty members of the choir took their seats in the choir stalls, together with four distinguished New York soloists, and the guests of honour began to appear.

Mrs Astor arrived first with Mr and Mrs John Jacob. She was escorted to her seat by Reginald Ronald and looked particularly splendid in a costume of grey completed by a velvet coat and a black toque with white satin rosettes. Mrs William Jay, close friend of Mrs Vanderbilt, followed closely behind with Colonel Jay, wearing one of the most 'effective' costumes of the wedding – a heavy black silk skirt, cut very full and with a bodice of yellow and crimson satin. A peal of church bells announced the arrival of the mother-of-the-bride, Alva Vanderbilt, with the bridesmaids. Miss Duer, Miss Morton, Miss Winthrop, Miss Burden, Miss Jay, Miss Goelet and Miss Bronson were 'all exceedingly pretty young women' gushed the reporter in charge of bridesmaids for *The New York Times*.[9] They wore gowns of ivory satin, with magnificent broad-brimmed velvet hats, topped with ostrich feathers, and around-the-throat bands of blue velvet with over-strings of pearls. All heads swivelled as Mrs Vanderbilt came up the aisle, accompanied by the bride's brothers, Willie K. Jr and Harold. She was wearing a gown of pale blue satin edged with sable, a hat of lace and silver with a pale blue aigrette, and had 'a decided look of satisfaction on her face'.[10]

Alva Vanderbilt's smug expression was not to last. Her arrival signalled the imminent appearance of the bride, escorted by her father, William K. Vanderbilt. A row of fashionable bishops appeared from the vestry room and took their places in the chancel. The Duke of Marlborough, who had slipped into St Thomas Church unnoticed with his best man, Ivor Guest, took up his position and waited, incidentally giving the wedding guests a good opportunity to stare at him instead. One of them told a newspaper that 'some of the guests partially rose in their seats to have a better view of the Lord of Blenheim. Some were asking where his relatives were and others answered that they were represented by the British Embassy.'[11] This was true – the groom's side was represented by the British ambassador, supported by two gentlemen from the embassy in Washington, Bax Ironsides and Lord Westmeath.

Along the New York streets and in St Thomas Church they waited. The ushers sauntered up to their stations, three on one side of the central aisle and three on the other – then sauntered back to the church porch again. Mr Damrosch, who had completed his concert programme, beat time with his baton in silence, his head turned round towards the church door. The Duke of Marlborough began to fidget nervously, and only regained some of his composure when he noted the English sang-froid of his best man. As the delay lengthened, the guests shuffled and whispered. Alva was observed looking uncharacteristically worried. And then decidedly strained. Five minutes passed . . . then ten . . . then twenty . . . and still the bride had not appeared.

PART ONE

I

The family of the bride

As the delay lengthened and nervousness grew, self-appointed society experts in the crowd had time to debate one important question: had anyone seen the Vanderbilt family, whose apotheosis this was alleged to be? There could be no dispute that Vanderbilt gold was a powerful chemical element at work in St Thomas Church on 6 November 1895. It had given the bride her singular aura; it had drawn a duke from England; and without it, Alva would not now be waiting anxiously for her daughter to arrive. Scintillae of Vanderbilt gold dust brushed everything on the morning of Consuelo's wedding, from the fronds of asparagus fern to the glinting lorgnettes in the crowd outside. It was remarkable, therefore, that apart from the father-of-the bride, its chief purveyors should be so conspicuously absent; and even more striking that this scarcely mattered because of the force of character of the bride's Vanderbilt great-grandfather, whose ancestral shade still hovered over the players in the morning's drama as if he were alive.

Cornelius Vanderbilt, founder of the House of Vanderbilt, lingered in the collective memory partly because he laid down the basis of the family's extraordinary wealth; and partly because of the robust manner in which he did it. He died in 1877, a few weeks before Consuelo was born, but he left a complex legacy and no examination of the lives of Alva and Consuelo is complete without first exploring it.

* * *

Fable attached itself to Cornelius Vanderbilt, known as 'Commodore' Vanderbilt, Head of the House of Vanderbilt, even in his lifetime. He generally did little to discourage this, but one misconception that irritated him was that the Vanderbilts were a 'new' family and he embarked on genealogical research to prove his point. However, he held matters up for several years by placing an advertisement in a Dutch newspaper in 1868 which read: 'Where and who are the Dutch relations of the Vanderbilts?', causing such offence that none of the Dutch relations could bring themselves to reply.[1]

More tactful experts later traced the Vanderbilts' roots back to one Jan Aertson from the Bild in Holland who arrived in America around 1650. A lowly member of the social hierarchy exported by the Dutch West India Company, Jan Aertson Van Der Bilt worked as an indentured servant to pay for his passage and then acquired a *bowerie* or small farm in Flatbush, Long Island. His descendants traded land from Algonquin Indians on Staten Island, starting a long association between the Vanderbilts and the Staten Island community of New Dorp. They also joined the Protestant Moravian sect, whose members fled from persecution in Europe in the early-eighteenth century and settled nearby. The Vanderbilt family mausoleum is to be found at the peaceful and beautiful Moravian cemetery at New Dorp on Staten Island to this day.[2]

In a development that goes against the grain of immigration success stories, the Vanderbilt family arrived in America early enough to suffer a downturn in its fortunes in the mid-eighteenth century. Just at the point when the Staten Island farm became prosperous, it was repeatedly sub-divided by inheritance and by the time the Commodore was born in 1794, his father was scratching a subsistence living on a small plot, and ferrying vegetables to market on a *periauger*, a flat-bottomed sailing boat evolved from the Dutch canal scow. Historians of the family portray this Vanderbilt of pre-history as feckless and inclined to impractical schemes, but he compensated for his deficiencies by marrying a strong-minded, hard-working, frugal wife of English descent, Phebe Hands. Her family had also been ruined, by a disastrous investment in

Continental bonds. They had nine children. The Commodore was their eldest son.

Circumstances thus conspired to provide the Commodore with what are now known to be many of the most common characteristics in the background of a great entrepreneur: a weak father and a 'frontier mother'; a marked dislike of formal education (he hated school and spelt 'according to common sense'); and a humble background.[3] A humble background is almost mandatory in nineteenth-century American myth-making about the virtuous self-made man, but it was a characteristic the Commodore genuinely shared with others such as John Jacob Astor, Alexander T. Stewart and Jay Gould. After his death the Commodore was accused of being phrenologically challenged with a 'bump of acquisitiveness' in a 'chronic state of inflammation all the time', but he was not alone in finding that childhood poverty and near illiteracy ignited a very fierce flame.[4] More unusually for a great entrepreneur, the Commodore was neither small nor puny. He developed enormous physical strength, accompanied by strong-boned good looks, a notorious set of flying fists and a streak of rabid competitiveness. Charismatic vigour, combined with a lurking potential for violence, made him a force to be reckoned with from an early age and even as a youth he developed a reputation for epic profanity and colourful aggression that never left him.

The Vanderbilt fortune was made in transportation. Its origins lay in the first regular Staten Island ferry service to Manhattan, started by the Commodore in a *periauger*, under sail, while he was still in his teens. From there his career reads like a successful case study in a textbook for business students. He ploughed back the profits from his first *periauger* ferry service until he owned a fleet. He expanded into other waters and bought coasting schooners. Then, when others had taken the risk out of steamship technology, he sold his sailing ships and embraced the age of steam, founding the Dispatch Line and acquiring the nickname 'Commodore' as he built it up.

The Dispatch Line ran safer and faster steamships than any of its competitors to Albany up the Hudson, and along the New

England coast as far as Boston up Long Island Sound disembarking at Norwalk, New Haven, Connecticut and Providence. Between 1829 and 1835, the Commodore moved easily into the role of capitalist entrepreneur, profiting from the impulse to move and explore as waves of immigrants fanned out and built a new country. By 1845 he began to appear on 'rich lists.' *The Wealth and Biography of the Wealthy Citizens of New York City*, compiled by Moses Yale Beach that year estimated his fortune at $1.2 million and added 'of an old Dutch root, Cornelius has evinced more energy and go-aheaditiveness in building and driving steamboats and other projects than ever one single Dutchman possessed'[5]. The size of the Commodore's fortune is particularly remarkable when one considers that the word 'millionaire' was only coined by journalists in 1843 to describe the estate left by the first of them, the tobacconist Peter Lorillard. In 1845, the *millionaire* phenomenon was still so rare that the word was printed in italics and pronounced with rolling 'rs' in a flamboyant French accent.[6]

It was only in the last quarter of his life, between the mid-1860s and his death in 1877, that the Commodore moved into railroads – the industry with which the Vanderbilts are usually associated (this came after an adventure in Nicaragua where he forced a steamship up the Greytown River to open a trans-American Gold Rush passage in 1850, fomented a civil war and took his bank account to $11 million by 1853). It took much to convince the Commodore that railroads were the future: 30,000 miles of railroad track had to be laid by others before he accepted that the argument was won. Once convinced, he divested himself of his steamships and began buying up railroads converging on New York in a spectacular series of stock manipulations or 'corners' at which he proved extremely inventive and adept.[7]

The Commodore's accounting methods may have been pre-industrial – he kept his accounts either in an old cigar box or in his head – but the enterprise he created made him a pioneer of industrial capitalism. He was also a master of timing. The American Civil War (1861–65) confirmed the absolute dominance of the railways at the heart of the growing US economy. On 20 May 1869, he secured the right to consolidate all his railroads into the New York Central, and recapitalised the stock at almost twice its previous market value. He was much abused for inventing the practice of issuing extra stock capitalised against future earnings, or 'watering' as it was known, but the practice has since become not only standard practice but a key instrument of modern finance.[8]

The first version of Grand Central Terminal in New York, which opened in 1871, had serious design flaws, rather like the Commodore himself. Blithely ignoring the impact on the local community, trains ran down the middle of neighbouring streets and passengers wishing to change lines had to dodge moving locomotives and dive across the tracks in all weather conditions (in old age it amused the Commodore to play 'chicken' in front of oncoming trains). But this was the first American railroad terminal and it encapsulated an extraordinary achievement. 'A powerful image in American letters,' writes the historian Kurt Schlichting, 'depicts a youth moving from a rural farm or small town to the big city, seeking fame or fortune . . . As the train arrives, the protagonist confronts the energy and chaos of the new urban society . . . Great railroad terminals like Grand Central provided the stage for this unfolding drama, as a rural, agrarian society urbanized.'[9] In constructing the first version of the Grand Central Terminal (his statue still stands outside the 1913 version), the Commodore constructed a metaphor for both his own life and the industrialisation of America.

That, at any rate, was the business history, the journey from ferryman to railroad king which his first biographer, William Croffut, believed should be held up as an inspiring example to the young. This is not the whole truth, however. He may have been a great entrepreneur, but the Commodore had some disconcerting domestic habits. He is alleged to have consigned his wife, Sophia, and his epileptic son, Cornelius Jeremiah, to a lunatic asylum when they stood up to him; he was rumoured to be a womaniser, especially with the notorious Claflin sisters who were eventually prosecuted for obscenity (though not with him); and he dabbled in spiritualism for advance news from the spirit world on stock prices. There were many who objected to his meanness, for his habit of

Dutch frugality was steadfastly extended to anything approaching a philanthropic gesture until very close to his death. 'Go and surprise the whole country by doing something right,' wrote Mark Twain in despair in 1869.[10] New York society preferred to keep him at a distance too. There was, it was felt, altogether too much of the farmyard about him. He was even rumoured to have spat out tobacco plugs on Mrs Van Rensselaer's carpet.

Alva later wrote that the Commodore was a charming old man. She came to know him in the last three years of his life, understood that he was a visionary and refused to be cowed by him which he always liked.[11] He was certainly forthright, as a love letter he penned to the young woman who would become his second wife demonstrated: 'I hope you will continue to improve for all time,' he wrote after she had been ill. 'Until you turn the scale when 125 pounds is on the opposite balance. This is weight enough for your beautiful figure.'[12] It may be untrue that the Commodore spat out tobacco plugs, but he was not the only industrial tycoon unwilling to tone down a forceful style for genteel drawing rooms. Occasionally he was gripped by the urge to show off to snootier members of New York society but even this lacked conviction. In 1853 he commissioned the largest ocean-going yacht the world had ever seen, the *North Star*, and took his first wife and family to Europe. But though they were greeted by grand dukes, sculpted by Hiram Powers and painted by Joel Tanner Hart, he sold the yacht to the United Mail when he went home, never repeated the experiment and returned to what he enjoyed most, which was making money and doing down his competitors.

Unsurprisingly, the Commodore's wealth inspired great jealousy as well as admiration. Some of the stories about his coarse behaviour came from his aforementioned competitors, and from embittered members of his own family contesting his will; he may also have played up to his image as a farmyard peasant when it suited him. Whatever the explanation, Frank Crowinshield could still write in 1941: 'The most persistent myth concerning the family was that they were all, if not boors exactly, at any rate unused to the social amenities. The myth was so pervasive that one may still hear it from the lips of decrepit New Yorkers who, in discreet whispers, recite the risks their fathers ran in crossing the portals . . . Such people still speak as though their sires had risked calling upon Attila, or visiting, without benefit of axe or bludgeon, the dread caves of the anthropophagi.'[13]

The force of the Commodore's personality was so great that it affected society's perception of his children and grandchildren. For his own part, he left no-one in any doubt that his sons were a disappointment to him, and he was much exercised about the best way to hand on his great fortune until he felt he had solved the problem in the 1870s. There was naturally no question of giving any kind of financial control to his daughters; his favourite son died of malarial fever during the Civil War; and Cornelius Jeremiah, who not only suffered from epilepsy but also an addiction to gambling, was regarded as beyond redemption. This left Consuelo's grandfather, William Henry Vanderbilt, who was treated with utter contempt well into middle-age, and was habitually addressed as 'blatherskite', not to mention 'beetlehead'. William Henry – or 'Billy' as he was known to his family – made matters worse by kowtowing to his father at every turn. Even on the *North Star* cruise, he responded to the Commodore's offer of $10,000 if he would give up smoking by refusing the money saying: 'Your wish is sufficient,' and flinging his cigar overboard. This tactic was so perfectly calibrated to irritate the Commodore that he slowly lit a cigar of his own and blew smoke in his son's face.[14]

William Henry was a far more careful, painstaking and methodical man than the Commodore, showing little of the latter's startling entrepreneurial flair – one of many causes of the Commodore's profound scorn. During his early career at a banking house, William Henry worked himself into a state of nervous collapse, attracting further contumely, and was promptly expelled with his wife Maria Kissam to work a small and difficult farm on Staten Island. (The Kissams were an old and distinguished family, and although Alva Vanderbilt later claimed to have propelled the Vanderbilts into society, this match could certainly have taken them into its outer circles if either party had been interested.)

On Staten Island, Maria Kissam Vanderbilt carried on the family tradition by producing a large family of her own – nine children in all. Three of her sons would later become Consuelo's famous building uncles: Cornelius II of The Breakers, Newport; Frederick of the Hyde Park mansion, New York; and George, who created the Biltmore Estate in North Carolina. The fourth – though not the youngest – was Consuelo's father, William Kissam Vanderbilt, often known as 'William K.'

In the family mythology, William Henry, the father of these sons, only finally won respect from the Commodore after many years with one of the double-crossing japes over a deal that nineteenth-century Vanderbilts seem to have enjoyed. It involved the definition of a scow-load of manure. William Henry offered to buy manure for his farm from his father's stables at $4 a load. The Commodore then saw him pile several loads on to one scow and asked him how many he had bought. 'How many?,' William Henry is said to have replied; 'One, of course! I never put but one load on a scow.'[15] Finally impressed that his son was capable of getting the better of him, the Commodore, who was a shareholder in the near-bankrupt Staten Island Railroad, decided to turn it over to William Henry to see what he could make of it. Within two years the Blatherskite had put the little railroad on a secure financial footing and proved his value in the only vocabulary the Commodore truly understood by turning worthless stocks into $175 a share.[16]

The Commodore then moved William Henry and his growing family back from Staten Island to New York, made him vice-president of the newly acquired Harlem and Hudson Railroad, and put him in charge of the daily operation of the lines. Once again, William Henry responded magnificently to the challenge, finding economies and efficiencies wherever he looked, whereupon his father made him vice-president of the New York Central after 1869. The Commodore remained in overall strategic control of the enterprise until the day he died, but increasingly left the day-to-day management to William Henry. In coming to trust his eldest son's managerial capabilities, the Commodore, always in the vanguard of entrepreneurial capitalism, grasped that the qualities needed to build a fortune were not the same as the qualities needed to maintain it. 'Any fool can make a fortune,' the Commodore is said to have told William Henry before he died. 'It takes a man of brains to hold on to it after it is made.'[17]

The difficult relationship that existed for so many years between the Commodore and William Henry would have repercussions for Consuelo: her father, William Kissam Vanderbilt (later one of the world's richest men) grew up in modest circumstances on Staten Island during the period when his parents were out of favour with the Commodore. *Munsey's Magazine* found this reassuring, thankful that the humble circumstances principle would hold good for another generation or two: 'The decline of ancestral vigour and the dissipation of inherited wealth, which sociologists claim is almost inevitable among the very rich, has doubtless been deferred for a very few generations, among the Vanderbilts, by the sturdy plainness in which William Henry had brought up his sons and daughters,'[18] it said pompously. This may have been true, but it also meant that William K. would spend much of his adult life having as little to do with sturdy plainness as possible, an attitude to life with considerable implications for his own children.

William K. was also raised in a very different atmosphere from his father, who was a kind and mild-mannered man, an affectionate husband and not in the least given to domestic tyranny. A charming painting of the William Henry Vanderbilt family by Seymour Guy in 1873 suggests a large family at ease with itself, and even allowing for polite obituarists and nineteenth-century sentimentality, there appears to have been none of the contemptuous atmosphere that blighted the youth of the Commodore's children. Maria Kissam came from a cultivated background and both she and her husband saw to it that their children were properly educated. Willie (as he was known) was taught by private tutors and his parents took the unusual step of sending him to Geneva in Switzerland for part of his education. According to architectural historians John Foreman and Robbe Pierce Stimson: 'Few Americans of the time possessed the means, let alone the inclination, to send their sons abroad to school. Willie became a true sophisticate at an early age. He was fluent in French, and a connoisseur of European culture, art, and manners. The scandal-mongering tabloids of the era loved to portray the Vanderbilts as coarse parvenus. However, the truth in the case of Willie's generation – and especially in the case of Willie himself – was precisely the opposite.'[19]

William K. Vanderbilt grew into an outstandingly good-looking young man who later became famous for his charm, hospitality and agreeable manners. Consuelo adored him. '[He] found life a happy adventure . . .,' she wrote. 'His pleasure was to see people happy.'[20] The problem for such a gregarious young Vanderbilt was that while his grandfather the Commodore was alive there was little possibility of making an entrée into New York society, or of enjoying a life of leisure. The Commodore's reputation as a vulgarian put paid to contact with New York's emerging social elite; and while his grandfather retained an iron grip on the family fortune, it was essential to behave as he wished. 'What you've got isn't worth anything unless you have got the power,' was one of the Commodore's favourite financial saws.[21] Even in 1875, two years before he died, he continued to strike fear into the heart of his relations. He believed that extravagance was a weakness, a sign that one was not responsible enough to inherit a cent. 'He's a bad boy,' he said of his son Cornelius Jeremiah. 'Money slips through his fingers like water through a sieve.'[22]

In 1868, William K., whom the Commodore liked, had no choice other than to join the family railroad enterprise alongside his brother Cornelius II and start learning the railroad business, some way down the hierarchy of the Hudson River Railroad. For a charming and sociable youth, this cannot always have been easy. For the time being, however, there was little alternative to assenting amiably to the Commodore's assertion that only '"hard and disagreeable work" would keep his grandsons from becoming "spoilt".'[23]

The person who would not only solve William K.'s problem but do much to change society's perception of the Vanderbilts was Alva, nee Erskine Smith, later Mrs Oliver H. P. Belmont and the mother-of-the-bride. The reasons why she was drawn to this challenge lay deep in her own background, about which there remain many misconceptions. She has variously been described – in even scholarly works – as the daughter of the wealthiest couple in Savannah, and so poor that she helped her father keep a boarding house in New York after the American Civil War.[24] Whereas Consuelo, her own daughter, thought that her mother was the daughter of a ruined plantation owner.[25] Some of this was Alva's fault for she often exaggerated to suit her purpose, particularly when it came to issues of status and power. Commodore Vanderbilt's disdain for New York society was particularly unusual; for many others nineteenth-century America was a time of straightforward struggle for social advantage. Alva was one of them, and she was not alone in claiming aristocratic genealogy to assist her case.

On her father's side, she maintained that her pedigree stretched back to Scotland, and the Earls of Stirling. She was named Alva after Lord Alva, a Stirling descendant, and she called her youngest son Harold Stirling Vanderbilt to underscore the connection. One of her Stirling forebears emigrated to Virginia, and married a Smith of Virginia. Her father, Murray Forbes Smith, was a descendant of this line. The antecedents of Alva's mother, Phoebe Desha, were much less hazy. She – rather than Alva – was the daughter of a plantation owner, the distinguished and powerful General Robert Desha of Kentucky, who won his rank in the war of 1812, and was twice elected to the House of Representatives. Her uncle, Joseph Desha, was a governor of Kentucky. Thus far, Alva's claims to a relationship with America's southern landed aristocracy appear to have been valid, but in her parents' generation they became diluted. Murray Forbes Smith had just finished training to be a lawyer in Virginia when he met Phoebe Desha. 'His entire career, like all women's but unlike most men's was upset by this marriage,'[26] Alva wrote later, for his powerful father-in-law persuaded him to abandon his fledgling legal practice and move to Mobile, Alabama to look after the family cotton interests. This made Murray Forbes Smith, in effect, a superior cotton sales agent working on behalf of the Kentucky Deshas.

While there may be some confusion about her background there is no doubt at all about the strength of Alva's personality, which impressed itself on everyone who ever met her. 'When convinced,' said one witness 'Not God nor the devil can frighten her off.'[27] 'When she speaks, prudent men go and get behind something and consider in which direction they can get away best,'[28] said another. 'Her combative nature rejoiced in conquests,' wrote Consuelo. 'A born dictator, she dominated events about her as thoroughly as she eventually dominated her husband and her children. If she admitted another point of view she never conceded it.'[29] If anything, Consuelo, anxious not to appear over-critical of her mother, always downplayed Alva's forcefulness. Alva, by contrast, seemed to take great pride in her own strength of character to the point of sounding puzzled by the strange impulse within her and writing in her (unpublished) autobiography: 'There was a force in me that seemed to compel me to do what I wanted to do regardless of what might happen afterwards . . . I have known this condition often during my life.'[30]

Alva explained her dominant personality by saying that she had been born forceful, that she was the seventh child and – according to an old saying – the seventh child was always the strongest and the mainstay of the family. Elsewhere she attributed it to her upbringing, particularly her mother. 'There is, I believe, no stronger influence on the development of character and personality than our early environment, and childhood memories,'[31] she wrote. It is difficult to dispute this. Her domineering character was given free rein by her strong-minded mother in childhood, and family circumstances which involved a weakened father in her teenage years conspired to emancipate it entirely.

The Smiths moved to Mobile in boom years for the cotton trade, when Mobile was a great cotton port. In 1858, Hiram Fuller described Mobile as 'a pleasant cotton city of some thirty thousand inhabitants, where the people live in cotton trade and ride in cotton carriages. They buy cotton, sell cotton, think cotton, eat cotton, and dream cotton. They marry cotton wives, and unto them are born cotton children.'[32] Life for the Murray Forbes Smiths was not entirely happy, however. According to one account, Phoebe was determined to show the cotton wives how things were done by grand families such as the Deshas of Kentucky. Sadly, Mobile society was first indifferent and then irritated. Alva omits to mention in any account of her childhood that Phoebe's attempt to conquer Mobile ended in abject failure, and would not have been pleased by a book which appeared some years later where her mother's social efforts in Mobile were pilloried. 'Some people ate Mrs Smith's suppers; many did not. There was needless and ungracious comment, and one swift writer pasquinaded her social ambitions in a pamphlet for "private" circulation. Then the lady concluded that Mobile was . . . unripe for conquest,'[33] commented Thomas De Leon in 1909.

Mobile must have been a difficult time for the Smiths in other ways. Alva was born on 17 January 1853, the seventh of nine children, of whom four died in infancy. Of the eight children born to the Smiths in Mobile, three are buried in Magnolia Cemetery – Alice, aged twenty months in 1847, one-year-old Eleanor in 1851, and thirteen-year-old Murray Forbes Smith Jr in 1857.[34] In her memoirs (which she wrote after her conversion to the cause which would dominate her later life – women's suffrage), Alva traced intense feelings of resentment towards men back to the death of this brother, Murray Forbes Jr. He died when she was four, and she grew up being made to feel that as far as her father was concerned, the death of his thirteen-year-old son and namesake was a far greater loss than her baby sisters. 'He was always kind to us, always generous in his provision and care, but atmospherically he made his daughters feel that the family was best represented in the sons . . . I didn't suffer with tearful sadness but with violent resentment.'[35]

In spite of this, Alva remembered the house in Mobile with deep longing. The Smiths were comfortably off and their house stood at the corner of Conception and Government Streets, one of the grandest and most distinctive houses in Mobile with crenellations round the roof, and a Renaissance suggestion to the porches. Memory of the dream house of early childhood would haunt her all her life, influencing the design of the remarkable houses she created as an adult: 'Always these houses, real and imaginary, reproduced certain features of the home in which I was born and where my early childhood was spent . . . It had large rooms, wide halls, high ceilings, with high casement windows opening upon the surrounding gardens . . . apart from the big house, also, was the bath house. The floor and bath were of marble, and marble steps led down into the bath which was cut out and below the level of the floor.'[36] When Alva was six, her parents decided to leave Mobile and go to New York. Quite apart from his wife's problems with Mobile society, Murray Smith sensed that success as a 'commission merchant' would be unsustainable if he stayed. His judgement (correct, as it turned out) was that the rapid spread of railroads would tip the balance from Mobile and the ports of the Gulf of Mexico in favour of New York. It was therefore the onward march of the American railroads that would ultimately bring the Smiths and Vanderbilts together.

The Smiths' move to New York shortly before the Civil War initially seemed well judged. They avoided the depression in the South that came with the Civil War, and profited from a property boom in Mobile. Several characteristics of nineteenth-century southern life moved with them. Like almost every other well-to-do southern antebellum family, the Murray Smiths had slaves, given to Phoebe by her father in lieu of a marriage settlement. These slaves went with the Smiths to New York. Alva had her particular favourite, Monroe Crawford, whom she adored. 'The reason Monroe Crawford and I got along without conflict for the most part was because I managed the situation, I wanted my own way and with Monroe I got it. I bossed him. It was a case of absolute control on my part,'[37] she told Sara Bard Field, the writer and poet to whom she dictated her first set of memoirs in 1917. Though Alva never said so, this early exposure to a system of human relations based on slavery may explain as much about her as Murray Smith's lack of interest in his daughters. She never entirely lost the habits of mind of a southern slave owner in relation to those she regarded as her inferiors: more profoundly such total control over another human being at such a young age can only have contributed to Alva's later obsession with power and control, and her almost phobic fear of losing her grip on it.

There were other aspects of her childhood that set the Smith children apart from middle-class New York. First, they were unusually international in outlook, partly, no doubt, as a result of their southern parents. Prosperous southerners were a familiar sight in London even in the eighteenth century and Phoebe Smith loved travelling. She began taking her children abroad when they were very young. One expedition included a babe in arms, a little dog, two maids, and a southern mocking bird in precarious health in a large cage. They all crossed the Atlantic in a wooden paddle steamer on a voyage which took fourteen days, and travelled to England, France, Germany, Austria and Italy. While it would clearly have been easier to leave her children at home, Phoebe Smith liked to broaden their minds and teach them to observe. This international outlook also extended to fashion. 'My mother, who loved the beautiful in dress as in all else, preferred the clothes made by European dressmakers, designed as they were by artists of an older civilisation, to those worn in her time by women in the United States. It was her custom, therefore, to order from Paris her own clothes, and later those of her children. Twice a year, from Olympe, a famous house of that day, would come a box containing clothes sufficient for our needs for the next six months.'[38] Alva would pass on this feeling for French style and couture to Consuelo, but in childhood she did not appreciate being a fashion pioneer. Parisian outfits were a major provocation to sniggering little New York boys, whom she claims to have pitched into the gutter.

One striking feature of Alva's account of her nineteenth-century upbringing in New York is the extent to which she presents herself as an aggressive, violent child. It was impossible to find a nurse to manage her. When she wanted to leave the nursery for a room of her own she smashed it up; and she particularly enjoyed thumping boy playmates when displeased. Any boy who teased her soon learnt better. 'I can almost feel my childish hot blood rise as it did then in rebellion at some such taunting remarks as "You can't run";

"You can't climb trees"; "You can't fight. You are only a girl,"' she once wrote in a letter to a friend.[39] Even at thirteen, passers-by had to pull her apart from a male tormentor in a fight so fierce that Alva boasted it was reported in the local newspaper. No-one has ever succeeded in tracing this report, but it is telling that it was a story Alva liked to recount about herself. 'I caught him and threw him to the ground. I choked him and banged his head upon the ground. I stomped on him screaming: "I'll show you what girls can do,"' she told Sara Bard Field.[40] It comes as no surprise that Alva had few girl playmates. She greatly preferred playing with the opposite sex, for she found boys' lives more interesting. 'I wanted activity and I could not find enough of it in the circumscribed and limited life of a girl. So I played with boys and I met them on their own ground. I asked for no compromise or advantage. I gave blow for blow.'[41]

It is perhaps surprising that although she detested the life conventionally lived by nineteenth-century American girls, Alva liked playing with dolls and designing imaginary houses – two activities that she regarded as closely connected. She told Sara Bard Field that she was unable to sleep if her sisters left their dolls sitting up with their clothes on: 'I loved dolls . . . I took them very seriously. I put into their china or sawdust bodies all my own feelings. They could be hot or cold. They could be weary, sleepy and hungry. Their treatment had to vary accordingly to these supposed conditions.'[42] She saw this as a childish manifestation of maternal instinct which she thought men had downgraded. She saw no contradiction between her love of dolls and her rebellion against the constricted life of a girl, claiming: 'It is because I must have felt then in an inarticulate way and feel now with a passionate conviction that the very fact of her maternity which men have used to lower woman's status, raises her to superior position. Thus my love for the doll children and my rebellion against the superimposed restrictions of a girl's life were bound up together'[43] – an insight which would have an impact on Consuelo later.

Phoebe Smith was ahead of her time in the amount of freedom she granted her headstrong daughter, allowing Alva to ride out alone all day when they holidayed in Newport. She had no hesitation, however, in whipping Alva when the boundary was finally crossed – when, for example, she took a horse from the stable and rode bareback in the garden (Alva maintained that the pleasure was worth the whipping, and her streak of physical daring was noted by others throughout her life). Alva expressed it to Sara Bard Field thus: 'My Mother found me the most difficult of all her children to train. The combination of rebellion and daring were difficult for her to meet. And in those days there was no Montissori [sic] methods and books on child psychology by which parents directed their training of children. The rod was the all-sufficient guide and to this my Mother resorted. There is a record in our family of my receiving a whipping every single day one year . . . but the end I desired was always strong enough to overcome the fear of the whipping . . . I was an impossible child.'[44]

Moving north before the start of Civil War, the Smiths appear to have made a smooth entrée into New York society. New York's elite at this period has often been seen in terms of a dichotomy between the old introverted 'Knickerbockcracy' – descendants of the original Dutch settlers, the Knickerbockers – on the one hand, and extrovert new money on the other. However, New York society was always more permeable than this suggests, and making one's mark on society was largely a question of becoming part of the right networks and (unlike the Commodore) demonstrating that one understood society's rules and wished to opt in. Genteel in values, tone and style, the Smiths fitted quite easily into New York's socially mobile elite before the Civil War, and there is every reason to suppose that without this unfortunate interruption, they would have soon felt well established. After a brief spell in a house at 209 Fifth Avenue, the family moved to a fine house at 40 Fifth Avenue, built for an affluent merchant in the 1850s by the well-known architect Calvert Vaux. New York City tax assessments of the house bear out Alva's assertion that the family was well-to-do at the time of the move north, for records put its value at between $25,000 and $39,000, making it one of the more valuable houses in the city.[45]

Nonetheless, some of Alva's assertions about the social position of the Smiths in New York during this pre-war period are simply wrong – saying much about Phoebe's own social anxieties since she was probably the source of the errors. Alva maintains, for example, that her mother was on the receiving line for the Prince of Wales at a famous ball held in his honour in New York in 1860. If true, this would have put Phoebe Smith at the pinnacle of society, but the ball took place on 12 October, seven days before Phoebe gave birth to her second daughter, Julia, in Mobile, Alabama.[46] Phoebe would not have stepped out of doors, let alone travelled to New York. Similarly, Alva suggests that the Smith's entrance into New York circles came through the department store owner Alexander T. Stewart, but Stewart had an uneasy relationship with New York's elite because he was a shopkeeper, and was suspected of vulgarity. The Smiths did not, as she claimed, have a box at the Academy of Music, though they may well have attended performances there; neither did they have a pew at Grace Church (one clear marker of membership of New York's inner circle), though they belonged to another fashionable Episcopalian church, the Church of the Ascension.

On the other hand, Murray Forbes was elected to the Union Club in New York in 1861. This was a significant social success since one of the most important developments in the emerging exclusivity of New York society was the expansion of gentlemen's clubs. Like gentlemen's clubs in London, New York clubs were, to quote the historian Eric Homberger, 'rooted in an ethos of exclusion'.[47] The Union Club was the first of the gentlemen's clubs from which many others emerged as a result of splits and disagreements. Membership was limited to a thousand members and lasted for life unless one chose to resign. By 1887 an observer noted that 'membership in the Union implies social recognition and the highest respectability'.[48] Founded purely for social (as opposed to political or sporting) purposes, the Union Club's early membership tended to favour merchants over 'gentlemen of leisure', but even here, Murray Smith was several steps ahead of the Vanderbilts. The Commodore had become a member in 1844, resigned and then rejoined only in 1863. William Henry Vanderbilt would not become a member until 1868, and William K. Vanderbilt was only elected to the club after his marriage to Alva and the death of his grandfather in 1877.

The outbreak of the Civil War, however, brought real difficulties for the Smiths. They were slave owners; Murray Forbes Smith did not believe that slaveholding was wrong, and took the view that emancipation was only possible if it happened gradually. As hostilities began, tension with northern neighbours escalated. One of the first places this manifested itself was in the Union Club itself. According to the club's historian: 'feeling rose high against the South in New York . . . Many Southerners, including Benjamin [the Confederate Secretary of State] and Slidell [the Confederate Commissioner to France] resigned, and more were dropped for non-payment of dues.'[49] Such clashes were not surprising in view of the fact that the Union Club's membership also included General Ulysses S. Grant, General William Sherman and General Philip Sheridan, as well as twenty-four Confederate major generals. The abolitionist views of the rector at the Smith's church, the Church of the Ascension, caused such offence to the southern members of the congregation that they all withdrew. Mounting tension affected the children directly too – this was a time when Jennie Jerome, later Lady Randolph Churchill, remembered pinching little southerners with impunity at dancing class.[50]

The turning point, according to Alva, was the assassination of Abraham Lincoln just after the end of the war on 15 April 1865. The Smiths felt obliged to sign up to the general mood of mourning, putting black bows of bombazine in their windows to avoid attack. By now, Alva recalled, 'feeling against southerners had risen from unfriendliness and suspicion to active antagonism and enmity,'[51] scarcely surprising considering that the city had lost over 15,000 men to the war and, in a last desperate throw of the dice, Confederates attacked New York itself by setting fire to ten hotels in November 1864.

Bows in the window, it turned out, were not enough to prevent unpleasantness. After the President's funeral, life became so difficult for the family that Murray Forbes Smith decided they should not remain in New York and sold their fine house on Fifth Avenue to a Mr McCormick of Chicago, inventor of the reaping machine. Social and business networks in New York once plaited closely together were torn apart by wartime antipathies. The cotton trade was disrupted by the war, and so was the transport system from south to north.

From 1866, Murray Forbes Smith based his business activities in Liverpool, the main English port for cotton from the southern states. That summer, when Alva was thirteen, the Smith family briefly took a villa on Bellevue Avenue in the resort of Newport, Rhode Island, where they probably met the Yznaga family for the first time. Mr Yznaga was Cuban and owned a cotton plantation in Louisiana that had been worked by over three hundred slaves before the Civil War. Mrs Ellen Yznaga was of New England stock but was thought 'fast' by some. These attributes were enough to disbar them from certain aristocratic New York households after the Civil War. At one point the Yznagas owned a house in New York on 37th Street, but in the post-war years their fortunes fluctuated so dramatically that they ended up living in Orange, New Jersey and the Westminster Hotel in New York. In the summer of 1866, however, they were still in a position to spend the summer in Newport. This was almost certainly the time of Alva's fight as a thirteen-year-old with a male tormentor, for her opponent was a Yznaga houseguest. She spent much of that summer fearlessly rolling down a hill that ended in a cliff face in the company of Fernando Yznaga, her future brother-in-law; and started a long and important friendship with Consuelo Yznaga, who was about three years her junior, and almost as high-spirited.

Soon after this, and quite possibly speeded on their way by some of Newport's matrons, Phoebe took her daughters to Paris, rented an apartment on the Champs Elysées and set up home. Although the Smiths kept smaller houses in New York throughout the period, they were based in Paris for much of the time between 1866 and 1869. Like other southern families who appeared in Paris during and after the Civil War, they were able to live well in reduced and uncertain circumstances. Apparently affluent, they were welcomed by the imperial court of Napoleon III and the Empress Eugenie, at a time when the Second Empire was at its most brilliant and glamorous. Precise gradations of wealth and social distinction of New York meant little to society circles in Paris. The Smiths were able to mix on easy terms with French aristocracy, equally untroubled as to how or when the latter had acquired their noble titles, some more recent than others. Lilian Forbes of the Forbes family, who had been a neighbour of the Smiths in New York, married the Duc de Pralin. Prince Achille Murat, distantly related to the Smiths by marriage, called at the house. The Marquis Chasseloup Loubat, who was Napoleon III's Ministre de la Marine, and married to an American, Louise Pelier, was particularly cordial in his invitations, inviting Phoebe and Alva's eldest sister, Armide, to select dances, and inviting the children to the Ministere de la Marine to watch processions. In Paris, Phoebe arranged a debut for Armide (who would never marry) and launched her into French society.

The impact on Alva of the move to Paris would have many consequences in the decades to follow: for American architecture, for the Vanderbilts and for Consuelo. Now in her early teens, she fell passionately in love with France, and above all with its history, art and architecture. In New York she had been just as resistant as the Commodore to attempts at formal education ('I could not learn from impersonal pages. I wanted the contact of mind with mind. I liked the friction of thought it engendered,'[52] she remembered later.) Now she responded to the clarity, rigour and competitiveness of French schooling which appealed to both ambition and pride; she particularly liked the French approach to learning history which she thought made sense. At one point she even demanded to go to a boarding school run by one Mademoiselle Coulon. She enjoyed this too, though she continued to prove a most difficult girl to handle and only stayed for about a year.

Much of Alva's French education, therefore, was a freelance affair in the hands of French and German governesses, with trips to places that appealed to Miss Alva Smith. Thanks to the ever-expanding French railway system, there were frequent visits to the great Renaissance chateaux on the Loire, and to Versailles. It is understandable, given her dominant personality and her love of history, that Alva would be drawn to the magnificent architecture of both the French Renaissance and of the Bourbons. It is easy to imagine her walking in awe through the Hall of Mirrors at the Palace of Versailles, or believing that the apartments of Madame de Pompadour really belonged to her, or sketching the Petit Trianon for the umpteenth time – doubtless followed by a breathless governess much relieved to have found a way of passing the time in such an acceptable manner.

At the height of the Second Empire there was much to grip the imagination of such a child: French history was invested with a magical quality of particular intensity. As Alistair Horne writes: 'The *haut monde* escaping from the bourgeois virtuousness of Louis-Philippe's regime had sought consciously to recapture the paradise of Louis XV. In the Forest of Fontainebleau courtesans went hunting with their lovers attired in the plumed hats and lace of the eighteenth century.'[53] The retrospective mood was set by Louis-Napoleon himself who loved to appear as a masked Venetian noble (masked balls being a particular feature of the Second Empire's illusory and fantastic world).

There were times, indeed, when the imperial court reminded observers of an endless Venetian carnival, with every ball outdoing the one before in dazzling display. One of the most extraordinary balls was given in 1866 by the Smiths' friend, the Ministre de la Marine, where the guests formed *tableaux vivants* of the four continents and 'a procession of four crocodiles and ten ravishing Oriental handmaidens covered in jewels' entered in front a chariot in which one English guest noticed the Princess Korsakow was seated *en sauvage*. Africa was represented by Mademoiselle de Sevres, 'mounted on a camel fresh from the deserts of the Jardin des Plantes, and accompanied by attendants in enormous black woolly wigs'; finally came America – 'a lovely blonde, reclined in a hammock swung between banana trees, each carried by Negroes and escorted by Red Indians and their squaws'.[54] Three thousand guests came to this ball which cost about 4 million francs. Although there were balls and assemblies a-plenty in New York before the

Civil War – indeed they were deeply embedded in the structure of society life – there was certainly nothing that came close to such a ball in terms of fantasy or expense until, that is, Alva threw one herself in 1883.

As a young lady protected from the seamier side of Second Empire life, Alva could see only enchantment in the Paris of Napoleon III. It seemed to embrace a great international vista, a future of scientific wonders as well as a magical past, encapsulated in the Great Exhibition on the Champs de Mars in 1867. The Great Exhibition was an extraordinary, opulent, dreamlike, awe-inspiring spectacle: As dusk fell, the Goncourts exclaimed that 'the kiosks, the minarets, the domes, the beacons made the darkness retreat into the transparency and indolence of nights of Asia'.[55] Alva particularly remembered the astonishing exhibits of Thomas Edison, and looking on in wonder from the windows of their apartment on the Champs Elysées at the great reviews held in honour of visiting kings and emperors by Napoleon III. 'The people seemed to worship their Imperial family,'[56] she later said wistfully. Alva may have spent her year at Mademoiselle Coulon's school in 1868–9, while her parents moved back and forth between houses in New York and Paris.[57] In 1869, however, Murray Smith decided that the whole family must return permanently to the United States. Sixteen-year-old Alva was utterly distraught. 'I was broken hearted that I must leave France. I was in sympathy with everything there. This musical language had become mine. I loved its culture, art, people, customs. Child that I was, America struck me in contrast to France, as crude and raw.' France, unlike America, was a 'finished product'.[58]

The New York to which the Smiths returned in 1869, was a very different city to the one before they left for France in 1866. Capital markets were already centralising in New York by the end of the Civil War in 1865. Commodore Vanderbilt's consolidation of his railroads in 1869 was a harbinger of things to come. The drive to expand the economy for military purposes had created a national market for the first time and war precipitated an almost limitless demand for goods that only increased with peace. This was the beginning of what Mark Twain termed 'The Gilded Age', the period spanning the final third of the nineteenth century that ended when Theodore Roosevelt became President in 1901, determined to control its worst excesses.

Twain's novel *The Gilded Age* satirised what he described as 'the inflamed desire for sudden wealth',[59] and came to define the period of about thirty-five years of economic boom centred on New York, characterised by rapid industrialisation, urbanisation and technological invention, harsh social inequity, grandiloquent, competitive opulence and by a relentless drive towards economic monopoly and big business. A new phenomenon, the industrial corporation, emerged quite suddenly, driven forward by intensely competitive individuals with energy as great as the Commodore's, whose corporate power was unrestrained and who were assisted by corrupt politicians and a regime of virtually non-existent taxation – inheritance tax expired in 1870, income tax was abolished in 1872 and tax on corporate profits did not exist. Labour costs were low, and workers had yet to organise themselves efficiently against exploitation. The potential for vast personal fortunes suddenly became limitless. Those who did well out of the war continued to fare very well after it was over. Wealthy men became richer; others suddenly acquired fortunes overnight.

The fact that Murray Forbes Smith had sold his Fifth Avenue house to new money from Chicago was significant. The newly rich flocked to New York, often accompanied by wives determined to partake of the delights of New York society. Before long, 'old' New York felt itself besieged by outsiders, an impression born out by the demographics of the period and a range of expressions for the new arrivals: 'social climber, men of new money, arriviste, bouncer, (as in the Yiddish *Luftmensch,* air man, someone who has arrived apparently from nowhere). The parvenus, objects of fierce social mockery, were assumed to be rich, crude, half-educated, and were seen as embodying the raw hunger for social distinction.'[60]

The Smiths had put down a marker in New York society between 1861 and 1865, but after absenting themselves for nearly four years, they came back to the city to find themselves at a remove from its inner circle. Worse, they discovered that there were far more people knocking on high society's door demanding admission and that the financial cost of re-entry had gone up sharply. In the early 1870s, according to Eric Homberger, 'New York was literally swirling with cash. Prices rocketed, but even inflated costs seem to have no effect upon the *ton* . . . The holdings of the New York banks had risen from $80 million in the early 1860s to $225 million in 1865. When the Open Board of Stock Brokers merged in 1869 with the New York Stock & Exchange Board, forming the New York Stock Exchange, membership increased from 533 to 1,060. There were many more millionaires in the city than there had ever been before.'[61]

In his memoir *Society As I Have Found It*, the southern gentleman Ward McAllister made the same point: 'New York's ideas as to values, when fortune was named, leaped boldly up to ten millions, fifty millions, one hundred millions, and the necessities and luxuries followed suit. One was no longer content with a dinner of a dozen or more, to be served by a couple of servants. Fashion demanded that you be received in the hall of the house in which you were to dine, by from five to six servants, who, with the butler, were to serve the repast.'[62] In an era of conspicuous consumption, this had an immediate impact on modish womenfolk. The newspapers began to notice, for example, that the cost of dressing fashionably was reaching breathtaking new levels. 'Ladies now sweep along Broadway with dresses which cost hundreds of dollars,' noted the *New York Herald*. 'Their bonnets alone represent a price which a few years since would more than have paid for an entire outfit. Silks, satins and laces have risen in price to an extent which would seem beyond the means of any save millionaires, and yet the sale of these articles is greater than ever.'[63]

As often seems to happen in periods of intense social mobility, a self-defined social elite emerged quite suddenly, as those who had been there longest, and felt they had the best claim to be top of the pile, pulled up the social ladder. It was the queen of this society resistance movement who swept up the aisle of St Thomas

Church as guest of honour at Consuelo's wedding on 6 November 1895. Mrs Caroline Schermerhorn Backhouse Astor, *the* Mrs Astor, was born into the Schermerhorns, an old Dutch family who were already entertaining and patronising the arts when the Commodore started the Dispatch Line. A generation older than Alva (there was a twenty-three year difference), Caroline married new money in the form of William Backhouse Astor in 1853. She immediately set about de-vulgarising Mr Astor, whose fortune was derived from furs, pianos and Manhattan slums. She persuaded him to drop the 'Backhouse' and 'Jr' and moved him north to 350 Fifth Avenue. This house famously had a ballroom into which she could squeeze 400 people, eventually giving rise to the idea that New York's elite comprised 'the Four Hundred'.[64]

Caroline Astor was essentially a conservative. Though she was aware that society needed new blood, she felt that New York social life should be conducted much as it had been by her great-grandmother a century earlier. Many in her close circle were descended directly from Dutch settlers. But even the 'Knickerbockers' could fall out of favour with Mrs Astor, however, and some simply refused to opt-in. It was Mrs Astor's considered opinion that New York society would be fatally undermined if vulgar wealth alone was allowed to dictate the social agenda. In her view, it was essential to harness the power of money, tame its owners, and show them how to behave if standards were to be maintained. Unharnessable individuals such as the Commodore, and by extension, his family, were not to be admitted to the Four Hundred. Indeed, Mrs Astor regarded the Vanderbilts as just the sort of people New York should rally round to exclude. At the same time, however, Mrs Astor was not immune to the effect of the new money swirling round New York. She was, after all, married to Mr Astor. The effect of this was that, with very few exceptions, wealth became a *sine qua non* for anyone wishing to participate in Mrs Astor's elite circle.

By 1870, when the Smiths had returned to New York, Mrs Astor's power was reinforced by a symbiotic relationship with Ward McAllister, a southern gentleman of quite remarkable fatuity who self-consciously modelled himself on Beau Nash, arbiter of society elegance in another period of intense social mobility in eighteenth-century England. Confronted by rows of post-war millionaires, this self-styled dandy took it upon himself to tell them quite explicitly how to stop living like vulgarians, acting as spokesman for Mrs Astor – who never pronounced in public (he referred to her as his 'Mystic Rose'). His advice extended to how to dress, what to eat, how to serve wine, how to provide suitable music, correct etiquette, and forms of address – in short, he provided 'a code of manners that would act as the constitution of upper-class social life in America'.[65]

For over a decade, from the early 1870s, this odd couple held extraordinary sway. As wealthy New York went through a social convulsion in the years following the Civil War, Ward McAllister and Mrs Astor acquired the power not simply to constrain but to exclude. They did this partly by setting up 'the Patriarchs' – a committee of twenty-five New York society gentlemen, whom Ward McAllister persuaded to draw up guest lists for three exclusive subscription balls at Delmonico's each year – balls that were modelled on those held at Almacks in eighteenth-century London. The Patriarchs' guest lists in turn defined New York's social elite. Behind the scenes, Mrs Astor almost certainly had power of veto over the names on Ward McAllister's committee, and dictated indirectly just who could be asked to the Patriarchs' balls. From 1872, for about two and half decades, membership of New York's elite was thus largely determined by Mrs Astor's family relationships, Mrs Astor's friendships, and those in the world of business Mrs Astor deemed suitable for membership.

The evolution of society and social life in this highly monopolistic direction created great difficulties not just for the younger Vanderbilts, but also for the Smiths as they returned to New York from France. Their presence in New York did not go back even one generation, and they were not acquainted with Mrs Astor. It was of little use that their main point of contact with rich New York circles was the department store owner A. T. Stewart, since Mrs Astor excluded Mr and Mrs Stewart from her drawing-room, remarking sniffily: 'I buy my carpets from them, but then is that any reason why I should invite them to walk on them?'[66] The Smiths could not rely on business connections either, for in spite of the fact that Murray Smith was one of 300 members of the Cotton Exchange in 1871, and remained a member of the Union Club, his southern background put him at a disadvantage after the Civil War and he may have severed many of his old links by conducting much of his business activity outside the US between 1866 and 1869.

Then, just at the moment when the financial bar to participating in the top drawer of New York society was raised to eye-watering levels, Murray Smith's business started to fail. Whereas Commodore Vanderbilt relished the atmosphere of economic boom in the 1870s and thrived, Murray Smith was defeated. Smith may not have been a particularly effective businessman in the first place; according to Alva, her father was never able to come to terms with the new dog-eat-dog spirit of mercenary capitalism abroad in the land. 'He could not stoop to the new methods which to him seemed underhand,' she wrote. 'Nor was he trained in the arts of clever manipulation by which big deals were put through. His inability to meet these changes resulted in a great change in our circumstances.'[67]

Murray Smith was forced to tell his children that the family must retrench. One outward sign was that the houses they rented went steadily downwards in terms of status from 1869 onwards. As New York's aristocracy moved north, the Smiths moved south and away from Fifth Avenue, so that by 1870 they were living at 14 West 33rd Street. In 1871, there was another terrible blow. Phoebe Smith, who had many of the characteristics of a frontier mother, would probably have been able to find a way of hacking through the social jungle and steering her daughters towards Mrs Astor. She was unable to withstand a severe attack of rheumatoid arthritis, however, and died at the age of forty-eight. Alva was eighteen and was left grief-stricken and feeling that she had lost the only person in the world who really understood her.

In her biography of Samuel Pepys, Claire Tomalin observes that families often select one child on whom they pin their hopes; but that sometimes the child who is not preferred in this way elects him or herself as a saviour of the family.[68] In the Smith family, hopes had been pinned on Murray Jr, the brother who died aged thirteen. Now, with her mother's death, Alva elected herself to save her family in his place:

> I remember as I stood by her coffin and gazed upon the being that had been everything in my life, to whom I had given so much trouble, and who had borne my follies with so much patience, interest and love, I felt I owed her a great debt ... But I knew, too, that if she could, she would ask that I should not content myself with grieving for her, that she would wish that grief should become on my part a determination to do as she would have me do if she were still living beside me ... And I solemnly vowed to do my best always to carry out the wishes I knew would have been hers.[69]

Although Alva was inclined to self-dramatisation, it seems that she now genuinely felt the need to step into the vacuum left by her mother's death – a vacuum that can only have expanded when her only surviving brother, Desha Smith, left home at seventeen, probably after a disagreement with his father (he may have never compared well with his dead brother). None of her sisters were doughty fighters like Alva. The eldest, Armide, who might have been expected to take on a maternal role after Phoebe's death, was a particularly gentle character. In the short term, however, there was little that even Alva could do to rescue the Smiths. Phoebe's death in 1871 was followed by a year of mourning which took the sisters out of society for many months. A visit to Smith relations in Virginia, as mourning ended, only reinforced Alva's lack of enthusiasm for proud poverty. Even in 1872, seven years after the end of the Civil War, the countryside was in ruins and some of her Smith relations had lost everything except a faintly obnoxious dignity.

On their return from this visit, the Smiths' financial position started to deteriorate in earnest. Murray Smith may have lost money in the Panic of 1873 and it has sometimes been suggested that the family was forced to run a boarding house. There is no evidence of this: Alva mentions in her memoirs that her father was in such an anxious state that he claimed they would be reduced to keeping a boarding house, but she makes it clear that this was merely a figure of speech – an indication of the desperation of a man faced with declining income and four uneducated daughters. She cites it simply to explain her own limited room for manoeuvre as the family's financial situation slid from bad to worse. 'Through change of circumstances he began not only to make no money but to lose it, so he notified us that we must move from 33rd Street to 44th Street. I could not understand the great worry and grief to my father because it did not seem to affect me. I remember hearing him say when he was very worried "we shall have to keep a boarding house" – at this my sisters would look dismayed but I would shout "If we do keep a b.h., I'll do the scrubbing". My father's anxiety was dreadful but it was perfectly justified by existing circumstances – 4 daughters never educated to do anything.'[70]

By her own account, Alva took the only option open to her. She put herself on the marriage market for two anxious years. Her social circle revolved round a younger group than Mrs Astor's. Within it, social demarcation as delineated by Mrs Astor was already breaking down. Alva described her set as an exclusive group based on family connections where the parents all knew each other and which was 'very exclusive and safe on that account'.[71] Some members of this younger set undoubtedly came from families already in the Four Hundred, while others would have been regarded by Mrs Astor as 'fast' or 'new'.

Edith Cooper and the Livingston sisters, for example, were from old New York families. Alva's Newport friend Consuelo Yznaga, on the other hand, did not meet Mrs Astor's exacting standards on account of her oscillating Yznaga fortune and flamboyant Cuban background. Minnie Stevens, a friend whom Alva met at Madame Coulon's school in Paris, was another of the younger set who suffered from the disdain of the Four Hundred. She was the daughter of Mr Paran Stevens, a hotel owner who collected hotels 'as assiduously as Commodore Vanderbilt collected railways'.[72] Although she was much more financially secure than either Alva or Consuelo Yznaga it was to Minnie's disadvantage that Mrs Paran Stevens was rumoured to have been a hotel chambermaid, a charge she most indignantly fought off, eventually becoming a successful hostess in New York in her own right.

For the young blue bloods in the group, this mix was part of its charm. In truth, society as constructed by Mrs Astor was often intensely dull. This younger set was fun precisely because it was far less concerned with keeping out arrivistes. Its members skated on Central Park and danced at ultra-exclusive Delmonico's too – not at the Family Circle Dancing Class, or even the first Patriarchs' ball, but at the 'bouncer's balls' which the press took great delight in describing as 'opposition' to Mrs Astor and Ward McAllister. The newspapers' suggestion at the time of wholesale exclusion from the Four Hundred is misleading, however. According to Eric Homberger: 'When we look at accounts of events labelled "bouncer's balls", such as a subscription ball held in New York in 1874, we find it under the management of an impeccable group of young blue bloods led by Charles Post, William Jay, and Peter Marie.'[73] William Jay would later marry Alva's great friend Lucie Oelrichs, another member of the same circle; and on 6 November 1895, Colonel and Mrs William Jay would walk up the aisle as guests-of-honour, just a little way behind Mrs Astor herself.

Although social demarcation lines were changing, it is possible to overstate the idea of 'permeability' too: entry to a circle such as this presented a formidable challenge. It seems likely that as a motherless girl, Alva was helped by the patronage of mothers of her friends who already knew the genteel Smiths well, and that once she was accepted she made sure her social behaviour was impeccably charming. Such mothers would have included both Mrs Yznaga, and Mrs Paran Stevens – who knew the family from the Smiths' Paris years, when Mr Stevens was a US commissioner to the Great Exhibition in Paris in 1867. Even if these mothers were not ideal patronesses from the point of view of the Four Hundred, the attitude of New York's social elite towards them was also being forced to change. Mrs Paran Stevens was particularly ambitious and frequently took Minnie to Europe in response to cold-shouldering by Mrs Astor. By 1874, New York knew that Minnie was a success in London society and had met with the approval of the Prince of Wales. Eric Homberger also suggests that Alva's circle would have interested her fellow southerner Ward McAllister, and that it probably benefited from his informal protection. His reactions often ran ahead of the intensely conservative Mrs Astor. A younger set comprised of old families, genteel southerners and energetic and pleasant newcomers that kept the truly rich at a distance was a development of which McAllister would have approved.[74]

This group also provided the entry point for William K. Vanderbilt into New York society. Its members did not hold the Commodore's reputation against him, whatever Mrs Astor had to say on the subject. William K. was handsome, charming, amusing, potentially rich, and keen to join in. It helped that he could spend time with these new young friends without attracting the Commodore's opprobrium, for the ethos of this group was not flashily vulgar. For example, parents put a stop to a custom that suddenly sprang up of young men sending bouquets to girls they admired on the night of a ball, so that favourites would go home loaded down with flowers, while others had nothing at all – on grounds of cost to the young men. William K. was introduced to the circle by Consuelo Yznaga, who also effected his introduction to Alva, bringing him to at least one social event at the Smith house at 14 West 33rd Street in 1873 before financial problems precipitated another downward move to a house at 21 West 44th Street.

In her memoirs, Consuelo wrote that she could not understand why her parents ever came to marry, but Alva Erskine Smith and William K. Vanderbilt had much more in common than either of them were later prepared to admit. They had both been educated in Europe, spoke fluent French and shared a more international outlook than many of their peers. The imbalance in good looks that came later was not so apparent when both were in their twenties. Around the time of her engagement Alva was a highly attractive young woman with a great mane of hair (she lopped it off after catching a rich husband). In those of her personal papers that have survived her drive and wit shines through – alongside other less attractive characteristics. Perhaps Consuelo was never permitted to understand the extent to which the Vanderbilts were anathematised by Mrs Astor and the extent to which William K. felt that he needed Miss Smith. In 1874, Alva appeared to be a young woman whose genteel background and energy would open doors – and for some considerable time it was Willie who was widely perceived to have the better part of the bargain, and was thought, however unfairly, to have been led out into the world by his socially accomplished wife.

Perhaps Consuelo was never allowed to grasp just how close her mother and aunts came to financial ruin when they were in their teens, either. Alva's experience of genteel poverty thus far had made her almost as 'inflamed' on the subject of money as the Commodore. Like him she was acutely aware of its power. Even in 1917, she tried to persuade Sara Bard Field to marry a rich man for the benefit of her children saying: 'You cannot help your children to advantages through sentimental romance but through money which alone has power.'[75] For over two years Alva had experienced at first hand the growing powerlessness that poverty brings, the terrifying insecurity of a family sliding ever closer to bankruptcy with nothing in the way of a safety net as it tried to preserve a refined front. She had known what it was to have slaves. She now sensed the enslavement of poverty for herself and its capacity to force her to the margins of her own existence. At best she faced a world of erratic kindness from her friends' parents, a life of constant gratitude. At worst, she could expect extended and difficult stays with patronising relations, a world of fading watering-holes and drab and grimy boarding houses.

In New York, true poverty was often close by. As the historian H. Wayne Morgan writes, this was a time of extremes, 'of low wages and huge dividends, of garish display and of poverty, of opulent richness in one row of houses and degrading poverty a block away'.[76] Somewhere beneath 14th Street, the 'other half' lived in a world of slum tenements and sweatshops, but it was not a sealed world. In Manhattan, said another writer, 'Wealth is everywhere elbowed by poverty.'[77] Genteel poverty was even closer in the streets near the Smiths; and to Alva, life as a woman on these margins was no life at all. 'It has seemed to me since that women and girls always play the part of spectators in the theatre of life while men and boys have the vivid action. And except to the serene gods there is nothing attractive in looking on,'[78] she once said. It was a theme on which she would play many variations throughout her life.

When Alva was twenty-one, her father's health started to fail as a result of acute financial strain, and possibly because of a drink problem. In the summer of 1874, Alva travelled to White Sulphur Springs in Virginia, a popular resort since the eighteenth century on account of its mineral waters, and a well-known spot for southern belles wishing to secure proposals of marriage. She knew that William K. Vanderbilt was also on his way there and she was under immense pressure. Although Alva may appear as coolly cynical as Edith Wharton's Undine Spragg in *The Custom of the Country*, her anxious manipulations in 1874 seem much closer to those of the orphaned Lily Bart in *The House of Mirth* – who knew that she had to 'calculate and contrive, and retreat and advance' if she were to succeed, and who 'hated dinginess as much as her mother had hated it, and to her last breath she meant to fight against it, dragging herself up again and again above its flood till she gained the bright pinnacles of success which presented such a slippery surface to her clutch'.[79] At twenty-one, it was clear that Alva already felt that her years were against her for she would lie to William K. about her age, telling him that she was a year younger than she actually was – an error which continued to appear in Vanderbilt legal documents long after his death.

Much later, *Town Topics* would recall the gossips' story about the way Alva netted her husband at White Sulphur Springs in 1874. According to the whisperers, she had decided that a grand gesture was required so she went to the village store and bought yards and yards of black tarlatan which she and her old nurse turned into a dress in the course of the afternoon: 'At dusk when everyone was upstairs dressing for dinner, the gay young girl and the old bent black woman took a mountain trail over the hill and half an hour later slipped back into the hotel laden with goldenrod.' Together, Alva and her nurse stitched tiny Dresden bouquets of yellow goldenrod flowers all over the black tarlatan dress to spectacular effect. 'The charming brunette, her beauty well framed in the black and gold, made her appearance in the ballroom. She was the sensation of the evening.'[80] When the group returned to New York in September, Alva Erskine Smith and William Kissam Vanderbilt were formally engaged.

2

Birth of an heiress

IN SPITE OF Murray Smith's failing health, the family rallied to give its saviour a smart wedding. Some press reports suggest that Murray Smith felt well enough to give his daughter away in marriage, but Alva was adamant in her memoirs that he was too ill to attend, telling Sara Bard Field: 'When I was all dressed up for the ceremony and about to leave the house he kissed me with great tenderness and told me I was taking a great burden off his mind and that he knew that if anything happened to him I would look after the rest of the family.'[1] This notion that she had rescued the Smiths – and had done as well as any absent son – was important to Alva's view of herself in later life, allowing her to think of herself as a heroine rather than a gold-digger. In a letter to her lover, Charles Erskine Scott Wood, Sara Bard Field remarked that Alva's 'terrible marriage to Mr Vanderbilt with its sordid selling of her unloving self but with the truly noble desire to save her Father' was much like life itself: 'a pathetic mixture of good and bad'.[2]

Whatever her reasons for marrying William K. Vanderbilt, Alva went to some lengths to ensure that her wedding on 20 April 1875 was impressively exclusive. It was reported as 'the grandest wedding witnessed in this city for many years',[3] and even in old age she was anxious to stress that Calvary Church was the most fashionable church in New York and that Dr Washburn who conducted the marriage service was the most fashionable divine. Her bridesmaids included Consuelo Yznaga and Edith Cooper, but Minnie Stevens was too ill to attend and had to be replaced at the last minute by Natica Yznaga. Alva's wedding dress from Paris failed to arrive in time (or so she said) and another was run up by Madame Donovan of New York using Phoebe's antique lace flounces. *The New York Times* remarked that the wedding guests included 'hundreds' of the 'wealth and fashion of the city',[4] although most of those it listed were in-laws to the Vanderbilts. Four policemen had to escort the bride through crowds from her carriage. Significantly, Alva was the first bride in New York to issue cards of admission to her wedding guests, a move guaranteed to bring crowds of the excluded flocking to the church door.

The account of her wedding that Alva dictated to her secretary, Mary Young, after 1928 suggests that she grasped early the Faustian bargain emerging between publicity and social success in Gilded Age New York – a development deplored by Henry James a decade later. 'One sketches one's age but imperfectly if one doesn't touch on that particular matter: the invasion, the impudence and shamelessness, of the newspaper and the interviewer, the devouring *publicity* of life, the extinction of all sense between public and private. It is the highest expression of the note of "familiarity", the sinking of manners, in so many ways, which the democratisation of the world brings with it,'[5] he expostulated in 1887.

In a democratised world with few other navigational aids, visibility was rapidly becoming the key to social success and it was largely in the gift of newspapers (now becoming big businesses in their own right), and later assisted by the invention of photography. Almost everyone with social ambitions had to come to terms with this. Alva and William K. were part of a younger group. They were already much less inhibited about using publicity as a weapon than their elders, though even they would find it difficult to manipulate by the 1890s. 'Unable to control the press, and unwilling to consider life without heightened visibility, the late nineteenth-century aristocrats were America's first celebrity-martyrs,'[6] writes Eric Homberger. If heiresses such as Gertrude Vanderbilt and Consuelo later complained that they hated being watched, they had good reason to blame their parents' generation for seeking out publicity twenty years earlier.

Two weeks after her wedding in 1875, however, Alva had to

attend to sadder matters than publicity, for her father finally died. His daughter's change in circumstances had come too late to help him. 'Had he died sooner, the whole course of my life might have been other than it was. But who is there living who cannot say that of some event in his or her life?,'[7] Alva remarked to Sara Bard Field. After Murray Smith's death she was shown great kindness by William Henry Vanderbilt who told Alva that he regarded her as a daughter and that she should turn to him for whatever she needed. The affection was mutual for Alva always held him in great regard. This relationship was not the problem however. Even after marriage, Alva continued to experience the effect that the power of money has on the powerless as she watched her in-laws tiptoe round the ageing Commodore. Though she always maintained fiercely that she was not overawed by him, Alva also took great care to avoid giving offence, for no-one knew how his fortune stood, nor what he proposed to do with it after his death.

Fortunately, the Commodore took to his pugnacious new granddaughter-in-law from the outset, perhaps divining qualities which were less apparent in his handsome grandson. When she expressed a fondness for country life, he shocked everyone by giving her the use of his old family home on Staten Island. 'Much to his surprise, and I believe also his interest and gratification, I took him at his word . . . I renovated the old house, which had been his home many years before, and went there one July intending to remain perhaps through August.'[8] The visit was an unqualified disaster and soon 'between mosquitoes, and chills and fever, I had quite enough of it'. The Commodore remained happily unaware of this. 'I never told the Commodore, leaving him under the impression that I stayed there longer than was really the case. It pleased him, and that was all that mattered.' For his part William K. also netted a significant success after becoming engaged to Alva: his name appeared as one of the sponsors of a bouncer's ball, marking the first appearance of the Vanderbilt name in a social column.[9]

After the wedding in 1875, there was a period of mourning for Alva's father. Mr and Mrs William K. Vanderbilt concentrated on settling in to their new brownstone house on West 44th Street and avoided taking any action that might unsettle the Head of the House of Vanderbilt. Alva could at least console herself with the knowledge that she had a fashionable home, a secure income, an amiable and handsome husband whose social standing was improving, warm relations with his rich family and excellent expectations. Compared to at least two of her closest friends, her position was enviable. In spite of her success with the Prince of Wales and her reputation as an heiress, Minnie Stevens stayed on the marriage market until 1878 by which time she had suffered considerable public humiliation. After the death of Mr Paran Stevens, his wife and Minnie spent much time in Europe on the look-out for an aristocratic husband. The Duc de Guiche proposed to her but broke off the engagement when the Duc de Gramont had a man go through her affairs who discovered that she was not worth as much as anyone had imagined. Lady Waldegrave, one of Miss Stevens' sponsors in London society, wrote to Lady Strachey: 'I must say I think this business very cruel, but at the same time I can't help thinking she deserved a snubbing as she told me she had £20,000 a year and would have more and she told me that sum in dollars as well, so there is no mistaking the amount.' Minnie Stevens finally married the titleless Arthur Paget in 1878, at the age of twenty five (though the story could be said to have one kind of happy ending for he eventually became a baronet, and she became Lady Paget).

Alva's oldest female friend, Consuelo Yznaga, meanwhile, caused a social sensation the year after the Vanderbilts' wedding by marrying Viscount Mandeville, heir to the 7th Duke of Manchester. Like Alva, Consuelo Yznaga brought very little money to the marriage, a disadvantage compounded by a growing family reputation for eccentricity, though it must be said that the bar for eccentricity was set low in late-nineteenth-century New York. Consuelo Yznaga's brother Fernando was later divorced by Alva's sister Jenny, 'because he never wore socks';[10] Consuelo Yznaga herself became famous for whipping out a banjo in London drawing rooms and playing popular songs to the assembled company. Viscount Mandeville's parents were deeply dismayed by the engagement because of his fiancée's inadequate dowry, but in the longer term it was their son who proved to be the libertine. In spite of a magnificent wedding in Grace Church attended by 1,400 guests, it was not long before the gossip columns were talking openly of the manner in which the Viscount was putting the Atlantic between a music-hall singer and his wife: and when he died young, in 1892, it was said of his widow that she had spent much of her married life as 'the pet of the spare bedrooms'.[11]

Later, Edith Wharton would use Consuelo Yznaga as the model for an unhappy, indebted, adulteress in *The Buccaneers.* In 1876, however, her transformation into Viscountess Mandeville looked like a pace-setting coup. It may have unsettled Alva and it certainly seems to have implanted an idea. Alva had already turned her attention to starting a family, becoming pregnant in June 1876. The William K. Vanderbilts' first child, a daughter, was born on 2 March 1877. The baby was immediately named Consuelo after her godmother, the only duchess-in-waiting that either of the Vanderbilts knew.

About the same time as Alva became pregnant with Consuelo, the Commodore was diagnosed with cancer. His strong constitution made death protracted. A miasma of disinformation floated over his deathbed as competitors circulated tales of his demise to undermine the Vanderbilt stocks. He is said to have thrown hot-water bottles at his doctors and yelled imprecations at waiting journalists, though his wife was encouraged that he simply paid off a noisy organ grinder beneath his window rather than threaten to shoot him.[12] The Commodore finally died on 4 January 1877, surrounded by large numbers of his family. It was claimed that he had enjoyed singing hymns on his deathbed, although the Reverend Henry Beecher spoilt the party by adding sourly: 'I am glad he liked the hymns, but if he had sung them thirty years ago it would have made a great difference.'[13]

The Commodore's obituary in *The New York Times* ran to several pages, and the flags in New York flew at half-mast. He was buried in a simple ceremony at the Moravian Cemetery at New Dorp. In one sense, the Commodore's story had come full circle – the farm boy from Staten Island returning to be buried as a titan of industry. In another sense, however, the arrangements the Commodore made for his fortune demonstrated his absolute determination that the history of the earlier Staten Island Vanderbilts should never be repeated.

When it was published, the Vanderbilt will caused uproar. The Commodore had left an astonishing $100 million*, making him the wealthiest man in America, twice as rich as John Jacob Astor and the department store owner Alexander T. Stewart. Even more surprising, however, was the manner in which he had bequeathed his fortune. Almost $90 million went to William Henry to keep the New York Central Railroad intact, suggesting that when the Commodore had said 'If you give away the surplus you give away the control,'[14] he meant it. A further $10 million was divided between William Henry's four sons, with the greater share assigned to two of them already working in the family enterprise – Cornelius II and William K. Even in death, the Commodore had flown in the face of convention. He had, in effect, transferred to the English system of primogeniture from the European principle of equal inheritance, setting up a Vanderbilt dynasty which would descend through William Henry, and, in the words of Louis Auchincloss, consigned the rest of his children and their descendants to life as 'nonkosher Vanderbilts'. When William Henry heard the news, he is said to have put down his head on the piano and wept.

There were few sounds of weeping from William K. and Alva, however. Their cautious strategy had paid off. William K.'s charm and application on behalf of the family enterprise (and possibly the Commodore's affection for Alva) netted him $3 million – without the responsibilities laid upon his elder brother Cornelius II. The industrious and conscientious Cornelius received a larger bequest totalling $5.5 million, but this signalled his position as head of the family designate and a clear understanding that he would eventually take charge of the business. Others of whom the Commodore

* approximately $13.9 billion today

approved also fared well, including William Henry's younger sons, Fred and George, while his widow had already agreed to $500,000 and the house in Washington Square as her dower settlement. The rest of the family were less than delighted. The Commodore's eight daughters were left $2.45 million between them, split unevenly and depending, it would seem, on relative degrees of spite towards his second wife. The unfortunate Cornelius Jeremiah was only awarded $200,000 in trust.

Just when William K. and Alva felt financially secure enough to launch themselves at the very pinnacle of New York society, Cornelius Jeremiah and two of his sisters determined to sue. This was a major setback. The will case dragged through the courts for months until March 1879. Allegations flew backwards and forwards of the Commodore's profanity, his aggression, his association with spiritualists and 'healing hands', and his cruelty to his afflicted son. Much to the disappointment of the press, the Claflin sisters did not produce the embarrassing testimony that was anticipated (possibly because William Henry paid them off) but the public was once again reminded of a most unfortunate association. It was alleged that the Commodore suffered from a form of mania when it came to money and his 'virility' was adduced to support this. In turn, the defence made much of Cornelius Jeremiah's drunkenness, gambling and indebtedness. Most of these arguments were rejected by the judge, and the trial was suddenly settled in 1879 when William Henry volunteered to hand over some of his fortune to his sisters. Nonetheless, the feud was never patched up and it is small wonder that in 1878, Mrs Astor still felt compelled to hold the line when it came to admitting Vanderbilts to her famous ballroom.

Although the trial was acutely embarrassing, William K.'s legacy of $3 million was not included in the contested part of the will. He and Alva continued to keep a modest profile as they set about their first important building project, a house on Long Island. If Alva's imagination had been ignited by French culture as a teenager, the model that entranced William K. was that of the English sporting gentleman with a house in the country. He was not alone. 'Wealthy Americans learned to drive fancy coaches, play polo, hunt with hounds, breed racehorses and pedigree livestock and took up yachting . . . ,' writes Eric Homberger. 'They collected Old Masters, oriental carpets, heirloom silver, and precious jewels. Americans began to describe themselves as "sportsmen". English taste and style, suggesting refinement, social position, and wealth were professedly aristocratic in the eyes of New Yorkers. They still are.'[15]

Soon after the death of the Commodore in 1877, William K. bought 900 acres of land near Islip on Long Island and asked the architect Richard Morris Hunt to build him a sporting retreat. Alva's vehement determination to control the story of her first marriage has disguised the fact that in their early years of wedlock, William K. was just as set on aristocratisation as his wife. At this stage, indeed, he led the way. The arrival of the railroad to Islip in the 1870s put an abundant supply of game within easy reach of New York; and it helped that Islip was secluded – the more the lives of the social elite were observed by the press, the more important privacy became. More significantly, however, the spot William K. selected for his country house was conveniently close to the first exclusive gentlemen's club that invited him to become a member, the South Side Club near Islip which he joined the year after his marriage in 1876. The South Side was a sporting club where pedigree and social connections mattered less than whether a chap was a good shot with pleasant sporting manners, making William K. a perfect candidate.[16]

The house, which was ready for occupancy in 1879 when Consuelo was two, was designed by Hunt in the fashionable 'Stick Style', an all-wood version of the English mock-Tudor method of half-timbering. Unlike the houses of England's landed aristocracy, however, it was conceived from the outset as a retreat from city life. The name it was given, 'Idle Hour', suggested a place of leisure, decided (it was said) on the toss of a coin with Mr Schuyler Parsons on the porch of the South Side Club, who then had to make do with 'Whileaway' for his own establishment nearby.[17]

Idle Hour cost the Vanderbilts a mere $150,000 out of the $3 million they had inherited. Throughout the 1880s they developed it to a point where the estate was almost entirely self-sufficient.

Eventually Idle Hour had amenities of which most English aristocrats, sporting or otherwise, could only dream, including an ice-house, a laundry, a water tower, a house for the superintendent, a house for the palm trees and a teahouse by the bay. Idle Hour played an important role in securing William K.'s membership of other smart clubs. Although he joined the Union Club in 1877, the most exclusive of them all – the Coaching Club – only capitulated after he invited all its members to stay at Idle Hour in 1883.[18] The building of Idle Hour was just as significant for Alva for quite a different reason: it brought her into contact with its architect Richard Morris Hunt. In Alva, Hunt found a visionary client in sympathy with his ideas; she encouraged daring and innovation, allowing him to find new levels of creativity and audacity that would make him the leading architect of the Gilded Age. For her part, Alva suddenly found a way of expressing herself.

In another world, at another time, it is perfectly possible that Alva might have been an architect. Some of those who knew her best, including Consuelo, thought she was always at her happiest when she was designing houses and rearranging landscapes. This was one of life's theatres where she ceased to be a spectator and became a paid-up member of the cast. When it came to designing Vanderbilt houses, she considered every detail and it seemed to calm her down. Hunt understood this instinctively. He described her as a 'wonder' to his wife, and gave Alva the use of a draughtsman in his office to help her work out her ideas. 'I spent many delightful hours in his office, working with the draughtsmen he placed at my disposal, always encouraged by him, and inspired alike by his kindness and great genius. He was my instructor and dear friend for many years, and the work we did together was for me always an endless delight, and a great resource.'[19] This is not to say they did not fight, but when they did they were well matched. 'Mr Hunt had a fiery temper . . . my own was not mild. We often had terrific word battles. With fiery intensity he would insist on certain things. I would, with equal eagerness, insist on the contrary. Once during the planning of this house we had had a long and heated argument over some detail of measurement. Finally he turned to me in rage and said "Damn it

Mrs Vanderbilt who is building this house?'' and I answered ''Damn it, Mr Hunt, who is going to live in [it]?'' '[20]

Richard Morris Hunt has the distinction of being one of the very few men Alva ever really loved, although there is no suggestion that the relationship was anything other than platonic and plenty to demonstrate that she was a rigorously demanding client. She called it one of the great companionships of her life. It gave her scope to fulfil a long-held ambition – to change the way New York looked, and to turn it, as far as possible, back into France. The prevailing architectural fashion was the brownstone house, symptomatic, in Alva's later professed view, 'of the lightly veneered crudeness of America'. When Alva and Richard Morris Hunt first met there was a meeting of minds on this issue: 'I told him how my taste trained in the European capitals had been shocked with what seemed to be a conspiracy of bad taste in American architecture and how willing and eager I was to break away from all precident [sic] and under his guidance build a thing of beauty . . . [I] determined that if ever the time came when I built a house I would profit by my contact with the architectural beauties of the Old World . . .'[21].

Richard Morris Hunt had found a kindred spirit. Although a generation older, he had spent nearly nine years studying in Paris. Like Alva, he was captivated by French art and architecture in his youth, and came to speak French so fluently that he was sometimes mistaken for a Frenchman. Moreover, he studied architecture with Hector-Martin Lefuel, official architect to the Second Empire which had so entranced Alva. Lefuel encouraged Hunt to enter the Ecole des Beaux-Arts in Paris, making him the first and only American architect of his generation to be trained there. Alva and Richard Morris Hunt were not, of course, the only people in New York who were fascinated by the opulent world of Second Empire France but between them they pioneered something new: the introduction of beaux-arts architecture to New York, a style that would define the Gilded Age and dominate the city's architecture until the First World War.

The core idea at the heart of the beaux-arts was the conviction that the architectural ideal was classical, embracing not just Greek and Roman architecture but the French and Italian Renaissance as well. However, beaux-arts theory also looked to the future with state-of-the-art construction techniques, using modern materials such as plate glass and iron.[22] Indeed the great beaux-arts buildings of New York were only possible thanks to the more remarkable inventions of the industrial revolution, such as the elevator and electric lighting, which allowed corporations to construct great edifices for large numbers of workers. One characteristic of these buildings was the way they dramatised space. According to the architectural historians Foreman and Stimson: 'Good beaux-arts buildings have a very calculated dramatic effect . . . Facades and entries were held to be crucial in establishing important initial reactions to the building's use and importance.'[23] In public beaux-arts buildings such as the New York Public Library the effect was democratic – anyone gliding down its great main staircase could feel stately. Applied to domestic architecture, however, the beaux-arts philosophy had quite the opposite effect. The style provided sweeping backdrops for America's new aristocrats in much the same way that Versailles dramatised the *ancien régime*. Alva would have agreed with Henry James that the secret of its appeal lay in a 'particular type of dauntless power'.[24]

The first beaux-arts collaboration between Alva and Richard Morris Hunt was a house in New York to replace the brownstone house on West 44th Street that she secretly greatly disliked. The design of the new house was undeniably the outward expression of social ambition. Alva was, in effect, pioneering vertically rather than horizontally, creating a space that the aristocrats of New York would find irresistible. However, 660 Fifth Avenue can also be understood as the first great example of Hunt and Alva's shared vision – a house designed to show American aristocracy what could be done if the great architecture of the European past was combined with the American gift for the modern in America's own 'Renaissance'. Alva's vision for the Vanderbilts went even further, for she felt the family should act like Renaissance merchant princes and become great patrons of the arts. The Medicis of Florence had built houses that were not merely beautiful private residences but an outward expression of the importance of the family. They had 'represented not only wealth but knowledge and culture, desirable elements for wealth to encourage . . .' If the Medicis could do it, so could the Vanderbilts. 'I preached this doctrine at home and to William H. Vanderbilt' she wrote later.[25]

Persuading her father-in-law, William Henry Vanderbilt, to behave like a Medici, turned out to be surprisingly easy. Liberated from both the Commodore and the will case after 1879, he went into action in an uninhibited manner which astounded New York society, only just coming to terms with his transformation from Staten Island farmer to railroad tycoon. He needed little persuasion that the Vanderbilts should build houses that reflected the family's wealth, and encouraged his elder son, Cornelius II, to follow suit. The settlement of the will case had the effect of a starting pistol: William H., Cornelius II and William K. all filed plans for houses along Fifth Avenue on the same day. Fifth Avenue north of 50th Street was at that time unfashionable, but by the 1880s the area would be known as 'Vanderbilt Alley', setting a tone for sumptuous development in New York for the rest of the century.

660 Fifth Avenue was not the largest of the three houses (the others were at 640 and 1 West 57th Street) but it was certainly the most audacious. Alva drew gasps from her in-laws when she presented her plans to them in 1879:

> I knew that they were more elaborate and would have a somewhat staggering effect on the family group. Nor was I mistaken. When the paper was unrolled and they all saw the pretentious plans of a house which would cover almost a city block there was a unanimous gasp from the assembly. With much elation I carefully explained the drawing, elaborating all the details and enjoying the effect on my audience. After a while my father-in-law said crisply: 'Well, well, where do you expect to get the money for all this?' 'From you' I answered instantly, giving him an affectionate slap on the back. The rest sat appalled at my temerity. To them it was like being families with Established Power. My father-in-law laughed and the money for the house came.[26]

660 Fifth Avenue, designed by Richard Morris Hunt for Alva and William K., would become a New York landmark for decades. Based on the sixteenth-century chateau of Blois on the Loire, out of Chenonceau, it also enjoyed fifteenth-century French Gothic accents, and marked a turning point for Morris Hunt as he finally slipped his American architectural moorings. Suggesting that Sleeping Beauty's castle had somehow landed from outer space on Fifth Avenue, its dominant feature was a three-storey tourelle by the entrance. There were gargoyles, flying buttresses and gables. The designs for this house, a masterpiece of aristocratic image-making, suggested something more complex than straightforward conspicuous consumption, or even aristocratic emulation – though both were an important part of its make-up. The most striking note of all was an unmistakeable flight from reality. In Richard Morris Hunt's conceptual watercolour for 660 Fifth Avenue, ghostly figures inhabited a fairy-tale palace; drawings for other rooms, such as the Supper Room, were peopled by tiny Renaissance princes.[27] In 660 Fifth Avenue, it was as if a deliberate decision had been taken to turn an aristocratic back on the drab, poverty-stricken world a few blocks away – a world into which one could fall so easily without a safety net. This sense of withdrawal to a magical past was a new departure for American architecture; it would make its own contribution to the growing sense of division between rich and poor in New York and it would be copied to the point of pastiche by the early-twentieth century.

Although New York's architects generally approved of 660 Fifth Avenue, and admired its originality, reaction from New York society was mixed. The pale Indiana limestone of its exterior marked a decisive break with the ugliness of brownstone houses. Every block of limestone was tooled – worked over by a hand chisel. The facades were covered with a riot of rich and decorative carving which caused great consternation: 'This radical departure from the accepted brown-stone front raised a storm of criticism among my friends,' wrote Alva. 'O these whippings from parents and society when the child or adult wishes to be a person and not a member of a mass.'[28] (Readers of Edith Wharton's *The Age of Innocence* may remember that Catherine Mingott's decision to build a pale, cream-coloured house also marked her out as a morally courageous eccentric.) While she was away in Europe, Alva received a stream of alarmed letters telling her that carvings 'of naked boys and girls' were appearing on the rooftop. 'They failed to see, as many, fatally tainted by Puritanism still fail to see, the exquisite beauty of the human form and of its significance in connection with the special period we were trying to represent,'[29] she commented.

Even to those who could take the psychological strain, the interior was almost overwhelming, dominated by spaces intended to dramatise the authority and economic power of the Vanderbilts. The dining room was 80-feet long, 28-feet wide and 35-feet high, had two colossal Renaissance fireplaces and a stained-glass window depicting a scene from the meeting of Henry VIII and Francis I at the Field of the Cloth of Gold. The sweeping grand stairway of Caen stone to the second floor was a tour de force of trophies, fruit, masks and cherubs. The entrance hall measured 60 feet and was lined with carvings and tapestries. The dominant theme may have been illusion and flight from reality but the translation of the remains of France's *ancien régime* to this new American interior was real enough. It had all been masterminded by the French firm, Jules Allard et Fils, who would come to specialise in importing architectural salvage, artefacts and paintings directly from the houses of ruined French aristocrats for the houses of plutocratic aristocrats in the United States. The William K. Vanderbilts' paintings not only included Rembrandt's 'Man in Oriental Costume', and Gainsborough's 'Mrs Elliot', but François Boucher's spectacular 'Toilet of Venus'. This came to Alva's boudoir – indirectly – from the boudoir of Madame de Pompadour; and at least one fine secretaire came from the apartments of Marie Antoinette herself.

As late as 1882, Mrs Astor was still refusing to acknowledge the Vanderbilts formally. Her attitude was increasingly irrational for leaving aside Alva's claims to southern gentility, Cornelius II had married Alice Claypoole Gwynn in 1867 (whose great-great-grandfather was Abraham Claypoole, a direct descendant of Oliver Cromwell), and in 1881 William K.'s sister, Lila, married William

Seward Webb, whose grandfather had been an aide to George Washington. Both William Henry and Cornelius II, head of the family elect, were building fine houses and lived lives of unimpeachable luxoriousness. However, Mrs Astor's strength of feeling on this matter may have been reinforced by two Vanderbilt upsets in the same year. One came about as a result of William Henry Vanderbilt remarking: 'The public be damned!' in answer to a reporter's question about running a Pennsylvania train for the public benefit. Some maintain that William Henry was simply defending the interests of shareholders as he had every right to do, but he was universally excoriated for this jest, and the image of a Vanderbilt as a boorish robber-baron was successfully dangled before the public once again by his opponents. The scandal surrounding the unfortunate Cornelius Jeremiah was worse. After the Commodore's death he became obsessed with funding his addiction to gambling. In 1882 he shot himself in the Glenham Hotel in New York, leaving debts of over $15,000. An undignified auction of his belongings compounded the disgrace of a family suicide.

Undaunted, Alva and William K. pressed on with their entrée to New York's social elite. A charming and energetic couple, about to take possession of a huge and dazzling house which would flatter the ambitions and pretensions of New York's *gratin*, they were already being asked to the best parties. In spite of family scandals they were invited to a Patriarchs' ball in 1882 and another early in 1883. As 660 Fifth Avenue neared completion, they started to plan a house-warming party of their own. The Vanderbilt ball, as it came to be known, has gone down in the annals of party history. In deciding to hold it in March 1883, and to send out 1,600 invitations, Alva and William K. must have calculated that to a very great extent, society's resistance to the Vanderbilts was already collapsing. They knew that the elite of New York was agog with curiosity over 660 Fifth Avenue; they made sure that society understood that the ball would be like no other in terms of expense and display; and Alva shrewdly reduced the social risk to invitees (and herself) by giving the party in honour of her old friend, Consuelo Yznaga, now Viscountess Mandeville, knowing full well that the presence of a real aristocrat would overcome residual hesitation – a manoeuvre she would repeat in the future. This left the problem of Mrs Astor.

The story goes that Alva used the ball to outwit Mrs Astor, who had not, in March 1883, been persuaded to relax her Vanderbilt-denying ordinance. This may have been because of recent scandals; possibly because she still thought the Vanderbilts remained a symbol of the dangers of vulgar wealth; and probably because she had anathematised them in the past and was in no hurry to back down. Her daughter, Carrie, on the other hand, was closer in age to the William K. Vanderbilts and enjoyed parties given by younger 'swells'. She looked forward to being asked to the Vanderbilts' house-warming ball, and even started to rehearse quadrilles with her friends. It then transpired that there could be no invitation because, according to the etiquette of the day, Mrs Astor had to call on Mrs Vanderbilt before Alva could invite Miss Carrie Astor to the ball. Such was the distress of Miss Carrie Astor that Mrs Astor's maternal love overcame her pride. She relented, made the call and an invitation was forthcoming.

This story has long been called into question. There is no doubt that the ball was planned with an element of calculated risk and that Alva wished Mrs Astor to grace it with her presence. There is no doubt that Mrs Astor only called on Alva for the first time shortly before the ball. However, Alva and Mrs Astor sat together on the executive committee of the Bartholdi Pedestal Fund,[30] and the Vanderbilts had already attended two Patriarchs' balls, which would have been impossible without Mrs Astor's implicit approval. It is more than likely that if there had been no ball, Mrs Astor would have called on Alva soon after she moved into her new house – at the moment when, as one wag put it, the Vanderbilts had finished Vanderbuilding. The ball simply acted as a catalyst for Mrs Astor's public acknowledgement as Alva hoped it might.

Once the invitations had been sent out, it is perfectly possible that Carrie Astor appealed to her mother to speed things up and that Ward McAllister sensed that it would be better for Mrs Astor to acknowledge the Vanderbilts formally if she wished to stay abreast of the zeitgeist and avoid looking foolish. The story that Alva deliberately outwitted Mrs Astor is too crude, however. In one sense she had done that long before when she started to plan 660 with Richard Morris Hunt. The end result was the same, however. It only took a brief glimpse of the interior of 660 Fifth Avenue to reassure the Queen of Society that Mr and Mrs William K. Vanderbilt were fine upstanding examples of the civilised 'money power'[31] which she and Ward McAllister so wished to encourage. 'We have no right,' she commented in 1883, 'to exclude those whom the growth of this great country has brought forward, provided they are not vulgar in speech or appearance. The time has come for the Vanderbilts.'[32]

Proust's remark that parties do not really happen until the day afterwards when the uninvited read about them in the newspapers is only partly true of the Vanderbilt ball. This party was a wild success before it ever took place. Not only did Mrs Astor finally capitulate, but the ball was the principal subject of discussion for weeks beforehand among the prospective guests. It was a fancy-dress ball, of course, in the spirit of make-believe and flight from reality that characterised the house; and the elite of society happily collaborated. 'Every artist in the city was set to work to design novel costumes – to produce something in the way of a fancy dress that would make its wearer live ever after in history,'[33] wrote Ward McAllister with a characteristic sense of proportion. Alva was deeply gratified by the time and energy expended by hundreds of guests on their outfits, which took weeks of work by New York's best dressmakers and couturiers. The degree of focus, effort and cost expended could only be seen as a compliment to the new generation of civilised Vanderbilts and marked out their elevation to the apex of society just as clearly as any endorsement from Mrs Astor.

In Alva's view, the male guests at the ball were, if anything, 'more brilliantly and perfectly turned out than the women'.[34] The invitation certainly sent some of them into a great sartorial tizzy. On the day of the party Ward McAllister was obliged to recruit extra helpers to get him dressed, 'two sturdy fellows on either side of me holding up a pair of leather trunks, I on a step-ladder, one mass of powder, descending into them, an operation consuming an hour'.[35] Another male guest, Augustus Gurney, never managed to resolve his outfit crisis. He went home in the middle of the ball and changed, disappearing as a Moldavian chieftain and re-appearing as a Turkish pasha.

It was, said Alva modestly, 'the most brilliant ball ever given in New York'.[36] It was certainly one of the more surreal. Don Carlos chatted away over supper with Little Bo Peep; Mary Stuart was seen in conversation with Neapolitan fishermen and a Capuchin monk; a plethora of Hungarian hussars mingled with several representatives of the French Bourbons; and the Cornelius Vanderbilts stood for both past and future with Cornelius as Louis XVI and Alice as 'Electric Light', in a costume that intermittently lit up, courtesy of batteries secreted in her pockets. Curiously, both Alva and Mrs Astor appeared as Venetian noblewomen, and were seen chatting amiably and publicly on the stairs. Alva's dress was made of white satin embroidered in gold, with a velvet mantle, and a diadem of diamonds. Many of the costumes, including Lady Mandeville's as Queen Maria Theresa of Austria, came posthaste from Paris. Perhaps most interesting of all, William K. was dressed as François I in doublet and hose, bearing a remarkable resemblance to the small princely figure whom Richard Morris Hunt once inserted into his earliest designs for the Supper Room.

That evening, the involvement of the guests in the success of the party went further than turning up in elaborate costumes and acknowledging that the Vanderbilts had 'arrived'. The other huge compliment paid to the hosts was the trouble taken over the quadrilles, which became the high point of the evening. Quadrilles were square dances in five movements which had become elaborate fixtures at society balls, for they were danced in costumes designed round a theme, and took weeks of organisation and rehearsal by teams of guests beforehand. The six quadrilles at the Vanderbilt ball exceeded anything that had ever been seen before, danced by over a hundred of the Vanderbilts' friends.

According to one authority, 'the chief attraction was the "hobby horse quadrille,'' for which the dancers wore costumes that made them look as if they were mounted on horses. The life-size hobby horses took two months to construct and were covered with genuine leather hides and flowing manes. Tails were attached to the waists of the dancers and false legs placed on the outside of richly embroidered horse blankets, giving the illusion that the dancers were mounted; ''the deception'', one observer enthused, ''was quite perfect''.'[37] Ward McAllister organised the Mother Goose Quadrille himself (another compliment to the hosts) which involved participation from Jack and Jill, Little Red-Riding Hood, Bo-Peep, Goody Two-Shoes, Mary, Mary Quite Contrary, and My Pretty Maid. He was forced to concede, however, that it was the Star Quadrille containing the 'youth and beauty of the city' which was the most brilliant, for all the young ladies wore electric lights in their hair which produced 'a fairy and elf-like appearance to each of them'.[38]

As Alva put it later, the 1883 Vanderbilt ball 'marked an epoch in the social history of the city'. As well as consolidating the position of the Vanderbilts, it marked a change of pace in two other ways. Alva, ever mindful of maximum visibility, was the first hostess to allow a full report of the ball to be syndicated to the newspapers through the *New York World* and to allow reporters to wander through the house earlier in the day. It was one of the *World*'s earliest society scoops and set a precedent for press coverage of similar events in the following decades. The paper calculated that the ball cost $155,730 for the costumes, $11,000 for the flowers, $65,270 for champagne and music, and $4,000 for hairdressers. This meant that Mrs William K. Vanderbilt had also set a vertiginous new standard for just the kind of social expenditure that had come so close to defeating the Smiths when they returned from France to America.[39]

When writing her memoirs in later life, Consuelo could recall very little of her early childhood. She remembered nothing of the ugly brownstone building where she was born. No-one registered her birth either, an oversight that subsequently caused a great deal of bureaucratic trouble. She moved into 660 Fifth Avenue with her parents in 1883, just before she was six, so the childhood she recollected began in surroundings of extraordinary affluence. She does not seem to have been present at the 1883 ball (unlike Cousin Gertrude who was two years older and went for part of the evening, dressed as a tulip). She remembered other parties, however: 'How gay were the gala evenings when the house was ablaze with lights and Willie [her younger brother] and I, crouching on hands and knees behind the balustrade of the musicians' gallery, looked down on a festive scene below – the long dinner table covered with a damask cloth, a gold service and red roses, the lovely crystal and china, the grown-ups in their fine clothes ... the ladies a-glitter with jewels seated on high-backed tapestry chairs behind which stood footmen in knee-breeches.'[40]

At other moments, there were distinct disadvantages to living like a princess in a neo-Gothic palace, which, like many houses built primarily for entertaining and display, could feel gloomy and frightening when no-one else was there. The fact that the stairway was carved in Caen stone was quite irrelevant when the princess happened to be cursed with a neo-Gothic imagination. 'I still remember how long and terrifying was that dark and endless upward sweep as, with acute sensations of fear, I climbed to my room every night, leaving below the light and its comforting rays. For in that penumbra there were spirits lurking to destroy me, hands stretched out to touch me and sighs that breathed against my cheek.'[41] Life in an urban chateau had its compensations, however. On the floor beside her bedroom there was a playroom big enough for bicycling with friends. There were horse-drawn sleigh rides in the streets of New York in winter, trips to the family box at the Metropolitan Opera to hear Adelina Patti sing, and weekly classes at Dodworth's Dancing Academy marking her out as a junior member of New York's elect from birth.

Alva always said that she loved motherhood. She remembered a sense of religious joy when she discovered she was to have her first baby. If it ever became fashionable to decry such feelings, she wrote, she would not join in. 'So long as the world endures there will be women who will quiver to these emotions . . . no matter what freedom of expression is finally attained.'[42] Consuelo's birth in 1877 was followed by the arrival of her brothers William Kissam II (known as Willie K. Jr) in 1878, and Harold Stirling in 1884. Alva prided herself on the fact that, unlike members of the English aristocracy, she did not hand her children over to the care of others. 'I dedicated the best years of my life to rearing and influencing and developing those three little beings who were my links with the future. I gave them an exclusive devotion. I considered their welfare before all else. I lived in their lives and cultivated no other apart from them for myself.'[43]

In 1909 Alva announced that she was writing a book about her child-rearing methods, and though it never materialised, she told the *New York City Journal*: '[My children] were not put away to sleep in a room with the nurse; they slept in my room. The nursery was next to my room, and when they were older they slept there, but with the door open to I could look after them, and the smallest one slept in my room. I nursed all my children, though I don't know that anyone is particularly interested in that.'[44] By 1917, however, she had come to believe that excessive pre-occupation with her children had been misguided, and that mothers should not sacrifice themselves as she had done. 'I want to say unhesitatingly that I believe this was wrong. I deplore the eternal sacrifice of women for another or others. Motherhood and Individuality should not conflict. Motherhood ought not to kill Personality in the mother and Personality in the mother ought not to injure the child.'[45]

However much Alva enjoyed motherhood she was also ambivalent about it – largely on the grounds that many women became mothers just at the moment they were finding themselves. 'It is a formative time for them so far as intellect goes . . . [A young mother is] in a sense a diamond already cut and ready to sparkle as she can find the light. Yet for the sake of developing the unknown quantity which her children are she gradually slips back into the darkness.'[46] Alva always felt that the equation between the perfect woman and virtuous female martyr was wrong. 'The whole history of most women's lives is summed in self-sacrifice. If it is not for a child whose future is uncertain then it is for an aged parent whose life is done. Again and again people have pointed out to me some splendid woman who was burying her talent under care for a decrepit relative. "Isn't her life beautiful!" they would exclaim. No, it is not beautiful. I think it is disgusting. I think it is wicked,'[47] she told Sara Bard Field.

In Alva's case, talk of immolating maternal self-sacrifice should be treated with caution. This was not modern hands-on motherhood. Like other affluent households in New York in the 1880s, 660 Fifth Avenue had nursemaids, nannies, housemaids, governesses and cooks. The fact that Consuelo's earliest years were so unmemorable has much to do with the disciplined and dull world of an affluent nineteenth-century nursery where the emphasis was on avoiding undue stimulation, building up the infant's strength and avoiding infection. Even when her children were very young, Alva was occupied with other matters: designing and decorating houses with Richard Morris Hunt, ensuring the Vanderbilts were behaving like Medicis, taking her rightful place at the apex of New York society, as well as the complex task of managing two large households.

There is also no sign that Alva's personality was in any way dimmed by maternity, though as each child left the nursery she certainly exercised an increasing degree of control over its life. Alva saw a direct relationship between building houses and building children: 'If one can judge of her own self I would unhesitatingly say that the two strongest characteristics in me are the constructive and the maternal. They are or ought to be associated.'[48] Children were, of course, the greater responsibility for here one was building character. Alva's view of maternal responsibility was first, that the mother was directly responsible for developing the character of each child; second, that each child should be treated as an individual with an independent mind; and third, that it was the parent's responsibility to 'guide' the child to the right course in life, based upon (and this was the rub) parental assessment of the child's individual characteristics.

This view of maternal responsibility was, in many ways, an

extension of the way Alva had described how she played with her dolls as a child ('I loved dolls . . . I took them very seriously. I put into their china or sawdust bodies all my own feelings.'[49]) She frequently expected Consuelo to behave with the submission of a doll, a 'china body' on to which Alva projected all her own feelings. Consuelo was to be the princess in Sleeping Beauty's palace. 'Gertrude and I were *heiresses*,' Consuelo once told Louis Auchincloss. 'There seemed never to have been a time when this was not made entirely clear.'[50] She was even dressed to stand apart by Alva, forced into 'period costume' for parties and sniggered at by other children. However, this often clashed with Alva's other view, which she held with equal conviction, that her children should be independent-minded individuals – like her, in other words. This contradiction at the heart of her approach to child-rearing was frequently irreconcilable and posed a very difficult conundrum for her offspring, especially Consuelo. Should they please her by submitting to her as doll-children? Or would Alva be more contented if they showed signs of independence? It was often very difficult to know.

In practice, submission to Alva's will generally took priority. It was, in any case, an age when inculcating obedience in children was widely considered a major parental responsibility, the first step in developing moral character. Childcare manuals of the period recommended that obedience training should start as early as twelve or fourteen months to encourage 'self-control and self-denial, and advancing a step towards the mastery of [the child's] passions'.[51] If obedience was important in boys, it was essential in girls. 'We were the last to be subjected to a harsh parental discipline,' Consuelo wrote. 'In my youth, children were to be seen but not heard; implicit obedience was an obligation from which one could not conscientiously escape.'[52]

Even by the standards of the day, however, Alva was a ferocious disciplinarian, administering corporal punishment with a riding-whip for the most minor acts of delinquency. When Alva was a child, her mother's whippings had had little effect. But a less head-strong personality like Consuelo could still feel the impact in old age. 'Such repressive measures bred inhibitions and even now I can trace their effects,'[53] she wrote later. Most difficult of all, perhaps, was the stomach-knotting tension induced by a mother with a volatile and ferocious temper: 'Her dynamic energy and her quick mind, together with her varied interests, made her a delightful companion. But the bane of her life and of those who shared it was a violent temper that, like a tempest, at times engulfed us all.'[54]

While Alva certainly took time to be with her children it was not quite the unalloyed pleasure for her offspring that she seemed to imagine. 'The hour we spent in our parents' company after the supper we took with our governess at six can in no sense be described as the Children's Hour,' wrote her daughter. 'No books or games were provided; we sat and listened to the conversation of the grown-ups and longed for the release that their departure to dress for dinner would bring.'[55] Alva lunched with her children almost every day for seventeen years, refusing (or so she later claimed) all social invitations in the middle of the day so that she could be available to her children. While she maintained that these lunches were the 'children's dining table', an 'open forum' at which 'everyone's opinion was gravely received' even when there were adult guests present, Consuelo remembered longing to express a view but invariably being repressed by a look from Mamma.

Having one's character developed by Alva could also be a brutal experience. 'Sitting up straight was one of the crucial tests of lady-like behaviour. A horrible instrument was devised which I had to wear when doing my lessons. It was a steel rod which ran down my spine and was strapped at my waist and over my shoulders – another strap went around my forehead to the rod. I had to hold my book high when reading, and it was almost impossible to write in so uncomfortable a position.'[56] Later, however, Consuelo attributed her famous straight back in old age to this dreadful device.

One result of Alva's passionate involvement in her children's upbringing was that, unlike cousin Gertrude who went to school, Consuelo was educated almost entirely at home so that Alva could oversee her doll-child's educational curriculum. Alva wanted to educate her sons at home too but lost the battle. 'I regretted very much the sending of my sons to preparatory schools. Personally I did not see the necessity of it. When parents have the intelligence required to guide and direct youth, I think it is better for children to stay at home as long as possible. I neither appreciate nor approve the theory held by many as to the value of outside influence in the rearing of children.'[57] In particular Alva objected to the 'one-size-fits-all' approach to education she felt had failed her badly as a child. It is likely that William K. was just as certain that only boarding school stood four square between his sons and total domination by their mother.

Consequently, Consuelo bore the brunt of Alva's educational experiments and maternal philosophy. Alva insisted on proficiency in foreign languages, an accomplishment that was also encouraged by William K. 'At the age of eight I could read and write in French, German and English. I learned them in that order, for we spoke French with our parents, my father having been partly educated in Geneva,'[58] wrote Consuelo. She was made to recite long poems in French and German to her parents every Saturday so that by the time she was ten she was capable of reciting 'Les Adieux de Marie Stuart' at a solfège class concert with such emotion that she burst into tears and was thrown a bouquet.

While instruction was given by tutors and governesses, Alva kept a very close eye on her curriculum, saying that she 'knew the books from which [Consuelo] was being mentally fed as I knew the food that nourished her body.'[59] Alva later told the *New City Journal* that Consuelo often had three governesses at any one time, but 'it was a great nuisance to have them around'.[60] At the same time, Consuelo's education as a linguist did represent genuine encouragement of individual talent, though it was along strictly approved lines. She showed an early talent for languages and everything was done to promote it; and when she occasionally did something well enough to please Alva, the praise was worth having.

Physical independence was also encouraged. At Idle Hour, no-one could have been less like a conventional nineteenth-century mother than Alva. The children crabbed, fished and experienced a taste of the autonomy Alva enjoyed as a child, though even here she could not resist instruction. She had a pond specially constructed so that they could learn to sail and she could dispense geography lessons:

> As the knowledge of navigation increased a mast and sail were added. The row boat, like a caterpillar, put on wings and became a butterfly of the water, a sail boat. With this craft and the pond we developed the Geography of the whole world. Now we were going from Dover to Calais on the choppy Channel. Now we were coming from New York to Liverpool on the perilous Ocean. William, the elder boy, by continuous exertion rocked the boat so successfully that we believed in storms and what they could accomplish for we were all pitched into the pond . . . no young friend who ever visited us met me at the luncheon table attired in her or his clothes.[61]

A governess was also pitched into this pond by the children, who promptly received one of Alva's more memorable thrashings. In spite of her impulse to control every aspect of her children's lives, Alva could be great fun, and courageous if things went wrong. At least once she prevented a serious accident when she jumped up and seized the bridle of a galloping pony as it bolted with Consuelo towards a water hydrant.

In a household where the children were waited on hand and foot, Alva thought it necessary to provide a play house where they could acquire some self-reliance. It was called 'La Récréation' and was one aspect of childhood which Alva and Consuelo later agreed had been a success. 'The German governess and my daughter made preserves there and did a great deal of cookery. In fact, they superintended the cooking while my eldest son was the carpenter and waiter. I and my friends often went there for afternoon tea. It was prepared and served by the children and was most excellent,'[62] wrote Alva. These hours in La Récréation gave Consuelo an early taste of the pleasure of running houses where she was in control. 'This playhouse was an old bowling alley, and when my mother handed it over to us she insisted as a matter of training that we should do all the housework ourselves,' wrote Consuelo. 'Utterly happy, we would cook our meal, wash the dishes and then stroll home by the river in the cool of the evening.'[63]

The children were also given a garden where they grew flowers and vegetables which they were encouraged to take to the nearby Trinity Seaside Home for convalescent children, Alva's first philanthropic undertaking. She told Mary Young that she started the home after watching her delicate eldest son grow into a robust boy and grasped the extent to which wealth had assisted his recovery from precarious health in infancy. Realising that poor mothers lost children because they could not afford the necessary care, Alva purchased land and built a home where convalescing children from poor homes were looked after by Protestant sisters. This was also Consuelo's first exposure to the lives of those less fortunate than herself.

Consuelo's nurse 'as near a saint as it is possible for a human being to be',[64] was another person responsible for drawing back the curtain a little further so that one of the most protected little girls in America had a glimpse of how other people lived. In conversation with a workman from Bohemia responsible for cutting the grass at Idle Hour, Consuelo discovered that he had a crippled child. Encouraged by her nurse, she loaded up her pony-cart with presents and went over to see the child, an experience which forced her to realise for the first time 'the inequalities of human destinies with a vividness that never left me'.[65] At other times, the children sold the vegetables they grew at La Récréation to their mother in an exercise in elementary capitalism: 'I know that they have grown up to profit by these lessons,'[66] wrote Alva. In one respect she was right. Behind her back her children gave themselves elementary lessons in gambling. 'My brother Willie, who was of an impatient nature, would pull up the potatoes long before they were ripe,' wrote Consuelo. 'Our earliest bets were made on the number we would find on each root.'[67]

In 1885 Consuelo's grandfather, William Henry Vanderbilt, collapsed mid-conversation with an old competitor at 640 Fifth

Avenue, and died. He was sixty-four. It is difficult not to feel sorry for William Henry. In addition to vilification as a result of 'The public be damned!' incident, he only lived to enjoy eight years of liberation from his father (or six, if one deducts the years spent attempting to settle the Commodore's will). In the short time available he made up for years of repression. He flung down a challenge to Mrs Astor on his own account as one of the founders of the new Metropolitan Opera Company, set up by a group excluded from the Academy of Music because they were born too late to acquire a box; and he indulged a passion for horseflesh which he inherited from his father. He particularly loved trotting horses, and was often seen driving his famous trotting teams up and down Fifth Avenue. His favourites, Maud S and Aldine, broke the record for a mile at the track at Fleetwood Park in 1883. His stables for the 'trotters' were renowned for having gas lamps with porcelain shades, and sporting pictures on the walls.

When his new house at 640 Fifth Avenue was completed, it was clear that William Henry had finally lost all inhibition when it came to shopping. It was stuffed with enormous pieces of Renaissance furniture (in line with proto-Medici thinking), suits of armour, marble statues, bronzes, mirrors, tapestries and oriental rugs. His front doors were an exact copy of the Ghiberti bronze doors in Florence. There were Japanese rooms, early-English rooms, Grecian rooms. The walls were hung with paintings by Alma-Tadema, by Fortuny, Millet, Munkacsy, Bonheur and Bouguereau, and his great favourite, Meissonier. Those who regarded this favourably saw it as 'regal magnificence'. Edith Wharton, on the other hand, described such Vanderbilt excess as 'a Thermopylae of bad taste'.[68] Amidst it all, William Henry is said not to have seemed entirely at ease, boxing himself into one corner of his library in his old rocking-chair.

In William Henry's hands the Vanderbilt fortune had continued to grow and multiply. Though assisted ably by Cornelius II and William K., as well as the Vanderbilt man of affairs Chauncey Depew, he found it hard to delegate and therefore much of the credit for this must go to him. He shepherded the railroads through a difficult period of unregulated competition, appalling accidents, organised protest at exploitative and abusive freight rates, and serious labour unrest (much of it justified). He improved Vanderbilt trains and managed to keep the Vanderbilt workforce largely on side during a violent railroad strike in 1877. As Alva put it: 'He lacked the commanding qualities of the Commodore who had founded the family fortune, but he had a quality of genuine kindness – almost an extreme kindness and a dogged persistence and thoroughness which father had either instilled or encouraged in him and which made [it] possible for him to handle the great Rail road business left in his care.'[69]

A few days after his death, William Henry Vanderbilt's will was the source of even greater astonishment than the Commodore's. In the short time that he had been responsible for the family fortune, dogged persistence and careful control had doubled its value from about $100 million to $200 million*. This made him the richest man in America and the poet-statisticians of New York's newspapers went into overdrive. In gold, the estate would weigh 500 tons and would need 500 strong horses to pull it down Wall Street; if paper, it would take a man eight hours a day for thirty days to count it. The *New York Sun* declared: 'Never was such a last testament known of mortal. Kings have died with full treasuries, emperors have fled their realms with bursting coffers, great financiers have played with millions . . . but never before was such a spectacle presented of a plain, ordinary man dispensing of his own free will, in bulk and magnitude that the mind wholly fails to apprehend, tangible millions upon millions of palpable money. It is simply grotesque.'[70]

William Henry altered his will nine times in six years, as he fretted over how best to bequeath such a legacy. He was determined to prevent the embarrassment of another will trial, and he felt strongly that the burden of such a fortune was too great for one man alone. 'The care of $200 million is too great a load for any brain or back to bear. It is enough to kill a man. I have no son whom I am willing to afflict with the terrible burden,' he is quoted as saying. 'I want my sons to divide it and share the worry which it will cost

* approximately $20.7 billion today

to keep it.'[71] At the same time, he appears to have been anxious to respect the Commodore's wish that the family fortune should remain intact. Within his own family everyone was treated generously. His daughters were all given the houses in which they lived, and each of his eight children received $5 million with a further $40 million in trust for them jointly with arrangements made for grandchildren. Maria Kissam Vanderbilt, his widow, received 640 Fifth Avenue, its contents and an annual allowance of $200,000, as well as a bequest of $500,000 which she used to help her Kissam relatives. There were donations to Vanderbilt University and a range of smaller bequests.[72] However, it was the two sons with the longest experience of managing the family enterprise who received the bulk of the estate between them: Cornelius II and William K. now discovered that they had inherited about $50 million each.

Because William Henry died prematurely, his sons and daughters-in-law were unexpectedly young when they inherited his fabulous wealth. Cornelius II was only forty-two, William K. was thirty-six, and Alva thirty-two. Under normal circumstances, they would all have had to wait at least another ten years before coming into such riches. But William Henry's early demise meant that Alva and William K. could now have whatever they wanted. The consequences for Miss Consuelo Vanderbilt were even greater. She became one of the world's greatest heiresses at the age of nine. This gave Alva plenty time to think about her daughter's future; and this made the impact of her grandfather's bequest on Consuelo's life almost incalculable.

Alva and William K. immediately reacted to the unexpected improvement in their circumstances by commissioning two new accessories: their own private yacht, the *Alva*, and a summer cottage in the fashionable summer resort of Newport, Rhode Island, which would become the backdrop to much of the drama ahead. At first glance it seems odd that the charming and refined colonial town of Newport, expressly founded on the principle of religious tolerance during the seventeenth century, should be the locus of titanic social struggles in the Gilded Age. The town manifests something of a split personality to this day, with elegant small colonial houses nestling together round the harbour and strenuously competitive nineteenth-century palaces on the slope above scarcely conceding existence to the throng below. The explanation for its singular history lies partly in its geography: a cool summer breeze which has always attracted visitors in search of a 'healthy climate' and a deep natural harbour which made it accessible to steamships from the south from the early nineteenth century. It is not surprising that Alva was brought to Newport as a child by her southern parents, nor that it should have been in Newport that she first made friends with the Yznaga family who came from Cuba and Louisiana.

For much of the nineteenth century, Newport was a holiday resort for writers, artists and intellectuals of modest means. After the Civil War however, Newport fell victim to the noisy arrival of the urban rich, 'quick to pick up the scent and take over the land, driving up prices to push out the eggheads'.[73] The transformation of Newport into the epicentre of social warfare at its most vicious was largely the work of two enterprising speculators: Alfred Smith and his associate Joseph Bailey. Spotting an opportunity in a manner of which Commodore Vanderbilt would have been proud, this duo acquired 140 acres of land on the slope to the north of the colonial town and began to develop terrain along Bellevue Avenue, creating large tracts of building land amid broad tree-lined streets in an informal exclusivity zone. This development paved the way for competitive snootiness unparalleled anywhere in America. In the hitherto smart resort of Saratoga, for example, society stayed and entertained in hotels, making it easier for those on its fringes to find a foothold. In Newport, on the other hand, rich families built their own 'summer cottages' on Smith and Bailey's land, while those who could not afford it were kept out.[74]

The summer cottages of Newport were, of course, nothing like cottages at all. Those that remain range from the elegant, to the ludicrous, to the very slightly mad. Henry James famously described them as 'white elephants . . . all cry and no wool . . . They look queer and conscious and lumpish – some of them, as with an air of brandished proboscis, really grotesque.'[75] Several of the most famous were designed by Richard Morris Hunt. Uncle Cornelius Vanderbilt II's house, The Breakers, had seventy rooms; the gardens of The Elms required twelve gardeners simply to keep them in order; and every Gilded Age cottage along Bellevue Avenue had a ballroom large enough to accommodate several hundred guests. This was the point of being in Newport in the first place. Even allowing for the appearance of the Casino (where one played tennis or croquet, rather than gambled) and swimming at Bailey's Beach, the focus of activity during Newport's short summer season was private entertaining by society figures, creating a vicious circle – or a virtuous one, depending on your point of view – of aristocratic exclusivity.

It only took the arrival of a few rich society families in Newport in the post-war years to attract others, turning Newport for a few brief weeks in July and August into New-York-by-the-Sea. By the end of the nineteenth century almost every wealthy family of the industrial age had established some kind of presence there. 'They were the Astors, the Vanderbilts, the Morgans, the Paran Stevenses, the Lorillards, the Oelrichses, the Belmonts, the Goelets, the Fishes, the Havemeyers, the Burdens . . . There were several hundred of them in Newport in any one summer season – a magical inner circle of those powerful few who called the social tune and those newly arrived families who desperately danced to it.'[76] Even for its aristocrats, Newport was anything but a holiday. For six short weeks the social competition of New York was transferred to the seaside, twisted, condensed and inflated. By 1890 the unwritten rules of competitive display required a twice-daily appearance in a phaeton on Bellevue Avenue in a different dress, a swim at private Bailey's Beach from one's own cabana (one of the least pleasant beaches in Newport by all accounts), luncheon on a yacht moored in the harbour, or a *fête champêtre* at a farm, attendance at the polo field, dinner and a further change of costume, then a ball at the Casino or, if one was of the elect, in a summer cottage on Bellevue Avenue. A season could require over ninety new dresses.[77]

For the hundreds of visitors who were not part of the inner circle and who arrived in Newport each summer with the hope of breaking through, it was far, far worse. 'It is an axiom of Newport that it takes at least four years to get in,' wrote society author Mrs John Van Rensselaer. 'Each season the persistent climber makes some advance through a barrage of snubs. The seasoned member of the Newport colony enters into the cruel game of quashing the pride of the stranger with great glee. Eventually, if he will bear all this, the candidate receives an invitation which indicates that he has finally been accepted by whatever particular set he has besieged. Then he turns about and snubs those remaining petitioners as harshly as he himself was snubbed. For the privilege of being a guest at certain houses and the license to affront those not yet in, he has spent perhaps a million dollars.'[78]

In commissioning Richard Morris Hunt to design Marble House as her Newport summer cottage Alva accelerated Newport's progress towards becoming the social capital of the Gilded Age. It has been remarked that the Vanderbilts only went to Newport in 1885 because of the Astors. Alva would have resented this deeply for though she undoubtedly set up camp in Newport near Mrs Astor's house, Beechwood, and then proceeded to outshine her architecturally, she had, of course, been to Newport for holidays as a child. She was now a leader of society herself; and by 1885 she would have regarded a Newport summer 'cottage' of her own as a matter of entitlement. Having acquired a plot of land on Bellevue Avenue, however, she and William K. set about taking the business of aristocratisation one step further than anyone else. While Cornelius Vanderbilt's Newport house, The Breakers, was modelled on the palazzo of a Medici merchant, Alva left the Medicis behind and addressed herself directly to the Bourbon monarchs of the *ancien régime*.

Often described as Richard Morris Hunt's masterpiece, Marble House contained allusions to the White House and the Petit Trianon at Versailles. It was certainly not lacking in ambition. In the memoir she dictated to Matilda Young, Alva also made mention of the Acropolis. For Richard Morris Hunt it was one of the great commissions of his life: he had unlimited resources, a client whose historical imagination and ambition matched his own and who had a sense of refinement and taste far more developed than any of his other clients.

Construction of Marble House began in conditions of great secrecy in 1888. By 1889, the contractor, Charles E. Clarke of Boston, had leased a wharf and warehouse in Newport harbour for materials which were brought in by ship. Artisans imported from France and Italy were quartered in separate lodgings and banned from communicating with each other on site. High fences went up round the building plot to hide it from the gaze of curious Newporters. It would take four years to complete. As drawings in the archives of Richard Morris Hunt show, it was very much Alva's project and she involved herself in every detail. 'This absolutely disapproved of by Mrs Vanderbilt' notes an anonymous hand on one drawing of a doorway. 'This is all wrong,' declares Alva in her own handwriting on another drawing. 'Will send photograph of marble to be adopted and each side of mantle to be solid marble panels and *no* columns on this end of the room.'[79]

While Marble House was under construction, the Vanderbilts kept themselves amused with their other new toy, the steam yacht *Alva.* Launched by Alva's sister, Jenny Yznaga, on 14 October 1886, the yacht was 285-feet long and 32-feet wide and had a tonnage of 1,151.27, making her the largest private yacht in America by a good 35 feet, beating J. P. Morgan's *Corsair* (165 feet),William Astor's new *Nourmahal* (233 feet) and Jay Gould's *Atlanta* (250 feet). In fact, the *Alva* was so large that the Turkish authorities once mistook her for a small cruiser and fired two shots across her bow in the Dardanelles. 'Mrs Vanderbilt, who is generally credited to be a lady of excellent taste, deems that elaborate and ornate furnishings are out of place on a yacht. She thinks that she is rich enough to afford simplicity in this instance,' reported *The New York Times*.[80] It was true that all the walls were simply panelled with mahogany, and the teak decks were simply covered with oriental rugs, but this principle was extended to a dining room which had a piano, a library with a fireplace, seven guest rooms, and a ten-room suite for the Vanderbilts (though the mahogany gave out below stairs in the accommodation for the crew of fifty-three).

While Alva's mother, Phoebe Smith, had once travelled with two maids and a southern mocking bird, the Vanderbilts found it necessary to take along a crew that included a master officer, a first officer, a second officer, a boatswain and a boatswain's mate, a storekeeper, four quartermasters, a ship's carpenter, twelve seamen, a chief engineer, first and second assistant engineers, six firemen, three coal passers, three oilers, a donkey engineman, an electrician, an ice machine engineer, a chief steward, a ward-room steward, a firemen's mess-man, a sailors mess-man, two mess boys, a baker and a doctor.[81] (The crew total of fifty-three does not include the French chef, family friends, household servants, or tutors and governesses for the children who were frequently present.) Labour unrest was dealt with in peremptory fashion. On 4 December 1887, the ship's log noted that men who had demanded better rates of pay and who refused to work were 'quickly landed'. Replacements were then picked up in Constantinople. The Vanderbilts and their guests, meanwhile, not only travelled in the lap of luxury but were treated as visiting dignitaries wherever they went, greeted by consuls, admirals, ambassadors, and kings. Even the Sultan of Turkey made recompense for the shots fired across the yacht's bows in the Dardanelles by granting William K. an audience and arranging a tour of his private palaces which included a visit to his harem; (the abject dependence of the women there would make a lifelong impression on Alva).

Cruises on the yacht between 1886 and 1890 took William K., Alva and the children to the West Indies, Europe, Turkey, North Africa and Egypt and often lasted several months. One voyage started in July 1887 and only ended on 31 March 1888, stopping at Madeira, Gibraltar and at Alexandria – where the party left the *Alva* and engaged one of Thomas Cook's steam dahabiyehs, the *Prince Abbas*, for a trip up the Nile. Alva later remembered that while they were in Cairo, one of their regular travelling companions, Fred Beach, became the object of Baroness Vetsera's attentions when they met her in Shepheard's Hotel – attentions to which he showed no objection at all, but allegedly did not respond. (The following year the Baroness Vetsera would be found dead of gunshot wounds alongside her lover, Crown Prince Rudolf of Austria, at Mayerling.) Alva shopped for furniture for Marble House during cruises on the *Alva*, and on at least one occasion left the children at Nell Gwyn's house in London while she went to look at potential purchases. On one cruise, the men of the party hunted deer at Guantanamo Bay in Cuba and stalked them in Scotland where the Vanderbilts took Beaufort Castle for the shooting season. Its owner, Lord Lovat, died while they were in residence and they witnessed a Highland funeral. This added to the general gloom of the experience, which Alva never wished to repeat: 'I always found the climate very trying in Scotland, and caring nothing for sport, found little to do there of interest.'[82]

The benefits of this style of international travel were not always clear to Consuelo. Extended cruising put paid to any chance of conventional schooling and it isolated Consuelo from the company of children her own age for months at a time. Life on the largest private yacht in America could be dull for a child, due in part to Alva's relentless emphasis on improvement. 'Heavy seas provided our only escape from the curriculum of work,' Consuelo wrote later, 'for even sightseeing on our visits ashore became part of our education, and we were expected to write an account of all we had seen'.[83] In spite of her magnificence, *Alva* was not a particularly seaworthy boat. There were extended bouts of seasickness (noted in the ship's log when they moored off Burntisland in Scotland in 7 August 1887), and life at sea could sometimes be positively frightening. 'Ship rolling a great deal and shipping quantities of water, which found its way below . . . Both large tables in forward and after saloons carried away,'[84] read an entry in the ship's log of an early cruise. During one storm an idiotic tutor maintained that it would only take seven huge waves in succession to sink the boat. 'Willie and I spent the rest of the day counting the waves in terrorised apprehension as the green water deepened on our deck,'[85] recalled Consuelo.

The cruises often ended in Nice, and the party travelled to Paris, where the Vanderbilts spent May and June with their retinue. As a child Consuelo fell in love with Paris just as Alva had done. Here she could ride on the carousel, watch Punch and Judy on the Champs Elysées and sail her toy boat in the gardens of the Tuileries. Like her mother, Consuelo came to associate Paris with liberation. After months on the yacht she could play with friends from New York on the same international circuit – Waldorf Astor who would marry Nancy Langhorne, May Goelet who would be her bridesmaid and later marry the Duke of Roxburghe, and Katherine Duer, later Mrs Clarence Mackay, already demonstrating that she had a bossy streak. 'She was always the queen in the games we played, and if anyone was bold enough to suggest it was my turn she would parry "Consuelo does not want to be Queen" and she was right,'[86] wrote Consuelo later. For several years in succession the early summer months were spent in Paris, followed by a brief return to New York; Newport in June and July; and a few weeks at Idle Hour in the early autumn before returning to New York for the confined world of the winter season.

When Consuelo reached her mid-teens, Alva finally allowed her to attend 'Rosa classes' when they were in New York. These were classes given by a Mr Rosa to a group of six young ladies in the home of one of the pupils – in Consuelo's case, the classes took place at the house of Mrs Frederick Bronson on Madison Avenue and 38th Street. Blanche Oelrichs attended the Rosa classes a year or two after Consuelo, and remembered Mr Rosa as 'a very stylish gentleman, with sideburns and a heavy watch chain, whose ambition to die in Rome was eventually gratified'.[87] The classes lasted from eleven till one while Mr Rosa fought to cram in as much English, Latin, mathematics and science as he possibly could. Consuelo preferred studying English and history and kept her early essays for Mr Rosa on the Punic Wars until she died. Two hours each morning with Mr Rosa were followed by French, German and music lessons with governesses, and an hour of exercise in Central Park.

None of this meant that Consuelo was gradually permitted greater independence. Instead, such freedom as she had enjoyed as a child was steadily curtailed in her teens, and gave way to a life that was increasingly controlled and introspective. Her brothers became more distant as they went away to boarding school and as she grew older she was forbidden to join in with their holiday activities. By the time Consuelo was sixteen there were 'finishing governesses' in residence, one French and one English. Since French and English views about finishing young ladies were sharply divergent if not contradictory (and probably still are), these governesses had to be handled with great tact. Alva spent many hours in the schoolroom supervising the curriculum and directing the finishing governesses. Unable to resist a competition, she sent off for the entrance papers to Oxford University and 'found that so far as [Consuelo's] equipment went she could enter with a condition in three live languages and one dead one'.[88]

Even by the standards of the day, Consuelo's teenage life was highly managed. It is striking that cousin Gertrude Vanderbilt was permitted far more independence, in spite of the fact that Uncle Cornelius and Aunt Alice were serious and strict. Gertrude's teenage diaries are filled with accounts of close female friendships, sorrow at leaving school, upsets about being too young to take part in 'tableaux', quarrels with her best friend and making-up. As Gertrude and her cousin Adele Sloane emerged from the schoolroom and into society, they were encouraged to form views about young men in the circle of aristocratic families in which they moved. Gertrude came home and analysed some of them: 'You have not enough go. You are trustworthy without being interesting.' [Mo Taylor]. 'If anyone ever looked out for No. 1, you are that person.' [Richard Wilson]. 'You mean well by people, but you will not take very much trouble to make yourself agreeable.' [Lewis Rutherfurd].[89] Adele was even allowed to go out riding with some young gentlemen, though she was never permitted to be alone with a man indoors ('Had nobody in the older generation read *Madame Bovary*?' asks Louis Auchincloss in astonishment.[90])

Alva would allow none of this. 'My mother disapproved of what she termed silly boy and girl flirtations . . . and my governess had strict injunctions to report any flighty disturbance of my thoughts.'[91]

There were moments when the doll-child found such micro-management truly insulting: 'I remember once objecting to her taste in the clothes she selected for me. With a harshness hardly warranted by so innocent an observation, she informed that I had no taste and that my opinions were not worth listening to. She brooked no contradiction, and when once I replied, "I thought I was doing right," she stated, "I don't ask *you* to think, *I* do the thinking, you do as you are told".'[92]

In America in the 1890s there were many constraints on the lives of well-to-do young ladies: few telephones, no motor cars, corsets, long skirts, hats fixed with pins, gloves and blouses with high whalebone collars. Even at Bailey's Beach at Newport, Consuelo bobbed up and down in the water in an outfit of dark blue alpaca wool consisting of a dress, drawers, stockings and a hat. It is perhaps not surprising that almost two pages of her memoirs are given over to a long list of the books she read in French, German and English. One German governess in her teens particularly inspired her with a love of German poetry and philosophy – to such an extent that after her marriage Consuelo considered translating *Also Sprach Zarathustra* into English, only to discover that there were twenty-seven translations already in existence. Meanwhile, she was inspired to secret but short-lived experiments in austerity by Plutarch's *Lives* (she spent a night on the floor, but caught a cold) and reached a 'real emotional crisis' when she found a copy of *Mill on The Floss* in the yacht's library. The picture Consuelo paints of herself as a somewhat sensitive, solitary and rather bookish teenager is reinforced by an entry in the diaries of the household superintendent, William Gilmour. On Thursday 2 March 1893, he wrote: 'Miss Vanderbilt's birthday, 16 years old. I went down to Wintons [Huttons] 23 St this morning and bought 3 vols Keats poems for Willie's present to his sister.'[93]

For many years, the marriage of Alva and William K. Vanderbilt had been propelled by shared ambition. They had conquered New York society together, paving the way for other Vanderbilts, particularly Cornelius II and Alice, to take their place at the apex of New York society. By the mid-1880s, William K. and Cornelius II were members of all the most exclusive gentlemen's clubs. Between them, the Vanderbilts had a row of magnificent houses on Fifth Avenue. Alva had undermined Mrs Astor's monopoly to such an extent that it had become a newspaper joke to talk about the 'Astorbilts'. Alva made her mark on New York's architectural history too, forging an important creative link with its greatest architect, Richard Morris Hunt. But these achievements came at great emotional expense. Even by 1885, when William Henry's death made the William K. Vanderbilts one of the richest couples in America, the glue of shared ambition had dried out. Consuelo's sixteenth birthday in 1892 may have been celebrated with a thoughtful present from her brother; but the next two years would be deeply scarred by the unhappiness already engulfing her parents.

3

Sunlight by proxy

WHEN SHE TALKED about the story of her early life in later years, Alva was only prepared to discuss the disintegration of her relationship with William K. Vanderbilt in general terms. She intimated to Sara Bard Field, however, that the start of married life had been dismal. Field, whose feelings about Alva were mixed (at best), wrote to Charles Erskine Scott Wood that Alva had stopped her in the middle of the lawn at Marble House, where no servant could eavesdrop, and had spoken of herself as 'a girl of barely seventeen who did not fully know the sex mystery'. Alva had alluded to an 'agony of suffering'. The memory brought 'tears from her hard heart to her eyes'. She refused to allow Field to write about this, saying that 'it was the sacred confidence of a woman's heart' and that 'the children would object . . . and the Vanderbilts'. Sara Bard Field suddenly found herself in tears too, partly because her own experience with Wood was very different and partly because she felt that 'a heart that could have been loved into beauty . . . has been steeled against its own finer and softer emotions. O, it is all fascinating what she is now telling me. Really, it is Life.'[1]

Leaving aside the fact that Alva was twenty-two and not seventeen when she married, it is possible that her wedding night did indeed come as a terrible shock. Her mother had died almost five years earlier, her elder sister Armide was unmarried and such 'innocence' was not uncommon. (One can only hope that Mrs Oelrichs, her chaperone at White Sulphur Springs, took it upon herself to have a quiet word.) The historians John D'Emilio and Estelle B. Freedman point out that there were also tensions in the sexual education of young men which did not help the process of marital adjustment. Many young men in New York in the 1870s had their first sexual experiences with prostitutes, 'a poor training ground for middle-class bridegrooms'.[2] In pioneering studies carried out in late-nineteenth-century America, middle-class women talked of finding sex pleasurable, but it depended on the behaviour of their husbands. Young men used to encounters with prostitutes would often 'bring to the conjugal bedroom a form of sexual expression badly out of line with what their wives might desire. On the other hand, some married men may have continued to visit the districts precisely because they could not find in their wives the kind of sexual availability, or responsiveness, they wanted.'[3] The problems caused by this kind of mismatch were often exacerbated by fear of contracting venereal disease. There is some evidence in the later part of Alva's life that she was familiar with this particular anxiety while married to William K. Vanderbilt.

For several years, the Vanderbilts found a way of resolving these early difficulties which cannot have been helped by the death of Murray Smith two weeks after the wedding. Until about 1885, however, the marriage had such forward momentum and such a triumphantly successful agenda, that both husband and wife ignored its disadvantages. Alva later hinted that the real difficulties set in after about ten years. 'Not many men are in love with their wives after ten or twelve years,'[4] she wrote. Elsewhere she remarked that 'sex passion' between man and wife generally lasts about ten years, and that after that time men of her class 'amused themselves elsewhere'.[5] In the case of William K. and Alva, however, ten years of marriage coincided with the death of William Henry in 1885. William Henry's fondness for Alva may have acted as a check on his son's behaviour. After his death, this impediment disappeared and William K., always a handsome man, found himself in possession of a limitless fortune and much less to do. By 1885 the Vanderbilts had achieved most of their shared objectives: their yacht, the *Alva* and Marble House may have kept them busy – but these were opulent extras, icing on a well-baked cake.

In the second set of memoirs that Alva dictated to her secretary,

Mary Young, after 1928, she suggests that having fought so hard to extract herself from the snares of genteel poverty, she now found herself faced with an even more pernicious form of exclusion. 'It was a time' according to Alva, 'when men of wealth seemed to think they could do anything they liked; have anything, or any woman, they, for the moment wanted. And so, as a matter of fact, they very nearly could, and did. If a man was rich enough and had enough to offer there were, unfortunately, women willing and waiting to throw themselves at their heads, women who were younger and more attractive to them than the wives of whom they had grown tired.'[6] Alva does not mention William K. by name when she talks of women insulted by their husbands' 'open and flagrant and vulgar infidelities', but she comments that the conduct of J. Pierpont Morgan, Colonel John Jacob Astor, and others was notorious. 'Col Astor's yachting parties were public scandals. He would take women of every class and kind, even chambermaids out of the hotels of the coastwise cities where the yacht put in, to amuse himself and the men of his party on these trips.'[7]

And what of the wives of these rich men? These men did not seek divorce for there was no need. They simply set their wives aside, leaving them 'to maintain the dignity of their position in the world, such as it was, and to care for their children, while they amused themselves elsewhere. That, they took it upon themselves to decide, was all that a woman was good for after they had finished with her in ten years or less of married life.'[8] No-one was prepared to challenge the convention by which a society woman in her prime ignored adulterous behaviour on the part of her husband and withdrew into a kind of half-life, while bravely maintaining a public front of domestic respectability. 'It was considered religious, dignified and correct for the wife to withdraw into the shadows while her husband paid the family respects to the sunshine . . . she was supposed to get her sunlight by proxy through the husband.'[9] It was, in Alva's view, an intolerable by-product of monopoly capitalism, a uniquely American form of purdah: the seclusion of cast-off wives enforced by rich men whose solidarity in the matter was perceived to be indestructible.

When she recalled working with Richard Morris Hunt on Marble House, Alva remarked that the period from 1886 and 1892 marked 'some of the saddest years of my life'.[10] It is possible that she welcomed long cruises on the yacht as a way of controlling her husband's infidelities. Later, the *New York World* recalled that she had looked unhappy for much of this time. 'She looked both weary and sad, and people wondered why it was. They said it was because she was naturally of a peevish and discontented disposition. They said it was because she had achieved every ambition possible to her, and was made wretched because there was nothing further to achieve . . . But gradually the truth crept out and it was known that Mrs Vanderbilt was wretched because her husband had broken his marriage vows, not once but over and over again.'[11]

The tension certainly affected sixteen-year-old Consuelo. 'I had reached an age when the continual disagreements between my parents had become a matter of deep concern to me. I was tensely susceptible to their differences, and each new quarrel awoke responding echoes that tore at my loyalties.'[12] On 16 July 1892, in an apt metaphor for the disintegrating state of the William K. Vanderbilt marriage, the *Alva* sank. Bound for Newport from Bar Harbor, the yacht was forced to anchor in dense fog off Monomoy Point where she was accidentally rammed by the mellifluously named freight steamer, *H. F. Dimmock*. William K. reacted by commissioning an even more luxurious – and rather more seaworthy – yacht, the *Valiant*.

While the *Valiant* was under construction, Alva occupied herself with the finishing touches to Marble House so that it was ready to receive its first guests in August 1892. There was plenty to amaze these visitors who were welcomed into the house through an elegant and elaborate bronze entrance grill (weighing 10 tons and made by the John Williams Bronze Foundry of New York). In the hall, warm and creamy Siena marble lined the walls, floors and staircase. Guests were then invited to admire rooms that have been described by one expert as a series of knowledgeable experiments in French decorative style.[13] The dominant theme was the art and architecture of Versailles. In the upper hall a bas relief of Richard

Morris Hunt faced a matching bas relief of the architect of Versailles, Jules Hardouin Mansart. The dining room was inspired by the Salon of Hercules, the Siena marble of the entrance hall giving way to walls lined with pink Numidian marble specially quarried in Algeria. A painting of Louis XIV attributed to Pierre Mignard, said to have hung in the Salon of Hercules at the time Alva visited the palace in the late 1860s, dominated one end of the room.

The dining room was only surpassed by the ballroom – the Gold Room – Alva's miniature edition of the Hall of Mirrors at Versailles, a riot of neo-classical exuberance with panels of Aphrodite, Demeter, Pan and Heracles suggesting a world of love, beauty, revelry and music sadly at odds with the lives of the proprietors. (Only a panel of Heracles aiming an arrow at Nessus who had made off with his wife comes close to reflecting emotional turmoil behind the scenes.) Above the marble mantelpiece, bronze figures bore vast candelabra, while cupids capered playfully and cherubs blew trumpets on the walls and ceilings. The Gold Room was dominated by wood panels gilded in red, green and yellow gold carved by the architectural sculptor Karl Bitter, its dazzling magnificence multiplied many times by vast mirrors hung over the four doors, above the mantelpiece, on the south wall, and by the south windows. Elsewhere in the house, Louis XV replaced Louis XIV in an outbreak of Rococo Revival: swags and garlands of flowers, masks, and somersaulting cherubs prevailed here and in Alva's bedroom an eighteenth-century four-poster bed stood on a very fine Aubusson carpet.

The anomaly was the so-called Gothic Room, probably inspired by the Bourges house of the great medieval merchant, Jacques Coeur, whom Alva greatly admired. Paul Miller, curator at the Preservation Society of Newport County, suggests that the Gothic Room may originally have been intended for 660 Fifth Avenue. In 1889 the Hunts and Vanderbilts met in Paris to discuss furnishings at a meeting that coincided with the publication of a *catalogue raisonné* of Emile Gavet's collection of European works of art from the thirteenth to the sixteenth centuries. The Vanderbilts bought half the collection, including a 'Madonna and Child' by Luca della

Robbia that now hangs in the Metropolitan Museum of Art in New York. Hunt's design for the Gothic Room was then transferred to Marble House to display *objets* purchased from the Gavet collection, though the room acquired American accents in the process: the foliate cornice around the room which was inspired by Coeur's house reappeared with crabs and lobsters to reflect the seaside setting.[14]

In 1892, those who knew Alva best might have detected her unhappiness in much of this design. She once described Marble House as her fourth child and its interior made few concessions to her husband, other than cartouches bearing the monogram 'WV' and a small study reflecting his sporting interests. Meanwhile, Alva's preoccupations could be found everywhere: on the ceiling painting in her bedroom where the paradoxical Goddess Athene reigned supreme, war-like but the goddess of fine craftsmen, and in many references to the French *ancien régime.* Even the use of marble suggested a fugitive memory of the Smith house in Mobile. If it is true that the best buildings of the Gilded Age dissolved almost entirely into make-believe, her greatest collaboration with Richard Morris Hunt had this quality in abundance. Even more than 660 Fifth Avenue, Marble House was characterised by a feeling of withdrawal from the world outside. But here there was a sense of unhappy withdrawal from a miserable marriage too, as if Alva has turned in on herself and back towards the world of the *ancien régime* she loved as a girl before the harsh compromises of adult life took their toll. To some, the Gold Room still stands as a symbol of the heartless, glittering emptiness of the Gilded Age; but it can also be seen as the most heartfelt room in Newport, an intense and private dream.

As far as Consuelo was concerned, however, Marble House was associated with sensations closer to nightmare, claustrophobia and control. It felt like a gilded cage. Even the gates were lined with sheet iron. 'Unlike Louis XIV's creation,' she wrote tartly, 'it stood in restricted grounds, and, like a prison, was surrounded by high walls.'[15] Consuelo was sixteen when Marble House was finished. In spite of this, Alva conceded nothing to her daughter's taste. In this instance her vision of the Marble House interior entirely overpowered the section of her child-rearing theory that involved independence. Still a doll in a doll's-house, Consuelo's bedroom was designed by her mother down to the last detail and furnished with objects which she scarcely dared to move. 'To the right on an antique table were aligned a mirror and various silver brushes and combs. On another table writing utensils were disposed in such perfect order that I never ventured to use them. For my mother had chosen every piece of furniture and had placed every ornament according to her taste, and had forbidden the intrusion of my personal possessions.'[16] It was this bedroom that inspired one of the most quoted passages about Alva from Consuelo's memoir *The Glitter and the Gold*: 'Often as I lay on the bed, that like St Ursula's in the lovely painting by Carpaccio stood on a dais and was covered with a baldaquin, I reflected that there was in her love of me something of the creative spirit of an artist – that it was her wish to produce me as a finished specimen framed in a perfect setting, and that my person was dedicated to whatever final disposal she had in mind.'[17]

When Marble House opened to widespread acclaim during the Newport season of 1892, Alva was less concerned with the final disposal of Consuelo than the state of her own marriage. 'Sunshine by proxy' was decidedly not for her. She was only thirty-nine. She refused to accept a scenario in which she tolerated her husband's philandering and retired to a virtuous life in the shadows. She particularly objected to the way in which rich husbands enforced their wives' powerless position by reminding them of their financial dependence. 'If a wife, hungering for love and with more spirit than most of her sex, asserted her right to a lover or to contacts with the outside world, the husband declared she was ruining his reputation along with her own and with the power of the bank resources at his command, bade her retire to the obscurity of respectability.'[18] Alva's reaction to this was spirited. She acquired a lover of her own.

Oliver Hazard Perry Belmont was the wayward son of financier August Belmont. Married to a socially pre-eminent wife of impeccable pedigree, August Belmont was of Jewish origin, though he had converted to Christianity, and represented the Rothschilds' interests in New York. He lived flamboyantly, introducing the first French chef to a private New York house, establishing a pace-setting example when it came to wining and dining, and causing wild gossip. He was another of Mrs Astor's principal *bêtes noires*, though her resistance to the next generation of Belmonts gradually dissolved.

Before his relationship with Alva, Oliver Belmont was often to be found in the Oelrichs household, charming Blanche Oelrichs as a child. She liked his 'slow urbanity, his face rutted with lines – from the hopes and disillusions of his life as a lover, I suspected. For certainly he must be a romantic man.'[19] The circumstances surrounding the collapse of Oliver Belmont's first marriage suggest that his behaviour was not always romantic. After a long courtship which was bitterly opposed by both his parents, Belmont married a beautiful socialite, Sara Whiting. On their honeymoon in Paris they were joined by Sara's domineering mother and two sisters, who moved in with the newlyweds and refused to leave. Oliver eventually marched out on the ménage – understandable perhaps had he not stormed off in the company of an exotic Spanish dancer, bad form at any time, but especially on one's honeymoon. On hearing that his new bride was pregnant he returned to Paris to attempt a reconciliation, only to find himself accused of heavy drinking and physical violence – allegations which he rebutted furiously. Sara Whiting later gave birth to a daughter, Natica, whom Belmont refused ever to acknowledge, while Mrs Whiting insisted on a divorce.

Oliver Belmont's parents were mortified by the publicity surrounding his first marriage. They had in any case long despaired of him: in spite of various attempts to find him gainful employment he appeared to have no greater ambition than to live as a gentleman of leisure. As early as 1888 they were concerned that he was joining a cruise on the *Alva*, fearing that Vanderbilt sojourns in resorts such as Monte Carlo would do nothing to raise his level of ambition and knowing that his friendship with Mrs William K. Vanderbilt was already a talking point.[20] Oliver joined part or all of subsequent Vanderbilt cruises in 1889 and 1890, however.[21] Indeed, his obstinacy and readiness to ignore society's opinion on this matter may have attracted Alva. Here was someone with strength of personality, someone to brace against, unlike William K. whom Alva later described as a 'weak nonentity'. It may also be true, as Louis Auchincloss has written, that Oliver was attractive because he represented a challenge. He had already caused offence. There was just a whiff of violence about him. He was a Belmont. 'One begins to suspect that the setting up of hurdles in order to jump them was her way of adding a bit of zest to the sameness of a social game that was already showing itself a drag to her lively spirit. And were not the Belmonts partly Jewish? Better and better!'[22]

Initially the relationship between Alva and Oliver Belmont raised few eyebrows for it was not unusual for the neglected wives of rich men to acquire 'walkers'. 'The Newport ladies of those days were trying hard to emulate their sisters in cosmopolitan Europe,' writes Blanche Oelrichs; 'and it would have been thought extremely "bourgeois" for attractive matrons not to have gentlemen about them who were "attentive".'[23] As the warmth of feeling between Alva and Belmont began to show, however, the gossips got down to work. 'I used to think Oliver Belmont one of the handsomest men at the Coaching Parade, with his dark eyes, clear-cut profile and slender, faun-like grace,' wrote Elizabeth Lehr, thinking back to her teens. 'Mrs W. K. Vanderbilt often sat at his side on the box behind the four famous bays, Sandringham, Rockingham, Buckingham and Hurlingham. The women glanced at her as she sat wide-eyed and innocent-looking, and whispered to one another.'[24] *Town Topics* also picked up Oliver's constant presence at Alva's side and talk persisted into later generations. In a delightful lecture about her childhood on Bellevue Avenue, Eileen Slocum remarked: 'Down the years I especially remember the gossip about Mrs William K. Vanderbilt's affair with Mr O. H. P. Belmont . . . Daddy was very critical . . . "Poor Willy K. drove up, unexpectedly, one day from the train in his carriage," Daddy said, "and entered his own house and ascended his own staircase and found Mr Belmont hiding in the closet of his own bedroom. Willy should have shot him."'[25]

It does seem perverse, therefore, that in the autumn of 1893, when their marriage was strained to the point of collapse, the Vanderbilts not only decided to go on a long cruise on the *Valiant* to India but invited Oliver Belmont to join them. It is just possible that Alva and Oliver were not yet lovers, for this would have put Alva, who was always political, at a disadvantage. Perhaps William K. welcomed Belmont's presence because he improved Alva's mood. Perhaps the expedition was William K.'s idea and Alva only agreed to go on condition she could take Oliver too. Consuelo later said that it was clear even to her that the cruise was a desperate last attempt to patch things up, one last effort to avoid 'the rupture which I felt could not be long delayed'. The expedition set off in an atmosphere of 'dread and uncertainty' with a party that included 'my parents, my brother Harold, a doctor, a governess and the three men friends who were our constant companions. Willie, being at school, remained at home. My mother, claiming that my governess gave sufficient trouble, refused to have another woman on board.'[26] The three men friends whose names appear in the ship's log were Oliver Belmont, Fred Beech and J. Louis Webb.

The cruise began on 23 November 1893 at 3.35 p.m. precisely with a total of eighty-five people on board, seen off by a crowd that 'surged and pushed and jostled on the pier like animated stalks in a bunch of asparagus'.[27] The *Valiant* arrived in Bombay just over a month later, on Christmas Day. On 30 December, the Vanderbilt party disembarked for a two-week overland journey by special train to Calcutta, while the yacht made its way round from Bombay to await them. Alva was pleased to discover that the Taj Mahal had been inspired by the spirit of a woman. Otherwise, much of what she saw in India appalled her. If Alva was taken aback by what she described as superstition and 'repulsive religious ceremonies',[28] Consuelo was frankly terrified by such unusually close proximity to humanity en masse, particularly when it rattled at the doors of the Vanderbilt sleeping cars and tried to force an entry. 'It was difficult to secure bath water and the food was incredibly nasty. We lived on tea, toast and marmalade . . . It was wonderful to find all the luxuries of home on the *Valiant* which had come round India from Bombay and lay anchored in the Hooghly.'[29]

What Consuelo did not know as she recuperated from this taxing journey, was that the stay in Calcutta would mark a turning-point in her life. While Consuelo, Harold, and the Vanderbilts' friends remained on board the *Valiant*, Alva and William K. were invited to stay by the Viceroy of India, Lord Lansdowne, at Government House in Calcutta. Sometimes described as 'the most neglected statesman in modern British history' Lord Lansdowne (or Henry Charles Keith Petty-Fitzmaurice, 5th Marquess of Lansdowne), had already had a distinguished career as Governor-General of Canada and would go on to become Secretary of State for War, Foreign Secretary, leader of the Conservative and Unionist peers, and a member of Asquith's wartime cabinet. At the time of the Vanderbilts' visit to Calcutta, however, his sojourn in India as Viceroy was almost at an end, and, worn out by his tour of duty, he was longing to go home. Nonetheless, Lord and Lady Lansdowne extended generous hospitality to the Vanderbilts with the result that just when Alva was feeling most vulnerable to a life of 'sunlight by proxy' she witnessed the life of the Vicereine, Lady Lansdowne, when the British Raj was at its zenith.

'We might as well be monarchs,'[30] wrote Mary Curzon when she arrived as Vicereine herself three years later. Even aristocrats such as Lord Lansdowne, accustomed to palatial space and waited on since birth, found Government House in Calcutta somewhat grandiose. 'Words cannot describe the hugeness of this place or the utter absence of anything like homely comfort . . . [The bedroom with its] colossal bed large enough for half a dozen couples . . . the ceiling which is so far up that one can scarcely see it,'[31] he wrote to his mother. Historian David Cannadine suggests that the grandeur was a deliberate political ploy: 'The British now saw themselves as the legitimate successors of the Mughal emperors, and came to believe that their regime should project a suitably "oriental" and "imperial" image. So they set out to construct a new ritual idiom for the government of India, partly based on the appropriation of what they believed were traditional Mughal court ceremonials, and partly invented and developed by themselves, through which they could express their own authority . . . The ceremonial surrounding the Viceroy, both in Calcutta and at Simla, and as he travelled round India, became increasingly splendid, ornate, elaborate and magnificent – far grander than the state in which British monarchs themselves lived at home.'[32]

The illusionists of the British Raj found a most appreciative audience in Alva, though even she was startled by the size of the Government House guest suite and the 'ten native servants who were assigned . . . in beautiful royal liveries of red embroidered in gold to serve us'.[33] What impressed her most, however, was the quasi-imperial role of both Lansdownes. 'The numerous house guests and outside friends assembled in an antechamber, and at a given moment the double doors were thrown open and the Viceroy and Lady Lansdowne were announced'. Even at lunchtime. Calcutta House had a throne room and on state occasions the Vicereine took her place on a throne on the dais beside her husband, receiving Indian princes in magnificent ceremony. Alva was even more impressed by the extent to which the British Vicereine made an important contribution in her own right, undertaking charity work in Calcutta and running much of the social life at Government House.

The Vanderbilts' visit coincided with plans for the handover of power to Lord Elgin and tributes were already flowing in to the departing Viceroy and Vicereine. Lord Lansdowne had been a popular viceroy and the view was frequently expressed that his tour of duty had enjoyed 'an almost unique popularity, to which the social gifts of Lady Lansdowne had largely contributed'.[34] Alva would also have been aware of the splendid formalities planned for the Lansdownes' departure, ceremonies which would acknowledge the contribution of them both, just as the ceremonies to welcome the Curzons in 1898 acknowledged Mary Curzon's American birth. There was no life in the shadows or sunlight by proxy for a Vicereine of India; and just as she had once pictured the Vanderbilts as Medicis, Alva could now visualise her daughter's future.

'My mother, whose habit it was to impose her views rather than to invite discussion, had already, on occasion, revealed the hopes she nourished for my brilliant future, and her admiration for the British way of life was as apparent as was her desire to place me in an aristocratic setting. These intentions, I am sure, crystallised during her visit at Government House,'[35] wrote Consuelo later. Worse, conversations between Alva and Lady Lansdowne revealed that there was a most interesting way of moving this vision forward. Maud Lansdowne had a nephew of the right age, with an interest in politics. He was already a duke – the young Duke of Marlborough. Consuelo thought later that it was during her parents' stay in Calcutta that 'the possibility of my marriage to him may have been discussed'. Even if the idea was not discussed explicitly, however, 'it is certain that it was then [my mother's] ambitions took definite shape; for she confessed to me years later that she had decided to marry me either to Marlborough or to Lord Lansdowne's heir'.[36]

It was possibly in a spirit of mutual inspection that Consuelo was invited to spend a day with the Lansdownes' younger daughter, Lady Beatrix, for Lady Lansdowne was fond of her nephew and knew that he had inherited a troubling financial burden in Blenheim Palace. The impression made by Miss Vanderbilt on the Lansdownes is not recorded but the serious-minded Consuelo was astounded (to the point of sounding quite priggish) by the ignorance and 'homespun education' of Lady Beatrix. On 19 January 1894 the captain of the *Valiant* recorded that 'the Viceroy & party from Government House were entertained on board'.[37] The *Valiant* left its moorings in Calcutta on the same day and headed back to Europe. It mattered not that when she played with her friends in Paris, Consuelo never liked being queen: Alva had decided what she wanted for her only daughter.

It was later claimed by the press that the *Valiant* cruise broke up in India after a final blazing row between the Vanderbilts; but according to both Alva and the ship's log, it continued as planned, sailing first to Ceylon, where the entry read: 'left a fireman behind at Colombo so we are one short'.[38] Apart from the fireman, the party remained intact, winding its way back to the Mediterranean by way of Alexandria, where the yacht was detained by rough seas. Near Rhodes, in another strangely symbolic incident, the *Valiant* lost its way – the captain took a local pilot on board who turned out to be incompetent. There is no doubt that relations between the Vanderbilts were strained to the limit and these setbacks can have done little to help matters. A visit to Delphi in Greece briefly acted as balm to Consuelo's troubled soul, but the break came by the time the yacht reached Nice. As the *Valiant* docked, Consuelo was told that her parents' marriage was definitely over.

Consuelo's initial feeling was one of relief 'that the sinister gloom of their relationship would no longer encompass me'.[39] It was only later that she realised how little she would now see of her father and the extent to which Alva would come to dominate her life. In the short term nothing changed. After their yacht moored at Nice on 24 February 1894, Alva took Consuelo to Paris, as she had so often done before. Both Vanderbilts remained in Europe for the rest of the summer, leaving the American press in something of a bother about where they were. *Town Topics* sneered derisively at newspapers alleging that the Vanderbilts were simultaneously in Newport, New York and Marseilles, asserting confidently that they had left America for three years and had leased a deer forest in Scotland. There was a calm interlude of several weeks before the press grasped what had actually happened.

Meanwhile, Consuelo's experience of Paris during the late spring of 1894 was happier than it had ever been. She and Alva moved into the Hôtel Bristol. 'I can still see the view over the Tuileries Gardens from our windows, still enjoy our walks under the flowering chestnuts of the Champs Elysées and our drives in the Bois de Boulogne in our carriage and pair. Every day there were visits to museums and churches and lectures at the Sorbonne, but the classical matinées at the Théâtre Français were my greatest pleasure.'[40] It was only with hindsight that she realised that her mother spent the early summer of that year preparing her for an aristocratic setting. Alva chose Consuelo's dresses from the great French dressmakers – Worth, Doucet and Rouff – and she arranged for her to have elocution lessons, in French, with an actress from the Comédie Française, where there was a long tradition of perfect diction. It seems likely that Alva arranged these lessons to prepare her daughter for a public life such as that of Lady Lansdowne's, where good voice projection was required when opening bazaars and returning speeches of welcome. 'Whatever her motive, the lessons produced a voice that carried,' said Consuelo. (Alva was later frustrated by her own fear of public speaking, brought up in a world where, in the rare event that a woman wrote a speech, she would hand it over to be read by a man.)

While they were in Paris, Alva also commissioned the portrait of Consuelo that now hangs at Blenheim, by Carolus-Duran. Alva's choice of artist was significant for Carolus-Duran was a fashionable painter particularly renowned for his portraits of aristocratic women. In an early exercise in branding, Alva requested that the background of red velvet which Carolus-Duran normally used should be replaced by a landscape in the classical style of the English eighteenth century, wishing Consuelo to 'bear comparison with those of preceding duchesses who had been painted by Gainsborough, Reynolds, Romney and Lawrence'.[41] On its completion, Alva arranged for it to be shipped to America and hung in the Gold Room at Marble House.

Consuelo made her Paris debut that summer at a ball given by the Duc and Duchesse de Gramont for their eldest daughter; she wore a dress of white tulle by Worth. 'It touched the ground with a full skirt, as was the fashion in those days, and it had a tightly laced bodice. My hair was piled high in curls and a narrow ribbon was tied round my long and slender neck. I had no jewels and wore gloves that came almost to my shoulders. The French dubbed me *La belle Mlle. Vanderbilt au long cou.*'[42] The party was a *bal blanc*, as parties for debutantes were known, where all the young women wore white. Elisabeth de Gramont remembered Consuelo as 'a tall girl whose small head with retroussé eyes like a Japanese, drooped languidly over her shoulder. She possessed great charm.'[43] Such evenings were misery for 'wallflowers' for whom any help from artifice was banned. 'Good girls were dressed in light, insipid colours and the poorest of materials, and all the touches that give "tone" – diamonds, powder, paint and perfume – were rigorously forbidden.'[44] The aces of the period, the grand 'marrying men', would sometimes look in briefly at these social gatherings, at the rows of nervous, perspiring debutantes lined up like cattle for their inspection. (On one occasion Elisabeth de Gramont heard one say: 'This place stinks of armpits, let's go to Maxim's.'[45]) There was little opportunity for conversation because permission to dance had to be sought from the young lady's chaperone and as soon as the dance was over, she was led straight back to her mother.

There was no shortage of partners for a seventeen-year-old American heiress, however, and by the end of June, Consuelo had received five proposals of marriage. 'When I say I had, I mean that my mother informed me that five men had asked her for my hand . . . She had, as a matter of course, refused them, since she considered none of them sufficiently exalted.'[46] Consuelo was only allowed to consider one: Prince Francis Joseph, a German prince who was the youngest of the four Battenberg princes, and at the centre of an intrigue to elect him ruler of Bulgaria. Confronted with the prospect of a royal crown rather than an English ducal coronet, Alva seems momentarily to have wavered from her original plan and Prince Francis Joseph was allowed to present his case to Consuelo. She was horrified both by the idea and by the Prince to whom she developed an immediate aversion. Alva too had second thoughts, unsure whether the intrigue would succeed. Nothing more was heard from her on the subject, though news of this potential engagement eventually reached *Town Topics* in New York who asserted (correctly this time) that: 'There is a general feeling that the report is not based upon facts, at this time at least.'[47]

In June, Alva took Consuelo to England. '[Alva] did not let her dally long in the drawing-rooms of Paris,' wrote Elisabeth de Gramont. 'She intended [Consuelo] for the English aristocracy, which she deemed more advantageous.'[48] Here Alva rented a house at Danesfield near Marlow and asked her old friend Mrs William Jay and her daughters to join them. The weather was so cold that they only went to Danesfield at the weekends and spent the rest of the time in the warmth of a London hotel. Consuelo described it as 'frowsty in the true English sense',[49] and thought with longing of their lovely hotel in Paris beside the Tuileries Gardens.

In England, Alva made use of her networks. The two people whose help she enlisted in the summer of 1894 were Consuelo Yznaga, now Duchess of Manchester, and Minnie Stevens, now Mrs Paget – pre-eminent figures in English society, favourites of the Prince of Wales and leading lights of his circle known as the Marlborough House Set. Consuelo did not care for Minnie Paget (later Lady Paget) one jot, however. 'Lady Paget was considered handsome; to me, with her quick wit and worldly standards, she was Becky Sharp incarnate ... Once greetings had been exchanged I realised with a sense of acute discomfort that I was being critically appraised by a pair of hard green eyes.'[50]

Such scrutiny was all too familiar. In an age when young women were commodities on the marriage market, they were forced to become accustomed to such analysis, which is not to say they enjoyed it.* 'I was particularly sensitive about my nose, for it had an upward curve which my mother and her friends discussed with complete disregard for my feelings,' wrote Consuelo. 'Since nothing could be done to guide its misguided progress, there seemed to be no point in stressing my misfortune.'[51] In London, Minnie Paget expressed her views forcefully. 'The simple dress I was wearing, my shyness and diffidence, which in France were regarded as natural in a debutante, appeared to awaken her ridicule. "If I am to bring her out," she told my mother, "she must be able to compete at least as far as clothes are concerned with far better-looking girls" ... It was useless to demur that I was only seventeen. Tulle must give way to satin, the baby décolletage to a more generous display of neck and arms, naiveté to sophistication. Lady Paget was adamant.'[52]

* In *The Buccaneers* Edith Wharton writes: 'A good many hours of Mrs St George's days were spent in mentally cataloguing and appraising the physical attributes of the young ladies in whose company her daughters trailed up and down the verandas ... As regards hair and complexion, there could be no doubt; Virginia, all rose and pearl, with sheaves of full fair hair heaped above her low forehead, was as pure and luminous as an apple-blossom. But Lizzy's waist was certainly at least an inch smaller (some said two),' pp. 4–5, p. 6.

Minnie Paget was once described by *Town Topics* as having 'watchful eyes ever on someone with money to burn',[53] and was rumoured to accept a fee for this kind of help. Having made over Consuelo to her satisfaction she arranged a dinner party to which she invited the young Duke of Marlborough. By now Alva's plan was becoming clear, even to her daughter. Minnie Paget placed the Duke to her right with Consuelo on his other side – 'a rather unnecessary public avowal of her intentions' Consuelo thought afterwards. 'He seemed to me very young, although six years my senior, and I thought him good-looking and intelligent. He had a small aristocratic face with a large nose and rather prominent blue eyes. His hands, which he used in a fastidious manner, were well shaped and he seemed inordinately proud of them.'[54]

They only met once during Consuelo's visit to England, and it seemed at the time that nothing would come of the matter, to Consuelo's great relief. Behind her back, however, English tongues were already wagging. Mrs Paget (later described by George Cornwallis-West as the worst gossip in London) was unable to keep quiet about the plan. On 19 July, the Duke's grandmother, Frances, Duchess of Marlborough, wrote to her daughter-in-law Lady Randolph Churchill that she was 'amazed at the news . . . [of] Marlborough's marriage. Mrs Paget has been very busy introducing him to Miss Vanderbilt and telling everybody she meant to arrange a marriage between them, but he has only met her once and does not seem to incline to pursue the acquaintance.'[55]

One reason that the introduction may have stalled was that the American press had finally picked up the scent of the Vanderbilts' separation. By 1894, the dark side of the Faustian bargain between the press and newer members of high society was all too obvious: socialites who had courted publicity now found themselves the captives of its machinery. It had become big business too. By the early 1880s most newspapers in New York responded to demand and carried social columns, while magazines devoted entirely to society matters began to appear. Both were aimed at two audiences. The first was a wider readership well outside the social elite, and included those who simply enjoyed society sagas as entertainment, nosey servants and those who worked in society's service industries for whom information was power, such as Mrs Heeney in Edith Wharton's *The Custom of The Country* (the 'society' manicurist and masseuse whose alligator bag was always filled with newspaper clippings). The second audience was high society itself and those who aspired to it. Here, the position of its members was reinforced and legitimised by constantly seeing their names, clothes and parties in print. 'If one's social goal was to force an entry into the most exclusive circles, half the satisfaction of achievement would have been lost if one's erstwhile acquaintances had not been able to read all about it,'[56] writes Ruth Brandon.

In some cases, newspaper editors were society figures in their own right, like James Gordon Bennett Jr of the *New York Herald*, or the society columnist George Wetherspoon who wrote for *The New York Times*. Though the social elite sometimes claimed to be irritated by comment in such publications, it generally remained on the right side of intrusive. Oddly, the two publications where it was most important to be 'seen' were the two which explicitly held the Four Hundred in the greatest contempt. One was the *New York World* after 1883, when it was bought by Joseph Pulitzer, who combined formidable liberal campaigning with a keen sense of the aspirations of his poorer female readership, and reconciled the two by covering the activities of high society in sensational and barbed detail while stopping just short of pouring unmitigated scorn. The other key publication was *Town Topics*, which changed the whole nature of society journalism after it was purchased by the piratical Colonel D'Alton Mann in 1891. When he took over ownership of the magazine that year he wrote: 'The 400 of New York is an element so absolutely shallow and unhealthy that it deserves to be derided almost incessantly'[57] – an editorial philosophy he pursued with great ebullience until a court case in 1905 exposed the seamier side of his methods. Colonel Mann paid for stories from a wide network of clubmen and other members of society down on their luck for his information, as well as servants and suppliers, which then became part of his weekly 'Saunterings' column. As a weekly magazine, *Town Topics* harassed society's elite week in, week out using a well-placed network of spies so that long-running plot lines emerged for the initiated, which often turned out to be accurate because his informants were so close to the heart of society. Colonel Mann was known to accept money from society figures in return for pulling unflattering stories; and it would later emerge that he had a group of eminent 'immunes' whom he blackmailed into handing over large sums of money in exchange for soft treatment.

One of Mann's favourite tricks was to place paragraphs in his column that described reprehensible behaviour on the part of anonymous individuals, giving the readership the fun of decoding his allegations (this was often easy because he frequently placed another paragraph describing quite innocuous activities by the named individual close by). On 19 July 1894, *Town Topics* leapt into print with a story of 'a most offensive liaison going on in high life between a man who has been conspicuous in society and . . . the wife of a millionaire that moves in the same set'. It had long been thought that this relationship would become a scandal. 'But with a great deal of manoeuvring some sort of treaty of peace was patched up.' Much to *Town Topics*' sorrow however, 'the shameful affair had continued without abatement', the lover in question was now in Europe with the married woman, and the husband's reputation had been 'recklessly besmirched'. The names of two honourable families were about to be 'dragged in the dust, all to gratify the passions of a pair that have renounced the thousand legitimate delights at their command to embrace the one that is forbidden and reprehensible'.[58]

But there was another twist to the story. It would appear that the husband in the case had inexplicably forsaken the moral high ground by taking up with an inamorata of his own in Paris, a demimondaine whom he was entertaining in 'the fashion of Lucullus of old'. By the following week *Town Topics* had stopped bothering to keep up the fiction. William K. Vanderbilt was in Paris flaunting his relationship with one Nellie Neustretter, a very grand courtesan – 'one of the prettiest and nicest of the high-class *horizontales*'.[59]

Alva seems to have decided to sit the publicity out in England, staying on after the London season and all suitable aristocrats had dispersed to the grouse moors of Scotland. It is unclear whether *Town Topics* was correct in maintaining that Oliver Belmont joined her, but it is quite likely. Alva and Consuelo returned to New York on 28 September 1894 on board the *Lucania*, arriving in Newport well after the season closed on 29 September. Alva now prepared to implement a three-point plan. She would divorce William K. for adultery, ensuring that she could have custody of the children; she would place Consuelo in an English aristocratic setting; and she would regularise her own position with Oliver Belmont. These three objectives would become intricately entangled in the months ahead.

After the amusements of Paris, Consuelo looked forward to a winter season in New York, well away from Europe and threats of international marriage. She and Alva settled back into 660 Fifth Avenue. William K. was banished to his club. (Dissatisfied with the configuration of space he called in workmen to knock down partition walls and redecorate. 'When at the club Mr Vanderbilt can entertain at dinner forty friends on the same floor upon which his rooms are and be sure of no intrusion,' insinuated *Town Topics* silkily.[60]) It was reported variously that his brother Cornelius Vanderbilt II had rushed to Paris in the summer for crisis talks and that the Vanderbilts had met for a family caucus in Boston. Whether or not these family conferences took place, the Vanderbilts now rallied firmly behind William K., because, according to *Town Topics*, Alva had condescended to them all in the most supercilious manner for years.[61] There was certainly tension. As far as Alva was concerned they were either with her or against her. She broke off relations with every one of William K.'s siblings and anyone else who failed to offer her unconditional support. As a result, Consuelo's hopes of a New York debut were dashed. 'During the following months I was to suffer a perpetual denial of friendships and pleasures, since my mother resented seeing anyone whose loyalties were not completely hers,'[62] she wrote.

Disliking scandal and controversy, William K. did his best to dissuade Alva from pressing for a divorce. However angry he may have felt, he was concerned that given the double standards of the day, disgrace would rebound on her alone. Well into the autumn, Alva's lawyer, Joseph Choate, did his best to dissuade her, pointing out that her close circle would regard her as a traitor for drawing scandalous attention to the lives of the ultra-wealthy. 'He saw immense fortunes in the hands of a privileged few. He knew the inevitable social unrest which would result from such a condition. If Wealth laid itself open to attack from any source its throne was weakened.'[63] When that failed to have any effect, Choate tried to warn Alva that by insisting on divorcing William K. Vanderbilt for adultery, she would be pitting herself against the vested interests of American male wealth. 'He knew better than I did the power and influence of wealth. He knew its sway over Courts of Kings and Courts of Law ... prelates and laymen ... even those who called themselves "friend".'[64]

Choate argued that the punishment meted out to women daring to challenge male hegemony would be so harsh that even Alva would not be able to withstand it. Reflecting on the episode, Alva once again presented her reaction as heroic: 'My argument in return was that I believed it was necessary for some woman to blaze the way for a just recognition of her own personality.'[65] Later, though, she also said that if she had known how difficult it would be, she might have thought twice about going into battle alone. The problem which Alva never mentioned was that it was one thing to sue for adultery (and this was courageous); but it was quite another matter to survive the battle when the world knew that she had a lover of her own whom she wished to marry. Once Joseph Choate assured her she would have custody of the children, however, Alva determined to press ahead regardless. 'The legalized prostitution that marriage covers is to me appalling ... If marriage is a protection for the woman against many wrongs, divorce is also an escape from many degrading evils,'[66] she said to Sara Bard Field.

Having surrendered on the divorce issue, William K. went back to Paris, where observant correspondents reported on his dalliance with Nellie Neustretter. A reporter for *Town Topics* thought that he looked wretched. 'There were large circles under his eyes, and he looked neither well nor happy.'[67] William K. arrived back in New York on 22 December 1894, and even the taciturn superintendent Mr Gilmour noted that the Christmas atmosphere was strained and tense. 'Willie and his father went out walking this morning. In the evening I went to the Knickerbocker Club, 32 Street to get Mr V. for Mrs V. but he was not at home. Mr Jay came in the evening to see Mrs V. I was called out of my bed to take a note to Mr V. 11 pm.'[68] On New Year's Day, Alva had a huge row with another servant: 'He was told to leave the house. He replied he would go when he felt so disposed.'[69]

The only person who did her best to ease the tension was seventeen-year-old Consuelo who treated her maid, her governess and Mr Gilmour to tickets for the opera on Boxing Day. In the middle of January 1895, William K. fled back to Europe amid mounting press speculation that the Vanderbilts were filing for divorce. On the day of his departure the *World* finally broke the story in prose breathless with excitement: 'Mr Vanderbilt came from Europe just one month ago. His stay has been almost entirely devoted to arranging his family affairs. There has been no reconciliation between him and Mrs Vanderbilt.'[70] One influential figure rallied to Alva's defence. On the evening of 16 January, Mrs Astor publicly supported Alva by inviting Consuelo to a party for her great-niece, Helen Kingsland. It was a kind gesture but one society reporter noted that Consuelo had a miserable and embarrassing evening as the gilded youth of New York tittered about the scandal whenever her back was turned.

From a Vanderbilt point of view, William K.'s precipitate departure to Europe was both unfortunate and misjudged, for it handed control of the story to Alva. When the divorce was finally granted on 6 March, the dam of publicity burst. Never a newspaper to understate matters, the *World* described it as 'the biggest divorce case that America has ever known. It is, in fact, the biggest ever known in *The World*.'[71] The paper saw it as its moral duty to provide the reading public with everything it wanted to know, while simultaneously lambasting the rich for lax moral standards. One striking feature of its reportage, however, was the extent to which it favoured Alva over William K., leading to the suspicion that she had managed to brief its journalists. Mrs Vanderbilt had not fled to Europe, like her husband wrote the *World*. She was determined 'to stay here until the divorce should be publicly announced; not to run away from the publicity which reflects only on her husband, who is pronounced guilty'.[72] A photograph of Nellie Neustretter was printed in what looked suspiciously like her underwear. Alva (though the report was not entirely complimentary) was presented as the unhappy victim, made peevish by her philandering husband; and Oliver Belmont was never mentioned at all.

It is possible that Alva arranged a deal. Oliver's name would be kept out of the *World*'s story in exchange for a most intriguing piece of information. On the morning of 7 March, the *World* produced a sensational piece of news. Nellie Neustretter was an elaborate sideshow, possibly just a decoy. The real object of William K.'s affections, and the true reason for Alva's implacable fury, was that her husband had been having a longstanding affair with her very old friend, Consuelo, Duchess of Manchester.

There is no means of establishing for sure whether this story is true. It was never formally denied by anyone involved, however, and it may have some basis. Years later Sara Bard Field told an interviewer that although Alva would not allow her to mention it in the memoirs, William K. 'had brought his mistresses right into the home' including 'poor women of the nobility of England'.[73] Consuelo Manchester, to all intents and purposes, disappeared from Alva's life after 1894, which is odd since she was not simply Consuelo's godmother and an English duchess, she was also a relation by marriage after Alva's sister, Jenny, married Fernando Yznaga. In her memoirs, Consuelo (Vanderbilt) makes very few references to her godmother.[74] Consuelo Manchester was also famously unhappily married. Her husband had been declared bankrupt in 1890, and had abandoned her in favour of a music-hall singer whom he escorted round London before his death in 1892. She was constantly short of money; her other lovers included the Prince of Wales. The *World* suggested the affair between Consuelo Manchester and William K. was well established (though not exclusive):

a 'titled American woman' and William K. had been linked eleven years earlier, in 1884. There was even one report that Alva had almost thrown a 'titled American friend' out of the marital home as early as 1879.[75] William K.'s inexplicable conduct with regard to Nellie Neustretter was now quite comprehensible, said the *World*. He was simply trying to deflect attention away from a scandal involving his mistress by flaunting a relationship with a *grande cocotte*.

Town Topics, peeved at its failure to uncover this story first, managed to keep it alive by downplaying it. 'According to the rumour most generally credited among those who know nothing on the subject, one of them is to marry a banker and the other a duchess.'[76] Whatever the truth, the warm relationship between William K. and Consuelo Manchester was to be highlighted in the most tragic fashion possible within days of this publicity firestorm. In a coincidence far more dreadful than the sinking of any yacht, Consuelo Manchester's daughter, who was named after Alva and who was only in her late teens, died in Italy ten days after the divorce was granted. Acutely distressed, Consuelo Manchester turned to William K. for help. On Saturday 16 March, he gave orders to the captain of the *Valiant* to sail to Civitavecchia in Italy. The captain wrote in the ship's log: 'In the afternoon we embarked the remains of the late Lady May Alva Montagu, accompanied by the Duchess of Manchester, Lady Alice Montagu, Miss Yznaga, Dr A. Muthie, Mr F. Yznaga, and servants, and at 6 p.m. sailed for Marseilles.'[77] According to the same log, William K. went up to Rome by the 4.50 p.m. train, possibly to assist with legal formalities, or to avoid making scandalous rumours worse. The Duchess of Manchester and her party sailed with Lady Alva Montagu's body back to Marseilles, and from thence to Paris. The *Valiant* then turned round and went back to Italy to pick up the rest of the party. William K. was back in Paris by 2 March.

This story of the liaison refused to go away for several months. It was noted that Alva did not attend her namesake's funeral. By April, *Town Topics* was reporting that the rumour mill had it that the death of Lady Alva Montagu had marked a turning point in the relationship between William K. and Consuelo Manchester, and that there was a persistent story 'that will seemingly not die down . . . to the effect that Mr Vanderbilt would have become the husband of the Duchess of Manchester had it not been for her bereavement in the loss of her twin daughter Lady Alva Montagu'.[78] By the middle of June, a consensus seemed to be emerging in the society press that Nellie Neustretter had indeed simply been engaged by William K. as co-respondent, though this does not wholly explain why he felt obliged to spend several months in her company.

The affair caught the attention of Henry James, also in Paris in the summer of 1895, who thought that William K.'s relationship with Nellie was part of a complicated strategy to force Alva into divorce, and that it had the makings of a short story: 'The husband doesn't care a straw for the *cocotte* and makes a bargain with her that is wholly independent of real intimacy. He makes her understand the facts of his situation – which is that he is in love with another woman. Toward that woman his wife's character and proceedings drive him, but he loves her too much to compromise her. He can't let himself be divorced on her account – he can on that of the *femme galante* – who has nothing – no name – to lose.'[79] This would become the starting point for James's novel, *The Special Type*, published in 1903.

Under the terms of the divorce, Alva kept Marble House, which had already been made over to her at her insistence, and refused William K.'s offers of both 660 Fifth Avenue and Idle Hour, which were 'rendered disagreeable by unpleasant memories'.[80] The terms of the divorce settlement were never made public, in spite of furious efforts by the press to find out, but Alva received a sum close to $2.3 million and an income of about $100,000 a year, with provision that specified amounts of the capital sum should be transferred to each of the children on marriage or at the age of twenty-eight.[81]

Predictably, Alva faced a harsh reaction from some elements in society, but as ever, she presented herself as having toughed it out: 'I did not fail myself at this stormy time. I got my divorce and just as in childhood days I accepted the whipping my mother gave me for taking the forbidden liberty, so I bared my back to the whipping of Society for taking a freedom which would eventually better them as well as myself.'[82] In spite of Choate's warnings about the viciousness of hegemenous males, society women were worse. 'Yes, and they put on the lash, especially the women, and especially the Christian women. When I walked into Trinity Church in Newport on a Sunday soon after obtaining my divorce, not a single one of my old friends would recognize me.'[83]

On Wednesday 13 March, Alva departed for Europe with Consuelo and Harold, seen off by William Gilmour. The *New York Tribune* reported that Alva travelled in her usual style with five maids, one man servant and seventy pieces of luggage.

By now, Alva had another compelling reason for sailing to Europe. Preoccupied by her divorce, she had failed to take seriously Consuelo's growing attachment to a man of thirty-three, which was threatening to undermine her plan to place her daughter in an aristocratic setting. It is impossible that Alva failed to notice the warmth between Consuelo and her American admirer since the indefatigable *World* had picked up the scent as early as the middle of February that year. On Valentine's Day, it chose to run the story as a romantic tale of shattered hopes: 'A young man, bearing an old family first name, prefixed with a prominent Boston family surname, has been all devotion to Miss Consuelo Vanderbilt and she apparently was most happy in his attentions. This joyousness must now be relegated to the saddest of "might have beens".'[84] Two days later the same paper explicitly linked Consuelo and the Duke of Marlborough asking: 'Is she to be a Duchess? It is quite generally recognised that the Duke must marry money if he is to keep up Blenheim. His income is only £8,000 ($40,000) a year and Blenheim costs £14,000 ($370,000) a year.'[85]

The young man who had been all devotion to Consuelo was Winthrop Rutherfurd, son of the eminently respectable Mr and Mrs Lewis Rutherfurd, a New England family of impeccable pedigree (Lewis Rutherfurd was one of the earliest Patriarchs in 1872). Through his mother, Winthrop Rutherfurd was a direct descendant of Peter Stuyvesant, colonial governor of New York, and John Winthrop, the first governor of Massachusetts. 'Winty' Rutherfurd was tall and famously good-looking. Though trained as a lawyer, he spent much of his youth playing polo and golf, for which he had something of a reputation. He was a member of the elite Newport Golf Club and has been described as suitable for Consuelo in every way.

As far as Alva was concerned, however, he was not suitable at all. The first problem was that he simply represented the wrong marital path. In America in the 1890s, there were two routes to dynastic marriage open to the new phenomenon, the American heiress. One was to marry into the network of American families enriched by industrial capitalism, further consolidating vast fortunes, creating an aristocracy of money but effectively embracing the 'new'. The other was to marry into one of the European aristocracies, depleting the industrial fortune but ennobling the American family through association with nobility and centuries of tradition, elegance and culture.[86] This trend had been started in Alva's generation by Jennie Jerome, who married Lord Randolph Churchill and by Consuelo Yznaga, though as it happened neither of them had huge dowries. By 1895, the European route to aristocratising one's family had become highly competitive. That year alone there were nine marriages between heiresses and English aristocrats* while Anna Gould's marriage to Frenchman Count Boni de Castellane set new standards for lavish New York weddings. By 1914 commentators calculated that over 500 American fortunes had been transferred to Europe through this route.[87] Alva, always ambivalent about the 'crude' and 'unfinished' nature of American life, embittered by the power structures of American society, drawn to those parts of European history where aristocratic marriages were arranged as a matter of course and a great admirer of British

* They were: Maud Burke to Sir Bache Cunard; Mary Leiter to George Curzon; Josephine Chamberlain to the 1st Baron Scarisbrick; Lily Hammersley to Lord William Beresford; Elizabeth LaRoche to Sir Howland Roberts; Leonora Van Roberts to the 7th Earl of Tankerville; Pauline Whitney to Almeric Paget; Cora Rogers to Baron Fairhaven of Lode; and Consuelo Vanderbilt to the 9th Duke of Marlborough.

aristocracy was, of course, determined that it would be the European and not the American route for Consuelo.

A further problem with Winthrop Rutherfurd, however, was that he was far too close – and far too similar – to William K. Vanderbilt, the 'weak nonentity' whom she had just divorced. According to Cornelius Vanderbilt Jr, 'the Rutherfurds lived well, dressed expensively, and did little else', though Winthrop's father, Lewis Rutherfurd, was a distinguished astronomer who took some early photographs of the surface of the moon. As far as Alva was concerned, Winthrop Rutherfurd was a fine example of the new breed of useless male now emerging, like her ex-husband, from three generations of plutocratic wealth. Alva also suspected him of being a gold-digger. American society had evolved to a point where it was impossible to participate without being very rich. Consuelo's dowry was a clear temptation to a young man from a good family with social ambitions but without great wealth. Alva, of course, took the view that almost all rich American men were serial adulterers who left the business of keeping up respectable appearance to their wives, while they romped like young colts in 'the world-wide field'.[88] In Consuelo's case there was a real danger that she would facilitate 'romping' by financing it. Alva always maintained that her divorce had no effect on her children's lives. In reality, the bitterness and cynicism engendered by William K.'s philandering profoundly coloured her plans for Consuelo's future.

For the moment, however, she dealt with her daughter's first love badly, in a manner guaranteed to encourage romance rather than stifle it. According to Consuelo, her first line of attack was contempt, 'reserving special darts for [the] older man who by his outstanding looks, his distinction and his charm had gained a marked ascendancy in my affections'.[89] Winty's response was to propose, and when the proposal came, it would not have been out of place in an Edith Wharton novel. It took place on Consuelo's eighteenth birthday on 2 March 1895, a few days before the finalisation of the Vanderbilt divorce. First, he sent her an American Beauty rose, her favourite. Later, he joined Consuelo, a group of other young people, and Alva, on a cycling expedition along

Riverside Drive. 'My Rosenkavalier and I managed to outdistance the rest. It was a most hurried proposal, for my mother and the others were not far behind; as they strained to reach us he pressed me to agree to a secret engagement, for I was leaving for Europe the next day. He added that he would follow me, but that I must not tell my mother since she would most certainly withhold her consent to our engagement. On my return to America we might plan an elopement.'[90]

Consuelo was not, in fact, due to leave for Europe for another fortnight. But there were to be no further meetings with Winty. A few days after Alva and Consuelo set sail for Paris, several newspapers also noted the departure for Europe of Winthrop Rutherfurd. If he hoped to see Consuelo he was to be disappointed. Alva regarded her daughter's glow of happiness with dark suspicion and did everything in her power to prevent a meeting. 'She laid her plans with forethought and skill, and during the five months of our stay in Europe I never laid eyes on Mr X, nor did I hear from him. Later I learned that he had followed us to Paris but had been refused admittance when he called. His letters had been confiscated; my own, though they were few, no doubt suffered the same fate.'[91] The happiness of the previous summer in Paris was a distant memory. Consuelo tried on new clothes 'like an automaton.'[92] Alva was intensely irritated by her daughter's air of adolescent 'martyrdom', and her complaints about it only served to deepen Consuelo's misery.

Alva later argued vigorously that she had only had her daughter's interests at heart in keeping her from Rutherfurd in this way. It should not be overlooked that in this period immediately after her divorce, Consuelo's interests were closely bound up with her own. Alva wished to marry Oliver Belmont. She did not, however, wish to abandon her position as a leader of society once she remarried, and thus retreat from the only theatre of life that was open to her. However high-minded Alva's reasons may have been for saving her daughter from life with an American plutocrat, Consuelo's marriage to Winthrop Rutherfurd would have done little to bolster Alva's position in America, however popular he might have been at Newport Golf Club. The Duke of Marlborough was another matter entirely. Consuelo later maintained that Alva ordered her wedding dress in Paris that spring, so sure was she about the successful conclusion of her plans. There is no evidence for this; but Alva certainly bought hundreds of expensive 'favors' – small presents – for a ball, as she now planned what would become a decisive manoeuvre.

As *Town Topics* put it: 'There has been little doubt in the minds of those who know Mrs Vanderbilt intimately, and consequently, understand her character and temperament, that she would return to Newport this summer and assert her position.'[93] Months in advance of her return to Newport, Alva fired her first shot by letting it be known from Paris that she would be giving a ball at Marble House the following August, and that she would construe acceptance of this invitation as a pledge of loyalty. By the middle of June, these reports were sending New York's elite into a frenzy, particularly in the absence of any signal from the Vanderbilt family whom nobody wished to offend. 'Small wonder it is that the approaching dilemma begins to assume tremendous proportions in the minds of not only those who are not yet absolutely sure of their position in the social world, and who feel they cannot afford to risk their chances by a false move in the start, but even, indeed, in those of the contingent of assured position, who have no prejudice or animosity toward Mrs Vanderbilt herself, who certainly feel kindly toward her daughter, and yet are on terms of friendship and even intimacy with the other members of the family,'[94] said *Town Topics* sagaciously.

Alva then moved Consuelo from Paris to London to participate in the London season of 1895. Here, she re-established contact with Minnie Paget who took the necessary steps. Consuelo was asked to a ball by the Duke and Duchess of Sutherland and, knowing almost nobody, was grateful to anyone who requested her as a partner by marking her dance card. Perhaps Aunt Lansdowne had had a word, for the Duke of Marlborough claimed several dances. To Alva's intense satisfaction, he followed this up by inviting them both, and Lady Paget, to spend a weekend at Blenheim Palace.

The party that travelled to Oxfordshire on 15 June was small, consisting of Alva, Consuelo, Minnie Paget, 'three young men' – including Lord Lansdowne's heir – and the Duke's two sisters, Lady Lilian and Lady Norah Spencer-Churchill. They all seemed 'lost in so big a house' wrote Consuelo, but she liked Lilian immediately, finding her unaffected and kind.[95] Saturday evening was spent listening to the Duke's organist, Mr Perkins, playing the organ in the Long Library, installed when his father the 8th Duke married 'Duchess Lily', a wealthy American widow to whom Blenheim also owed the installation of central heating and electric lighting.

The following day, Alva's usual rules of chaperonage were conspicuous by their absence for no obstacle was placed in the way of the Duke showing Consuelo round part of the Blenheim estate. They drove together to pretty outlying villages where 'old women and children curtsied and men touched their caps as we passed'.[96] Although enchanted by the countryside, the feudalism on display made Consuelo feel uncomfortable, and in Alva's absence she was quick to say so. 'That Marlborough was ambitious I gathered from his talk; that he should be proud of his position and estates seemed but natural; but did he recognise his obligations? Steeped as I then was in questions of political economy – in the theories of the rights of man, in the speeches of Gladstone and John Bright – it was not strange that such reflections should occur to me.'[97]

According to Consuelo – and we only have her side of the story here – the Duke of Marlborough seemed to find these remarks amusing rather than tiresome, and made up his mind that very afternoon that he would set aside his feelings for an English girl with whom he was in love and marry Consuelo. It seems more likely, given his subsequent caution, that the Duke of Marlborough simply decided that marriage to Consuelo was a possibility that could reasonably be explored. Even if her notions were a trifle outlandish, she was intelligent and thoughtful; and the intervening year had given this young duke ample time to discover that both his sense of obligation to Blenheim and his political aspirations required substantial financial resource. As far as Alva was concerned, however, the weekend at Blenheim and his pleasant attentions to Consuelo made it easy for her to extend an invitation to her ball at Marble House in August. The Duke immediately accepted, giving out that he had never visited the United States, and would come to Newport as part of a longer tour.

This was a major coup for Alva. By late June, the society press were lying in wait in Newport to await her return. The *World* even sent detectives – an early form of paparazzi – to Newport to watch every move both Vanderbilts made and report back. Once again, there were multiple narrative lines. How would the Cornelius Vanderbilts, who would be opening their house The Breakers that August, react if they met Alva? How would society as a whole respond to the invitation to her ball? There was also the delicious extra twist of Oliver Belmont's arrival and the news that he too would be giving a house-warming ball at his Newport house, Belcourt. 'The housewarming of this new mansion will probably be one of the chief social events of the Newport season, and may, if reports be true, also be the opening gun in the *Montague and Capulet* warfare that is still a menace to the peace of the season and looms like a dark cloud on the horizon,' reported *Town Topics*.[98] It was all feverishly exciting.

When Alva finally arrived with Consuelo in Newport in July, she soon put Newport society out of its misery by unleashing a secret weapon in the diminutive form of the Duke of Marlborough. The attention paid by the Duke to Consuelo had been noted by *Town Topics*, but stories of an engagement were dismissed on the grounds that the divorced status of Mrs Vanderbilt would present an obstacle to such a match. Now, Alva let it be known that there was no obstacle whatsoever for the Duke of Marlborough had accepted an invitation to attend her ball and would be coming to stay with her in Newport for several days. Suddenly, the much anticipated drama ebbed away. Realising they had been wholly outflanked, the denizens of Newport reached for their pens and their blotting paper, thanked Mrs Vanderbilt for her kind invitation through gritted teeth, and told her they would have much pleasure in accepting.

* * *

Consuelo faced a much more serious problem. She felt that she was being 'steered into a vortex'.[99] She considered herself secretly engaged to Winthrop Rutherfurd, and after the weekend at Blenheim she was certain that she did not wish to marry 'Sunny' Marlborough. 'Homeward bound, I dreamed of life in my own country with my Rosenkavalier. It would, I knew, entail a struggle, but I meant to force the issue with my mother.'[100]

Once they reached Newport, however, even making contact with Winthrop Rutherfurd became very difficult and with the Duke of Marlborough's visit less than six weeks away, Consuelo became anxious and despondent. Marble House stood in a prominent but isolated position on Bellevue Avenue, where every move was scrutinised by the summer colony and by the press; assignations were impossible, and all her post was monitored. 'On reaching Newport my life became that of a prisoner, with my mother and my governess as wardens. I was never out of their sight. Friends called but were told I was not at home. Locked behind those high walls – the porter had orders not to let me out unaccompanied – I had no chance of getting any word to my fiancé. Brought up to obey, I was helpless under my mother's total domination.'[101]

Was this melodramatic? Probably not, for by now the stakes for Alva were very high. It was essential to the success of Alva's manoeuvres that nothing should prevent the Duke from honouring her invitation. She had no intention of letting her daughter undermine such a careful campaign with a misjudged teenage crush, and she may have feared that an obstinate but desperate Consuelo would somehow arrange an elopement. (One fictional account of Alva's life even has her turning this period into a test of Winthrop Rutherfurd's strength of feeling, which is not implausible either.[102]) Quite apart from Rutherfurd's intrinsic unsuitability, Alva would be the laughing stock of America and her chances of protecting her own position in the aftermath of divorce would be greatly diminished.

In spite of every difficulty being placed in their way, however, Consuelo and Winthrop Rutherfurd eventually met once more at a ball. They had one short dance before Consuelo was taken away by Alva, but he had time to tell Consuelo that his feelings had not changed. That evening, matters came to a head in the most famous mother-daughter row of the Gilded Age. Following an ominous silence on the drive home, Consuelo went to Alva's bedroom and informed her mother that she felt that she had a right to choose her own husband, and that she intended to marry Winthrop Rutherfurd. 'These words, the bravest I had ever uttered, brought down a frightful storm of protest. I suffered every searing reproach, heard every possible invective hurled at the man I loved. I was informed of his numerous flirtations, of his well-known love for a married woman, of his desire to marry an heiress.'[103] Alva went on to declare that there was madness in the Rutherfurd family, and that he could never have children (this was certainly inaccurate). Consuelo, by her own account, stood her ground. Alva argued back that Consuelo was far too young to make the choice herself, and that her 'decision to select a husband for me was founded on considerations I was too young and inexperienced to appreciate'[104] – sentiments Alva would later repeat almost word for word herself to Sara Bard Field.[105]

Alva had prided herself in bringing up independent-minded children, but when her doll-child finally showed some signs of independence, mother and daughter collided with force. For the first time in her life, Consuelo stood her ground and argued back. 'I still maintained my right to lead the life I wished. It was perhaps my unexpected resistance or the mere fact that no-one had ever stood up to her that made her say she would not hesitate to shoot a man whom she considered would ruin my life.'[106] Shouting that she would shoot Winthrop Rutherfurd was characteristic of Alva at her most impulsive, and it would give anyone who knew her a moment's pause for thought. When Consuelo's cousin, Adele, indicated she might want to marry her old roué of an uncle, Creighton Webb, her mother Emily – a far kinder and more subtle character – simply replied that she would rather see Adele in her coffin first, and that that was the end of the matter.[107]

What followed went far beyond the firm but well-meant line taken with Adele by Aunt Emily. The next day, the house was ominously quiet, and no-one came to see Consuelo. She was told that her mother was ill and that the doctor was on his way. Even her calm and collected English governess seemed harassed. Eventually, her mother's friend, Lucie Oelrichs, now Mrs William Jay, came to see her. Aunt Jay condemned Consuelo's behaviour. She may have pointed out that what Consuelo wanted to do was potentially very damaging to Alva. Most seriously, Aunt Jay gave Consuelo to understand that her mother had had a heart attack 'brought about by my callous indifference to her feelings. She confirmed my mother's intentions of never consenting to my plans for marriage, and her resolve to shoot X should I decide to run away with him. I asked her if I could see my mother and whether in her opinion she would ever relent. I still remember the terrible answer, "Your mother will never relent and I warn you there will be a catastrophe if you persist. The doctor has said that another scene may easily bring on a heart attack and he will not be responsible for the result. You can ask the doctor yourself if you do not believe me!".'

The precise details of this scene may have been embellished over time, but much of what Consuelo maintained took place is consistent with Alva's later behaviour at other times and in different places. Alva's crude attempt to translate the question of Consuelo's marriage into one about her own health and happiness is typical behaviour of a highly controlling personality in a very anxious state. Unlike Aunt Emily, Alva was the first to claim that when crossed, her instinct was to head straight for a tremendous fight and an outright win. In this instance she was fighting three battles at once: to stop Consuelo from marrying Winthrop Rutherfurd; to prevent Consuelo from doing anything which might stall the Duke's visit; and to protect her own social position. Consuelo's determined reaction may have taken her by surprise. Perhaps her daughter's unprecedented display of strength of character did indeed make Alva feel so powerless that she fell ill. Who can tell? Whatever the truth, being told that she would kill her mother if she persisted had the desired effect on Consuelo as Alva must have known it would. 'In utter misery I asked Mrs Jay to let X know that I could not marry him.'[108]

The short period between this terrible row and the Duke's arrival was marked by a time of intense introspection when Consuelo felt compelled to keep her feelings to herself. She wrote that friends who had been rebuffed no longer called; her brothers meanwhile were too young and too preoccupied with their own affairs. What is perhaps more shocking to the modern sensibility is that no adult intervened. This was because they either shared Alva's view of Consuelo's best interests, were too frightened of Alva to protest, or, like Mrs Jay, had a vested interest in the Duke's arrival in Newport. Remembering the gossip of previous generations, Eileen Slocum remarks that no-one in the wider summer colony could believe that Consuelo would hold out against such an advantageous match for long. It soon became clear that Winthrop Rutherfurd would not be attempting a dramatic elopement. A kind interpretation is that he simply took Consuelo at her word and did not wish to force the issue; a less charitable view is that the prospect of a fight with Alva which might damage a wedding settlement caused him to back off sharply, and he seems to have spent the rest of the Newport season in the background, pottering about on the golf course.

William K. Vanderbilt, meanwhile, was even less help. Even though the *Valiant* was moored in Newport harbour (and was not 'away at sea' as Consuelo thought in her memoirs), he felt out of reach. Consuelo adored her father too much ever to describe him as a weak man but this is the inescapable conclusion: 'his gentle nature hated strife,' she wrote. Even while her parents had been married, the children knew it was pointless appealing to him in any struggle with their mother. 'He played only a small part in our lives . . . he was always shunted or side-tracked from our occupations . . . with children's clairvoyance we knew that she would prove adamant to any appeal our father made on our behalf and we never asked him to interfere.'[109] The Commodore's first biographer, who met him, thought William K. showed signs of a 'morose disposition', and a rare interview in later life does indeed suggest that however charming and gregarious, William K. also had a melancholic, passive streak. 'My life was never destined to be quite happy,' he told the journalist. 'It was laid along lines which I could not foresee almost from earliest childhood. It has left me with nothing to hope for, with nothing definite to seek or strive for.'[110] On this occasion, passivity may have led him to fail his daughter.

It is also possible that the idea of Consuelo becoming a duchess appealed to him. Here indeed was the apotheosis of the Vanderbilts; here at last was the final symbol of the family's rise to the highest echelons of international society; and here was splendid protection from any untoward consequences of his divorce from her mother. In fairness, Consuelo later admitted that she had kept her feelings to herself, and that she knew there was little point in involving her father in a struggle which would 'only involve him in a hopeless struggle against impossible odds and further stimulate my mother's rancour'.[111] The log of the *Valiant* during the Newport season of 1895 suggests that though William K. had no need to protect his social position as Alva did after her divorce, he was equally determined to consolidate it with an on-board entertaining schedule that culminated in a luncheon for the Duke of Marlborough, giving rise to a dark suspicion that he may even have colluded with Alva on this issue.

Meanwhile *Town Topics* reported that Oliver Belmont would also be entertaining the Duke when he arrived in Newport, and that he was planning his own splendid ball to take place shortly after the one being given by Alva. So many people had a vested interest in the success of the Duke's visit that eighteen-year-old Consuelo must indeed have felt that the forces ranged against her were overpowering and that the whole situation was too difficult to fight. The only person to whom she confided her fears was her English governess, Miss Harper, of whom she was very fond. In Edith Wharton's novel *The Buccaneers*, the governess sacrifices her own happiness to secure the happiness of her charge. Miss Harper chose a more pragmatic approach. 'How wisely she spoke of the future awaiting me in her country, of the opportunities for usefulness and social service I would find there, of the happiness a life lived for others can bring. And in such gentle appeals to my better nature she slowly swung me from contemplation of a purely personal nature to a higher idealism.'[112] It was just as well for the news soon arrived at Marble House that the Duke was on his way to New York aboard the *Campania* and would be in Newport in just a few days.

4
The wedding

THE DUKE OF MARLBOROUGH was not the star passenger as he left Liverpool on the Cunard steamer *Campania* on Saturday 16 August 1895. This slot was reserved for Keir Hardie, leader of Britain's emerging Independent Labour Party who was on his way to the United States for a lecture tour, and who was seen out of the harbour by waving supporters in a tug boat, *The Toiler*, complete with bunting, a band, and fluttering socialist mottoes.

Keir Hardie noticed the 'haughty aristocrat' immediately he boarded the *Campania* but refused to be intimidated. 'There are dukes and archbishops and bishops and State Senators on board; but the I.L.P. passengers were the only ones who could command a crowded tug boat by way of a farewell,' he wrote in tones of satisfaction a week later.[1] One of the ship's waiters told Keir Hardie that the Duke of Marlborough had been most interested in his presence, though he did not attend any of Keir Hardie's impromptu on-board talks about socialism, where Hardie drew on the relationship between the *Campania*'s cabin accommodation and the British class structure to illustrate his point.[2]

When they disembarked in New York on Friday 23 August, however, it was the Duke who was greeted by the New York press, in a manner for which he was wholly unprepared. He was followed to the Waldorf; he was observed eating breakfast at 10 o'clock; he was joined by Captain A. H. Lee, a fellow passenger on the *Campania*; he took a stroll down Fifth Avenue; and he was called on by Creighton Webb (the same old roué who had tried to marry Cousin Adele). He then travelled in a reserved seat in a parlour car on the

5 o'clock train to Newport on the following day, Saturday. 'Lookouts from some of the great housetops on the Cliffs are already watching for his Grace's arrival,' said the *New York Herald*; 'and should he come he may expect a charge such as his famous ancestor, John Churchill, never met.'[3]

The charge soon came. The news that the Duke had been seen with the Vanderbilts at Trinity Church in Newport on Sunday morning spread fast. That afternoon Alva held open house for Newport society and was promptly mobbed. It was clear that an in-house duke had eviscerated all scruples. 'Mrs Vanderbilt has been informally "at home" on Sunday afternoons ever since her arrival at Newport, and a few of her friends have dropped in there for tea and a chat,' reported *Town Topics*. 'But on Sunday afternoon last – the morning newspapers having announced that the young Duke had arrived at Marble House – the huge iron gates were swung open to admit the entrance, during the afternoon, of almost every member, with the exception of the Vanderbilts, of the Newport summer colony.'[4]

The following day the *New York Herald* reported: 'Everyone in Newport today was running around saying to everyone else "Have you seen the Duke?" And then all strained their necks to find a man who looked like a duke, however a duke may look.'[5] Those who felt confused should have bought copies of the *Newport Mercury* which reported that those who had called at Marble House 'did not find, as many expected, a big strapping Englishman with a loud voice, whose grip you would remember with pain for hours after, but instead a pale-faced, frail-looking lad, with a voice devoid of that affected drawl peculiar to the English, and as soft as a debutante',[6] who looked amused by all the excitement he was causing.

On Monday, those who had not called at Marble House for tea on Sunday crowded into Newport Casino to catch a glimpse of the Duke of Marlborough near the tennis courts. That evening Richard T. Wilson Jr gave a calico party (where all the favours were made of calico) at the Golf Club for 300 guests. 'The Duke of Marlborough was present, of course, and that meant that all of the cream of the elite set would attend,' wrote the *Newport Journal*. On Tuesday 27

August, the newspaper estimated that about 5,000 people went to the Casino to watch tennis in the hope of catching a further glimpse, but were disappointed. On Wednesday 28th, the day of Alva's ball, the Duke of Marlborough demonstrated that he was a passable tennis player himself and 'played two sets on the casino grounds, with Mr P. M. Lydig'.[7] More significantly, William K. assisted Alva on the day of her ball (to which he was naturally not invited) by entertaining the Duke to lunch on board the *Valiant* with the cream of Newport society.[8]

Alva's long-planned ball, it was generally agreed, was the highlight of the Newport season. One newspaper called it 'The Most Beautiful Fête Ever Seen' – which would have pleased Alva because from the outset she had been determined to outdo all previous entertainments. Every invitee accepted. As if working to a plan (and he probably was), 'Mr W. K. Vanderbilt steamed away on *Vigilant* [sic] just at sunset.'[9] From early evening, scores of onlookers gathered at the gates to catch a glimpse of the guests. This was not a fancy-dress ball, but the party had an *ancien régime* flavour in the spirit of the house. A small army of servants was dressed in the style of Louis XIV; there were nine French chefs; and 'the grounds were illuminated by thousands of tiny globes of different colors, just as they used to be in Versailles when Louis strolled across the broad terrace of Versailles with his court'.[10]

The world of Louis XIV and Versailles was particularly noted by the party correspondent of the *New York Herald* who thought he had been thence transported until woken from his reverie by the strains of an Hungarian polka. Alva was dressed 'in a superb costume of white satin, with court train and wonderful diamonds, and looked as if she might have stepped out of one of the old court pictures in Versailles. Her daughter Miss Consuela [sic], becomingly arrayed in white satin and tulle, stood beside her.'[11] Lotus flowers and water hyacinths filled with tiny globes of light floated in a fountain in the hall; on every table there were orchids, ferns and pink hollyhocks tied with illuminated pink ribbons; and – in a touch that was a talking point of the evening – tiny humming birds swarmed amid the flowers.

Partly because of the heat, Newport balls started late. That night, guests danced to three different orchestras and supper was served at midnight (one course alone included 400 mixed birds) before breakfast appeared at 3 a.m. Richard T. Wilson Jr led Consuelo in the cotillion where she distributed the favours bought by Alva in Paris earlier in the year to those who had not been fortunate in winning them for themselves. These included 'genuine bagpipes made by French peasants',[12] as well as ladies' silk sashes, etchings and fans of the Louis XIV period, work baskets, mirrors, watch cases, ribbons and bells, and white 'Marble House' lanterns. One newspaper reported that the favours were so fine that they 'occasioned an immense amount of heartburning, envy and jealousy, and led to a deal of petty thievery. I am told that some of the women . . . stole favours from each other whenever they could.'[13]

Alva left nothing to chance and a good deal to shameless suggestion where her central campaign was concerned. The portrait of Consuelo in duchess mode by Carolus-Duran hung above the fireplace in the Gold Ballroom. The Duke of Marlborough stood beneath it, beside Mrs Jay, 'viewing the pretty women with interest',[14] the only barbed note in reports of the evening's entertainment. The *Newport Mercury* thought that Mrs Vanderbilt and the Duke were the 'cynosure of all eyes'.[15] 'It was a perfect night', Alva told Mary Young, 'and the house and grounds [looked] lovely in the moonlight, provid[ing] a setting of almost unreal beauty for one of the most beautiful balls I have ever seen.'[16] It was less amusing for William Gilmour. The ball ended at 5 a.m. and according to his notebooks, 'some had to be taken home as their navigation was somewhat uncertain, especially the gentler sex'.[17] He finally went to bed about 6 a.m., 'tired out', though he was luckier than the policemen stationed at the gates whose cab at daybreak collapsed after just a few yards, compelling them to walk home.

Alva, meanwhile, was almost certainly lying in bed, staring up at the Goddess Athene on the ceiling and basking in triumph. *Town Topics* concluded that the ball had been just as significant a social event as the great Vanderbilt ball of 1883. 'The Marble House ball of 1895 put the seal of fashionable approval upon that lady and all her doings, and was in its way, quite as remarkable and significant an entertainment as the fancy ball. The presence of the Duke of Marlborough – if not an acknowledged suitor for the hand of Miss Consuela [sic] Vanderbilt, certainly a suspected one – was of itself a successful stroke of diplomacy on Mrs Vanderbilt's part, and when was added to this a dance marked by the richest and most beautiful favors bestowed at an entertainment in years, and every appointment that taste could suggest or wealth provide, the success of the entertainment as a whole may be easily imagined.'[18]

Immediately after the Marble House ball, there was a momentary setback when Oliver Belmont collapsed from exhaustion. This meant that his much-anticipated house-warming ball at Belcourt had to be postponed until the following Monday, whereupon it clashed magnificently with a musical evening at the Cornelius Vanderbilts'. Faced with this social emergency – a gap in the collective party schedule – Mrs Robert Goelet rallied nobly and threw a 'surprise' party, which, of course, gave her an opportunity to entertain the Duke too. When it finally took place, Oliver's Bachelor's Ball (which simply meant that he received his guests alone) demonstrated the architect Richard Morris Hunt's ability to follow his clients into a marked degree of eccentricity when required.

Inside an exterior inspired by a Louis XIII chateau, Oliver, who was famous for his love of horses, had instructed the architect to build palatial stable accommodation on the ground floor. 'It is a most singular house,' wrote Julia Ward Howe to her daughter, 'with stalls for some thirteen or more horses, all filled, and everything elaborate and elegant. Oh! To lodge horses so, and be content that men and women should lodge in sheds and cellars!'[19] The residential part of the house was on the first floor, in Gothic style, but even here Oliver had had two of his favourite horses preserved by a taxidermist and placed at one end of the large salon.

The Bachelor's Ball was another splendid event, where the favours included small riding whips to reflect the masculine tenor of the invitation. Consuelo, Alva and the Duke of Marlborough were also entertained by the John Jacob Astors on their yacht the *Nourmahal*, and by the Goelets on the *White Lady*, though the Duke

– who must have been feeling the pace by now – declined further invitations to cruise on the grounds that he was a bad sailor. 'How leisurely were our pleasures!', wrote Consuelo later. 'In the mornings, with my mother, we drove to the Casino in a sociable, a carriage so named for the easy comfort it provided for conversation. Face to face on cushioned seats permitting one to lean back without the loss of dignity, we sat under an umbrella-like tent. Dressed in one of the elaborate batistes my mother had bought for me in Paris, with Marlborough opposite in flannels and the traditional sailor hat, we proceeded in state down Bellevue Avenue.'[20] Oliver Belmont was often in attendance: 'Sometimes he drove us to the Polo Field, where the young Waterbury boys were giving early proof of the dash and skill that later placed them in the team known as the Big Four.'[21]

The Duke of Marlborough had been invited to stay for the America's Cup races, but everyone knew that this was not the real reason for his visit. The days passed, and then a week, but there was still no announcement of an engagement, although the Duke was frequently seen having tennis lessons at the Casino. The social campaign at Marble House, meanwhile, continued unabated. On Saturday 31 August, Alva gave a dinner 'in honour of the Duke of Marlborough . . . among her guests being Mr and Mrs John Jacob Astor, Mr and Mrs Victor Sorchan, Miss Burden, Miss Post, Mr and Mrs T. S. Tailer, Miss Wilson and Mr Sidney Smith.'[22] This guest list also included Edith and Teddy Wharton, who were on the outer fringes of Alva's social circle and had based themselves largely in Newport since their marriage in 1885. The next day, Consuelo gave a party of her own at Marble House. Mr Gilmour noted: 'Sunday September 1st/2nd Miss V. had a huge reception in the afternoon. 3 Hindoos performed tricks for the guests.'[23] The *New York Herald* called it: 'the society event of the afternoon',[24] (which was hardly top billing), and mentioned that two new English arrivals were present, the daughters of Lord Dunraven, whose boat the *Valkyrie* would shortly compete for England against America in the America's Cup race.

On Thursday 5th and Friday 6th September, there was a sudden exodus from Newport to New York where the America's Cup races were to be held. Just as suddenly, society's focus swivelled away from Consuelo and the Duke of Marlborough to the race itself, leaving Newport 'as if stricken with a pestilence' and at the mercy of a few 'hen' dinners organised by women in desperation at having been left behind.[25] According to William Gilmour's notebooks, Alva, Consuelo and the Duke of Marlborough joined many other spectators on the 1.20 train to New York on Thursday 5 September, and watched the races from the Astor's yacht the *Nourmahal*.

In the event, the America's Cup of 1895 became mired in one of the more acrimonious controversies in the history of the race. Although William K.'s yacht, *Defender*, won the America's Cup with a 3–0 victory over the *Valkyrie*, it only won the first race on water. At the start of the second race, the *Valkyrie*'s boom hit *Defender*'s topmast stay and broke it. Although *Defender*'s crew made emergency repairs, they were unable to overcome the handicap, and the race committee reversed *Valkryrie*'s win by disqualifying her. Lord Dunraven, patron of the *Valkyrie*, reacted furiously and defaulted from the third race to challenge the decision that he had lost the second. He blamed the large fleet of small spectator boats crowding the starting line, until it was pointed out that this had affected *Defender* too. Then he alleged that *Defender* had been illegally ballasted. His protest was disallowed, but he continued to make it so indignantly that he was stripped of his membership of the New York Yacht Club, causing such a breach that England made no further official challenge for the America's Cup until 1934.[26] The controversy had serious implications for Consuelo too, for just at the moment when she might have found an opportunity to talk to her father, William K. was caught up in the furore which threatened to bring the America's Cup race to a premature end, and a row which called into question the honour of his captain, and his own.

Unlike many other members of society, Alva, Consuelo and the Duke of Marlborough returned to Newport on Sunday 8 September. There had been rumours that the Duke was planning to proceed from New York to Lenox, a smart resort in the Berkshires in Massachusetts, but the press noted with interest that this plan had been set aside. The days came and went. Nothing materialised. Impertinent speculation continued. 'The lingering of the Duke of Marlborough at Marble House must mean something,' thought *Town Topics*, 'and his daily drives with the fair daughter are, in the minds of Newport gossips, convincing proof that America will have another Duchess, and a reigning one of the house of Spencer-Churchill at that.'[27] By now, the magazine was explicitly linking the presence of the Duke of Marlborough to Alva's relationship with Oliver Belmont. The Duke of Marlborough, the magazine remarked, 'seems to be the exclusive property of the Marble House Vanderbilts and the Stone Stable Belmonts'.[28] Moreover, every step was being taken by the aforementioned working in tandem to give the two young people time alone together. 'While the Duke and Miss Consuela [sic] are driving, you may meet any morning, and again in the afternoon, Mrs Willie and Mr Oliver Belmont wheeling or walking.'[29]

As speculation reached fever pitch, another week passed. It is possible that the Duke of Marlborough, who had an obstinate streak, may have disliked the idea that he was being pressured into a proposal and refused to be rushed; by now he may have noticed that the ferocious Mrs William K. Vanderbilt always succeeded in getting her way and suspected that he was being used as a weapon in her armoury. He may have felt that it was undignified, given Consuelo's wealth, to propose to her too quickly; and on closer inspection he may have found Miss Consuelo Vanderbilt most difficult to read. For what is one to make of Consuelo?

There were no newspaper reports that she looked unhappy during this courtship, though public sulking and failure to rise to the occasion would have been regarded as almost as insubordinate as eloping with Winthrop Rutherfurd. There were no reports of her looking radiantly happy either, however. In fact there was very little discussion of Consuelo's demeanour at all. Her name was often misspelt, even by newspapers that had spent weeks tracking every move. There were philosophical debates such as 'Why Do Women Crave Titles? Are They by Nature Imperialists and Enemies of Democracy?'[30] Otherwise, it was as if Consuelo scarcely existed. Perhaps that is what she felt too, for Consuelo later spoke of being frightened of risking her mother's displeasure and of being 'disciplined and prepared'[31] for the Duke's arrival. For his part, the Duke may have found her so inscrutable that he began to doubt whether they were remotely compatible, for even marriages of convenience require a degree of mutual understanding to make them work. The difficulty was that the longer he stayed, the more awkward the position became.

Years later, Alva gave evidence to the Rota – the Catholic court in Rome – that she had precipitated the engagement by announcing it in the newspapers. The Duke of Marlborough told his hosts that he intended to depart during the week beginning 16 September and Alva may have applied pressure by issuing a formal denial of an engagement knowing that he was about to go. 'New York papers insist upon the engagement of Miss Consuelo Vanderbilt to the Duke of Marlborough. Mrs Vanderbilt said to a reporter of the *Daily News* that evening "Miss Vanderbilt is not engaged to the Duke of Marlborough. I regret that the papers so often see fit to connect her name with different friends of ours",' wrote the *Newport Daily News* on the morning of Wednesday 18 September. This could have had the desired effect on the same evening, for according to William Gilmour's records, Alva made a dash to New York on Thursday 19 September, returning the following day.[32] She almost certainly went to New York to put matters in hand for the formal announcement of Consuelo's engagement on Friday 20 September.

When it came, the proposal itself was undramatic. After dinner on the night before he was supposed to leave, the Duke of Marlborough took Consuelo into the Gothic Room at Marble House, which she famously described as 'propitious to sacrifice', and asked her to marry him, saying that he hoped he would make her a good husband. Consuelo ran upstairs to break the news to her mother. 'There was no time for thought or regrets,' she said. 'The next day, the news was out.'[33] These two short sentences may disguise a moment of real maternal cruelty by Alva however. It is possible that when she told her mother of the Duke's proposal, Consuelo was still hesitating over whether or not to accept. It is equally possible that Alva simply ignored her daughter's obvious doubts, and chose to regard the engagement as a fait accompli, leaving for New York as soon as she could to arrange for the announcement. The reaction of Consuelo's twelve-year-old brother, Harold, did not help. He looked at her calmly and said: 'He is only marrying you for your money.' 'With this last slap to my pride,' wrote Consuelo, 'I burst into tears.'[34]

As soon as the engagement was announced a fierce battle began for control of the narrative. The formal announcement on Friday 20 September 1895 immediately triggered a convulsion of publicity. Initially, much of the coverage was deferential, although Alva's successful direction of the matter attracted some snide remarks. The 'short but decisive campaign of General Alva', was congratulated by *Town Topics*.[35] 'It was a Famous Victory,' crowed the *World*, quoting Southey on the triumph of the 1st Duke of Marlborough at the Battle of Blenheim, in an unexpected outbreak of erudition.[36] Newspapers dedicated several column inches to profiles of the young Duke, 'Blenheim Castle' and the Spencer-Churchills, and apart from remarking that he did not seem particularly clubbable, the commentary was superficially polite. Consuelo was described as sweet and cultivated though her 'youth' and 'simplicity' were consistently underlined. Sometimes this went a little far. As the *Newport Journal* remarked: 'Her picture as a girl of ten or twelve years old, wearing a tucked guimpe and a childish gown of white muslin and lace with a baby sash is made to do duty in a full page reproduction as "The Fiancée of the Duke of Marlborough" . . . and some of the papers are indulging in ill-natured criticism.'[37] Suspicion about the Duke's motives was never far from the surface either, even in those newspapers who claimed to despise such cynicism. 'All doubt as to what impelled the Duke's visit to this country is dissipated by the announcement of his engagement,'[38] opined the *New York Herald*, immediately planting seeds of doubt by mentioning it, and failing to explain why becoming engaged to an heiress in any way cleared the matter up.

Once excitement about the engagement subsided, public

attention turned to the scale of the deal. There was a short delay until the Duke of Marlborough's lawyer arrived from England. 'The marriage settlements gave rise to considerable discussion. An English solicitor who had crossed the seas with the declared intention of "profiting the illustrious family" he had been engaged to serve devoted a natural talent to that end,'[39] wrote Consuelo. Although wild sums were discussed in the press – $10 million according to one source, plus an additional $5,000 to pay off the Duke's creditors – the eventual settlement to the Duke was $2.5 million in $50,000 shares of capital stock of the Beech Creek Railway Company, on which an annual payment of 4 per cent was guaranteed by the New York Central Railway Company, giving him an annual income of $100,000*. This income, which was very similar in structure and total to Alva's divorce settlement, was the Duke's for life, and was guaranteed even if his marriage to Consuelo ended. In a most unusual arrangement, however, which may have reflected some unease on the part of William K. about the motives of his daughter's fiancé, a comparable sum was settled on Consuelo. William K. agreed to pay her $100,000 a year in four equal quarterly instalments, a sum which almost certainly took account of $50,000 already paid to Alva annually for Consuelo's upkeep which was now transferred to her on marriage under the terms of the divorce settlement.[40]

There was some delay in the negotiations until Consuelo proposed that the final sum should be split between them 'in equal shares, at my request'.[41] It is interesting to note, in view of the later charges of coercion, that Consuelo herself came up with the proposal that finally unlocked the problem, for failure to find a compromise could have resulted in the engagement coming to a premature end. She may have felt, however, that matters had proceeded too far for her to back out. Later, Blanche Oelrichs remembered Newport servants gossiping that Consuelo cried all night at the conclusion of the settlements between the Duke and her father. 'What were these settlements that tied people up in

* approximately $7.1 million today

them against their will? For what did they barter this mysterious something which they cared for enough to cling to with tears? I put a few leading questions to my sister, a great friend of Consuelo's . . . to be angrily told that if I went on playing with "street children" I would never get "anywhere".'[42]

After the first flush of enthusiasm, the attitude of the American press became much more ambivalent, as if the editors were responding simultaneously to a sentimental desire to see the engagement as a love match and to widespread cynicism about the Duke's motives. In the *World,* which had both a political agenda and a wide female readership, both interpretations of the story appeared on the same page. Consuelo wrote later that the Duke went off on a tour of America shortly after the engagement was announced, but this is not true. Soon afterwards, Oliver Belmont arranged a coaching trip to Tuxedo for Consuelo, the Duke, Alva and the Jays, which lasted a few days. While the Duke was still – to all intents and purposes – Alva's guest, criticism by the flock of journalists who followed him was muted. On the return of the coaching party to New York, however, this changed when the Duke took rooms at the Plaza Hotel. From the moment that he ceased to be Alva's house guest the press declared open season. The Duke was quite inexperienced in dealing with this kind of publicity, accustomed to a far more deferential press in England. At the same time, however, he clearly lacked Alva's instinctive grasp of publicity as an instrument of social power. Shortly after the engagement was announced, for example, he let it be known that the marriage had been 'arranged by his friends and those of Miss Vanderbilt',[43] a most unfortunate turn of phrase which would be held against him for a very long time.

The Duke cannot be held responsible for all criticism, however, for some of it was politically motivated. Joseph Pulitzer at the *World,* in particular, had a longstanding objection to the manner in which 'our vulgar moneyed aristocrats' were prepared to buy 'European gingerbread titles'[44] for their daughters. He thought it was deeply unpatriotic and objected just as strenuously to the European nobles who came hunting for American bounty. The day after the coaching trip, when the Duke was joined from England by his cousin Ivor Guest (who would be his best man), they departed for a short excursion to look at the famous blood stock of Kentucky, followed by a bevy of reporters. Such a trip does not seem wholly unreasonable given that the Duke had been staying with the Vanderbilts for several weeks and that Mrs Vanderbilt was on the point of moving into the new house at 72nd Street and Madison Avenue from which Consuelo would be married. Indeed, he may have felt that his presence would have been a burden at such a time.

The Duke must soon have regretted the decision to strike out alone, however, for the *World* in particular was determined both to poke fun and to show him in the worst possible light. In common with other newspapers, it particularly objected to the fact that he measured just over five foot two inches and that Consuelo stood taller than him at five foot eight. He was accused of discourtesy at a Kentucky racecourse when he picked up a glove; he showed excessive enthusiasm for Kentucky whisky; and in Louisville he was spotted with various sporting friends, at a performance of a high-class comedy "The City Club of Gay Paree" at the Buckingham Theatre. At this point, the *World*'s reporter thought he had a scoop, maintaining that the Duke had been spotted giving 'the glad hand and the cheerful word' to Miss Sophie Erb who had played the role of Tottie Coughdrops.[45] Miss Sophie Erb told the reporter that she had been ogled throughout her appearance as a living picture in 'The Birth of Venus' by a sporty-looking man who later sent word that he was the Duke of Marlborough and asked her for supper – an invitation she indignantly refused saying that she didn't care if he was the Prince of Wales. Since 'sporty-looking' is an adjective that no-one else has ever applied to the 9th Duke of Marlborough, it seems likely that he was the victim of a prank, but the Tottie Coughdrops incident was soon picked up by other more sober newspapers including the *New York Tribune*.[46]

The *World* then proceeded to print an exceedingly unflattering profile of the Duke, describing him as 'no credit to his tailor ... hollow-chested ... with queer hats ... very short of stature and some people say of money',[47] and was unable to understand why this had no influence over certain young ladies who, after his return to New York, took to hanging round the foyer of the Plaza Hotel. 'They want to speak to the Duke, to touch him, to cut off a piece of his coat tails or in some other way to obtain a souvenir of the affianced husband of Miss Vanderbilt. The faithful attendants with difficulty preserve the amiable and ingenious duke from their clutches,' wrote the *World* on 15 October. Almost certainly acting on Alva's advice, the Duke sent for a reporter from the *World* on his return from Kentucky, his trip having lasted no more than five days. In an attempt to set the record straight he said rather plaintively: 'They've told so many lies about me that really I hardly know myself any more. I've become a sort of stranger to myself don't you know . . . You Americans seem to like to amuse yourselves at the expense of the English, isn't that so?' He offered to tell the reporter anything he would like to know about arrangements for the wedding; but his hazy grasp of the wedding details did little to help his case. He was then reported as having asked the journalist: 'Why are you people are so fond of interviews with Englishmen? I suppose your American men never give interviews?' When told that, on the contrary, they were very fond of being interviewed, the Duke was said to have replied incredulously: 'No, really? They can't be such flats.'[48]

No sooner had the readership put the Tottie Coughdrops affair behind it than the Duke of Marlborough was arrested for 'coasting' on a bicycle in Central Park *with his feet on the handlebars*. This might be considered rather to his credit, but not by Policeman Sweeny. Said by his admiring colleagues to be capable of 'arresting anything', Sweeny had already ordered the Duke off the grass and moved him on, when, to his horror, the felon re-appeared 'scorching' down Block House hill, his feet elevated on the handlebars of his bicycle at a rate of at least twenty miles an hour. Policeman Sweeny marched the Duke to the police station, where he confessed his ignorance of park regulations and pointed out that there was no sign warning innocent scorchers that they were in breach of the law. There was considerable embarrassment when the Duke's identity was discovered, but since a crowd had gathered, 'it was too late to recede'.[49] The distinguished visitor was reprimanded, 'discharged', and proceeded on the offending bicycle back to the Plaza. This time the story appeared in *The Times* in London, though all mention of feet on handlebars was respectfully omitted.

As these stories appeared, there was a counterblast in different mode. The Sunday edition of the *World* began to print a weekly 'Diary of the Most Interesting Couple in America'. The newspaper was watching every move made by the Duke and Consuelo and was perfectly capable of fabrication, but there is also a strong possibility that it was being fed information by Alva in an attempt to manage criticism of her future son-in-law. Alongside impolite press coverage of the Duke, a different voice stressed his painstaking attentions to Consuelo, described by the *World* as 'in many ways more entertaining than one of Ouida's novels of high life and far more instructive to aspiring duchesses – for it is fact'.[50] He showers her with roses; she carries three of them to a soiree; she wears a fetching gown of white *mousseline de soie* with a jewelled buckle; she breakfasts late with her mother; she steals a rest on a veranda as she receives the congratulations of friends; when he hears that Miss Vanderbilt is slightly indisposed, he sends more roses; she selects the most beautiful and wears it in her hair, etc.

On Saturday 18 October the 'Diary' gave way to extensive coverage of Alva's new house on 72nd Street, complete with elaborate descriptions of the interior, including Consuelo's boudoir described as 'the lovely little rooms she will leave behind when she becomes mistress of Blenheim Castle'. As the *World* pointed out: 'The happy dwellers in it do not have to spend weeks in hanging pictures, living in one room at a time and so forth, as ordinary mortals do when they move into a new house,'[51] but Alva certainly had much else to think about, and this included protecting her future son-in-law from the raw energy of New York's newspapers. For a few days the tactic seemed to work. The Duke accompanied Consuelo and Mrs Vanderbilt to church on Sunday morning and on the Monday it was announced in the papers that the wedding would take place on 6 November (for some inscrutable reason the Duke refused to be married on Guy Fawkes Day); Walter Damrosch would direct the music; an orchestra of sixty players had been engaged. Letters appeared in the press saying that the Duke's 'arrest' had been ridiculous and inhospitable. On Monday 21 October, he enjoyed a good day's hunting with the Monmouth Hunt Club in New Jersey. And there, perhaps, matters could have rested.

By Tuesday 22 October, however, the papers were in full flow again, this time because, in a serious public relations blunder, the Duke had refused to pay duty on family jewellery and on wedding presents for Consuelo sent from England. On the face of it, this was not an unreasonable reaction from a man accustomed to making economies. The presents would only be in the States for a very short time before travelling back to England with the bride and groom. But he was about to marry one of the world's richest heiresses, there was great sensitivity about his motives and his instinctive reaction appeared curmudgeonly, mercenary and mean-spirited, particularly since it was also reported that he had bought four expensive white Kentucky mules which were being shipped back to England (a purchase he later denied). The *World*, needless to say, ran this as its front-page story, complete with a dramatic account of frantic attempts by 'fuming and perspiring' British officials to sort out the matter which ended with one Colonel Phelps of the New York Customs House suggesting in despair that the entire matter could easily be resolved if only the bride and groom were prepared to marry in a bonded warehouse.[52]

On Saturday 26 October there was a publicity counter-attack which could only have come from one source. The *New York Herald* devoted several columns to the wedding details, including a drawing of Miss Consuelo's wedding dress, a full list of bridesmaids, the number of invitations sent out, and police deployments. Contrary to claims by Consuelo in her memoirs, all press reports, including those in *Vogue*, maintained that her wedding dress was being made by Alva's dressmaker, Madame Donovan in New York, and that it incorporated old lace flounces that Alva had worn on her wedding day. One plausible reason why Consuelo may have been confused was that she says she was not consulted about either her dress or her trousseau: 'Ordering my trousseau, always an exciting event in a girl's life, proved of slight interest since I had very little to say about it, my mother not troubling to consult the taste she claimed I did not possess.'[53] The bride's underwear excited particular interest. Even the *New York Herald* recorded that: 'The bridal corset is made of white satin brocaded with tiny white carnations, and trimmed at the upper edge with a deep-pointed border of Valenciennes lace. The clasps, the large hook, and the buckles on the attached stocking supporters are all of solid gold.'[54]

By 31 October, *Town Topics* was purporting to be startled by the extent of the personal information now being circulated in the daily newspapers though it would have printed the same details immediately had its informants been privy to the information in time. *Town Topics* was convinced that much of this information was being supplied to the newspapers by Alva herself: 'There is a good deal of curiosity as to whether Mrs Vanderbilt has secured a press agent or not, following the plan of the Goulds at the Countess of Castellane's wedding. I know there have been several applications for this position. It is possible, however, that Mrs Vanderbilt prefers to be her own press agent, and thus far I have not heard that any of the applications for the post have been successful.'[55]

In suggesting that Alva was her own press agent, *Town Topics* may well have been right. It is quite clear, for example, that it was Alva who authorised *Vogue* to run an article about the bridal lingerie, and permitted samples to be delivered to *Vogue*'s offices so they could be sketched by a team of in-house illustrators. Edna Woolman Chase, who worked for the circulation department of the magazine at the start of her *Vogue* career, recalled the excitement that surrounded the arrival of these 'entrancing garments' as they appeared in the office, and the manner in which the publisher, Mr Turnure, shouted on the phone to a hard-of-hearing fashion editor: 'I said underclothes, Miss Vanderbilt's underclothes. We have some of her trousseau here in the office. I want you to come in and draw it.'[56]

The founder and first editor of *Vogue*, Mrs Josephine Redding, regarded this as the greatest coup of her career, and devoted the greater part of one issue to the trousseau which confirmed that most of the information printed in the newspapers had been accurate: the lingerie had indeed cost in the range of $3,000; the bridal corset did indeed have solid gold clasps; a copy of Consuelo's signature was embroidered on her nightgowns; her rose-coloured drawers were indeed caught up on the knee with jabots of lace; although it was incorrect that the gold clasps on Consuelo's garters were studded with diamonds.[57] As Alva busily publicised Consuelo's underwear, the mortified bride-to-be squirmed with embarrassment, wondering how she would ever live down such publicity.

Confident that in the face of such press attention Consuelo was most unlikely to create a scandal, Alva now finally relaxed her vigilance. She no longer confiscated Consuelo's letters which meant that she was inundated with proposals from gentlemen anxious to rescue her from what was widely perceived to be an unromantic marriage. 'Rendered sceptical by recent experiences, I viewed these offers with less enthusiasm than did their begetters,'[58] said Consuelo later. Her saddest would-be knight was a gentleman who called himself Sir Oliver de Gyarfres, of Bart Leczfaera, who claimed he was the Prince of Teck. Fortunately for all concerned he handed himself in to the police on Sunday 27 October and was later declared insane at Bellevue Asylum. There were other signs of concern about Consuelo's safety. Not content with simply hanging Consuelo's portrait by Carolus-Duran above the Duke's head when he came to Marble House, Alva now loaned it to a portrait exhibition in New York as if determined to parade her foresight. She and Consuelo were present on the opening afternoon, along with many other society figures, 'thus furnishing a splendid opportunity for those present to compare the picture with the original'.[59] The *World* remarked on the presence of two detectives to protect Miss Vanderbilt's portrait, leading the newspaper to comment that the counterfeit Miss Vanderbilt was almost as well guarded as the lady herself. Alva became concerned when proposals of marriage from strangers turned to threats of violence from those who objected to the idea of Americans marrying foreigners and Consuelo's fortune leaving the United States. When she spoke to the police she discovered that they had been discreetly protecting both Consuelo and the Duke from the time that their engagement was announced.

In the final days before the wedding, there was so much information available about the wedding arrangements that it seemed as if the Duke might now escape further persecution. To those outside New York's inner circle, the list of wedding presents in *The New York Times* must have read like something from a fairy tale. It was considered impolite to imply that Blenheim lacked furniture or works of arts, so many of the gifts were jewellery or small *objets*. Ivor Guest, the best man, gave the bride a blue enamel watch, set with diamonds and fastened to a golden chain with a true-lovers' knot. There were pearls and diamonds from her aunt, Armide Smith; a purse of gold mesh set with turquoise and diamonds from Mrs Astor; a fan of point lace with medallions by Watteau from Mr and Mrs William Duer. There were silver bon-bon boxes, silver loving cups, silver snuff boxes, silver mirrors, combs set with pearls, crystal vinaigrettes with stoppers set with aquamarine and diamonds; and – allegedly – a pair of antique candlesticks from a certain Mr Winthrop Rutherfurd. Vanderbilt wedding presents, much to Consuelo's distress, were returned unopened, for Alva made her return them 'without excuse or thanks'.[60]

Crowds of good-natured spectators continued to follow the couple wherever they went – even when they travelled to see Bishop Littlejohn for lunch and a pre-marital chat at Garden City. On Sunday 27 October, the *World* reported that the Duke had had a 'good day', as if it were a matter for comment that nothing had gone wrong. The Duke and Consuelo, accompanied by Alva, went to the theatre; he rode his bicycle without being arrested; he stayed for two days in Washington with the British ambassador, Sir Julian Pauncefote, who would represent his relatives at the wedding, and returned to New York in time to sign the all-important marriage settlement. Then, at the eleventh hour the Duke, clearly misunderstanding the importance of the wedding rehearsal in America, announced that he would not be attending, saying: 'That sort of thing is good enough for women.' 'He spent the hours of the rehearsal shopping,' said *The New York Times* pointedly, printing the whole sorry story on its front page.[61]

As the crowds, guests, conductor, orchestra, choir, bridesmaids, husband-to-be and anxious mother waited amid the spectacular flowers at St Thomas Church on 6 November 1895, what had become of Miss Vanderbilt herself? Consuelo wrote later that she spent the morning of her wedding in tears and alone. 'A footman had been posted at the door of my apartment and not even my governess was admitted.'[62] Brides can often find themselves more or less alone in the final hours before a wedding when everyone else is fully occupied. It is possible that the footman was there as a security measure; and that it was felt that Consuelo should be left in peace at the beginning of a most demanding day. But it does appear that this was one more example of the carelessness with which her feelings were treated.

According to the *World*, Consuelo's wedding dress cost \$6,720.35.[63] Made from cream-white satin it had graduated flounces of point lace and trails of orange blossom; a 15-foot train, embroidered with pearls and silver, fell in double box pleats from the shoulder. But by her own account, much of this was lost on the bride on her wedding day: 'Like an automaton I donned the lovely lingerie with its real lace and the white silk stockings and shoes. My maid helped me into the beautiful dress, its tiers of Brussels lace cascading over white satin,'[64] and fitted the veil of lace – which fell down to her knees – with a wreath of orange blossom. Consuelo wrote later that her bouquet from Blenheim did not arrive in time, but it was widely reported that flowers from Blenheim had arrived some days beforehand, so it is possible that once again some bitterness distorted her recall. It also seems unlikely Alva did not check the appearance of the bride in her wedding dress before she left for the church, though she may have been so tense and preoccupied that she failed to notice just how upset Consuelo had become, or she simply chose to ignore it.

After Alva had departed, however, there was a scene of some

distress. Consuelo's father appeared at the house as soon as Alva left, as instructed. When she saw him Consuelo broke down and cried uncontrollably. As the world waited, there were frantic efforts to sponge the tears from her face which delayed their departure for a full twenty minutes. Alva is reported as saying that Consuelo later told her they were detained by reminiscing about old times, but if she did, Consuelo can only have intended irony. Given the waiting crowds and the military scale of the arrangements, such absentmindedness is quite implausible. It was probably one of the worst moments of William K.'s life, for there was clearly no option other than to proceed and to persuade his weeping daughter to go through with it. When they finally appeared outside the house, Consuelo on her father's arm, it was obvious that all was not well. *The New York Times* noted that William K. was far from his normal merry self, for: 'There was a grave look upon his face as if his short talk with the future Duchess has been of a serious nature.'[65] The *Herald*'s correspondent focused on Consuelo. 'She looked sad and appeared to have been crying – a natural proceeding with brides,' he remarked callously.[66]

As the carriage containing the bride and her father moved off down the street, women ran along beside it trying to catch a glimpse of Consuelo but were quickly pushed back by the police. Inside St Thomas Church the guests – and Alva – heaved a sigh of collective relief as a distant cheer signalled that the bride was now just a few blocks away. Outside in the street a journalist for *The New York Times* reported that from 'far away you could see a wild fluttering of white handkerchiefs which spread and came nearer like the white crest of an incoming wave. "Hurrah!" the women cried. "Here she comes, look, look." They looked, looked, looked and looked and out of the 5,000 women who were there not 100 saw her face.'[67]

As soon as the cheers started, according to a man from the *World*, apparently caught up in the middle of it all, 'the women heard the rumble of heavy wheels and imagining that the bride was coming seemed to become possessed of demons. They struggled like so many drowning persons and there being such a tremendous pressure behind them they pushed the police line further and further towards the church. "Here," cried Acting Inspector Courtright [sic], "that won't do. Push those women back, every one of them – back they go. Quick, now." And laying his hands upon the shoulders upon the big fat woman who was trying to squeeze through between two policemen, he set his men an heroic example by pushing that woman back into the crowd until her ribs ached.'[68] In the end, even those who had waited for hours saw very little of the bride who descended in a haze of wedding finery, asked her father to straighten her dress and hurried into the church without looking right or left.

As the orchestra started up the Wedding March from *Lohengrin*, every head swivelled to look at the bride, and there was a noise of clicking lorgnettes as she walked up the aisle, squeezing her father's arm to slow him down. 'So many eyes pried my defences,' wrote Consuelo, 'I was thankful for the veil that covered my face.'[69] In spite of the veil, the *World* reported that it was possible to see that Consuelo looked pale with two strange black lines under her eyes, and that the veil accentuated a curiously enigmatic expression.[70]

The bridesmaids came up the aisle first. As they reached the chancel Miss Duer, Miss Jay, Miss Post and Miss Edith Morton filed to the left, and Miss Bronson, Miss Goelet, Miss Winthrop and Miss Burden filed off to the right. The Duke stepped out to meet his bride and lead her up to the chancel steps while the fifty-voice choir started the first hymn. The clergy were out in intimidating force: Bishop Littlejohn, Bishop of Long Island, who had baptised and confirmed Consuelo; the highly fashionable Bishop Potter, Bishop of New York; the Reverend Dr John Wesley Brown, Rector of St Thomas; the Reverend Dr J. H. Rylance, Rector of St Mark's Church, New York; the Reverend Dr Ralph W. Brydges, Rector of St Mark's Parish, Islip, Long Island, and (somewhat outnumbered) the Reverend Waldo Burnett of Southborough, Massachusetts, who for many years was Chaplain at Blenheim, according to one report. 'The usual hymns glorifying perfect love were sung, and when I glanced at my husband shyly I saw that his eyes were fixed in space,' said Consuelo.[71] The Duke later said that when he looked at Consuelo she seemed 'much troubled'.[72]

The marriage ceremony complete, the wedding party passed into the vestry for the signing of the register, accompanied by the British ambassador and the British consul-general. Once his signature was on the documents, William K. slipped out of the vestry door, into a waiting carriage and went off (it was said) to the Metropolitan Club, his part in the ceremony now over. While the signing of the register was still in progress, the bridesmaids walked down the central aisle distributing pink and white posies to guests, prompting a small eruption of avidity. 'There was great disappointment when Miss Duer and Miss Morton, thinking there was not a sufficient quantity of flowers to supply the guests, appeared to think it wise not to distribute them to the guests in the last six or eight stalls. There were enough flowers, however, and there was great rejoicing when at the last moment the flowers were supplied to all.'[73]

As the Duke and Duchess of Marlborough appeared back at the chancel steps, the orchestra struck up the march from *Tannhauser* and the bridesmaids lined up behind the young couple as they processed down the aisle. The crowd, or at least those nearest the church, waited patiently during the wedding ceremony, entertained by intermittent bursts of music and peals of bells. As the bride and groom emerged the *Herald* was able to report that all signs of sadness seemed to have disappeared – as had patience on the part of the crowd, who now surged forward as women tried to snatch flowers from Consuelo's bouquet. The sight of the bride on the Duke's arm brought out something less pleasant in the crowd too, for amid the cheers and pealing bells there were, according to Consuelo, 'less friendly sallies'.[74]

When the wedding carriage moved off towards 72nd Street for the wedding breakfast, all restraint broke down. 'Hundreds of women pushed the policemen aside and ran to the carriage waving their handkerchiefs and screaming with delight,' wrote the *World*'s reporter. 'The horses were going rapidly and none of the women could keep up with the carriage but their places were immediately taken by others and so the carriage moved through a narrower lane of excited women whom the police could not hold back and who waved their hands and cried, "Good luck. God bless you." And then turned round and abused the policemen who were trying to save them from being run over.' Some of the women tried to force an entry into the church after the guests had left to take flowers as souvenirs, but their way was barred by policemen who only returned to their precincts at around 2 p.m., 'worn out in mind and body'.[75]

At Alva's house on 72nd Street, the crowd also waited – though not always peacefully. The street corner nearest Alva's house was a special point of attack where 'small but lively battles were waged between the police and over-curious women for several hours'.[76] The latter were soon rewarded by the sight of the bride and groom, Governor and Mrs Levi P. Morton, the British ambassador, Mrs Astor, and a total of 115 guests representing the highest circles of New York society. It was the first chance for many of these distinguished visitors to admire Alva's new house; even here the floral display was an extraordinary achievement, featuring ferns ten-feet high and thousands of specially imported orchids, not to mention a bower of white roses in which the bride and groom received their guests.

Mrs Astor, as a guest of honour, sat beside Sir Julian Pauncefote at Alva's table, together with Governor and Mrs Morton and Bishop Littlejohn. Other guests were placed at tables in the two drawing rooms and in the hall. Another table displayed specially commissioned white boxes for the wedding cake and buttonholes for both men and women. Background music was provided by a Hungarian band. *The New York Times* reported that (far from failing to arrive) the flowers from Blenheim in the large bouquets carried by both the bride and the bridesmaids were one of the talking points at the wedding breakfast. William K. may have been sent back to his club, but there was help on hand from Oliver Belmont, the *World* reporting that: 'In addition to Mrs Vanderbilt's own dexterous house servants, four of Oliver Belmont's head men had come down from Newport and assisted in serving the guests.'[77] The Vanderbilt and Belmont liveries were remarkably alike, consisting of claret-coloured coats, knickerbockers, silk stockings and patent leather pumps, and all the servants had their hair powdered.

As the crowd outside waited, the first to depart after the speeches was Governor Morton – without his wife and daughters – who gave much pleasure by making use of public transport in the form of a Madison Avenue horse car. Shortly afterwards, the Duke and Duchess appeared. The Duke was still wearing the frock-coat that had been made for him in America for the wedding, but Consuelo had changed into a dark travelling dress. In time-honoured fashion, the guests now threw rice and tokens of good luck at the bridal couple. Aim, generally speaking, was poor. An over-enthusiastic usher caught the owner of Blenheim in the back of the neck with a handful of rice, so that he 'ducked his ducal head, and followed his wife into the carriage with more speed than grace'.[78] A blue slipper almost knocked off the coachman's hat; and a grey-haired gentleman threw a pink slipper that landed on top of the awning. Another usher led a round of cheers for the newlyweds as they prepared to leave. There were further cheers as the carriage set off down Madison Avenue escorted by an impromptu escort of a dozen cyclists. As the bridal couple drove away, both the hawk-eyed correspondent of the *New York Herald* and Consuelo noticed that Alva was crying as she watched from an upstairs window, trying to hide her tears behind the curtains.[79]

The newlywed Duke and Duchess of Marlborough travelled by private boat to Long Island City, and then by private train (decorated with a further profusion of flowers) to Oakdale to spend the first days of their honeymoon at Idle Hour. Sitting in the observation car as the little train chugged along a specially cleared track, Consuelo had her first introduction to British class-consciousness and reacted with a pang of disappointment. Sunny, as she now called him, read through congratulatory telegrams from England 'handing them on to me with the proper gestures of deference or indifference the senders evoked . . . Unfortunately, there was no silver platter on which to present Her Majesty Queen Victoria's missive, but it was read with due respect, and a sense of her intimidating presence crept even into that distant railway car'.[80] If Consuelo expected charming compliments and reassuring words of affection from her new husband she was disappointed.

Consuelo was relieved to arrive at Oakdale even though there were more gaping crowds and the first real hitch of the day occurred when the carriage failed to appear. Confronted by thick fog and the mute stares of the crowd, the couple decided to set off to Idle Hour on foot, much to the delight of local reporters who thought they had a big story. 'It was an odd spectacle, the solitary figures of the Duke and Duchess trudging along the muddy country road in the half darkness of a misty twilight and followed by a motley cortège of bicyclists, pedestrians and carriages of all descriptions.'[81] The couple were almost halfway to the lodge gate when the Vanderbilt carriage came dashing down the road drawn by a pair of spirited bays, driven by Johnson the coachman. The Duke and Duchess climbed in and were whisked away, leaving the crowd behind.

As the gates of Idle Hour shut behind her, Consuelo remembered feeling overwhelmed with homesickness – not for the home that she had just left, but for a happier time in childhood 'when my father and mother were united and I had my brothers as companions', she wrote later. It was not long before homesickness was compounded by fear. 'Here my mother's room had been prepared for me and my room next-door for Marlborough. A sudden realisation of my complete innocence assailed me, bringing with it fear. Like a deserted child I longed for my family. The problem created by the marriage of two irreconcilable characters is a psychological one which deserves sympathy as well as understanding. In the hidden reaches where memory probes are sorrows too deep to fathom.'[82]

This was an age when sentimental behaviour was common, but Alva's tears on her daughter's departure puzzled Consuelo who remembered thinking (a little too neatly perhaps) 'she has attained the goal she set herself, she has experienced the satisfactions wealth can confer, she has ensconced me in the niche she so early assigned me, and she is now free to let ambition give way to a gentler passion'.[83] The day after the wedding, press opinion too was divided as to the true story. The *New York Herald*, opined, for example, that: 'There is not one unkind or suspicious word to be said about this international episode. The Duke met the young lady in his own country surrounded by his own home influences. He followed her to America, pleaded and won. It is a love match, pure and simple.' In spite of this, the newspaper could not resist a sly dig at the use to which American money would now be put, while purporting to pay the Duke a compliment. 'There are hundreds of thousands of Americans who recognise the Duke of Marlborough's possibilities in the Senate of his country and who expect to applaud the career so plainly marked out for him by his marriage with an American heiress. It would be going too far to say that American dollars have more than once furnished capital for British political preference. They have perhaps been the humble means of renewing activities in certain expensive directions . . .'[84]

The New York Times, on the other hand, was far more critical of what it saw as ducal greed – in spite of the fact that it had published one of the fullest and most lavish accounts of the wedding, doubtless boosting its circulation figures in the process. 'In this country ninety-nine marriages in every hundred are love matches, and in all the English-speaking countries the pretence that a marriage is a love match is sedulously maintained. In this case it has been rendered difficult by the frankness of the bridegroom who, in admitting his engagement, was careful to add that the marriage had been arranged by his friends and by those of the young lady . . . Since the public has been taken into the confidence of the high contracting parties in this case, the public will naturally form and express opinions upon the nature of the contract, and we observe with interest that on both sides of the water the groom is esteemed to have the better of the bargain.'[85] The newspaper deplored what it saw as the growing tendency to arranged marriages in 'a small section of American society numerically very insignificant but pecuniarily very important'.

Alva, however, consistently rebutted this kind of criticism even after her conversion to the suffrage cause. In 1917, Sara Bard Field was most curious to see how Alva would justify her role in Consuelo's marriage. Any expectations that her employer might now claim it had been a love match were dashed, for Alva did nothing of the kind: 'I am perfectly willing to admit my own part in consummating Consuelo's marriage,' she told her, in an unsettling turn of phrase. 'To leave marriage to chance is a sin.' While she was 'perfectly aware of the feeling that exists in certain classes of America about alliances of rich American girls with the foreign nobility,' and was 'just as ready to admit that there is abundant cause of much of the censure of such alliances', she refused to accept that Consuelo should have had the right to choose her own mate. 'Happiness is the most uncertain creature in the world,' and it was a major parental responsibility to prevent mistakes for 'nature will have her way among any group of young people thrown together'. For this reason, she had been most anxious to prevent Consuelo from coming into contact with too much 'nature': 'I was careful that my daughter should not meet men for whom she might have a youthful and passing fancy that would lead her into a marriage where there was no opportunity for self-growth through public ministry.'[86] This was easier in Europe, said Alva, for the young men came to her to discuss their chances with her first.

Sara Bard Field was fascinated by these conversations with Alva. 'There are phrases vivid and I believe universal,' she wrote to Charles Erskine Scott Wood. 'She herself little knows the import of much she is uttering. In that respect she is like the Oracle of Delphi.'[87] Field seems to have understood that when large transfers of property took place on marriage, it was normal for parents to play an important role in the choice of a child's spouse, though she certainly did not approve.

In this case, however, she thought that much of Alva's compulsion to marry off Consuelo was driven by her 'will to power'.[88] This is also what Alva implied herself when discussing her relationship with Consuelo at the time of her engagement. 'Consuelo's marriage was of scarcely less importance to me than it was to her. She was my oldest child and my only daughter . . . To her mental and spiritual equipment I had given the most earnest attention and in her own future my own was in a sense wrapped.'[89] The decision about a suitable husband was based first and foremost on Alva's analysis of Consuelo's character. 'She was far less rebellious and covetous of the freedom which I had taken at any cost to myself. She was docile and tractible [sic] and rather conservative in her attitude toward life,' said her mother. 'She had a good mind and loved learning. When she reached young womanhood she was tall and slender and her face radiated certain spirituelle graces which were not belied in her manner or her approach to people. She had character and culture, a high moral sense of service and mentality to direct it.' Given these characteristics, Alva argued, it was essential that Consuelo should be placed in a marriage where 'she was given the widest possible field for serious activity . . . I did not want it dissipated in an environment that asked only frivolous amusement of a woman.'[90]

Like Lily Bart in Edith Wharton's *The House of Mirth*, Alva was all too aware that marriage to a rich American in the Gilded Age meant nothing more than life inside a 'great gilt cage in which they were all huddled for the mob to gape at'.[91] Alva's own experience of marriage to a rich man had taught her that marriage into an American 'dynastic' family brought frivolous amusement and nothing more, and that such a wife ceased to have any important role from the day of her wedding. 'What an ash heap of too early discarded females there is . . . strong women at the very prime of life with empty hearts and vague bitterness for the meagre return life gives them, sit and knit out their useful years,'[92] she said.

Alva's analysis of the position of the wealthy American wife was echoed by Edith Wharton some years later: 'It is precisely at the moment when her experience is rounded by marriage, motherhood, and the responsibilities, cares and interests of her own household, that the average American woman is, so to speak, "withdrawn from circulation" . . . On her wedding-day she ceases, in any open, frank and recognised manner, to be an influence in the lives of men in the community to which she belongs.'[93] Alva herself was the first to admit that her first marriage had brought 'bitter disillusion' after the first ten years. 'By the time she [Consuelo] was grown I had learned that happiness is seldom found through personal romance but through practical usefulness and I was eager to see Consuelo placed where the heart emptiness which comes when romance fails, as it generally does, could immediately be refilled by the stream of engrossing thought for others.'[94]

Given this embittered analysis, it was hardly surprising that Alva was horrified by the idea of Consuelo marrying Winthrop Rutherfurd. Here, however, she may have had a point. Having failed to secure Consuelo's hand, Rutherfurd waited seven years before marrying another heiress with social cachet, Alice Morton (whose father was Levi P. Morton, vice-president of the United States from 1889–93, Governor of New York and guest of honour at Consuelo's wedding). When Alice Morton died, Rutherfurd married a second wealthy and socially prominent woman, Lucy Mercer. On 18 January 1902, the year of his first marriage, the society magazine *Town and Country* described him as 'a graduate of Columbia '84; a member of the Union, Raquet, Meadow Brook, and Westminster Kennel Clubs', and noted that he owned 'the Rutherfurd Kennels at Allamuchy, Hackettstown, one of the most picturesque places in New Jersey. The estate covers several thousand acres, is beautifully laid out, and preserved with many kinds of game.'[95]

When Winthrop Rutherfurd died in 1944, his obituarist struggled to find something interesting to say, and was reduced to pointing out that Rutherfurd had been the 'owner of fox terrier kennels known the country over', a leader of society and member of a noted family.'[96] Alva may have been right in feeling that Consuelo deserved more, and that if she allowed her to proceed she would be simply consigning her daughter to life inside 'the great gilt cage' where 'most of the captives were like flies in a bottle, and having once flown in, could never regain their freedom'.[97] After many years of international travel, Alva thought that the kind of female influence she sought for her daughter was only possible in Europe, and above all England where 'position entailed such obligations and responsibilities on the wife as American women of the leisure class never know'. Unlike the aristocracies of France or Germany, the English nobility enjoyed stability, great prestige and political power vested in the House of Lords. In America, social responsibility was optional for the rich. In England, it was expected: 'An old family whose history has been part of the national life for many centuries and whose present actions are to make the recorded history of the future cannot run away from the duties of their office.' Alva then went on to say: 'My idea was that Consuelo should take a place in a life of whose firm establishment there could be no question. It was such a position that the House of Marlborough offered her through the present Duke at that time a young man of promise.'[98]

If this sounds like self-serving justification invented for the benefit of Sara Bard Field, a fellow feminist, it should be noted that Alva's defence of her role in arranging Consuelo's marriage was known to *Town Topics* as early as 1906, two years before Alva took any interest in female suffrage or began to clothe her arguments in feminist language. In 1906 the marriage of the Duke and Duchess of Marlborough was once again the focus of public attention and an anonymous correspondent was able to summarise Alva's reasons for arranging her daughter's marriage to the Duke in precisely the same vein: 'Three or four years ago the mother of the Duchess of Marlborough – who has, in her day, been considered a great general in social campaign – gave out for public reading her reasons for marrying her daughter, Consuelo Vanderbilt, to the greatest eligible title in England. She was reported to have said "that the American husband – the one who selects a rich wife – is not as he is represented; that he expects his wife to sit down and admire him; that he lives from New York to Newport and back again on her money;" that she preferred a career for her daughter'. As reported in *Town Topics*, this was a career of 'going to public functions; going to court; going to the laying of cornerstones, and having estates and tenants to think about, and responsibilities and children, and a life so full there could be no time for mischief'.[99]

In the end, it seems, Alva arranged for Consuelo what she would really have liked for herself: a great house by a great architect, a position of power, in Europe rather than America, where there was no question of becoming a 'spectator in the theatre of life' and where there seemed to be abundant 'vivid action'.[100] Alva, did not deny that she arranged her daughter's marriage. On the contrary, she was proud of empowering her. She had a Utopian streak and a taste for the role of fairy-tale heroine throughout her life. All would be well, she seems to have told herself, if she could only arrange marriage with a Vanderbilt, or build a fairy-tale palace, or marry Consuelo to a duke. In all three instances, the fairy tale was closely bound up with exploiting others. Alva certainly used the Duke of Marlborough's presence in Newport and his subsequent engagement to Consuelo as a means of restoring her social pre-eminence after a scandalous divorce.

Alva's motives were, as Sara Bard Field might have said, a 'pathetic mixture of good and bad' but social ambition, as commonly understood, was not her most important impulse. She had no intention of aristocratising the Vanderbilts, and did everything in her power to prevent her daughter's marriage to the Duke from being interpreted in this way by excluding them from the wedding. Conventional social ambition could be more appropriately ascribed to William K. Vanderbilt, born on a Staten Island farm, but now one of the richest men in America. Indeed, his apparent delight in such an advantageous match (shared by most of New York's elite) may have deterred Consuelo from making her reservations known. It is possible that Consuelo, who adored her father, sensed how much pain it would cause him if she confessed her true feelings. It was perhaps significant that William K.'s wedding present to his daughter was a particularly fine tiara.

A question remains, however, about Consuelo's own feelings, for she has been accused of distorting what happened in her memoirs, written many years after divorce and annulment proceedings. The story of her marriage to the 9th Duke of Marlborough has, of course, been distorted and exaggerated by others. Consuelo is a shadowy figure throughout this period and scraps of evidence are open to interpretation. At least one writer, Elizabeth Eliot, was told that Consuelo was 'mad for the Duke' but that this came from someone who may have wanted Winthrop Rutherfurd for herself.[101]

Given the extent to which it was Alva and not Consuelo who controlled communication with the press, it would be dangerous to rely on newspapers for accounts of her emotional state, but there is little to suggest that she was in love or even that she regarded the business of snaring the Duke as some kind of game (unlike her bridesmaid May Goelet a few years later when she became engaged to the Duke of Roxburghe). Consuelo admitted that she proposed the solution that unlocked a deal-breaking problem in the marriage settlement negotiations, but there was also servant gossip that she cried. It emerged later that there was enough gossip about coercion at the time to make the Rector of St Thomas Church seriously concerned. When he made checks he was reassured, but we do not know when he made them or to whom he spoke.

When Consuelo's cousin, Adele, wanted to marry a man her parents thought unsuitable, there was a complex emotional negotiation to persuade Adele such a marriage would never make her happy; while Adele would never have married anyone of whom her parents violently disapproved. In the end, however, Adele was allowed what might now be called 'ownership' of her decision to marry James Burden. Consuelo was never permitted ownership of the decision to marry the Duke. There was no complex emotional negotiation here, for it was not in Alva's nature. After the battle over Winthrop Rutherfurd, Consuelo found herself isolated from anyone prepared to probe her real feelings; surrounded by those who shared her mother's view of an advantageous match; and was made to feel that the happiness of others, her mother above all, was her responsibility. After Alva won, Consuelo was simply too intimidated to discourage the Duke and too young and disoriented to seek help. As soon as she did what everyone wanted and accepted the Duke's proposal, there was no way back, for the publicity that engulfed them both made sure of that.

The period from the Duke's arrival in Newport on Saturday 24 August to the announcement on the 20 September was very short indeed by modern standards and the wedding itself took place six weeks later. Consuelo cannot have been exaggerating when she spoke of being steered into a vortex. At best, her feelings can only have been mixed. The prospect of life with Alva after refusing the Duke of Marlborough may have seemed unpalatable. Marriage may have represented some kind of escape from her mother's overbearing will, a chance to run her own version of the playhouse, and to fill a role of social service which certainly interested her. There may have been moments when she was caught up in the excitement. There were other moments when she was beset by doubts. It seems perfectly possible that she both helped to break deadlock over the marriage settlement *and* cried all night when it was finally resolved. Attributing consistent feelings to anyone facing an arranged marriage is perhaps unwise. Even when she was at her most optimistic, Consuelo could still see the union for what it was – a marriage of convenience, which had been devised by her mother and which she would not have chosen for herself.

Alva would almost certainly have tried to arrange a great marriage for Consuelo whatever her own circumstances. Such a campaign appealed to her highly competitive instincts. She would always have been determined, in an age of international marriages, that her daughter should win a great prize. The impact of slavery on Alva's view of human relations, the fear of exclusion and loss of control that had fuelled many of her actions since her teenage years and her deeply ambivalent feelings about late-nineteenth-century America all played a part. Under these circumstances, Alva would have settled for nothing less for Consuelo than a position of power and influence. It was Consuelo's misfortune, however, that she reached marriageable age in 1895, when Alva's recent divorce had left her with such a dismal view of marital happiness that she felt it was barely worth attempting. It was even more unfortunate that the Duke of Marlborough was available in 1895, for they both became pawns in Alva's campaign to reassert her social power. After the visit to Lord and Lady Lansdowne in India in January 1894, Alva may have felt compelled to move quickly because such an eligible young man would not stay unmarried long; but the multiple pressures on her that year tipped her instinct for control into a kind of compulsive frenzy.

So why did Alva cry? Consuelo knew she could not be weeping

tears of regret. But perhaps she underestimated the extent to which her mother's life had been bound up with her own. As Alva had put it to Sara Bard Field: 'To her mental and spiritual equipment I had given the most earnest attention and in her own future my own was in a sense wrapped'[102] – so wrapped, perhaps, that she was ill-prepared for the emotional impact of Consuelo's departure. The wand had been waved. The princess had suddenly disappeared with her prince. It is one of the ironies of this story that as her frightened daughter prepared for her wedding night and a powerful social position, Alva turned for comfort – and love – to Oliver Belmont; and withdrew to a life of extravagant vacuity in Newport – where a stream flowed down the centre of one banqueting table and 'vivid fish swam pleasantly',[103] and guests took little sterling silver shovels and dug desperately in sand for favours made from rubies, sapphires, emeralds and pearls.

PART TWO

5

Becoming a duchess

HISTORY HAS NOT ALWAYS been kind to the stranger to whom Consuelo now found herself married. Some who met Charles Richard John Spencer-Churchill, 9th Duke of Marlborough, maintained that, detached from his title, he was a person of little significance and considerable ill-humour. The historian, Maurice Ashley, who worked as Winston Churchill's research assistant soon after taking his Oxford degree, described him as 'an inspissated little man' with a truly appalling attitude to the 'lower orders'. 'I remember standing with Churchill and the Duke in one of the courtyards where they were discussing the rising unemployment figures. The Duke said disagreeably that he hoped they would reach two million. Churchill saw me visibly blench and hastened to assure me soothingly afterwards when the Duke had gone, that his cousin had not really meant what he said.'[1]

The title of the immediate heir to the Duke of Marlborough is the Marquess of Blandford; the heir's son is known by the courtesy title of Earl of Sunderland; and guides at Blenheim are not the only people to remark that the 9th Duke was known to his circle as 'Sunny' because it was short for Sunderland and not because he was blessed with a cheery disposition. A typical verdict can be found in Jerry E. Patterson's book *The Vanderbilts*: 'The Duke's only distinction was his ancestry. He was surly, critical, suspicious and without intellectual qualities.'[2]

There are other voices, however. A much more sympathetic portrait of the Duke emerged from a study of the Churchills by A. L. Rowse, who wrote: 'Fastidious and fussy – for at bottom he was an aesthete, with a cult of perfection, whether in riding or architecture, buildings, landscape, dress or women. In him the taste, the connoisseurship of his Spencer ancestors burned bright and clear.'[3] Even his father, the 8th Duke, who was not generally noted for paternal sensitivity, seemed to have noticed that his son was clever and sent him to Winchester rather than Eton. By the time the Duke arrived in Newport in 1895 he had already published an article in *Pall Mall Magazine* entitled 'Blenheim and Its Memories', which attracted considerable attention. Drawing extensively on letters between the Marlboroughs in the Blenheim archives, it was certainly not the work of an ill-educated boor. Neither was the care and attention he gave to Blenheim throughout his life, for, as Winston Churchill commented in his funeral address, the state in which the 9th Duke left his palace was a great deal finer than the state in which he found it.[4]

He could also be a sensitive and hospitable friend. Those who were fond of him included F. E. Smith (later first Earl of Birkenhead) and writer Sir Shane Leslie. His first cousin Winston Churchill remained one of his staunchest supporters. Mary Soames relates how her father, Winston Churchill, invited her mother, Clementine, to Blenheim for the weekend with a view to asking her to marry him. Before retiring to bed he made an assignation with her to walk in the rose garden the following morning after breakfast: 'Clementine came downstairs with characteristic punctuality, and was much discomfited by Winston's non-appearance. While she was eating breakfast, she seriously turned over in her mind the possibility of returning immediately to London. The Duke observed that Clementine was much put out, and took charge of the situation. He dispatched a sharp, cousinly note upstairs to Winston, and, deploying his utmost charm, suggested to Clementine that he should take her for a drive in his buggy. He whirled her round the estate for about half an hour, and upon their return there was the dilatory Winston anxiously scanning the horizon.'[5] It was only thanks to this kindly intervention that Winston and Clementine were engaged by the end of the afternoon.

As Consuelo observed the unfamiliar figure of her new husband in the fog-bound confines of Idle Hour in November 1895, she was almost certainly aware that he had inherited Blenheim Palace – one of Europe's great baroque buildings – at a low point in its fortunes. She was already acquainted with the early history of Blenheim, its dukes and its first duchess. (If she needed reminding she only had to read her fiancé's article on Blenheim which was faithfully reprinted in the *New York Herald* when their engagement was announced.) The royal manor of Woodstock, and the dukedom of Marlborough had been given to the 1st Duke – the general John Churchill – by Queen Anne as a gift from a grateful nation for a series of military victories over the French. Of these, the Battle of Blenheim, named after a small village near the Danube where it was fought in 1704, was one of the most important and spectacular. The 1st Duke's victories put paid to the French king's expansionist ambitions in Europe, saved much of Continental Europe from his absolutist regime, and were instrumental in preserving the legacy of England's Glorious Revolution. Queen Anne had awarded the 1st Duke of Marlborough title after title as one victory followed another, so that Consuelo found herself married not simply to a duke but a prince to boot: he was Duke of Marlborough, Marquess of Blandford, Earl of Marlborough, Lord Churchill of Eyemouth, Lord Churchill of Sandridge, Prince of the Holy Roman Empire and Prince of Mindelheim in Swabia. (This did not mean that the 1st Duke's reputation remained unscathed, but Consuelo would have had to read Macaulay for herself to discover that there were those who persisted in regarding his pursuit of power and accumulation of wealth as disreputable and unscrupulous.)

Even the 9th Duke, whose view of his ancestry was understandably partial, was clear that Blenheim Palace itself was a source of conflict from the outset. The playwright and architect Sir John Vanbrugh was commissioned to design a palace intended not simply as a magnificently theatrical celebration of a great general and a great queen, but as a building that would rival the vanquished Louis XIV's palace at Versailles. (It is little wonder that Alva admired Blenheim, for its design not only drew on the spirit of the Bourbons but competed with them.) Sarah, the formidable Duchess of the 1st Duke and the Queen's favourite, hated Vanbrugh's designs from the beginning saying: 'I was always against the whole design of Blenheim, as too big and unwieldy, whether I considered the pleasure of living in it, or the good of my family who were to enjoy it hereafter.'[6] Long before building was finished, however, the Duchess fell from royal favour when Queen Anne finally tired of her 'teasing and tormenting' and replaced her with the tactful and emollient Abigail Masham. This compounded the difficulty that the precise extent of the Queen's gift had never been clear in the first place. The Duchess persevered with building and furnishing the palace after the Marlboroughs were restored to favour by George I. After the death of the Duke, she dedicated herself to completing the palace she believed her husband would have wanted, though she caused Vanbrugh to resign in fury and issued no fewer than 401 lawsuits against erring workmen in the process.

After the Great Duke's death, family history became more complicated. He had no surviving sons so the dukedom passed by a series of manoeuvres to Sarah's grandson, Charles Spencer, 5th Earl of Sunderland, who became the 3rd Duke of Marlborough in 1731 – thereby linking the Marlboroughs to the Spencers of Althorp. (The family name now became Spencer rather than Churchill until it was changed by act of Parliament to Spencer-Churchill in the patriotic atmosphere that followed the Battle of Waterloo.) The 3rd Duke of Marlborough began a long history of extravagance. His son, George, the 4th Duke, continued it to great effect over a period lasting nearly sixty years, for he employed Sir William Chambers and Capability Brown to enhance Vanbrugh's original designs, building the little temples of Diana, Flora and Health and transforming High Lodge into a small Gothic castle. Capability Brown's great contribution to Blenheim was the creation of the lake in the park – two lakes in fact – which greatly enhanced Vanbrugh's beautiful bridge. On the other hand, the 4th Duke also tampered with Vanbrugh's original design for the gardens, grassing over the Great Court, and the parterre to the south of the palace.

At this point, Sunny's account of family history for the benefit of his new bride may have become a little vague. The improvements to Blenheim and the cost of the upkeep of the palace were already proving a drain on expenditure by the time of the 4th Duke's death (he lived out the end of his life as an embittered recluse). Both the 5th and 6th Dukes made matters considerably worse. The 5th Duke, though a man of taste, a connoisseur and a bibliophile, was a spendthrift who was later forced to sell many of the treasures he had bought and retreat to live in one corner of the palace. The 6th Duke made no contribution to Blenheim at all, other than a much publicised row about the admission price for visitors which he was forced to reduce to a shilling shortly before he died in 1857. His chief claim to fame was a libel case involving allegations of bigamy.

The accession of the 7th Duke marked a change in pace for he stopped the family's growing reputation for degenerate behaviour in its tracks. His formidable widow Frances, known as Fanny, was still alive in 1895, and the high-mindedness of this couple was forcefully impressed upon Consuelo. The 7th Duke may have been a 'complete, full-blown Victorian prig',[7] but he led a worthwhile existence as a Privy Councillor, Lord Steward of the Household and was the author of the Blandford Act which resulted in an essential but stupefyingly dull sub-division of extended parishes. He too was forced to make stringent economies and started selling family treasures to make ends meet, including a precious collection of antique cameos and intaglios known as the Marlborough Gems and the priceless Sunderland Library.

Duchess Fanny, wife of the 7th Duke, and dowager duchess when Consuelo arrived in England, was the redoubtable daughter of the 3rd Marquess of Londonderry. She undertook important fundraising for famine relief when her husband became Viceroy of Ireland. Indeed, she had a keen sense of the duties of a duchess which she impressed on her descendants to the point of ennui. Her American daughter-in-law, Lady Randolph Churchill, (and Consuelo) cordially disliked her, but Hugh Montgomery-Massingberd comes to her defence: 'In fairness to Duchess Fanny it could be said that she was obliged to run the palace on a tight rein for reasons of economy. The Marlboroughs' income of £40,000 a year was by no means great by contemporary standards, certainly not with Blenheim to maintain. Rather than retreat to a corner like the spent-out 5th Duke, they managed to keep up appearances, dutifully entertaining the Prince of Wales and other notables of the day, even if the fare was somewhat sparse. Sir Alexander Cockburn, the Lord Chief Justice, complained in the Visitor's Book that while he was prepared to share almost everything in life, he drew the line at half a snipe for dinner.'[8]

The serious-minded and industrious atmosphere that surrounded the 7th Duke and his duchess would not endure for they produced two of the most erratic and wayward sons in the family's history. Their younger son was Lord Randolph Churchill, father of Winston Churchill. The elder, the heir, was Sunny's father, the 'wicked' 8th Duke. It would take Consuelo a good deal longer than the first week of her honeymoon to appreciate the havoc wrought by this man, who caused his parents great grief while he was still Marquess of Blandford. He continued the process of asset-stripping begun by his much worthier father so that by 1886, he had sold eighteen paintings by Rubens, two by Rembrandt, one Breughel, eight Van Dycks, not to mention works by Watteau, Stubbs, Reynolds, Poussin and a couple of Titians, the rest of the library and much of the porcelain. This caused a bitter quarrel between the 8th Duke and his brother, Lord Randolph Churchill, who kept finding family paintings as far afield as St Petersburg. In 1888, discovering that he had little of value left to sell, the 8th Duke went to America where he found a rich widow, Lilian Hammersley, and married her, a move not lost on the American press as they reported on Consuelo's engagement to his son. More interested in science and agriculture than in art, the 8th Duke spent money on equipping the estate farms, building hot houses and a scientific laboratory for himself, in which he died of heart failure 'with a terrible expression on his face' at the age of forty-eight.

The 8th Duke was guilty of much more than selling the family silver however. David Cannadine describes him as 'one of the more disreputable men ever to have debased the highest rank in the British peerage . . . rude, erratic, profligate, irresponsible and lacking in self-control.'[9] Even A. L. Rowse, who takes a more favourable view of Marlboroughs in general, remarked that he was a 'clever man whose cleverness ran all to seed'.[10] He was expelled from Eton as a youth. Immediately after Sunny's birth in Simla in India in 1871, he told his wife to make her own way back to England with a tiny baby and an ayah while he explored Kashmir. He was persistently unfaithful, miring the family in deep disgrace as a result of his affair with Lady Aylesford. In a wholly misjudged move to keep his proposed elopement with Lady Aylesford quiet, he and his brother, Lord Randolph Churchill, attempted a crude blackmail of the Prince of Wales, threatening to make public admiring letters the latter had written Lady Aylesford unless the Prince persuaded his crony, Lord Aylesford, to abandon plans for divorce. An acceptable apology from Lord Randolph to the furious Prince of Wales had to be drafted by the Lord Chancellor. The Queen became involved. The Prime Minister, Disraeli, finally dispatched the unhappy 7th Duke of Marlborough and his duchess to Ireland where he made the Duke Viceroy and told him to take Lord Randolph Churchill and his American wife, Jennie, out of England. The Prince of Wales refused to enter any house that entertained the Spencer-Churchills and denounced Blandford as 'the greatest blaggard alive'. None of it had any effect. His first wife Albertha finally divorced him after he fathered a son by Lady Aylesford in 1881. He went on to figure in the sensational divorce case of Lady Colin Campbell. It did nothing to help the family's reputation that when the 8th Duke died at the age of forty-eight, he left a substantial bequest to Lady Colin Campbell, and nothing to Lady Aylesford.

It would also take Consuelo time to understand just how dysfunctional her new husband's upbringing had been at the hands of such a father. The idea that dukes can suffer from unhappy childhoods may induce snorts of derision in a more democratic age, but there is little doubt that the damaging impact of the childhood on Sunny was recognised long before such explanations were fashionable. A letter from his aunt, Lady Lansdowne, to Sunny's second wife, Gladys Deacon, in 1921 remarked: 'Up to ten-years old he was one of the most charming boys I ever met & most joyous; after that his spirits seemed to have vanished & he quite changed,'[11] though she may not have mentioned this to Alva when proposing him as a potential suitor for Consuelo. As his marital affairs lurched from bad to worse, the 8th Duke seems to have taken it out on his eldest legitimate son. Sunny told Consuelo that his father had bullied him so badly as a child that he found it quite impossible to exert parental discipline himself, and humiliating ill-treatment of a different kind continued. The 8th Duke kept his family so short of funds that in 1891, when Sunny was still at Cambridge, his father ceased to pay him any form of allowance, and his mother was forced to apply to the courts, unsuccessfully, for it to be reinstated.

Consuelo's mother-in-law was born Albertha Hamilton, a daughter of the Duke of Abercorn. She was known as 'Goosey' because though she was beautiful she was generally felt to be dim. She was also kind and affectionate but this was not enough to compensate for her husband's cruel and damaging behaviour. She had a marked penchant for heavy-handed practical jokes which drove those around her to distraction, but which were probably a sign of deeply repressed and seething anger, for, according to Consuelo, 'Goosey' too had been forced to stand down the man she loved and had been made to marry her duke. Ink pots were placed on top of doors; guests fell ill after eating soap at dinner; one of her best japes was placing a pink toy baby under a silver dish for her husband's breakfast just after his mistress, Lady Aylesford, had given birth to his son.[12]

Consuelo later maintained that her husband ended up greatly preferring his aunt, Lady Lansdowne, to his mother. All this came on top of one dynastic trait for which even the 8th Duke cannot be blamed – a family tendency to depression observable from the 4th Duke onwards and which famously afflicted Sunny's cousin Winston. When Consuelo first arrived at Blenheim she found the 8th Duke's depressive tendencies still manifest in enigmatic but gloomy inscriptions. The mantelpiece of her bedroom read: 'Dust. Ashes. Nothing.' In another room he had inscribed: 'They say. What say they? Let them say.'

When his father died in 1892, the 9th Duke inherited several intractable problems. The family name was tainted by scandal. Blenheim had been stripped of many of its treasures. An agricultural depression had reduced rents, and in any case, the Marlborough wealth from land was modest compared with other British dukedoms. They were not great London landlords like the Dukes of Westminster and Bedford. Nor were they great landowners like the Duke of Buccleuch or the Duke of Devonshire. The Duke of Buccleuch derived an income from annual rent rolls of £217,000 from 460,000 acres. The Duke of Devonshire received £181,000 from 200,000 acres. The Duke of Marlborough, on the other hand, derived a mere £37,000 from 24,000 acres.[13]

At the same time, Sunny believed passionately that aristocracy was the cornerstone of the polity. Like most English patricians in the 1890s, he believed that society was unequal by God-given design, that his position was his birthright, and that his responsibility was to maintain an unbroken link with England's ancient and traditional structures. Though his friends teased him that he had a feudal mind, it is perhaps unsurprising since the Marlboroughs were ranked ninth in an aristocracy that was the most exclusive and prestigious in Europe – an elite of wealth and hereditary political power that had seemed to serve Britain well for centuries. But this was more than a conventionally aristocratic self-serving view. On account of his very difficult background, it is hard to avoid the conclusion that his feudal mind and passionate crusade to restore the Marlboroughs and Blenheim to their former glory was closely bound up, at a deeper level, with a crusade to restore himself.

Unlike his father, and several of his forbears apart from the 7th Duke, the 9th Duke believed that rank had its obligations. Disraeli, Salisbury and even his uncle, Lord Lansdowne, had all turned down dukedoms because they felt they 'lacked the resources to support the dignity'.[14] It was Sunny's misfortune that he felt this sense of obligation so acutely when the first real challenges to the aristocratic status quo were beginning to appear, and when the rules of what constituted ducal 'dignity' were being changed by royalty itself. The problem for those at the pinnacle of aristocratic society in England in 1895 was that the tone was now being set by the Prince of Wales who presided over an alternative court to that of his mother, Queen Victoria – a virtual recluse since the death of Prince

Albert. It was not simply that the Prince was easily bored, that he enjoyed gambling for high stakes, racing, elaborate meals and colossal sport – though several of his friends claimed to have beggared themselves by keeping him amused. More pernicious was the manner in which the cosmopolitan and metropolitan Prince was embracing opulent style and equating it with rank.

In the short term, the Vanderbilt dowry would do much to relieve these tensions. Many reports of the Duke's more disagreeable behaviour stemmed from later in his life, and include Consuelo's own memoir written from the perspective of a far happier second marriage. In 1895, the problems caused by his difficult legacy were much less obvious. Alva would have noted and approved of his emerging connoisseurship. She may have taken the sympathetic view that he was an artist whose canvas was Blenheim as hers had been 660 Fifth Avenue and Marble House. She later wrote that he was 'a young man of promise'[15] with political ambitions – which would have pleased her as she thought of the grandeur of the life enjoyed by the Viceroys of India.

Beneath the surface, however, there was a legacy of bad-temper, bullying, erratic behaviour, humiliation and scandal, a mother he had learnt to despise, and a family tendency to depression. Behind a polite veneer, the Duke was a hyper-sensitive, anxious and self-involved young man determined to restore Blenheim and the tarnished family name regardless of personal cost. None of this would make him an easy companion. In 1895 he also had the disadvantage of youth and inexperience. The 9th Duke of Marlborough came into the title in 1892 and found himself shouldering responsibility for Blenheim Palace before he turned twenty-one. When he married his eighteen-year-old Duchess, he was only twenty-three.

During 'the week's seclusion custom has imposed upon reluctant honeymooners'[16] – as Consuelo later described it – the press continued to hang about the Oakdale area reporting sightings. The couple were kept indoors by Long Island fog in the mornings, but were observed 'chatting gaily', with the Duke acting as his own coachman, when they went out driving in the afternoons. They went cycling and made tours of the nearby Southside Club estate and Islip itself, where the Duchess was reported as looking 'pretty'.

The coaching drives rather annoyed the residents of nearby Sayville and caused frustration for the Islip Cornet Band which was quite determined to give the Duke and Duchess a surprise concert. The original intention had been to turn up at Idle Hour uninvited until a member of the band panicked and suggested they should seek permission; after which the plan was abandoned. One afternoon, however, the Duke and Duchess seemed about to make an unexpected drive through Islip and the players scrambled to produce a tune – but the Duke, possibly suspecting that he was about to be ambushed by a cornet band, turned the coach round and headed back for Idle Hour before it could organise itself. On Sunday 11 November, the Duke and Duchess attended church at St Mark's in Islip. When they returned home they found that William K. had arrived to keep them company for the evening; he was doubtless relieved to see his daughter looking more composed.[17]

Travelling back to New York, the Duke and Duchess were recognised by other passengers on the ferry who confined their interest to 'sly glances'. Unfortunately, a group of teamsters were not so discreet. 'They peered through the glass doors and indulged in a number of rude witticisms at the Duke's expense, which could plainly be heard in the cabin. The Duke tried to appear unconcerned, but both he and his wife were so evidently embarrassed that they won the sympathy of their fellow passengers. Finally several men sitting near the doors went out and remonstrated with the teamsters and the annoyance ceased.'[18] The same intense interest was evident when they went with Alva to the Horse Show, a fashionable society event. Crowds mobbed their box and had to be moved along by the police. *Town Topics* had no sympathy. Given Mrs Vanderbilt's publicity campaign 'it would have been ungrateful and churlish in the highest degree for the people to fail to manifest their interest in articles so extensively advertised'.[19]

Before the Marlboroughs left for Europe, the Duke had an unexpected visitor in the form of his cousin, Winston Churchill, on his first visit to America. He was en route to Cuba in the company of fellow subaltern Reggie Barnes, having been given permission by his colonel to observe the rebel insurgency against Spain. Churchill and Barnes had originally intended to stay in New York for three days, but were enjoying themselves so hugely that they extended their visit to a week. 'What an extraordinary people the Americans are!' Churchill wrote to his mother on 10 November. 'Their hospitality is a revelation to me and they make you feel at home and at ease in a way that I have never before experienced.'[20] It was a very great country, he told his brother Jack in another letter on 15 November. 'Not pretty or romantic but great and utilitarian. There seems to be no such thing as reverence or tradition.' He had seen Sunny the night before and would be dining with the Vanderbilts that evening, he wrote, adding that Sunny 'is very pleased with himself and seems very fit. The newspapers have abused him scurrilously.'

Churchill could thus witness the hysteria in New York surrounding his cousin and the new Duchess (on whom, from the evidence of surviving correspondence, he passed no comment), and it both interested and repelled him. 'The essence of American journalism is vulgarity divested of truth. Their best papers write for a class of snotty housemaids and footmen & even the nicest people here have so much vitiated their taste as to appreciate the style.' On the other hand he also thought that American vulgarity was a sign of strength. 'A great, crude, strong, young people are the Americans – like a boisterous healthy boy among enervated but well-bred ladies and gentleman . . . Picture to yourself the American people as a great lusty youth – who treads on all your sensibilities, perpetrates every possible horror of ill manners – whom neither age nor just tradition inspire with reverence – but who moves about his affairs with a good-hearted freshness which may well be the envy of older nations of the earth.'[21]

Winston Churchill may have been interested and perplexed by what he described as the 'irreconcilable conflicts' of America, but by now the Duke of Marlborough had had enough. As far as he was concerned, it was his first and last visit. 'There was in his sarcastic comments on all things American an arrogance that inclined me to view his decision with approval,'[22] wrote Consuelo later. But after the experience of the Horse Show she too was ready to leave the United States for 'the glare of publicity that was focused upon our every act' was most trying and she would leave the city 'with few regrets'. The Duke and Duchess sailed for Italy on Saturday 16 November. The Duke of Marlborough had put in hand changes to the private apartments at Blenheim which made an immediate return to England difficult. Less well travelled than Consuelo, he now wished to see something more of the world, so he suggested that they should spend three months on an extended honeymoon before returning to Oxfordshire in the spring. Their eventual destination was to be Egypt.

The Duke and Duchess arrived at the pier at Hoboken shortly after 10 a.m. and went at once to their quarters aboard the steamship *Fulda* where a party came on board to bid them farewell. 'The Duchess was attired in a blue serge steamer costume, lined with yellow satin, the collar of the jacket being trimmed with fur. Her cap was a jaunty-looking turban-like affair with fur trimmings. The Duke wore a black derby hat, a grey mixed suit, blue Newmarket coat with velvet collar and cuffs and tan shoes,'[23] and they both carried bunches of violets. Much to the excitement of the press, the farewell party included both Alva and William K., as well as Miss Duer and Miss Post, two of Consuelo's bridesmaids. 'Mr and Mrs Vanderbilt met at the side of their daughter, and in answer to Mrs Vanderbilt's greeting, her former husband merely raised his hat. They exchanged no words,'[24] wrote the *New York Herald*. It reported that William K. Vanderbilt left the ship early and returned to New York. Mrs Vanderbilt and the young women went to the end of the pier where they waved farewells as the ship was warped out. When the *Fulda* cleared the end of the pier the Duke and Duchess were seen forward on the hurricane deck, and, according to one reporter, handkerchief signals were waved from ship to shore. On the pier, members of the Vanderbilt party did not leave their places until the ship was well on her way down river.

* * *

As soon as the Duke and Consuelo left New York, Alva turned her attention to marrying Oliver Belmont, and immediately encountered an obstacle in the Episcopal Church. It was explained that it was permissible for Alva to marry again because she had been granted a divorce on the grounds of her husband's adultery. The problem lay with Oliver Belmont, who had been divorced by his first wife for desertion, a state of affairs which the Church refused to recognise. 'Neither of us had even thought of it as having any possible bearing on our own marriage,' Alva said later; 'But the Church thought otherwise. No Episcopal clergyman could marry us.'[25] In the end a solution was found. The ever-reliable Colonel Jay talked to the Mayor of New York who agreed to perform the ceremony at Alva's house on 11 January 1896.

Not everyone was pleased. The Belmonts, as they now became, were much criticised for announcing their marriage the day after the death of the sixteen-year-old daughter of Florence Twombly, William K. Vanderbilt's sister. *Town Topics* was adamant that this was Alva's own doing, insisting that she had given the news herself to the *New York Herald* and the *World*: 'Mrs W. K. Vanderbilt and Mr O. H. P. Belmont organs [sic] during the last year.'[26] Oliver's brothers also reacted furiously. They were not concerned by Oliver's decision to marry a divorced woman (his brother Perry would do the same thing in 1899); what really angered them was Alva's insistence on holding on to her alimony from William K. Vanderbilt, which was unprecedented. William K. was said to be quite untroubled by the matter – *Town Topics* thought this was because he was anxious to get rid of Alva at almost any price and 'would think a long while before engaging in any proceedings that might disturb his present delightful sense of freedom and holy and halcyon calm'.[27] There was probably some truth in this: the Vanderbilt divorce settlement included complex arrangements for the support of their three children until each was independent, and William K. would have had no interest in starting negotiations all over again.[28] The Belmonts, on the other hand, regarded Alva's decision as such a humiliating blow to family pride that they refused to attend Oliver's wedding. He then added insult to injury by giving Alva all his

Newport property as a wedding present. This included both Belcourt and the Gray Crags estate valued at over $1 million. For many years, Oliver's brothers refused to have anything to do with him or their new sister-in-law.

Alva's decision to fly in the face of all convention by retaining her alimony was part of a determined attempt never to find herself in a position of financial dependence again. She and Oliver were marrying for love. They would be equal partners financially, and there would be no repeat of the humiliating dependence of wife on husband and the lack of financial power she had suffered with William K. In old age she described her relationship with Oliver Belmont as deeply happy. 'It was a marriage of mature people who had mated not on the basis of a passing attraction but on the basis of like interests, like choices, like hopes and aims . . . It means that each is always eager to be better and greater in order to have more to give . . . It is the only sort of union which can possibly endure.'[29]

On top of this, Consuelo's marriage had the desired effect. It was clear that Alva's position as a leader of society was secure in August 1896 when *le tout Newport* accepted an invitation to a Belmont ball. Personal happiness calmed Alva down. Her marriage to Oliver Belmont would turn out to be the least active phase of her life. From 1896 to 1908, when her life changed again, Alva was content to derive most of her power from her new-found status as Mother of the Duchess of Marlborough.

However well-intentioned, an extended honeymoon made for a difficult start to married life. Consuelo was far from intrepid; the Duke was a poor sailor and the *Fulda* was a simple ship.[30] The captain gave up his cabin and slept near the hurricane deck, but his quarters provided a 'minimum of comfort' according to Consuelo, who found herself cooped up on a gloomy boat with a seasick husband for whom it was difficult to feel sympathy given that he was so very sorry for himself. 'Sea-sickness breeds a horrible pessimism, in which my husband fully indulged, and it took all the optimism I possessed to overcome the depressing gloom of that voyage,'[31] she wrote later. By the time the *Fulda* reached Europe, all optimism had more or less evaporated. The Duke's seasickness was so severe that the couple changed their plans, disembarked at Gibraltar and arranged to travel through Spain.

This turned out to be almost as disagreeable as sailing on the *Fulda*. The weather was freezing. The great cathedral of Seville struck Consuelo as both dismal and gaudy, its decorated Madonnas shocking to a New England Protestant. The galleries in Madrid were unheated, pierced by an icy wind and filled with second-rate paintings badly hung. Tension was never far from the surface. Three weeks into their honeymoon it erupted in a furious quarrel that was etched forever on both their memories. Consuelo informed her new husband that she had been forced to give up the man she loved by her mother and had been ordered to marry him. Sunny told Consuelo that he had renounced the woman he loved because he felt obliged to save Blenheim.

It was probably the moment when they both suddenly realised what they had just done. Stripped of any tattered illusions of romance, depressed, and only just coming to terms with her new role as a wife, Consuelo may have been a trying travelling companion while Sunny was ill-equipped to deal sympathetically with a homesick girl of eighteen. Even in Spain, the clash between American respect for democracy and patrician respect for tradition loomed large and the formal demands of life as an English duchess were never far away. The Marlboroughs spent a pleasant evening in Madrid at the home of the British ambassador, Sir Henry Drummond-Wolff, where Consuelo was introduced to Lord Rosebery, leader of the Liberal Party in the House of Lords. Rosebery had been prime minister briefly from 1894 to 1895, when his main achievement had been to fulfil a dream of winning the Derby – twice – while prime minister.

Consuelo described Lord Rosebery as possessing an aristocratic arrogance in spite of a bourgeois appearance, with a keen sense of the ridiculous (her reactions were not always as democratic as she liked to think). She warmed to him, however, and was much relieved when he told her that he would be sending a good account of her to Sunny's formidable grandmother, the Dowager Duchess of Marlborough. This was uncharacteristically decisive on the part of Lord Rosebery who was famous for being unable to make up his mind about anything. The evening also had more stressful consequences. When Sunny heard that Lord Rosebery had been received by Queen Christina, then Regent of Spain, he pressed for a similar introduction. Etiquette at the Spanish royal court was known to be extremely strict. Consuelo managed to make the three ceremonial curtsies and acquit herself as protocol demanded, but she resolutely refused to enjoy it.

A visit to the racier atmosphere of the French Riviera in the company of her new husband was no more acceptable, though here Sunny also seems to have displayed a lack of sensitivity. Confronted by blue skies and azure sea, Consuelo's spirits rose as they made their way to the Hôtel de Paris in Monte Carlo, but she soon found herself eating among an unfamiliar crowd of exotic people, many of whom Sunny seemed to know. 'When I asked him who they all were I was surprised by his evasive answers and still more startled when informed that I must not look at the women whose beauty I admired. It was only after repeated questioning that I learned that these were ladies of easy virtue whose beauty and charm had their price. It became increasingly complicated when I heard that I must not recognise the men who accompanied them, even though some of them had been my suitors a few months before.'[32]

Two of the great demimondaines whom Consuelo observed from afar were La Belle Otero, 'a dark and passionate young woman with a strong blend of Greek and gypsy blood' who was later named by heiress Anna Gould in her divorce proceedings from Boni de Castellane, and Liane de Pougy, 'who looked the *grande dame* she eventually became by her marriage to the Rumanian Prince Ghika'.[33] These women, often more beautiful than the ladies of the Faubourg, had their own elegant houses, smart carriages and stunning jewellery, and entertained the intelligentsia of Paris in their salons. There was great rivalry between them and wild bets were placed on the value of their jewels and other less decorous statistics. Consuelo considered it most unfair that these women were allowed to wear make-up, while respectable ladies had to be very discreet about it. 'How different was this life from the prim monastic existence my mother had enforced. The goddess Minerva no longer sat enthroned. Beauty rather than wisdom appeared to be everyone's business.'[34]

The blinkers were further ripped from her eyes, to borrow Consuelo's phrase, when she was introduced in Monte Carlo to another bride, a young Englishwoman who was astounded by Consuelo's lack of worldliness. 'I listened to her vitriolic gossip with mounting concern. Seeing with her eyes those I was to meet, gauging as she meant me to the enmity of certain of my husband's friends, the future loomed complex and difficult. Finally, emphasising the need for fine clothes, with rich jewels and a lavish expenditure she added, "With our money, our clothes, our jewels we will be the two successes of the coming London season, and all the women will be jealous of us." '[35]

Though Consuelo was deeply dismayed by these remarks, and by Sunny's increasingly apparent taste for display, both expressed the prevailing spirit of the *belle époque*. The onward march of industrialisation had created a new international class of fabulously rich and assertive plutocrats in Europe as well as America. International wealth had given birth to its own ostentatious, international style, reflected in the glittering jewels of the demimondaines, in ocean-going yachts, dresses by Worth and Doucet, in racehorses and new casinos. In architecture, the European millionaire drew on a mixture of French classicism, French renaissance, the Jacobean and the Italianate – a style which had a trajectory of its own in England from the 1860s onwards. The international playground of the new plutocrats was Paris, a city more tolerant of plutocrat pleasures than anywhere else. By the time the Marlboroughs were on their honeymoon, the new opulence was dominant, sweeping aside the anti-vulgarian stance of the British landed classes. It was a sign of the times that on his one and only attempt to enter the Casino at Monte Carlo, the patrician Lord Salisbury was refused entry because he was unsuitably dressed, to the very great glee of his family.

One reason for this change in fashion was that the de facto

leader of English society, the Prince of Wales, was captivated by wealth himself to the extent of insisting that ostentatious display was now an aristocratic duty. 'Marlborough House [his London home] was essentially cosmopolitan, its habitués as much at home at Auteuil, Longchamps, Chantilly as at Newmarket, Ascot, Goodwood; the Prince and Princess of Wales were accustomed to spend several months of the year abroad, their regular routine including Biarritz, Homburg, Marienbad, with frequent appearances in Paris or at Monte Carlo',[36] wrote A. L. Rowse. Sunny was therefore right to believe that creating an opulent setting for the Marlboroughs would do much to win royal approval and restore the tarnished family image.

His point of view may now meet with little understanding, and at the time it never made much impression on his new bride whose enthusiasm for her husband's Marlborough restoration project was lukewarm at best. Since she hailed from one of the richest families of plutocrats in the world, it may seem odd that Consuelo chose to regard her new husband's taste for display as a symptom of crass European materialism quite at odds with her more innocent set of values. Part of the explanation may lie in her growing antipathy to the man she had married, who was also evincing a marked tendency to see things entirely from his own point of view (famously summed up by a postcard he once sent to an estate worker at Blenheim from the Riviera which read: 'Pray press on with the haymaking while this glorious weather lasts.'[37]) Consuelo's view of herself as an innocent Jamesian heroine abroad in the corrupt world of old Europe also stemmed partly from her autocratically sheltered childhood. As a child 'her prim monastic existence' kept her at one remove from her mother's campaign of aggrandisement, and she was never allowed to make purchases for herself. Moreover, Alva never quite set aside her own thrifty, puritanical streak. She often challenged tradesmen when she thought they were overcharging and turned down great works of art if she thought they were overpriced. When she discussed it at all, Alva chose to express her enthusiasm for spending Vanderbilt money in terms of a high-minded mission to bring European culture to American vulgarians. The Duke was rather more clear-sighted. Even in 1921, he would urge his second wife, Gladys Deacon, to keep in touch with her wealthy cousin, Eugene Higgins, 'because everyone in England from the Royal Family downwards was impressed by money'.[38] Consuelo may have found such attitudes distasteful, but in 1895 the 9th Duke of Marlborough was right.

By Christmas Day the Marlboroughs had arrived in Rome. Consuelo remembered the rooms of the hotel as high and bare, and felt acutely homesick. Concerned by her persistent listlessness, Sunny chose this moment to insure her life. A 'pompous Roman doctor' examined her and promptly informed them that she had less than six months to live. This alarming news led Sunny to summon a specialist from London posthaste. The specialist's diagnosis was that she was physically well but had 'outgrown her strength'. A modern doctor might well have diagnosed homesickness or even reactive depression and queried whether extended travelling was simply making matters worse. The news that the Duke of Marlborough had been so quick off the mark to insure his new wife's life and thus protect Blenheim against the loss of her part of the marriage settlement reached *Town Topics* by 2 January 1896. It took a dim view of the matter: 'After the hardships and perils incident to the successful hunt for "bigger game" in America, the Duke of Marlborough, like a worthy descendant of the thrifty founder of his house, has no fancy in being done out of his "bag" by a random shot from the "gray" horseman,' it commented sourly.[39]

Consuelo was ordered to rest while Sunny ransacked the antique shops of Rome, remarking that her appearance only served to drive up prices. This was probably an indirect compliment, but if so Consuelo failed to notice and simply felt excluded from the great project of setting Blenheim to rights for future Marlboroughs. Indeed, enforced rest proved most difficult of all: 'In the solitude of those long hours I began to realise what life away from my family in a strange country would mean. At the age of eighteen I was beginning to chafe at the impersonal role I had played in my own life – first, as a pawn in my mother's game and now, as my husband expressed it, "a link in a chain". To one not sufficiently impressed with the importance of ensuring the survival of a particular family, the fact that our happiness as individuals was as nothing in this unbroken chain of succeeding generations was a corroding thought; for although I greatly desired children, I had not reached the stage of total abnegation regarding my personal happiness. Nevertheless, to produce the next link in the chain, was, I knew, my most immediate duty and I worried at my ill-health.'[40]

Two weeks later, spies for *Town Topics* had better news to report. In spite of hundreds of visiting cards left for them, the Marlboroughs preferred to visit monuments like 'bourgeois tourists' and received no-one. The Duke was even reported as saying gallantly: 'I am here only for my wife, and I only take interest in her.' This was just as well, thought Colonel Mann in his weekly column, for 'during their stay in Monte Carlo, I understand that everyone remarked upon the woe-begone aspect of the two turtle doves.'[41]

In view of concerns about Consuelo's health and ability to conceive, it was perhaps rather tactless for Sunny to leave his new bride in the ruins at Pompeii and Herculaneum while he went with a guide to look at risqué paintings and 'statues erected to the worship of Priapus, the god and giver of life'. Indeed, it clearly caused another row since she later recommended honeymooners to avoid these sites which, she said, could only provoke discord. It was perhaps equally maladroit, when sailing on the Nile, for Sunny to summon nautch girls on board to dance. 'Not yet hardened to such exhibitions, I retired below and was not altogether sorry to hear that one of them had fallen into the river, from which she was fished none the worse for her immersion.'[42]

In Egypt, Sunny explicitly aped the Prince of Wales (who had returned from a visit to Egypt in 1869 with a ten-year-old Egyptian page called Ali Achmet) by acquiring an Egyptian pageboy for his duchess and taking him back to Blenheim as a human souvenir. It was not a wise decision. Consuelo soon found the unhappy pageboy's attentions trying, and he developed a violent streak. Gerald Horne, a hallboy at Blenheim in 1896, remembered that though he did not speak a word of the language on arrival, the Egyptian page could swear like a trooper in English within six months. 'Mike', as he was nicknamed, appealed to Sunny's sense of ceremonial display, and he often stood beturbaned and dressed in a yellow or scarlet Egyptian costume behind the Duchess's chair at dinner. 'He looked very picturesque, I'll say that much, but he was a dangerous customer all the same,' remarked Horne.[43] ('Mike' eventually returned to Egypt).

It was with feelings of considerable relief that Consuelo found herself back on the familiar terrain of the Hôtel Bristol on the Place Vendôme in Paris in February 1896. She had tasks of her own in Paris, for William K. had told her to buy whatever she needed to complete her trousseau, well away from the searchlights of New York's press. At this point, Consuelo suddenly realised how inexperienced she was. Until now, everything she owned had been selected by her mother. Her new husband promptly stepped in to take Alva's place, a move that was far from welcome since Consuelo felt he was primarily motivated by display, particularly when it came to her clothes. 'Marlborough took it upon himself to display the same hectoring rights [my mother] had previously exercised in the selection of my gowns. Unfortunately, his taste appeared to be dictated by a desire for magnificence rather than by any wish to enhance my looks. I remember particularly one evening dress of sea-blue satin with a long train, whose whole length was trimmed with white ostrich feathers. Another creation was rich pink velvet with sables. Jean Worth himself directed the fittings of these beautiful dresses, which he and my husband considered suitable, but which I would willingly have exchanged for the tulle and organdie that girls of my age were wearing.'[44]

Sunny's own shopping in Paris astounded his new bride who was surprised by the 'excess of household and personal linens, clothes, furs and hats' he ordered. He was also anxious to buy jewellery for Consuelo, for the lack of family heirlooms was a source of embarrassment. Marie Carola, Lady Galway noted in her late-nineteenth-century autobiography that it was the custom for a bridegroom to offer a *corbeille* to his fiancée, and that: 'In the case of the great and the rich, there were family heirlooms to which modern additions were made, and much pride and kudos was involved,'[45] but after the depredations of his ancestors the Duke had little to offer that was suitably grand. The purchase of jewellery for Consuelo in Paris only illustrated how poorly this young couple understood each other, even after weeks of travelling together. Once again, Consuelo felt that Marlborough's purchases were motivated first and foremost by a desire to convey the splendour of the family, a cause for which she felt little sympathy. The Duke, meanwhile, simply could not understand why she should object to wearing beautiful jewels and enhance the standing of the family into which she had married.

Both Sunny's desire to buy jewellery for his new duchess and Consuelo's resistance to it becomes clearer in the context of nineteenth-century ostentation. Historians of jewellery have noted a marked increase in sales of gold jewellery in England as well as America from the 1850s onwards, a trend that extended far beyond the aristocracy. Peter Hinks points out that even Charles Dickens's character, Mr Merdle, 'wanted something to hang jewels upon' and therefore 'got himself a wife.'[46] Nancy Armstrong has noted that this tendency became noticeable from the end of the eighteenth century when men stopped wearing jewels themselves. In the hierarchy of display, pearls and diamonds sat at the top until the successful development of spherical cultured pearls in the first decade of the twentieth century sharply damaged the value of real pearls.[47] Before that, pearls were often more valuable than diamonds and in 1908 George Kunz could still write: 'If one is a wearer of jewels, pearls are an absolute necessity; indeed, they are as essential and indispensable for the wealthy as are houses, horses and automobiles.'[48] Pearls were swung in huge loops round necks in a visible display of wealth. Mrs Astor even developed an odd habit of draping pearls down her back. On marriage, Alva gave Consuelo the magnificent string of pearls bought for her by William K., said to have belonged first to Catherine the Great and later to Empress Eugenie. These almost certainly came into Alva's possession after 1887, when a famous auction of most of the French crown jewels was conducted by Tiffany and Co. on behalf of the government of the Third Republic. The value of the jewels owned by American society women was frequently recorded and compared by the press. Comment in England on the jewels worn by Princess Alexandra and her court was only marginally less explicit.

It is not surprising, therefore, that the Duke was keen to ensure that the new Duchess of Marlborough had beautiful jewellery. Consuelo's resistance to wearing new jewels was unexpected, however, and pointed towards an emerging distaste for submission. Some commentators have suggested that there was a connection between male gifts of jewellery and acceptance of servitude by the female partner, and point out that as middle- and upper-class wives became increasingly divorced from economic production, their jewellery became more fetter-like.[49] The most fashionable piece of jewellery in Edwardian England was the pearl dog-collar, the most fetter-like style of them all, which usually consisted of a dozen or so strands of small, ungraduated pearls or diamond open work. Portraits and photographs of society ladies of the late nineteenth and early twentieth centuries show its near universal appeal. The fashion was led by Princess Alexandra, though Consuelo's nineteen-row pearl dog-collar, designed for her celebrated swan-like neck, became the most famous. But she hated wearing it, complaining that the diamond clasps rasped her neck and that wearing a heavy tiara gave her a violent headache. This may have sounded extraordinarily spoilt to the Duke to whom she doubtless complained, but if one accepts a subliminal association with submission, Consuelo's dislike of her famous dog-collar is much more understandable. It was, as Marian Fowler has said, 'a fit symbol for a young wife to be brought instantly to heel'.[50]

In spite of tensions with her husband, Consuelo would later write that she dreaded their honeymoon coming to an end as they prepared to travel to London. 'I realised that he would from now on be surrounded by friends and distractions that were foreign to me and that the precarious hold I had during our months alone secured in his life and affections might easily become endangered,'[51] she wrote. One clue that he did indeed hold her in some kind of affection comes in a letter to the Duke from the Marquess of Dufferin and Ava in Paris: it seems that Sunny had approached him to write a tribute to Consuelo for her birthday, knowing that the Marquess was already the author of some verses entitled 'Transatlantic Letters'. 'My dear Duke,' replied the Marquess:

> I would esteem it an honour and a privilege to pay any little tribute at the feet of the Duchess, but I am afraid it is quite impossible for me to do so on the present occasion. In the first place, I have not the practical faculty, though in early life I strung together a few very bad rhymes, and whatever is written about the Duchess ought to be a little gem; and in the next place, my daughter Flora made the same request to me for her picture. For a long time I refused, simply from the sense of my inability to write anything suitable. At last, on being further pressed, I sent her a couple of foolish verses which I had written years and years ago to please my sister-in-law who had instigated a book of 'definitions'. The subject of the verses was 'letters'. They have nothing whatever to do with Flora and are really worthless, but they were the only thing I could think of. To put some kind of meaning to them I added 'Transatlantic Letters'. A second equally mediocre effort in the same book by the same person would never do. I have written this long rigmarole in order to make you understand how vexed I am at not being able to take advantage of what would otherwise have been a most pleasant opportunity of evincing my genuine feelings of admiration for your wife. Believe me,
>
> Yours very sincerely,
>
> Dufferin and Ava.[52]

Shortly after her nineteenth birthday on 2 March 1896, Consuelo braced herself for introduction to a new life in England, suitably 'bejewelled and bedecked'. 'London looked immense as the train slowly wound through endless dimly-lit suburbs. They seemed drab to me, but the streets were clean and the little houses had gardens. There was a general air of homeliness. In those days there was little discontent – England was prosperous and only the intelligentsia ventured to discuss socialism.'[53]

To all appearances this was true. Keir Hardie's presence on the *Campania* and his lectures on the British class structure did not, as yet, pose any threat to the patrician elite. This was an era when, as Churchill would later write, the old world still existed in a 'glittering and it seemed stable framework'.[54] Aristocratic government reached its apogee between 1895 and 1905. The Conservative government that came to power in 1895 was dominated by England's great landowning families. The Prime Minister, Lord Salisbury, was a marquess. So was the Secretary for War. The Lord President of the Council and the Secretary for India were both dukes. The government also included one viscount, and at least three barons. Other than Joseph Chamberlain, even the commoners were more aristocratic than they first appeared: Arthur Balfour came from a Scots landowning family and was Lord Salisbury's nephew (a coincidence behind the expression 'Bob's your uncle'). In 1896, however, when Consuelo arrived in England, some of the glitter of British aristocratic dominance could already be imputed, in A. L. Rowse's ringing phrase 'to the phosphorescence of decay'.[55]

In the view of David Cannadine, the rot had already set in by the 1880s. It was not simply that the style of the patrician elite was being challenged by the opulence of plutocrats. Across Europe, the agricultural base of the economy was being adversely affected by the rise of industrial society and an influx of cheap foodstuffs from the Americas and the Antipodes. Agricultural depression meant lower agricultural rents. 'The result was that the rural sector was simultaneously depressed and marginalized, and the consequences for the essentially agrarian elite of European landowners were inevitably severe.' In Britain there was a more specific problem. The Third Reform Act, passed between 1884 and 1885, had altered the balance away from dominance by the landed classes, from 'notables to numbers'.[56] As a result of this extension of the franchise there had already been several challenges to the great landlords across Britain: a tithe war in Wales, a crofter rebellion in Scotland, a demand for Home Rule in Ireland. The patrician society to which Consuelo was about to be introduced may superficially have seemed all-powerful – rich, prestigious, and with political power vested in the House of Lords – but it was already on the defensive. The consequence was arrogance, insularity, a preoccupation with minute issues of status and precedence, and a complex – and not always amiable – reaction to a young American plutocrat bride with a very large fortune.

Consuelo would experience all of this and more as she came to know her new in-laws, some of whom now lined up to greet her on the platform at Victoria station in London. The reception committee included some familiar faces: Sunny's sisters, Lilian and Norah, whom she had met at Blenheim the previous summer; Ivor Guest, the Duke's best man; and cousin Winston whom she had probably met briefly before leaving New York the previous November. There were new faces too: her mother-in-law, Lady Blandford; Lady Sarah Wilson, an aunt whom Consuelo mistrusted on sight; and Lady Randolph Churchill, born Jennie Jerome, the first American bride to marry into the Spencer-Churchill family. 'I felt the scrutiny of many eyes and hoped that my hat was becoming and that my furs were fine enough to win their approval. They all talked at once in soft voices and strange accents which I knew I should have to imitate, and I felt thankful that I had no nasal twang.'[57]

There was a great deal to absorb that first evening in London. Lady Blandford, Consuelo's mother-in-law, 'had the narrow aristocratic face of the well-bred, with a thin slightly arched nose and small blue eyes that were kind and appraising'. Consuelo immediately realised that 'Goosey' was far from dim however. 'Her outlook was limited, for she had received an English girl's proverbially poor education, but she possessed shrewd powers of intuition and observation, and that she liked me I immediately realised.'[58] At dinner that evening, however, Consuelo was taken aback by some of her mother-in-law's more startling remarks 'revealing that she thought we all lived on plantations with negro slaves and that there were Red Indians ready to scalp us just round the corner'.[59] This was not an uncommon experience: Lady Randolph Churchill wrote in her memoirs that the young American woman 'was looked upon as a strange and abnormal creature, with habits and manners something between a Red Indian and a Gaiety Girl'.[60] Such treatment was also satirised by Edith Wharton: 'I am constantly expecting them to ask Mrs St George how she heats her wigwam in winter,' remarks Sir Helmsley Thwarte in *The Buccaneers*.[61]

It quickly became apparent to Consuelo that Sunny's aunt Lady Sarah Wilson was far less kindly than his mother, and that there was a feud between the Churchill side of the family and the Hamilton side, represented by 'Goosey'. When Lady Blandford regretted that she had been unable to come to New York for the wedding – because Sunny had refused to pay her passage – Lady Sarah tittered and remarked: 'But the Press did not spare us one detail' – making the new arrival feel 'that the word "vulgar" had been omitted but not its implications'.[62] Lady Sarah seemed hard and sarcastic to Consuelo, and plainly showed that she considered Alberta Blandford a fool. 'To me she was kind in an arrogant manner that made me grit my teeth, for I had no intention of being patronised,' Consuelo wrote later.[63]

She found the half-American side of the family much more appealing. Cousin Winston, a 'young red-headed boy a few years older than I', struck her immediately as 'ardent and vital and seemed to have every intention of getting the most out of life', although she wondered how he and his mother would react if he were displaced as heir to Blenheim. Lord Randolph Churchill had died a terrible and protracted death from syphilis fifteen months earlier in January 1895. His widow, Jennie, remained 'a beautiful woman with a vital gaiety . . . Her grey eyes sparkled with the joy of living, and when, as was often the case, her anecdotes were risqué it was in her eyes as well as in her words that one could read the implications. She was an accomplished pianist, an intelligent and well-informed reader and an enthusiastic advocate of any novelty. Her constant friendship and loyalty were to be precious to me in adversity.'[64]

There were three further calls to be made before the Marlboroughs proceeded to Blenheim. One of them was to Lansdowne House, the London home of Sunny's Aunt Lansdowne, whom Consuelo had met while she was still Vicereine of India in 1894. Sister of Lady Blandford, and unwitting role model, Lady Lansdowne now joined the ranks of those determined to lecture the former Miss

Vanderbilt on how to behave like an English duchess. She must remember that she should not walk alone in Piccadilly or Bond Street; or sit in Hyde Park unless accompanied; or take a hansom cab; or travel in anything other than a reserved compartment; or visit a music hall; or dance more than twice with the same man at any ball. It was essential to know her place in the social hierarchy, and learn the ramifications of both the socially prominent families into which she had married. 'One must, in other words, memorise the *Peerage*, that book that with the *Almanach de Gotha* in Europe and the *Social Register* to a lesser degree in America establishes pedigrees and creates snobs. I found that being a duchess at nineteen would put me into a much older set and that a measure of decorum beyond my years would be expected of me. Indeed my first contact with society in England brought with it a realisation that it was fundamentally a hierarchical society in which differences in rank were outstandingly important. Society was definitely divided into castes.'[65]

However, this lecture paled beside the one delivered by Sunny's intimidating grandmother Frances, the Dowager Duchess of Marlborough, wife of the 7th Duke of Marlborough. Duchess Fanny was 'a formidable old lady of the Queen Anne type' – by which Consuelo meant that she could be arrogant and familiar by turn and in baffling succession. The Duchess sat in an armchair in the corner of her drawing-room in Grosvenor Square, dressed in deep mourning and armed with an ear-trumpet. After bestowing a kiss 'in the manner of a deposed sovereign greeting her successor', she inspected Consuelo closely, before embarking on her own lecture about good behaviour, expressing the view that it was now Consuelo's duty to restore Blenheim to its former glory and to uphold family prestige. She then cut to the chase: '"Your first duty is to have a child and it must be a son, because it would be intolerable to have that little upstart Winston become Duke. Are you in the family way?" Feeling utterly crushed by my negligence in not having insured Winston's eclipse and depressed by the responsibilities she had heaped upon me, I was glad to take my leave,' wrote Consuelo.[66]

The Duke of Abercorn, who was Lady Blandford's brother and Sunny's uncle, was almost as bad, looking Consuelo up and down and saying: 'I see the future Churchills will be both tall and good-looking.' She decided to take this as a compliment, however, since Sunny was pure 'Hamilton', similar to his uncle in that he was small, fragile, and restless with many of the same fussy mannerisms. The Duke of Abercorn was also quite eccentric. 'He insisted on removing my coat, which was of green velvet entirely lined with Russian sables. "What a wonderful coat, what priceless furs!" he exclaimed. "I must send for my sables to compare them." Whereupon he rang the bell and had his valet bring his coat. To his deep concern, it did not equal mine!'[67]

This fabulous garment played a walk-on role the day that the Marlboroughs finally arrived at Blenheim. 'How fortunate that the day we left for Blenheim was cold since Marlborough had decided that I must wear my sable coat,'[68] wrote Consuelo bitterly. The Duke can hardly be blamed for this, however, for he had some sense of what lay ahead and wanted her to look her most magnificently elegant in front of crowds of interested spectators. (Neither of them would have been very pleased that a reporter from the *Oxford Times* described her coat as 'fox', though he was much more taken with Consuelo's pearls which, he wrote, were 'as large as small marbles'.[69]) On the afternoon of Tuesday 31st they left London from Paddington station in a special saloon carriage attached to the 1.30 express train. When the train arrived at Oxford station an hour and a half later, the Duke and Duchess alighted onto the platform to be greeted by Mr Davis the stationmaster and others, and received the first of a number of pleasant surprises.

Blenheim Palace is close by the small town of Woodstock in Oxfordshire, about seven miles from Oxford. The Woodstock branch engine, *Fair Rosamund,* now chugged into view specially decorated with evergreens and Stars and Stripes, bearing the words 'Welcome Home' on its front. *Fair Rosamund* was then coupled to the private saloon carriage and as the train pulled out of Oxford station bearing the Duke and Duchess, it transpired that a *fête de joie* of detonators had been placed on the track which exploded noisily as it departed, to the great pleasure of the many cheering spectators. In fact there were cheering spectators all the way along the line with large groups massing to wave at Walton Well Bridge and Wolvercote as the train passed on its way to Woodstock. *Fair Rosamund* stopped again at Kidlington station. Here, the Duke and Duchess not only found themselves being greeted by another unexpected reception party of railway officials but that the down platform had been artistically decorated for the occasion. Consuelo was given a bouquet of roses by Florence Cooke, the stationmaster's daughter, while the Duke was presented with a 'handsome red morocco case' designed to hold the local train timetable. As the train moved off on the small branch line from Kidlington to Woodstock, more spectators emerged from their farmhouses and cottages to wave, and flags fluttered from trees the whole way down the track and 'reminded the Duchess that she was at last nearing home'.[70] She was also nearing a welcome for which nothing could have prepared her.

At 3.13, according to *Jackson's Oxford Journal*, the ringing of the bell in the stationmaster's office indicated that the train had left Kidlington and five minutes later it steamed into Woodstock station amid cheers from local school children assembled on the opposite platform. There were more loud explosions as mortars were discharged in the park to announce the train's arrival, whereupon their graces were welcomed by Mr R. L. Angas (steward at Blenheim), the Mayor and Mayoress, the Town Clerk, the Rector (Reverend J. Farmer), the Reverend Dr Yule (chaplain at the Palace) and Mr Higgs. Salutations were exchanged and Miss Nellie Mabel Clarke (daughter of the Mayor) presented the Duchess with a magnificent bouquet of pink roses. His worship, the Mayor, was wearing a new scarlet robe specially presented to him for the occasion by members of the Woodstock Town Council. Consuelo received another bouquet of red roses and lilies of the valley from four local school children (two girls, two boys) as the Duke raised his hat to the crowd.

To Consuelo's dumbfounded astonishment, the horses were then unhitched from the Duke's carriage and amidst scenes of great excitement, the ducal carriage – bearing the Duke and Duchess – was dragged from the station to the town hall by several able-bodied men of Woodstock, escorted by the Witney troop of the Queen's Own Oxfordshire Hussars. The same thing had happened on Lady Randolph Churchill's first visit to Blenheim as the bride of Lord Randolph in 1874, and she thought it was rather fun. Although Consuelo's 'democratic principles rebelled',[71] she did her best to play the role of duchess in a fitting manner, and smiled and waved at the cheering crowds. The warmth of the welcome was undoubtedly most touching.

There was a large crowd at the town hall where discipline was maintained by the Blenheim and Woodstock fire brigades. Those who lived nearby watched from windows. The carriage was pulled up to a platform outside the town hall to the strains of 'Hail Columba' from the Oxford Volunteer Band. The Town Clerk read a speech of welcome from the Corporation. Then the Mayor of Woodstock had his turn. 'Every inhabitant of the ancient town of Woodstock was anxious to welcome the Duke and Duchess back,' he declaimed. 'It was a self-governing body forty years before America was discovered, and 200 years before the first settler set foot in New York.'[72] The rest of the speech wishing them many years of wedded happiness was rather lost on Consuelo, who was already sensitive to being patronised after only three days in England, and now seethed inwardly.

The Duke then rose to make a speech of thanks. He had no idea when he left Blenheim the previous summer that he would be gone so long nor that when he returned it would be 'the occasion of such good feeling and so kind a reception on their part'. He went on to say that one of the first sentiments Consuelo had expressed to him 'was that she might be able to become the friend of the people among whom she was going to dwell, and might succeed in endearing herself to their hearts'.[73] He felt sure that she had never anticipated such a warm reception, which he was sure was chiefly designed for his wife rather than himself. The Duke finished his speech of thanks with a splendidly feudal flourish. The *Oxford Times* reported: 'He thought he might say without assuming any tone of exaggeration, that his aim had always been to ameliorate the condition of Woodstock and its population – (hear, hear) – and he would conclude, thanking the Mayor for the kind address which he had presented to him, by expressing the hope that what had taken place that day might be the means of binding them more closely together in a common bond of sympathy, and that the good fellowship and esteem of all classes might be united (hear, hear).'[74] This was greeted by prolonged cheering.

Meanwhile the procession to the Palace was forming in Market Street. Elaborate floral arches that had taken days of work had been erected at intervals down the street. The procession now passed beneath them, watched by hundreds of people who had come from Oxford and the surrounding villages. The Blenheim and Woodstock fire brigades were followed by the local band playing 'Welcome Home'. They were followed by a deputation from the National Fire Brigades Union, the Olivet Friendly Society, the Foresters, estate employees, estate tenants, the Mayor and Corporation in carriages, and finally the Duke and Duchess in a landau, escorted by the Oxfordshire Hussars yeomanry. Visitors from the surrounding villages were already lined up along the road in the park, and school children from Bladon were mustered at the gate, where little Ethel Goodall presented Consuelo with bouquet number four.

When the procession finally arrived in front of the Palace they were received by Mr Angas, the Chief Steward, (who must have departed swiftly from the station) and Mrs Ryman the housekeeper to the strains of 'Home Sweet Home' from the band. Consuelo was presented with bouquet number five. Tenants were grouped on one side and employees on the other with a further crowd of ticket-holders stretching back to Blenheim Park. Mr Angas told the Duke that it was his fervent hope that he would never be away so long from them again. 'He estimated that the Duke had travelled during those seven months 15,000 miles by land and water, and he assured his Grace when storms were blowing at sea they thought of him. It was a happy moment when he saw that the steamship *Fulda* had touched at Gibraltar.'[75] Consuelo was assured they would all do their best to make her happy. Mr Nash, the Clerk of the Works, endorsed every word that Mr Angas had said on behalf of the employees; Mr Scroggs was deputed by the tenantry to say it all again; and Mr Gamble of the National Fire Brigades Union presented them with another address in pen and ink. The prize for unctuousness, however, went to the Reverend Farmer, rector of the parish of Bladon and Woodstock. 'The Duke had presented him with the living ten months ago, and he could safely say that in every single house in Woodstock and Bladon his Grace's name was as ointment poured forth, that was to say, it always led to the expression of some words of kindly goodwill and love from men of every creed and every phase of politics.'[76]

The Duke thanked them all, on behalf of them both. If Consuelo felt that the proceedings thus far were feudal, her husband's speech of thanks to Mr Nash who had spoken for the employees would have made her gasp. 'I feel sure that it will always be your wish to endeavour to promote the welfare and happiness of everyone connected with Blenheim,' said his Grace. 'And I think I may say on this occasion that the employees on the estate may look to you as their head to represent their interests as well as the interests of your master.'[77] But the chances are that no-one else noticed, for according to Blenheim hallboy Gerald Horne, they were all entranced by the new Duchess. 'It was a great thing for me to stand there and see them brought home in such style and when the Duchess stepped down from the carriage you might almost have heard us gasp at how young and how beautiful she was,' he said later. 'And she was as good as she was beautiful too.'[78]

She may have been good and beautiful but deep down, the Duchess felt close to tears. 'As I stood on the steps listening to the various speeches, I realised that my life would be very strenuous if I was to live up to all that was expected of me. My arms were full of bouquets, the fur coat felt heavier and heavier, the big hat was being blown about by the winds, and I suddenly felt distraught, with a wild desire to be alone.'[79]

Once she stepped indoors, Consuelo remembered that 'my

maid was waiting for me, a tea gown of satin and lace laid out, a hot bath ready, and I dressed for the ritual of dinner such as Marlborough, the chef and the butler had decreed it to be'.[80] There were further celebrations starting with a champagne luncheon for 350 people involved in the welcome in the audit room, with overflow tables in the laundry. The Duke paid them an impromptu visit while it was in progress, and thanked everyone again, saying that he had never seen anything better done. That evening there was a grand display of fireworks, which the Duke and Duchess watched from a tent in the park, and afterwards they drove through the streets of Woodstock and admired the illuminations. Though it had been an ordeal, Consuelo rallied. 'Everyone was charmed with the gracious and winning manner of the Duchess, who won the hearts of all who had the privilege of seeing her,' said the *Oxford Times*.[81] Peals were rung in churches for miles around, money from their graces was distributed to the poor of the parishes of Charlbury, Coomb, and Stonesfield, and in Wooton the villagers gave themselves a holiday. That evening special trains were laid on from Woodstock to Oxford, taking home more than thirteen hundred onlookers.

In the last quarter of the nineteenth century, David Cannadine suggests, British patricians were 'highly conscious of themselves, their families, and their order *in time*. More than any other class, they knew where they had come from, they knew where they were, and they hoped and believed they were going somewhere. This is what Edmund Burke meant when he spoke of "partnership not only between those who are living, but between those who are dead, and those who are yet to be born".'[82] Sunny expressed precisely this view when he talked of being a 'link in a chain', and after arriving at Blenheim, Consuelo understood better what this involved. 'When Marlborough spoke of a link in the chain he meant that there were certain standards that must be maintained, whatever the cost, for what was a generation but such a link? – and to him it was inconceivable that he, given the greatness of his position, should fail to uphold the tradition of his class. The English countryside was still rural, the farmers and labourers loyal to their landlord, the standard of living possible for those whose needs were elementary. It was not for me, with my more democratic ideals, to upset the precarious balance. I should have to adopt the role expected of me by my marriage and fulfil its obligations as conscientiously as possible.'[83]

6

Success

AT NINETEEN-YEARS OLD, Consuelo's only experience of running a house had come during playtime in La Récréation at Idle Hour. Otherwise, she had barely been permitted to move her own hairbrushes. As soon as she arrived at Blenheim, however, Consuelo was required to take charge of a palace where she never succeeded in counting the number of rooms, though an inventory prepared for the 4th Duchess suggests there were almost one hundred and seventy, with forty on the bedroom floor alone. She also became responsible for an indoor staff of more than forty servants. 'Marlborough had given me the supervision together with the financing of everything pertaining to the house, while reserving the administration of the estate for himself,' she wrote, a division of labour for which she should perhaps have been more grateful since the outdoor staff in Blenheim's self-contained world included a hunting department, stable staff to take care of forty grey horses, a fire brigade, lodge keepers and a cricket coach. Consuelo's new husband did not provide her with much guidance in her daunting and unfamiliar task, and disappeared to London soon after they arrived to make arrangements for the season. 'Unfortunately he was more inclined to criticise than to instruct and I had to trust to observation to ensure the continuity established by past generations of English women,'[1] she wrote.

There was plenty of continuity to observe. The servants' hierarchy was just as rigid as that above stairs, and almost more congealed. The butler, or house steward, came first. 'He was addressed as Mr So-and-So by the other servants, and his chief concern was to keep everyone, including himself, in his place.'[2] Only the two in-house electricians, who were regarded as men of science, equalled him in status. The Duke's valet was also very important. Like the butler, he wore tails and striped trousers and his role was almost as prestigious. Next in the chain came the groom of the chambers, one of whose tasks was to keep guest bedrooms supplied with pens and writing paper, sometimes removed by visitors who then absent-mindedly used it for their thank-you letters. Then came the under butler and three or four footmen, as well as 'oddmen' whose job it was to do odd jobs at the butler's bidding – such as carrying coal for over fifty grates and washing windows (there were so many that they were only washed once a year). Consuelo quickly learnt that it was easy to offend staff over protocol. 'On ringing the bell one day I was answered by the butler, but when I asked him to set a match to an already prepared fire he made me a dignified bow and, leaving the room, observed, "I will send the footman, Your Grace," to which I hastily replied, "Oh don't trouble, I will do it myself." '[3]

The French chef had his own staff of four. There were frequent rows between him and Mrs Ryman, the housekeeper, over serving meals and Consuelo was soon called on to adjudicate. Common sense might have suggested that cooking one part of breakfast in a kitchen several minutes away from the room where the rest of it was being prepared made co-ordination difficult, but there was much resistance to change. Consuelo had her own maid, Rosalie, selected for her by Lady Blandford after Sunny decided that the maid who accompanied them on their honeymoon was insufficiently *au fait* with English ways. Sunny may have later regretted this decision, for Rosalie died in service to Consuelo, whom she adored, while 'disliking men in general and my husband in particular'.[4] The housekeeper ran her own department. Consuelo thought Mrs Ryman was very understaffed with only six housemaids, five laundresses and the ever-popular still-room maid (who cooked breakfasts and teas); and that the maids, who lived up in 'Housemaids' Heights' – a tower without running water – were not well treated. But housemaids had lived like that at Blenheim for over

Consuelo at about ten years old, *c*.1887.

Above Consuelo's father, William Kissam Vanderbilt, *c.*1900.

Left Alva in costume for the Vanderbilt Ball of 1883.

Below Harold, Consuelo and Alva in the Bois de Boulogne, Paris, *c.*1894.

Woodstock welcomes home the Duke and Duchess of Marlborough after their honeymoon, 31 March 1896.

Left Consuelo and Lord Ivor Churchill (probably in a christening gown) and the infant Marquess of Blandford, early 1899.

Below The Duchess in coronation robes, 1902.

Above Consuelo and Winston Churchill at Blenheim, 1902.

Right The 9th Duke of Marlborough in uniform during the Boer War.

Left A drypoint etching of Consuelo by Paul-César Helleu, *c.*1901.

Below Consuelo in her electric car at Blenheim, a present from Alva, *c.*1904.

Opposite The Duke and Duchess of Marlborough, separated by a friend, aboard the P&O liner *Arabia*, en route for the Delhi Durbar, December 1902.

Consuelo at the Delhi Durbar, December 1902–January 1903.

two hundred years and she quickly learnt that improving invisible arrangements for the servants was not one of Sunny's priorities.

Blenheim differed from other great country houses in that the upper servants had their own dining room, but they followed the widespread practice of sitting down to meals in accordance with their own hierarchy. 'Visitors' servants made quite as much work as the gentry themselves,'[5] according to Gerald Horne, for when there were house parties the servants arranged themselves at table according to the rank of their masters and mistresses. (One visitor later reported that he had to lend his valet a dinner jacket when staying at Blenheim because blue serge was not acceptable in the steward's room in the evenings.) This practice was rigidly observed until 1939, a Blenheim housekeeper told James Lees-Milne. The only exception was that the 'valet of the eldest son of the house always sat on the housekeeper's right, taking precedence even over the valet of the most distinguished guest.'[6] In Vita Sackville-West's novel *The Edwardians* the date of creation of the peerage was taken into account when ranks coincided in the servants' hall and a copy of Debrett's was kept by the housekeeper to resolve disputes over placement below stairs.

In keeping with the ostentatious spirit of the age, the physical appearance of the Blenheim house servants was of the utmost importance. Quite apart from the exotic presence of 'Mike', the pageboy from Egypt, the six Blenheim footmen wore splendid liveries of a maroon coat and matching plush breeches, waistcoat with silver braid, flesh-coloured silk stockings and patent shoes with silver buckles. To qualify as a footman in 1896 a man had to be at least six-feet tall – and prepared to powder his hair. This was a horrible job according to hallboy Gerald Horne (who was called 'Johnny' while he worked at Blenheim because 'Gerald' sounded too upper class). 'You had to powder every day and that meant washing your hair with soap, combing it out, getting it set in waves and then powdering it. The powder you mixed yourself, buying violet powder from the chemist and mixing it with flour . . .' When the footmen took their place on the ducal carriage they had to wear hats which spoilt their hair and meant they had to powder it all over again. When it was dry, the powder set like cement. 'Honestly,' said Gerald Horne later, 'I don't think the gentry realised what you went through doing this kind of thing.'[7]

Consuelo soon discovered in those early weeks at Blenheim that there was one important difference between the wives of American plutocrats and those of English aristocrats, a difference which Alva had failed to grasp. In America, rich women were excluded from public life but had more or less total control over their households. (Even now, the great 'cottages' of Newport continue to be associated with their original female owners: Alva Vanderbilt, Mrs Astor, Mrs Oelrichs, and Mrs Fish.) An English duchess had no such power. In Consuelo's case, her work was subject to periodic tests and it did not help that some of the Duke's methods for testing the efficiency of indoor servants were reminiscent of his mother's more sadistic practical jokes. Consuelo once had to intervene to stop a distraught housemaid from resigning after the Duke hid a small china box to test whether anyone would notice, for the maid thought she was being accused of stealing. Sunny also made a monthly tour of inspection that everyone learned to dread. 'The Duke was all right but he could certainly make one tremble and when he came round with the steward on his monthly inspection you'd most likely hear a roar and think that what was coming down the passage was a giant,'[8] said Gerald Horne.

Everywhere she turned during those first few weeks, Consuelo found British class-consciousness in action. Introductions to country neighbours, with one or two exceptions, were almost as wearing as the first calls to Marlborough's relations in London. Etiquette demanded that Consuelo stay at least twenty minutes, but most neighbours – and her coachman – wanted her to remain much longer, so that they could give her tea and show off their houses. Here, it was not simply a matter of the English gentry patronising an American arriviste; some county families were just as anxious to patronise the Marlboroughs. 'It was apparent that the older families whose roots were embedded in Oxfordshire regarded the Churchills, who moved there in the eighteenth century, rather as the Pilgrim Fathers looked upon later arrivals in America. Perhaps also to impress me, they stressed their ancient lineage, seeming to imply that lives lived in a long-ago past conferred a greater dignity on those lived in the present.'[9] She greatly preferred her visits to the tenant farmers on the estate, in the company of Mr Angas, the steward, and came to respect them both as farmers and as loyal friends.

Soon after she arrived at Blenheim, Consuelo moved quickly to establish a rapport with the poorer and less fortunate tenants on the estate and 'domain towns', local villages on the edge of Blenheim's park. The day began with chapel at 9.30, before breakfast was served. 'At the toll of the bell housemaids would drop their dusters, footmen their trays, house-men their pails, carpenters their ladders . . . laundry maids their linen, and all rush to the chapel in time.'[10] The Duchess would often have to scramble too, particularly when she overslept. After the service the curate would tell her who was in need of a personal visit that day. This is where she had always felt she might have a role and her attentions were welcomed. 'There were old ladies whose complaints had to be heard and whose infirmities had to be cared for, and there were the blind to be read to. There was one gentle, patient old lady whom I loved. She used to look forward to my visits because she could understand every word I read to her while sometimes with others she could neither hear nor follow and was too polite to tell them so. I grew to know the Gospel of St John by heart because it was her favourite.'[11]

The saccharine flavour of such accounts may grate a little now, but there is no doubt that Consuelo's interest in the less fortunate was quite genuine; that it went right back to her childhood; and that it was an aspect of her daughter's character that Alva felt should have expression after marriage. Consuelo's kindness may partly have started in a reaction to the excessively self-interested flight from reality lived out by her parents, but stories about it became almost legendary. 'She would go out of her way to be kind to everyone,' said Gerald Horne 'and of course she was idolised . . . She was a great lady.'[12] 'Consuelo came to be adored,' writes David Green, Blenheim's principal historian. 'Naturally, when the coach-and-four was heard in the distance cottagers popped in for a spotless apron, to pop out in time to drop a curtsey as the ducal equipage passed; but personal visits by the beautiful American Duchess of course made red-letter days. As an old Long Hanborough woman told me: "I had a dream last night . . . I dreamt our Duchess visited me, and do you know what she said? She said mine was the cleanest cottage she had ever been in." Such visits and such dreams were treasured in a way which nowadays might easily be despised; yet treasured they were and made for happiness and self-respect.'[13] Years later, when *The Glitter and the Gold* was published, an admirer from England wrote and told Consuelo that she was still remembered as 'the angel of Woodstock'.[14] Her kindness was so marked that it was occasionally abused by the unscrupulous who put their children to bed and ordered them to look ill when they thought that the Duchess was on her way, but there is no sign that she minded. For her part, Consuelo enjoyed the visits for the strength they gave her, and for the sense of mutual obligation and commitment, which she had never before experienced.

This was all in sharp contrast to the stories told about her husband. The Earl of Carnarvon recalled that when he was staying at Blenheim one Boxing Day and looking forward to a day's shooting, the butler came in and announced nervously that the head keeper was ill, but had delegated his responsibilities. The Duke's reply was: 'My compliments to my head keeper; will you please inform him that the lower orders are *never* ill.'[15] Such differences in perspective meant that it was hard for a new duchess to make changes, but Consuelo quickly introduced one alteration for which the poor could only have been grateful. 'It was the custom at Blenheim to place a basket of tins on the side table in the dining-room and here the butler left the remains of our luncheon. It was my duty to cram this food into the tins, which we then carried down to the poorest in the various villages where Marlborough owned property. With a complete lack of fastidiousness, it had been the habit to mix meat and vegetables and sweets in horrible jumble in the same tin. In spite of being considered impertinent for not conforming to precedent, I sorted the various viands into different tins, to the surprise and delight of the recipients.'[16]

In spite of the warm welcome she was given, Consuelo was never reconciled to Blenheim as a dwelling-place. In fact, she lined up with those who denigrated it. 'It is strange that in so great a house there should not be one really liveable room,' she wrote. She would not have minded sacrificing comfort if the rooms had been elegant or beautiful, but in her view they were neither. 'We slept in small rooms with high ceilings; we dined in dark rooms with high ceilings; we dressed in closets without ventilation.'[17] A happy life in such a splendid but unforgiving house may have only been possible for a duchess who adored her duke. Since Consuelo did not, it was difficult to love a house so intently devoted to his forbears. And it made matters worse that Consuelo's sympathy was never enlisted for the Duke's attempt to reverse the decline of the Marlborough fortunes, his working assumption being that she was lucky to assist.

While her husband deeply admired the vision of his ancestors, Consuelo sided with those who saw only triumphant egotism. 'Having been confronted with Marlborough's victories in the tapestries that adorned the walls, having viewed his household in the murals painted by Laguerre, his effigy in silver on our dinner-table, his bust in marble in the library, his portrait over mantels, his ascent to celestial spheres on the ceilings, my feelings as I faced the funeral monument [in the Chapel] were akin to those of the Bishop of Rochester who says in a letter to Alexander Pope, speaking of the first Duke's funeral . . . "I go tomorrow to the Deanery [of Westminster] and I believe I shall stay there till I have said Dust to Dust and shut up that last scene of pompous vanity," '[18] she wrote.

After Consuelo had spent a few weeks settling in at Blenheim, she travelled to London for another challenge – her first season as Duchess of Marlborough, when she would be introduced to London society. 'I might almost say for my coming out, for there had been little gaiety in my previous life . . . with no time for friendships or even understanding.'[19] Before she could take her place in London's elite, however, she had to be presented at court. By now Sunny had bought a crimson state coach, a grandiose object whose coachman wore a livery of crimson cloth and silver braid stamped with the double-headed eagles of the Holy Roman Empire of which the Duke was a prince. This imposing equipage transported Consuelo and her mother-in-law, Lady Blandford, to Buckingham Palace.

Lady Blandford presented Consuelo at court at an afternoon event known as a 'Drawing Room', presided over by the Prince and Princess of Wales who now deputised for Queen Victoria at such functions. In keeping with tradition, Consuelo wore her wedding dress cut low, its long court train spread out on the floor behind her as the presentation began. A Drawing Room was a major opportunity for everyone to show off their jewellery, starting with the Princess of Wales, whose sloping shoulders, breasts and arms were particularly well suited to displaying her glittering jewels. Consuelo wore the tiara her father had given her, the infamous pearl dog-collar and a diamond belt given to her by Sunny.

Years later, she could still remember the thrill of excitement at the sound of the drum roll and the national anthem that accompanied the entrance of the royal procession into the ballroom, preceded by the Earl of Lathom and the Earl of Pembroke walking backwards before the Prince of Wales. It all went off perfectly. Consuelo made the requisite number of curtseys with dignity, feeling that it was her patriotic duty as an American to manage this without looking foolish. The only slight contretemps came afterwards when Lady Blandford breezily assured her that she had done so well that no-one would have taken her for an American. This time, Consuelo rebelled and asked her how she would feel if she thought that no-one would take her for an Englishwoman. '"Oh that is quite different," she answered airily. "Different to you, but not to me," I countered, laughingly.'[20] Even Lady Blandford refrained from making such remarks thereafter. As they left Buckingham Palace, the band of the Household Cavalry played. Crowds had gathered on the Mall to cheer the young women who had just been presented at court and Consuelo later remembered feeling like Cinderella after her pumpkin changed into a coach.

The London season at this time lasted from early May until late

July, when it made way for the opening of the grouse shooting season in the middle of August. Unlike the American plutocracy, the life of the English aristocracy was essentially rural. The social season was the metropolitan exception, a period lasting several weeks in the summer when the aristocracy based itself in London. Young unmarried women came out as debutantes early in May and met suitable young men at balls and parties. The end of the season was traditionally marked by a ball at Holland House in Kensington, and many marriages were settled in its gardens before society dispersed again. During these weeks aristocratic families stayed in the family town house, where there was one, returning to the country for large parties at the end of the week – a practice made easier by the advent of the motor car. Since the Marlboroughs had no London base, they took a 'tiny house' in South Audley Street for the London season of 1896, where they were close to the London town houses of the Lansdownes, Devonshires and others, in the streets and squares of Mayfair and Belgravia.

'Those who knew the London of 1896 and 1897,' wrote Consuelo, 'will recall with something of a heartache the brilliant succession of festivities that marked the season . . . To me it appeared as a pageant in which beautiful women and distinguished men performed a stately ritual.'[21] Although Lady Diana Cooper cautions against regarding all balls of this era as 'fairytale', remarking that they were frequently nothing of the sort, Consuelo looked back with some nostalgia to the days when one still danced quadrilles, with polkas and Strauss waltzes played by Viennese orchestras. The great beauties of those seasons were Lady Helen Vincent (later Lady D'Abernon), Lady Westmorland, and her sisters Lady Warwick and the Duchess of Sutherland; the great hostesses, Lady de Grey and Lady Londonderry.

Lady de Grey championed the opera at Covent Garden and made attendance almost mandatory for fashionable persons. In spite of Lady Lansdowne's warnings, the Prince and Princess of Wales broke the embargo on music-hall entertainment at the Gaiety Theatre. There were plays at the Empire Theatre, polo matches at Ranelagh, Roehampton and Hurlingham, and attendance at debates in the House of Lords where the oratorical power of many new relations by marriage was frequently on display. The Duke liked to go out in his new mail-phaeton with his own 'tiger' (a diminutive groom), and sometimes asked Consuelo to accompany him. Such elegance required that she wore her best clothes to match her husband's grey swallow-tailed jacket and high grey hat, a white gardenia in his buttonhole. Aristocratic ladies lined up at Grosvenor Gate to see the Princess of Wales pass by as she bowed right and left. 'We dined out nearly every night and there were always parties, often several, in the evening. Indeed, one had to exercise discretion in one's acceptances in order to survive the three months.'[22]

Amidst it all, Consuelo's debut in London 1896 was a great success – the 'rage of the season and everybody makes a fuss of her', according to Chicago heiress Mary Leiter, who had recently married George Curzon. 'Everybody raves about Consuelo,' she wrote in letters home to Washington. 'She is very sweet in her great position, and shyly takes her rank directly after royalty. She looks very stately in her marvellous jewels, and she looks pretty and has old lace which makes my mouth water. I never saw pearls the size of nuts.'[23] Consuelo – and her jewellery – were once again closely observed from different angles. 'We had a very nice view of the American Duchess of Marlborough (Vanderbilt)' wrote the diarist Lady Monkswell. 'She is not a beauty but nice looking, very tall and slender with a little retroussé nose and dark eyes . . . she had a splendid dress of turquoise velvet. A little open-work diamond crown, perched as they now do on the top of her head, quantities of strings of pearls round the neck, and a band of diamonds at least 3 inches wide, round her waist.'[24]

Back in Oxfordshire, Consuelo's impact was equally striking, even on children. Viscount Churchill (scion of a collateral line at Cornbury Park) recalled that while Blenheim suffered from 'a certain lack of colour in spite of the pomp and circumstance', the exception was Consuelo who 'passed completely through the barrier separating child from grown-up. The fact that Consuelo Vanderbilt was young, new to her surroundings and very watchful, just as a child is watchful, may have had something to do with it.

She was also beautiful and a child responds to beauty.'[25] Undergraduates at Oxford fell for her charm too. One, Guy Fortescue, saw her frequently in the distance and he and his friends fell 'instantly in love with her *piquante* oval face perched upon a long slender neck, her enormous dark eyes fringed with curling lashes, her dimples, and her tiny teeth when she smiled'.[26] In spite of Lady Monkwell's reservations, Consuelo quickly came to be regarded as one of the great beauties of the Edwardian era. It helped, no doubt, that 'it was the slim, tight look that counted',[27] and that Consuelo came to embody it. Stories of her beauty caused an equerry to exercise royal prerogative at a ball at Grosvenor House, and interrupt while she was dancing to introduce her to the old Duke of Brunswick. He was blind, but so anxious to see what the new Duchess of Marlborough looked like that he asked 'if I would object to his running his fingers over my face, since only so could he know what I looked like. It was an embarrassing procedure, but I felt too sorry for him to refuse.'[28]

Consuelo may not fully have appreciated the extent to which being Duchess of Marlborough gave her an easy entrée into English society. Mary Curzon had a far more lonely and difficult time because George Curzon's obsessional approach to his work as Under Secretary at the Foreign Office (and Member of Parliament, Privy Councillor and leading authority on Asiatic affairs) made it hard for her to go out in society. They had refused Ascot, she wrote home in June 1896. 'G's work defeats all forms of amusement.'[29] This was another difference for American women. In New York they exercised almost total control over the social domain, as they did over their homes. In London, participation of women in society was largely a function of their husband's status, and men, from the Prince of Wales downwards, were acutely aware of society's protocols which they actively maintained. Mary Curzon was prevented from enjoying herself at Ascot by George's work; Consuelo, on the other hand, was required to attend Ascot Week by Sunny.

For Ascot they took a small house where the Duke invited friends, and the chef complained (rightly, in Consuelo's view) that he was overworked. He retaliated by ordering quails at 5 shillings each and ortolans (buntings) – which cost even more – and serving them for breakfast, causing Consuelo to blush with shame at such nouveau-riche extravagance. The aspect of Ascot that troubled Consuelo most, however, was that it offered up yet another opportunity for 'blatant display' by her husband. 'The racecourse lay only fifty yards across the road from our house, but Marlborough had our coach-and-four sent on to Ascot simply so that he could drive onto the course. It was, moreover, a drive fraught with danger, since there was a sharp turn out of a narrow gate on to a main thoroughfare. A groom had to be sent ahead to hold up the traffic, and fresh horses over crowded roads provided a daily and unpleasant emotional experience.'[30]

As far as Consuelo was concerned, the activities of the season were interesting and novel, but a very great effort since everyone had to be classified and arranged in order. The rules of precedence meant that she frequently found herself sitting next to the same ancient noblemen on every occasion. On one visit to Althorp she sat between Earl Spencer and the Brazilian ambassador at every meal for four days. Fortunately, Earl Spencer was a most entertaining host, and kept her amused with a stream of anecdotes about the past. Eventually, her problems with arranging her own seating plans were solved when she found a 'Table of Precedence' with a number beside the name of each peer. 'I was glad to know my own number, for, after waiting at the door of the dining-room for the older women to pass through, I one day received a furious push from an irate Marchioness who loudly claimed that it was just as vulgar to hang back as to leave before one's turn.'[31]

It was felt that the best way to introduce Consuelo both to members of the family and the more important members of English society was at house parties at Blenheim during the London season. This was intimidating for such a young and inexperienced hostess, who would find herself arriving back at Blenheim at the end of the week, just ahead of twenty-five or thirty guests, to be greeted by the problems of feuding upper servants. Alva, Oliver and twelve-year-old Harold were guests at one party on 14 June 1896, the first record of a reunion between Consuelo and her mother since the departure of the Duchess for England the previous November.[32] Two weeks later, the list included the Londonderrys, Daisy, Princess of Pless, the Curzons, Henry Chaplin and Viscount and Viscountess Churchill. Guests had to be seated correctly, rooms allocated, menus approved. A duchess had to do much by hand herself in the way of writing cards and replying to invitations, for it was considered ill-bred to delegate correspondence to anyone else.

There were very few bathrooms at Blenheim (some reports put the total at one), and the six housemaids had to provide jugs of hot and cold water for over thirty baths during large house parties, while Consuelo shuddered at the 'toilet aids' on display on account of the lack of bathrooms. Meals ran to five or six courses – two soups, one hot, one cold, followed by a choice of hot or cold fish, then a sorbet, then a meat course (game in winter, quail or ortolans in summer), an elaborate desert, a hot savoury dish with the port and finally 'a succulent array of peaches, plums, apricots, nectarines, strawberries, raspberries, pears and peaches' grouped in large pyramids among bowls of huge pink malmaisons. Sundays were interminable. Games were forbidden, but Edwardian guests favoured promenades in the form of a tête-à-tête. Even this was competitive and ladies who did not have a cavalier were known to hide in their rooms. 'One never knew where one's duties as hostess would end,'[33] wrote Consuelo.

Consuelo soon realised if she were to succeed as chatelaine of Blenheim she would have to assert herself. Fortunately, Alva's contradictory maternal messages had not been without effect. Alva may not have allowed much independence in practice, but she approved of it in theory, and even encouraged it on an occasional basis at Idle Hour when she thought she could remain in control. An inner voice approving independent behaviour can only have helped Consuelo when it came to asserting herself against certain members of the Churchill family. She had been conscious on arrival in London of the enmity directed at her by one of Sunny's aunts, Lady Sarah Wilson, who does not emerge with credit from letters and diaries of the period. She was famous for her withering wit. In a letter to Winston Churchill, George Cornwallis-West described her tongue as 'becoming even more vicious than it used to be'.[34] In another letter, Pamela Lytton talked of her 'evil eye' shining 'brighter and harder'.[35] Consuelo's problem was that Lady Sarah had frequently acted as hostess for Sunny at Blenheim before his marriage and now resented being displaced by his new bride.

'At one of my first dinner parties,' wrote Consuelo, 'I found the ladies rising at a signal given by my husband's aunt, who was sitting next to him. Immediately aware of a concerted plan to establish her dominance, and warned by my neighbour Lord Chesterfield's exclamation "Never have I seen anything so rude; don't move!", I nevertheless went to the door and, meeting her, inquired in dulcet tones, "Are you ill, S?" "Ill?" she shrilled, "no certainly not, why should I be ill?" "There surely was no other excuse for your hasty exit," I said calmly. She had the grace to blush; the other women hid their smiles, and never again was I thus challenged!'[36]

Coming on top of other examples, such as the firm line with Lady Blandford about her American background, and her insistence on separating out the food that was given to the poor, it was becoming clear that Consuelo was not all that she seemed. She looked shy and graceful, and was only nineteen, but her demeanour was misleading. For now that Consuelo was finally out of Alva's orbit, the doll-child was giving way to the independent-minded daughter.

Consuelo's social success was sealed by royal approval. During the summer of 1896 an invitation from Queen Victoria to 'dine and sleep' at Windsor Castle arrived at twenty-four-hours' notice, as was often the case. The 'dine and sleep' invitation was in the nature of a royal command and was another ordeal for Consuelo, who would be formally presented to the Queen for the first time. On arrival at Windsor the Marlboroughs were greeted by one of the Duke's great-aunts, Lady Edward Churchill, who told Consuelo exactly what she would have to do. There would be a small number of guests. Consuelo must only speak when spoken to by the Queen, and should confine her remarks to answers to Her Majesty's questions, since only the Queen could initiate a subject. When she was presented she should kiss the Queen's hand. The Queen would then observe the correct protocol for a peeress by imprinting a kiss on Consuelo's brow.

No-one warned Consuelo that Queen Victoria was tiny, however. 'I almost had to kneel to touch her outstretched hand with my lips. My balance was precariously held as I curtsied low to receive her kiss upon my forehead, and a diamond crescent in my hair caused me anxiety lest I scratch out a royal eye.' Apart from the obvious affection demonstrated by the Queen towards her prime minister, Lord Salisbury, dinner was 'a most depressing function' where conversation was conducted entirely in whispers out of deference to Her Majesty. After dinner they were led to a small, cramped corridor where the guests waited before being led one by one to be introduced to the Queen, who addressed a few words to each of them in turn. 'I found it most embarrassing to stand in front of her while everyone listened to her kind inquiries about my reactions to my adopted country, which I answered as best I could. I was, moreover, haunted by the fear that I might not notice the little nod with which it was her habit to end an audience, having heard of an unfortunate person who, not knowing the protocol, had remained glued to the spot until ignominiously removed by a lord-in-waiting.'[37] Although Consuelo found the experience discomfiting and Windsor dismal and gloomy, the visit appears to have been regarded as a success by the Queen, for the Marlboroughs were invited back again the following summer.

The next royal communication was thoroughly daunting. On 2 August 1896 – less than six months after Consuelo arrived at Blenheim – Lord Knollys wrote to the Duke from the Royal Yacht at Cowes that the Prince and Princess of Wales wished to come and stay at Blenheim. They proposed arriving on Monday 23 November and remaining there until the following Saturday, bringing members of the royal family with them. The Prince of Wales, who did not bear grudges from one generation to the next, had already put his quarrel with the 8th Duke behind him and had invited

Sunny to Sandringham in the summer of 1894. A visit by their royal highnesses to Blenheim had been anticipated in the local press for some time, but as far as both the Marlboroughs were concerned the setting of a precise date now meant weeks of apprehensive planning. First, the proposed guest list had to be submitted and approved. Then work started on plans to make the visit as memorable as possible. It was not simply Consuelo's first royal visit – it was also her first big shooting party.

On top of everything else, the first phase of refurbishment at Blenheim had to be completed in time for the Prince's arrival, which included restocking the library and rehanging and adding to Blenheim's collection of pictures and tapestries. The Duke had also been much impressed with the opulent version of French classicism that he had observed in the great Newport mansions with their gilded *boiseries*, which he was eager to emulate. 'By the end of the century this very nouveau-riche mode had emerged as a style fit for courtiers and cosmopolites, from the D'Abernons at Esher Place to the Marlboroughs at Blenheim; for anyone indeed who aspired to the magic circle of the Prince of Wales,'[38] writes J. Mordaunt Crook. According to Paul Miller, curator of the Preservation Society of Newport County, fabric by Prelle of Lyon used at Marble House can be found covering one of the Blenheim bergères; and the *boiseries* installed at Blenheim after 1896 are so similar to those in Newport that they may even have been done by Jules Allard himself, particularly those holding the Carolus-Duran portrait of Consuelo which came from Marble House to Blenheim after the wedding. The Duke would later regret what he regarded as a lapse in taste saying that he was 'young and uninformed' when he put French decoration into state rooms with English proportions. 'The result is that the French decoration is quite out of scale and leaves a very unpleasant impression on those who possess trained eyes,'[39] he wrote. At the time, however, both he and Consuelo believed they were providing a magnificent new setting in the style of Versailles, closely in line with royal tastes and their own.

The sartorial question came next. The Prince of Wales was a stickler for protocol and adored dressing up. He thought that aristocratic women should be beautifully dressed, and was disconcertingly observant. 'To the remarkable Lady Salisbury, who had a mind above such things, he one day said reprovingly: "Lady Salisbury, I think I have seen that dress before." "Yes, and you'll see it again," replied that lady, undaunted but most improperly,' wrote A. L. Rowse.[40] Lady Salisbury was one of the very few women in England who could get away with such a reply however. Consuelo knew that protocol demanded four changes of clothes on each day of the royal visit, and had to buy sixteen new dresses in preparation for his stay, at vast expense (for some royal visits, twenty-six changes of clothes were required).

The Prince of Wales was equally concerned about what he would wear himself. A county ball was proposed as part of the entertainments, triggering a letter from Lord Knollys asking whether it would be in uniform 'as if Yeomanry Officers do come in uniform [the Prince of Wales] would propose wearing the uniform of the Gloucestershire Hussars of which regt he has recently become Honorary Colonel'.[41] On being told that uniforms would be worn, Lord Knollys wrote again a few days later saying that the Prince thought that though officers in the army, militias and yeomanry might come in uniform, he thought volunteer officers need not wear it, and reminded the Marlboroughs that civilians should wear ordinary evening dress, as 'knee breeches are only worn in country houses where the sovereign is present'.[42] In the event, the ball was cancelled. Consuelo's grandmother, Maria Kissam Vanderbilt, died in early November, shortly before the royal visit, and as a concession to mourning it was decided that the ball should be replaced by a concert. 'I expect the American press will be very nasty – but as they always are it does not matter,' wrote Lady Randolph Churchill to Winston in India on 13 November.[43]

On the first day of the royal visit, 23 November, the citizens of Woodstock demonstrated their loyalty (for the second time in a year) with illuminated triumphal arches in the market square, decorations at the station, and welcome banners in the town. The *Fair Rosamund* was garlanded once again. The royal party and the guests who travelled with them ('rather cross most of us', according to Arthur Balfour[44]) were greeted by the Duke at Woodstock station, where the Blenheim fire brigade formed a guard of honour and church bells rang out. *Jackson's Oxford Journal* noted that the crowds were smaller than at the homecoming of the Duke and Duchess, and its reporter thought there was less cheering too. Fortunately 'the waving of handkerchiefs from the windows, which were filled with spectators, and the respectful demeanour of the crowd made up in some measure for this omission' and the Prince was not in a position to make comparisons. In fact he was delighted by what he saw, especially an illumination in the form of a 'gas-lit device of the Prince of Wales's plumes, with the letters A.E. in variegated lamps', and he was heard to remark that the town was very beautifully decorated.[45]

There were over a hundred people in the house while the shooting party lasted. Apart from the Prince and Princess of Wales, it included their daughters Princess Victoria (whom Consuelo came to like very much), Princess Maud and her husband Prince Charles of Denmark; Mr and Mrs George Curzon; Arthur Balfour; the Londonderrys; the Duke's sister, Lilian; the Earl of Chesterfield and Lady Randolph Churchill. The Marlboroughs made over their rooms on the ground floor to the royal party and slept upstairs.

The women spent most of their time dawdling, chatting and changing. 'To begin with, even breakfast, which was served at 9.30 in the dining room, demanded an elegant costume of velvet or silk. Having seen the men off to their sport, the ladies spent the morning round the fire reading the papers and gossiping. We next changed into tweeds to join the guns for luncheon, which was served in the High Lodge or in a tent. Afterwards we usually accompanied the guns and watched a drive or two before returning home. An elaborate tea gown was donned for tea, after which we played cards or listened to a Viennese band or to the organ until time to dress for dinner.' The Marlboroughs had imported Herr Gottlieb's Viennese orchestra, which played at teatime while the ladies played cards. The musical programme for the week was designed by the distinguished Blenheim organist, Mr Perkins, who gave recitals to the guests. After tea the ladies changed again into evening dresses and 'a great display of jewels'.[46] In retrospect it all struck Consuelo as a tremendous waste of time.

The men were rather more active for they had come to shoot (the only exception being Arthur Balfour who detested both shooting and dawdling, and explored the park on a bicycle instead.) On Tuesday morning the Duke escorted his guests to part of the estate known as High Park where eight guns shot over two thousand rabbits. On Wednesday the party went to North Leigh where eighty beaters were on hand in light brown Holland smocks and red caps to assist with the bagging of over a thousand birds. By the end of the week it was reported that eight guns had bagged an average of two thousand a head, though this did not come near the record set in Consuelo's time of around seven thousand rabbits bagged by five guns in one day. In every one of the outlying villages visited by the shooting party there was bunting, arches of evergreen and peals of bells when the ladies joined the guns for lunch. On Thursday morning Consuelo took the Princess of Wales and other ladies of the party to Oxford where they toured the Bodleian Library and Magdalen College, before going to the deanery at Christ Church for lunch. When they returned to Blenheim the royal party planted commemorative trees with spades, which, like the table linen, had been specially commissioned for the occasion.

On Thursday evening there were public celebrations. Extra trains brought thousands of people to Woodstock to watch the fireworks. The arrival of a procession of thirty cyclists carrying Chinese lanterns in front of the palace triggered the lighting of a huge bonfire up at the Monument, so large it lit the countryside for miles around. While the royal party watched from the terrace along the north front of the palace, the cyclists took up positions round the great courtyard dressed as Robin Hood, Buffalo Bill and 'Country Milkman'. The waiting crowd was then treated to the 'largest pyrotechnic display ever produced outside the precincts of Crystal Palace'. Aerial maroons, bombshells filled with stars, rockets fizzing in tens, fifties, hundreds and thousands, Roman candles, electric spray, tourbillions and diamond dust lit the night sky 'some of the effects exceeding in brilliance the most powerful electric light'.[47]

Afterwards a procession of 800 torchbearers carrying Bengal lights of red, white and green snaked its way down from the Monument, across the bridge and up to the north front of the palace, an effect 'pretty and fantastic beyond description'.[48] On their arrival they were given refreshments and Woodstock rang with sounds of merriment until the small hours. It is not impossible, as Marian Fowler hazards, that there were sounds of tiptoeing and creaking floorboards in aristocratic bedrooms too.[49]

On the last night of the Prince's stay there was a reception followed by a concert to replace the county ball. There were over five hundred guests including masters and presidents of the Oxford colleges, as well as local gentry and aristocracy. The Duchess was said to look particularly beautiful in a costume of white velvet trimmed with sable, the front of the bodice embroidered with magnificent black lace. 'A diamond belt encircled her waist, while in her hair was a tiara, the jewels of which, together with those of the necklace she wore, sparkled and glittered under the glow of the incandescent lamps.'[50] The reception rooms were filled with malmaisons and orchids, in 'rare china and golden vases'.[51] The concert began at 11.30 p.m. At the special request of the Prince of Wales it included glee singing from lay clerks of the Oxford colleges.

As discussed with Lord Knollys, many of the gentlemen wore military uniform that evening – scarlet tunics, gold epaulettes, bright sashes and swords, stalwart figures arrayed in the uniform of the Hussars and Highlanders and Dragoons and Royal Artillery. Some came in hunting costume, 'scarlet and white-faced coats predominating'. The dinner beforehand was an unforgettable sight for Gerald Horne, who was allowed to peep down from the balcony. 'There it was, all gleaming with wealth. I think the first thing that struck me was the flashing headgear of the ladies. The Blue Hungarian was playing and there was the Prince himself looking really royal and magnificent in military uniform. The table was laid of course with the silver gilt service, the old silver duke busy writing as usual in the very middle of it all and the royal footmen waiting side by side with our own.'[52] Even *Town Topics* commented on the scene, reporting in a 'circumstantial' account that the Prince of

Wales was so stout that he made the Duke look even more of a slip than usual. It reserved its praises for Consuelo, who, it said, looked beautiful but pale. 'Everyone was saying how charmingly she bore herself. Amid many praises, the quietly spoken encomium of the Princess of Wales did her, perhaps, the greatest honour. "She is a sweet girl, I like her," said our rather undemonstrative Princess; and the more vivacious Princess Maud chimed in with "She's a dear." '[53]

To the great relief of their nineteen-year-old hostess, the royal party departed on Saturday 28 November, leaving by carriage for Oxford station. The only hitch of the visit occurred in Oxford when one of the Duke's horses took fright at the cheering crowds and slipped at Carfax, causing Prince and Princess Charles to enter another carriage, and slightly injuring the postilion. The incident created a considerable sensation but Oxford's Town Council was held to be responsible for its ill-surfaced roads rather than the Duke of Marlborough. (His postilion was wholly exonerated by the royal party though they were much more concerned about the horse.[54])

The whole week had been a ferocious ordeal for Consuelo. On top of constant anxiety about arrangements she was obliged by protocol to have the Prince of Wales as her neighbour at dinner every evening, and she worried about boring him 'since he liked to discuss the news and to hear the latest scandal, with all of which at that age I was unfamiliar'.[55] The Princess of Wales was a great deal easier. In spite of everything, however, the royal visit impressed on Consuelo that 'the Crown stood for a tradition that England would not easily give up'.[56] The success of the visit was a great achievement for a nineteen-year-old American newcomer obliged to entertain distinguished guests who were not only royal, but twice her age. Lady Randolph Churchill wrote to Winston in India that everything had been 'wonderfully done' and that Sunny and Consuelo had been 'quite at their best'.[57] 'The whole business appears to have been very satisfactory – and bound to do Marlborough a great deal of good,' Winston wrote back on 23 December from the Continental Hotel, Calcutta where he had been scanning newspaper reports of the visit. 'There is always a great name to be made by the judicious application of wealth – and he is just the person to do it.'[58]

Hard on the heels of the visit of the Prince and Princess of Wales came Consuelo's first Christmas at Blenheim. The party included the Dowager Duchess Frances, and Consuelo's least favourite aunt by marriage, Lady Sarah Wilson. In the best tradition of family Christmases, the Dowager Duchess proved 'somewhat of a trial' by constant carping. It was particularly irritating when Consuelo was trying to do her duty in other ways, 'seeing to it personally that everyone in the villages belonging to Blenheim had a blanket or a pig or a ton of coal or whatever they wanted', according to Gerald Horne.[59] But while he may have remembered her generosity to everyone on the estate, it cut little ice with the Dowager Duchess who characterised Yuletide generosity as extravagance and thought that Consuelo's behaviour was insufficiently duchess-like. The criticisms still rankled over fifty years later. 'Trailing her satins and sables in a stately manner, she would cast a hostile eye upon my youthful figure more suitably attired in tweeds, and I would hear her complaining to my sisters-in-law that "Her Grace does not realise the importance of her position." She did not perhaps realise that a little relaxation was necessary after my lengthy conversations with her, which were rendered difficult by being conducted through her ear-trumpet.'[60]

Consuelo was not alone in finding the Dowager Duchess extremely difficult. On a freezing cold Christmas Eve another house guest, Lady Randolph Churchill, wrote to Winston in India that the Dowager Duchess was 'not making herself pleasant to me and we have not exchanged a word – but I do not mind – & perhaps it is as well. To the world we can appear friends – anything of the kind in private is impossible.' Otherwise the party was pleasant enough and Sunny and Consuelo were 'charming in their own house'.[61]

From the family's point of view, however, Consuelo's greatest achievement was to become pregnant that Christmas, though the enforced lack of activity which followed meant that she had far too much time to contemplate the gloom of her English surroundings. Early in 1897, she went with Sunny to Sysonby Lodge near Melton Mowbray for the hunting. 'Whenever there was a frost Marlborough went off to London or to Paris, but since it was considered inadvisable for me to travel in my condition I remained alone. From my window I overlooked a pond in which a former butler had drowned himself. As one gloomy day succeeded another I began to feel a deep sympathy with him.'[62]

Being pregnant meant that she could not ride with the hunt, but she was sometimes driven to the kill by the Master of the Quorn, Lord Lonsdale, in a buggy. She found this manner of hunting fun but uncomfortable, and possibly just as nausea-inducing. Consuelo decided to discontinue these excursions until she could hunt in person and amused herself by hiring a teacher from London with whom she could read German philosophy. Such behaviour was regarded with great suspicion by the hunting set of Melton Mowbray, who then consigned her to the category of bluestocking. Observing the reaction of this circle, it appeared to Consuelo that Alva had greatly over-educated her for the job of English duchess. 'I realised that I had shown more courage than tact in advertising my preference for literature. Only this interest, however, got me through the first depressing winter, when my solitary days were spent walking along the high road and my evenings listening to the hunting exploits of others.'[63]

Advancing pregnancy meant that Consuelo's involvement in the great season of 1897, the year of Queen Victoria's Diamond Jubilee, was less frenetic than it might otherwise have been, though she appeared tightly laced at the Duchess of Devonshire's ball at Grosvenor House – a ball as symbolic of English aristocratic ascendancy as Alva's Vanderbilt ball had been of plutocratic change in 1883. The Duke went to Jean Worth in Paris to have his costume made. Worth acceded after some protestations and 'got to work on a Louis XV costume of straw-coloured velvet, embroidered in silver, pearls and diamonds . . . Every pearl and diamond was sewn on by hand, and it took several girls almost a month to complete this embroidery of jewels . . . When I came to make out his bill, I was almost afraid to begin it. But at last when I got it totalled it came to 5,000 francs.'[64] Once again, however, *Town Topics'* London spies reserved their praise for Consuelo. 'The young Duchess of Marlborough is orienting her way through the maze of English society in a manner that commands admiration and that is astonishing in so young a girl. Everyone has a good word to say for her absence of "side", her quiet dignity, her simplicity, and her talent for saying the right thing.'[65]

Alva came to England in September, staying with Consuelo at Spencer House by Green Park, which had been taken by the Marlboroughs for the confinement. A bedroom was created for Consuelo in a corner drawing room, where she sometimes felt a sudden cold draught 'as if a presence had glided through the room'. Alva claimed to have seen a ghost, which could hardly have made for a calm atmosphere since, unexpectedly, she had a neurotic fear of the supernatural. (Alva's enthusiasm for building new houses was partly ascribed by Elizabeth Lehr to her fear of living in a house where anyone had died.) Despite Alva regarding the obstetrician who attended Consuelo as unutterably inept, the Duchess gave birth to a healthy boy, who was named John Albert Edward William but was known as Blandford, in line with family tradition. For almost a week after the birth, Consuelo slipped in and out of consciousness, giving considerable cause for alarm; but she soon recovered, and found herself basking in universal approval. Even the Dowager Duchess was pleased for she lived to see the 'little upstart Winston' officially displaced. The news was broken to him in a most casual fashion in a letter from his mother on 21 September 1897, in which she remarked en passant: 'Duchess Rose – Consuelo Marlborough – has had a son – If you write to Sunny you might congratulate him – He is very fond of you.'[66] If this gave Winston Churchill a pang of disappointment at the time (and there is no sign that it did) it freed him for a great political career, and it has been pointed out that as a nation, Britain ought to be grateful to Consuelo for this if nothing else.

Back at Blenheim the news that there was another 'link in a chain' resulted in a celebration remarkable even then for its feudalism. 'You can imagine what a day it was then when we heard that an heir – Lord Blandford – had been born in London,' said Gerald Horne. 'The steward and his staff at once climbed to the palace roof and fired a salute; and at night a ball was given for the servants and the people of Woodstock and the rest. The menservants wore dress clothes with special buttonholes. I was in my morning suit (but I managed the buttonhole all right; I doubt if any man's was larger) and danced with the maids, who looked very nice and graceful in their long black dresses . . . We danced to the organ as well as to a string band. All the elaborate refreshments were prepared in the palace kitchens and then passed from hand to hand by a row of waiters reaching from the kitchen to dining-room. Free beer, free everything flowed like milk and honey.'[67]

The weeks and months after the birth of Blandford in 1897 were the high-water mark of the Marlborough marriage. Sunny had now assured both the future of the Marlborough dynasty and put in hand a new Golden Age for Blenheim. He had brought home a beautiful duchess who had charmed society and was idolised by the people of Woodstock and the Blenheim estate. Consuelo would have seemed happy too, for even she admitted that the happiness the baby brought her 'lightened the gloom that overhung our palatial home'.[68] She may have shouldered much of the work of visiting the poor, but during this period there were numerous glimpses in the local press of Sunny rushing about on ducal visits, opening technical schools, presenting prizes to boys in Burford, and even showing duke-struck lady journalists round the improvements at Blenheim for an article in *Woman and Home*.

After the uneasy years of the 'Wicked Duke' and the impulsive Lord Randolph Churchill, the Marlboroughs basked in the sunshine of royal approval. In another demonstration of esteem, the Prince of Wales offered to be the baby's godfather (hence the name of 'Albert', which Consuelo writes that they 'vainly tried to eschew'.[69]) He smiled with 'gracious urbanity' as the christening took place in the Chapel Royal, St James's, where the 'sun streamed through the oriel window, touched the gold vessels on the altar, the white lilies round the font and the scarlet tunics worn by the royal choristers'.[70]

William K. Vanderbilt, godparent as well as a grandfather, must have reflected happily on the distance travelled from the farm on Staten Island as he stood beside the Prince of Wales at the font. On a rather different note everyone tried not to laugh at the consternation of Lady Blandford's sister, the Duchess of Buccleuch, as she puzzled over the identity of a mystery lady – (Mrs Ryman, the housekeeper) who had been ushered (accidentally) into the same pew, it being inconceivable to the Duchess that anyone of lesser rank should be seated beside her.

There was more joy when another son, Ivor, was born in October 1898. Consuelo was exceedingly fortunate in producing an 'heir and a spare' so quickly (a phrase she is said to have coined) for she would otherwise been obliged to continue becoming pregnant until she succeeded. Instead, she was given credit for giving birth to two sons in such a professional manner. This time, the Marlboroughs took Hampden House from the Duke of Abercorn for the confinement, and as Consuelo lay recovering, Lady Blandford swept in and remarked: 'You are a little brick! American women seem to have boys more easily than we do!' The only tussle took place in the class-conscious nursery where the head nurse argued forcefully that she should be allowed to continue looking after the eldest, rather than the new baby, because he was a marquess and Lord Ivor was not.

Although Consuelo often wrote of 'gloom' in relation to life at Blenheim, her account of life there belies the assertion that it was always wreathed in miserablism. Each year, the Oxfordshire Hussars spent three weeks under canvas training in High Park and there was often fun to be had: 'I remember an exciting paper-chase which I won on a bay mare, thundering over the stone bridge up to the house in a dead heat with the adjutant.'[71] She was a good horsewoman, and loved her daily gallop across the park in the company of Mr Angas, the estate steward. 'Those were wonderful days,' wrote George Cornwallis-West. 'Taxation and the cost of living were low; money was freely spent and wealth was everywhere in evidence.'[72] Consuelo's success – as a society beauty, a hostess, and a mother of sons – vastly increased her confidence, and Vanderbilt cousins who stayed at Blenheim thought that there was much about being a duchess that Consuelo enjoyed. When visitors arrived they would be received in the Italian garden where tea tables had been laid, and they then strolled through the gardens and park until it was time to dress for dinner. 'We had an Indian tent set up under the cedars on the lawn, where I used to sit with our guests. We always brought out *The Times* and the *Morning Post* and a book or two, but the papers were soon discarded for conversation,' she recalled. 'Sometimes we played tennis or rowed on the lake, and in the afternoon the household played cricket on the lawn. The tea table was set under the trees. It was a lovely sight, with masses of luscious apricots and peaches to adorn it. There were also pyramids of strawberries and raspberries; bowls brimful of Devonshire cream; pitchers of iced coffee; scones to be eaten with various jams, and cakes with sugared icing. No-one dieted in those days.'[73]

There were plenty of interesting visitors, some more agreeable than others. Some of the least interesting arrived as 'glorified tourists'. One such was the German Emperor, brought down for the day by his uncle, the Prince of Wales. 'His conversation was self-centered, which is usual with kings' and his desire to impress was quite comic. He was delighted to see the Imperial standard on the Blenheim flagpole and by the hastily arranged recital of German music given on the organ by Mr Perkins. Consuelo's two little boys were brought downstairs by their nursemaids, 'just the sort of occasion Nanny enjoyed'.[74]

A much more congenial group of visitors was the political circle that gathered round Sunny's cousin Winston Churchill. In the absence of other forms of entertainment, there was constant conversation with Winston – the greatest talker of them all. Consuelo may have complained in a letter to a friend that his talk was tiring but he also 'represented the democratic spirit so foreign to my environment, and which I deeply missed. Winston was even then, in his early twenties, tremendously self-centred and had a dynamic energy.' He also possessed, it seemed to her, a near-photographic memory. No sooner had he read Taine's *History of English Literature* – at Consuelo's suggestion – than he was able to recite from memory pages he had barely scanned, or so it seemed. But even without Winston, the conversation never dried. 'We talked morning, noon and night, but we also knew how to listen. There was so much to be discussed. Politics were interesting, but so also were the latest novels of Henry James and of Edith Wharton – Americans who had the temerity to write of England and the English.'[75]

With Blenheim's recent misfortunes reversed, Sunny finally felt able to think about building his own political career. He was, of course, a Tory (Conservative Unionist). He had been made a Privy Councillor in 1894, and had made his maiden speech in the House of Lords in August 1895, shortly before he left for Marble House. He created a good impression in the view of Lord Lansdowne, who read his speech beforehand. 'You got through capitally – that was the general verdict,' he told his nephew by marriage. 'If you had done much better, you have done too well for a beginner.' He had some criticisms, but 'with these exceptions the speech was quite a sweep.'[76] The Duke now sought a role in government and in January 1899, soon after Ivor's birth, he was invited to become Paymaster General by Lord Salisbury. His cousin Winston wrote to congratulate him remarking: 'I am afraid I am vy ignorant of its duties and powers and their scope. But in any case it is a position of great dignity and will be a prelude to positions of greater responsibility.'[77] Sunny held the position, which nominally gave him responsibility for the government's cash reserves, until 1902. However, his responsibilities were not so onerous that they prevented him from going out to South Africa on Lord Robert's staff soon after the start of the Boer War in 1900.

Daisy, Princess of Pless, who knew the 9th Duke well, astutely remarked that he was too thin-skinned for politics.[78] His political career would be eventually cut short by the Liberal landslide of 1906. In the meantime Consuelo offered loyal support, at least in public. Though the political outlook of Duke and Duchess later diverged sharply, she can be seen in photographs on the platform at the great Unionist rally held by the Duke at Blenheim in 1901, for instance, where thousands of supporters gathered in the courtyard to hear speeches from the Duke, Arthur Balfour, and Joseph Chamberlain, leader of the Liberal Unionists. As the Duke pointed out later, rallies in the days before mass ownership of motor cars were difficult to organise, and it was no small achievement to gather crowds of thousands in the courtyard, who then had to be addressed without the benefit of microphones. Lunch was provided for 100 MPs in the great hall, while 3,000 delegates ate in tents.

When Consuelo, Balfour and Chamberlain entered the tents they were greeted with loud cheers, though it was Chamberlain who received the greater ovation and who seemed not to mind the rough and enthusiastic greeting he received, in spite of his dapper appearance. From the platform Consuelo could see in the distance the statue of the 1st Duke of Marlborough, standing aloft on his 'Victory Column': '[As] we listened to my husband and Mr Balfour's addresses I could almost detect a satisfied smile on John Duke's countenance. It was somewhat different when Mr Chamberlain spoke of social measures that in the distant future would still leave the Duke on his column but might drive his heirs from the palace.'[79]

Sunny's developing political career also made it imperative that the Marlboroughs finally resolved the question of a permanent establishment in London. Each year, finding a house to rent became more of an irritant and an apparent barrier to political progress. 'I only had to mention our wish for my father to promise its fulfilment,'[80] wrote Consuelo. The problem was that most of Mayfair and Belgravia was already owned by the great landlords like the Duke of Westminster or Lords Portman and Cadogan. The Marlboroughs finally found a solution by buying the freehold of a site in Curzon Street by Shepherd's Market and building their own establishment, which they would eventually call Sunderland House – though they had first to demolish a chapel on part of the site, which caused some controversy and was also considered unlucky.

* * *

'The last quarter of the nineteenth century and the years before the First World War witnessed a remarkable flowering of ceremonial and spectacle . . . the aim of those who stage-managed them was to create feelings of security, cohesion and identity, in an era of anxiety, uncertainty and social dislocation,' writes David Cannadine.[81] Consuelo's position as Duchess of Marlborough gave her a ringside seat at many of these great Edwardian pageants. Queen Victoria's funeral at St George's Chapel, Windsor, in 1901, was particularly enjoyable. The service itself was magnificent. The stalls of the Knights of the Garter were occupied by the German Emperor and a dazzling array of kings, queens, ambassadors extraordinary, Indian princes, Colonial dignitaries, generals, admirals and courtiers. Consuelo wore the prescribed deep black mourning and crepe veil, which rather suited her, and it had the effect of extracting what she describes as a 'rare compliment' from her husband who remarked: 'If I die, I see you will not remain a widow long' – a conceit which suggests that he was more of his father's son than he cared to acknowledge.

Consuelo later reflected that the funeral of Queen Victoria was a moment when it truly appeared that no other country in the world had an aristocracy so magnificent, nor a civil service so dedicated, which is precisely what was intended. The great doors were flung open as the royal cortege mounted the steps, a boom of distant guns and clanging swords the only sound other than the funeral march, until Margot Asquith broke the reverential silence with a quip. Consuelo thoroughly enjoyed herself at the reception in the Waterloo Chamber afterwards too. Her beauty bloomed in her twenties after the birth of her two children, and so did a streak of vanity which surfaced in her memoirs years later. 'I found myself so sought after by many public men – Arthur Balfour, George Wyndham, a man of great personal charm who was then Secretary of State for Ireland, St John Broderick, George Curzon, Mr Asquith and others – that I felt that Marlborough's compliment had perhaps been deserved.'[82]

Between the death of Queen Victoria and the Coronation of Edward VII, the Marlboroughs travelled to Russia at the invitation of Tsar Nicholas II to take part in the court celebrations that traditionally ushered in the Russian Orthodox new year. The Duke, who 'had a weakness for pageantry' (by which Consuelo meant dressing up, showing off and swaggering about) flew into something of a panic about the extent to which the Marlboroughs would stand comparison with the celebrated magnificence of the Russian court: 'Every detail had to be subjected to his exacting scrutiny,' she wrote. 'Court uniforms had to be refurbished, and in Paris I bought some lovely dresses. A diamond and turquoise dog-collar was ordered as a special parure to be worn with a blue satin gown. We had heard much about the fabulous furs of Russian nobles; it is true my sable coat was fine, but I had only one.'[83]

An entourage of magnificent persons was assembled to go with the clothes and jewels. To 'ensure an added prestige', the beautiful and cultured young Duchess of Sutherland was invited to accompany them, as well as Count Albert Mensdorff, then attached to the Austro-Hungarian embassy in London, and another friend, Henry Milner. The rest of the retinue was made up by valets, maids, and a security officer to safeguard the jewels. In letters home to her husband, Millicent Sutherland wrote that the Marlboroughs were good, if formal, travelling companions. The Duke always wore his top hat, and Consuelo constantly changed dresses – all lovely – even on the train.

Consuelo's first impressions of Russia in 1902 were far from favourable. The white plains in the moonlight were depressing; the hotel in St Petersburg reserved for visiting foreign dignitaries did not compare to the great hotels of Paris; the windows were sealed shut in winter making her feel imprisoned; the wind-swept avenues were lined with modern buildings in doubtful taste. But gradually, the hospitality of the great aristocratic Russian families won her over. They were entertained by Countess Olga Shuvalova in her palace with its own private theatre; they drove in open sleighs to islands on the frozen Neva where they danced to Tzigane music into the early hours of morning; they went to the ballet, where, according to tradition the *danseuse-en-tête* was the Tsar's ex-mistress, while the other ballerinas were distributed round the Grand Dukes 'as part of their amatory education'. Consuelo felt that they had taken a step backwards into an eighteenth-century society far less rigid than turn-of-the-century England.

The grandest occasions in St Petersburg were the great court functions and both duchesses dressed for the huge New Year ball in the Winter Palace with some apprehension. The stairs of the Winter Palace were guarded by Cossacks, with hundreds of footmen in scarlet liveries. 'I have never in my life seen so brilliant a sight,' Millie Sutherland wrote home. 'The light, the uniforms, the enormous rooms, the crowd, the music, making a spectacle that was almost barbaric in splendour . . . They seat at supper nearly four thousand people!'[84] When Consuelo was asked to dance a mazurka by the Tsar's brother, Grand Duke Michael, it bore no resemblance at all to the version taught at Mr Dodworth's classes for the children of the *ton* of New York. '"Never mind," he said when I demurred, "I'll do the steps."'[85] (Grand Duke Michael would be shot by the Bolsheviks in 1918.)

At a second more exclusive ball, the Bal des Palmiers, both duchesses felt the contrast between the wealth indoors and the poverty outside acutely. Inside the palace, the jewellery worn by aristocratic Russian women made even a Vanderbilt heiress feel inadequate. Outside, it was difficult to avoid the sight of long queues of hungry people, and beggars freezing in the streets, 'all the want and penury of the peasants and this strange show to keep up the prestige of the autocracy of one gentle, quiet little man',[86] wrote Millie Sutherland in amazement. Both Marlboroughs thereon teased her gently by dubbing her 'D.D.', for 'Democratic Duchess'.

At the Bal des Palmiers, Consuelo sat beside the 'gentle, quiet little man' at dinner (though she only set down her conversation with Tsar Nicholas much later when she wrote her memoirs). It was protracted, comprising 'soups, caviar and monster sturgeons, meats and game, pâtés and primeurs, ices and fruits, all mounted on gold and silver plate fashioned by Germain'. She noted the Tsar's close resemblance to his cousin, the Prince of Wales (later George V); she was struck by his youth; and she was impressed by the enormous difficulties he faced. 'When I asked him why he hesitated to give Russia the democratic government that was so successful in England, he answered gravely, "There is nothing I would like better, but Russia is not ready for democratic government. We are two hundred years behind Europe in the development of our national political institutions. Russia is still more Asiatic than European and must therefore be governed by an autocratic government."' Consuelo gathered that he saw his ministers every day but separately, to prevent any one of them becoming too powerful. He also seemed afraid of the Russian people, their 'ignorance, their superstition, their fatalism'.[87]

Both Consuelo and Millicent Sutherland thought that the Tsar was a good man, but weak. After the Russian Revolution, Consuelo also thought his analysis of why Russia was not ready for democracy was fundamentally correct, and that the person who was more to blame for the end of tsarism was the Tsarina, Alexandra Feodorovna (Queen Victoria's great-granddaughter), who rudely failed to grant an audience to the Marlborough party. The conversation also took a slightly more alarming turn when the Tsar asked Consuelo why Millie Sutherland had visited the writer Maxim Gorky when he was in temporary internal exile. The English party may have been honoured guests, but that did not protect them from being watched by the secret police.

Both Marlboroughs were good at ceremony. As the years went by, they came to find the Prince of Wales's Marlborough House set dull, and slowly withdrew. In spite of this, Consuelo was asked to act as a canopy bearer for Queen Alexandra at the coronation of Edward VII, a great honour which set the final seal of approval on her performance as an English duchess. Indeed, the photograph of Consuelo in her velvet robes trimmed with miniver, her dog-collar of pearls and diamond belt has come to stand as a symbol of the great international marriages of the Gilded Age. The Duke had also been invited to act as Lord High Steward, which meant that he would have the critical task of carrying King Edward's crown into Westminster Abbey.

The coronation, which had been postponed once when the Prince of Wales was struck down with appendicitis, was held in a shortened form on 9 August 1902. The Duke and Duchess both arrived as instructed at 9.30 a.m. at the west door, having travelled to the Abbey in the Marlborough state coach and been cheered through the streets of London by crowds who evidently mistook their crimson livery for royal scarlet. They separated at the west door, since the Duke had duties elsewhere, leaving Consuelo to process up the aisle alone, in a solitary but splendid moment. There was a very long wait, though Consuelo's hunger pangs were relieved by the bar of chocolate she had hidden in a pocket, and much amusement was provided by the rows of peers who sat opposite the peeresses, many of whom had not bothered to check beforehand whether their ancestral robes and coronets actually fitted or simply did not care. In some cases the robes were too big or too long, and there was much uneasy male hitching of hemlines; and when the solemn moment came for the King to be crowned and the peers placed their coronets on their heads, at least one slid downwards till it rested on an aristocratic chin.

The image of Consuelo in her coronation robes may have come to stand for a generation of American heiresses married to English aristocrats, but during the coronation itself she found herself overwhelmed by the pageantry to the point where all sense of her American self left her. 'The trumpets were blaring, the organ pealing and the choir singing the triumphant hosannas that greeted the King and Queen. The long procession was in sight – the court officials with their white wands, the Church dignitaries with their magnificent vestments, the bearers of the royal insignia . . . [which included the Duke bearing the King's crown on a velvet cushion], the lovely Queen, her maids-of-honour holding her train, and then the King, recovered, solemn and regal. I felt a lump in my throat and realised that I was more British than I knew.'[88]

But the feeling was short lived. Two events pierced the solemnity of the occasion. The first was the anguished look Queen Alexandra shot at Consuelo when the Archbishop of Canterbury anointed the Queen with sacred oil. Some of it ran down her nose and there was nothing either of them could do about it. The second was the great shout of acclamation from the assembly at the moment when the King was crowned which seemed downright shocking. It may have been an ancient rite, binding crown and aristocracy, but at that moment divine majesty appeared to have given way to something more earthly. 'But then I was not English,' she wrote later, 'and could not feel the same pride in the tradition of unbroken lineage the act of crowning symbolised.'[89]

At the end of 1902, the Marlboroughs sailed to India to attend the greatest Edwardian pageant of them all: a durbar in Delhi organised by George Curzon, now Viceroy of India, to mark the accession of Edward VII. They travelled out on a P & O liner with sixty other guests, all friends of the Curzons, for a ceremony which came to epitomise imperial pomp. The first durbar was held in 1877 by Viceroy Lord Lytton to mark Queen Victoria's new title as Empress of India. Curzon, who was convinced that the 'oriental mind' was susceptible to 'barbaric splendour', was determined that his ceremony would surpass Lytton's in every way. His meticulous planning, insistence on controlling every detail and profound sense of the past made him by far the greatest of all Edwardian impresarios. He saw the ritual of the durbar in terms of 'visual advocacy; the opportunity to make a case, to impart a message, to impress an audience, to reinforce a sense of identity and of community, to cement those links between past and present about which he cared so dearly'.[90] Even by Curzon's exacting standards, the Delhi Durbar of 1902–3 was a masterpiece of its kind, though it came at a cost of over £200,000. 'From the moment of our arrival in Bombay, where Marlborough and I were guests of the Governor, events as glamorous and gorgeous as those narrated in the tales of *The Arabian Nights* enchanted us,'[91] wrote Consuelo.

The durbar lasted almost two weeks, involved more than 150,000 people and hundreds of Indian princes from all over the subcontinent. An infrastructure of more than seventy camps was created around an amphitheatre which was the focal point of the ceremonies, requiring 'railways, water supply, sanitation, lighting, telephones and postal services to be installed; the grounds had to be grassed and ornamented, and the amphitheatre constructed. Sofas, arm chairs, tables, book cases, pianos and pictures had to be provided.'[92] A special train brought the Viceroy's guests to this great camp. 'A double row of beautiful tents lined a central avenue. We had a salon as ante-room, two bedrooms behind, and a smaller room which held a round tub. Marlborough's valet and my maid lived near-by and a native servant brought hot water for our ablutions, and breakfast hot and delicious at any hour.'[93]

There were two great ceremonies. One was the state entry into Delhi, a procession three-miles long with heralds and trumpeters, led by Lord and Lady Curzon riding an elephant lent by the Maharajah of Benares and covered in a cloth of gold embroidered with lions rampant and a silver boat-shaped howdah. This was followed by another elephant carrying the King's brother, the Duke of Connaught and his duchess, and a great procession of Indian chiefs. 'The mere bringing together of people from the Chinese frontier of Tibet and Siam, Burma, Bootea, Nepal, Gilgit, Chitral-Swat, Beluchistan, Travancore, and Kathiawar, has been the most marvellous object-lesson. Chiefs from the outer fringes of civilisation, who for years had been turbulent, gasped, "Had we known we were fighting *this* we would have remained at peace,"'[94] wrote Mary Curzon to Lady Randolph Churchill in England. The effect of them all in procession struck the correspondent for *The Times* as 'a succession of waves of brilliant colour, breaking into foams of gold and silver, and the crest of each wave flashed with diamonds, rubies and emeralds of jewelled robes and turbans, stiff with pearls and glittering with aigrettes'.[95]

The focal point of the two weeks was the Coronation Durbar on 1 January 1903. Sixteen thousand people, including the Indian ruling princes, assembled in the amphitheatre to hear the proclamation of King Edward VII as Emperor of India read aloud to a fanfare of trumpets. Thirty-one guns were sounded as Curzon entered and took his place on a throne as the King's representative, before each of the Indian princes was presented in turn. Around these two great ceremonies there was ceaseless activity – polo matches, fireworks, a huge ball at the Red Fort, and a military review of British and Indian troops. For Consuelo it was punctuated by memorable moments: Curzon at his grandest, leaving no-one in any doubt that it was he, and not the King's brother, the Duke of Connaught, who represented the King-Emperor in India; Mary Curzon bewitching, the cynosure of all eyes in a dress with a design of peacock feathers embroidered with semi-precious stones; a purdah party to meet the maharanees and princesses which only women could attend but profoundly reinforced Consuelo's sense of just how forcefully this 'bevy of imprisoned children' depended on their husbands;[96] an early morning gallop across a great plain as part of a falcon hunt . . .

After it was over, Curzon was congratulated by the King, the British press and politicians from both sides of the British political divide. 'The best show that ever was shown,'[97] said Arthur Balfour. 'The most gorgeous pageant that has ever been devised by the imagination and ingenuity of mortal man to point a moral or adorn a tale,'[98] said one newspaper. But not everyone approved of the moral that was pointed or the tale that had been adorned. There were objections to the contrast between the extravagance of the durbar and the poverty of India which, it was alleged, the Viceroy's guests never saw.

Consuelo saw plenty on that visit, however, and was as repelled as she had been during the *Valiant* cruise of 1894, by funeral processions with the dead on litters, by the bloated corpses of babies floating on the Ganges and by the mutilation of lepers. Unlike those guests in the party who saw only a breathtaking display of imperial magnificence, she also noted subversive rumblings – Maharajahs who felt that the Viceroy disdained them; a moment during the Coronation Durbar itself when one of them refused to bow to Curzon (who paled, but chose to ignore it). As far as sections of the Indian press were concerned, the Delhi Durbar was nothing more than a wholly inappropriate display of British imperial pomposity that was deeply humiliating to India. Curzon had gone too far, it was said, and one Indian newspaper even thought that 'the Durbar marks the beginning of Lord Curzon's decline, and has made it impossible for him to end his term of office on the same high note with which it began'.[99] It was a perceptive remark, which turned out to be true. But the Delhi Durbar did not simply mark the start of a decline in Curzon's career, it also coincided with further erosion of British aristocratic authority. And if a way of life was coming under pressure, so was the Marlboroughs' marriage. Exposure to the lifestyle of the Vicereine of India had set the seal on Alva's determination to marry Consuelo into the British aristocracy during the *Valiant* cruise of 1894; but by the time the Marlboroughs returned from the Delhi Durbar, the cracks in their own ceremonial facade were beginning to show.

7

Difficulties

WHEN HARPER & BROTHERS accepted *The Glitter and the Gold* for publication in 1951, Consuelo was asked to write more about the causes of her disagreements with the Duke. She refused. It would be 'necessary to place him in a light that will be painful to his children and to his cousin Winston Churchill', she wrote to her editor, Cass Canfield. 'In order to satisfy the curiosity of magazine readers I am unwilling to depart from a silence I have hitherto observed on a subject which after all is no one's business than my own.'[1] It was a curious protest. By the time the memoir was published in 1952, it was spattered with so many clues that those who were fond of her first husband protested, saying that she should not have blackened the name of a man who had died in 1934 and was thus in no position to defend himself. Piecing together his side of the story of the breakdown of their marriage is more difficult, for it is accessible only through a few letters and the remarks of those close to him. As is often the way, friends sometimes found it difficult not to take sides.

Later, *Town Topics* would remark that 'American mothers, if they must have titles for daughters, should see to it that hearts are not left out of the bargain'.[2] The obvious explanation for their later unhappiness is that this couple never loved each other in the first place and that Consuelo never recovered from feeling compelled to marry the Duke. Yet arranged marriages do sometimes end happily, even when the contracting parties have little choice. In this case, however, there was antagonism from the outset. In Consuelo's memoirs, written many years later during a second much happier marriage, she implied that even by the end of their honeymoon there were already many serious difficulties which were never resolved. By the time she met her new in-laws in London, she was already bristling at the Duke's arrogant and churlish dislike of all things American; his love of display; his 'hectoring rights' over her person which represented little change from life with her mother; and his deference to monarchy, aristocracy and the English class system. On top of a deeply upsetting quarrel, when all illusions of romance on both sides were ruthlessly dispatched, there had been sharply divergent points of view over the importance of presentation to Queen Christina of Spain; insensitive exposure to the vulgarities of *belle-époque* Monte Carlo; cack-handed timing of life insurance arrangements; serious disagreements over matters of taste; exhaustion and depression; and mutual episodes of deep gloom. Perhaps one of the most ominous signs during those first few weeks was that mealtimes began to assume enormous importance. 'We seemed to spend hours discussing the merit of a dish or the bouquet of a vintage. The maître d'hôtel had become an important person to whom at meals most of my husband's conversation was addressed.'[3] The Duke, in other words, was finding that after several weeks together, he had little to say to his new bride.

Settling into life at Blenheim did nothing to improve her sense that they had, *au fond,* very little in common and had fundamentally incompatible temperaments. Her account of long evening meals when the servants shut the door and left them alone make one feel for both of them:

> He had a way of piling food on his plate; the next move was to push the plate away, together with knives, forks, spoons and glasses – all this in considered gestures which took a long time; then he backed his chair away from the table, crossed one leg over the other and endlessly twirled the ring on his little finger. While accomplishing these gestures he was absorbed in thought and quite oblivious of any reactions I might have. After a quarter of an hour he would suddenly return to earth, or perhaps I should say to food, and begin to eat very slowly, usually complaining that the food was cold!

> And how could it be otherwise? As a rule neither of us spoke a word. I took to knitting in desperation and the butler read detective stories in the hall.[4]

This is only one testimony among many to the Duke's tendency to self-absorption, which Consuelo found herself unable to puncture. She later described him as 'a beast' to Louis Auchincloss, and it is clear from others that he had a domineering streak and a 'demoniacal temper' which meant that life with him was altogether too similar to life with Alva. Such was the damage wrought by the Duke's own childhood that it would have taken very great confidence, insensitivity, or submissiveness to live alongside him happily.

Consuelo was neither enormously confident nor insensitive, and she quickly became less submissive. At the same time, she had had an isolated and sheltered childhood from which she had been precipitately thrust, and the necessity of constantly surrendering to Alva's autocratic demands gave her little experience of emotional negotiation, or of articulating her feelings calmly when something was wrong. Instead, her antagonism towards her husband appears to have manifested itself in lethargy or festered until it finally erupted in a bitter and damaging row. Antagonism and antipathy towards his person, his house and his ancestors were much more destructive in the 9th Duke's case than in those of more robust emotional temperament.

They probably both hoped that once Consuelo was settled into English life a modus vivendi would emerge and some of these problems would settle down. Instead, more damage was done in the early months in England when Consuelo felt unsupported in finding her way through the mysterious maze of English high society. Sunny was insensitive to her anxieties while she often misunderstood his. Exceedingly fastidious, he had the critical eye of the acutely anxious man which made him more inclined to criticise than praise. His compliments were so rare that Consuelo could remember them years later. Conversely he probably felt her need for encouragement was insatiable, was unwilling to pander to her vanity and felt she should regard becoming chatelaine of Blenheim as a privilege.

Like other American women who married into the English aristocracy, Consuelo faced the constant unspoken assumption that she would, at any minute, do something vulgar. As Lady Randolph Churchill later wrote: 'If (the American woman) talked, dressed, and conducted herself as any well-bred woman would, much astonishment was invariably evinced.'[5] Consuelo, confronted with the ineffable certainties of the English upper classes, found her new acquaintance in general and her husband in particular unthinkingly arrogant and carelessly anti-American. She was understandably indignant when they tried to patronise her, by the implicit assumption that she was fortunate to have become a Marlborough, and that as a duchess she would work to glorify the family name without question. Even where she could see that change was necessary, she often found that the forces ranged against her made it difficult to implement. There is no sign, moreover, that she was ever allowed to participate in planning the refurbishment of Blenheim, though she was – at least indirectly – expected to finance it. For, as Lady Randolph Churchill also remarked of the American wife: 'Her dollars were her only recommendation . . . otherwise what was her *raison d'être*?'[6]

Consuelo later described herself and her husband as 'people of different temperament condemned to live together'.[7] From her in-laws' point of view, Consuelo may have been more difficult and less victimised than she gives us to understand. She was charming, elegant and wealthy, with a dowry that enabled her husband to carry out many changes to Blenheim of which he had long dreamed; but away from her mother, she was proving 'touchy' and 'oversensitive', characteristics that Consuelo herself recognised as clouding an otherwise amiable disposition, and which she thought stemmed from her confined and introspective childhood. There was also Churchill irritation at her lethargy in the face of many of the new experiences which now confronted her. Given that she described these, even in her memoirs, as 'boring' or 'tiring', she must sometimes have appeared spoilt and infuriatingly passive. In the

Blenheim Christmas tableaux and theatricals burlesque of 1897, for example, Consuelo was teased by her in-laws for feeling tired all the time.*

Today, these outbreaks of fatigue and lassitude would be better understood. They may well have had a physical basis. She was faced with a huge and demanding adjustment at the age of eighteen, followed by two pregnancies in quick succession before she was twenty. Many of the new experiences she was supposed to enjoy were imposed without explanation; the risk of failure was impressed on her constantly; she was criticised when she made mistakes and relations who found her lassitude tiresome may well have been jealous of the social success which appeared to cause it. Sunny's depressive tendency was also misunderstood by his new wife, however, who was more inclined to view his love of Blenheim as the psychological prop of an inadequate man, while his agenda was inaccessible to all but his own charmed circle. It was, at best, a fragile emotional construct and it soon came under new pressures.

The first of these, ironically, was success. For a few years, the challenge of reversing decline, launching Consuelo into English society and the birth of two sons acted as a distraction from underlying incompatibility. By 1899, however, the 'heir and the spare' had not only been produced but were being looked after by nannies, another English system that Consuelo found she could do little to change. She had done her duty while the Duke had done his by ensuring another 'link in the chain'. This gave both of them more time for introspection and discontent. Consuelo's success as a

* *She played the role of the Countess of Clondyke and sang:*
'The season's over, all the lights are blown out for the present;
I feel a perfect wreck, of course, a feeling not unpleasant . . .
And well I may, for not a ball was held, nor candle lighted
In Palace or in marble hall, where I was not invited.

And: I feel so tired
Terribly, awfully tired
I think I shall die
I don't know why
Except – I'm tired.'

duchess gave her confidence, however, and came partly because she learned to assert herself. As the independent-minded Consuelo gained the upper hand, she was more inclined to regard her former submissive self with disfavour and resent high-handed treatment by her mother and husband. But the conflict between the two sides of her nature did not make her any easier to deal with when she was angry. 'I know from long experience that with a child-like appearance and demeanour she is as dangerous as she can be,'[8] wrote her husband in exasperation to Winston Churchill in 1906.

As Consuelo conquered English society and made friends she found it all less intimidating. By the same token, its pretensions became much more apparent; its insistence on protocol increasingly irritating. After the death of Queen Victoria in 1901, for example, duchesses were required to wear black. When Consuelo wore white gloves with her black outfit to the races in Paris she found herself accused of showing insufficient respect to the memory of the dead Queen by, of all people, the Duchess of Devonshire – the famous 'double Duchess' who had a raffish reputation as a leader of the 'fast set' and added to her louche aura by wearing a great deal of make-up. On another occasion, she was scolded by the Prince of Wales for wearing a diamond crescent to dinner rather than a tiara like the Princess of Wales, forcing Consuelo to say that she had come direct from a charitable function and that the bank where she kept her tiara had been closed when she arrived in London.

It was not long before this reaction against stifling protocol became something more profound than irritation at the 'over-importance attached to the fastidious observance of ritual'. Even her engagement with the poorer tenants on the Blenheim estate opened her eyes to a different side of the English class system. It became increasingly difficult to enjoy life in a resplendent carriage when the linkman who opened the carriage door and bowed as one stepped out onto the carpet looked so drab and impoverished. Constant protection from everyday life was constricting and frustrating. 'The realities of life seemed far removed from the palatial splendour in which we moved and it was becoming excessively boring to walk on an endlessly spread red carpet,'[9] she wrote.

Although Consuelo supported her husband in his political activities while she lived under his roof, her instincts were always more liberal and she subscribed to the Whiggish view that it was the duty of those with wealth to alleviate suffering caused by poverty. She also remained an unreconstructed plutocrat. 'The accident of one's birth had always appeared to me no adequate reason for personal pride; though it is pleasant to realise that cause for shame in one's forebears is non-existent, still the achievements of others lend one no special glory.'[10] She did not regard herself as a snob. Her husband regarded her lack of snobbishness as a personality failing.

On at least one occasion their diverging views caused trouble. During the winter of 1898–9, while she was staying in Melton Mowbray once again, Consuelo heard that there were serious social and economic problems in the countryside round Blenheim, and took a small step to tackle it.

> Reports from our agent at Blenheim told of unemployment and of its accompanying train of hunger and misery. When I announced my desire to provide work for the unemployed it was labelled as sentimental socialism; but unable to reconcile our life of ease with the hardships of those who, although not our employees, were yet our neighbours, I dispatched funds to institute relief work. Unfortunately the men, grateful for the help given to them, sent a letter of thanks to my husband, who to his indignant surprise discovered that the roads on his estate had been mended and his generosity exalted. It was only then that I discovered how greatly he resented such independent action and that had I committed *lèse majesté* it could not have been more serious.[11]

With so little room for manoeuvre, it is hardly surprising that Consuelo found herself faced with the problem of ennui that afflicted other duchesses before and after her. Alva had been convinced that the life of an English duchess would provide a role and a career. In time, however, Consuelo came to find the routine opening of bazaars and the presentation of prizes easy, noting that the mere fact of her appearance was more important than anything she said for 'the cinema star had not yet eclipsed the duchess and archaic welcomes were still in line ... There were school treats for the various villages to arrange. There were cricket matches, which I never learned to appreciate. There were mothers' meetings, and women's organisations to address. I even wrote a sermon for a young curate who was shy and pressed for time.'[12]

Political discussions at Blenheim, particularly those with Winston Churchill in the eye of the storm, made her long for something beyond what she described as traditional but superficial public duties. After being told one day by a vicar presiding over a bazaar that she put him in mind of his favourite fruit the strawberry (found in a ducal coronet) Consuelo decided it was time to do something more ambitious and accepted an invitation to speak about technical education to a club of blind men in Birmingham. Cousin Winston went to great trouble to help her with this speech, telling her that no professor could have done better. Unlike her husband, who had to brace himself for contact with the 'lower orders', Consuelo warmed to her audience in an instant. 'I immediately felt in touch with them. And when, at the end, they greeted me with a storm of applause it made all the nervous anticipation worthwhile and encouraged me to continue.'[13]

This tentative step into public life was one of several strategies by which Consuelo distracted herself from the unhappiness of her marriage. Freedom to be alone was difficult to arrange. 'With a page in the house, a coachman or postilion to take me for drives and a groom to accompany my rides, my freedom was quite successfully restricted.' There were two ways of escaping endless surveillance. One was long walks in the park, which she could take unaccompanied. 'I loved to wander through the bracken among the great oaks with the lake shimmering below and to day-dream of past centuries and of the persons who had then haunted those green glades.'[14] The other was the electric car sent across from America by Alva. Because no protocol had grown up around ladies and cars, there was no objection to Consuelo driving out in it alone.

Another strategy, pursued by both Marlboroughs, was to spend increasing amounts of time apart. In Consuelo's case this sometimes meant going to Paris to see her father who was playing an ever greater part in French horse-racing, had an apartment on the Champs Elysées and moved to France permanently after he married Anne Harriman Sands Rutherfurd in 1903. She would go with him to race meetings at St Cloud and Maisons-Lafite and she loved the cosmopolitan life and relaxed atmosphere of Paris. William K. continued to look so young and handsome that on one occasion the Duke of Marlborough was refused admission to a private dining room where father and daughter were eating together by an over-zealous maître d'hôtel who feared a scene. 'When speeding homewards over the golden wheat fields to the white town of Calais I invariably felt sad at leaving a people whose civilisation I believed had truly assessed the values of life,'[15] wrote Consuelo later.

Early in 1900 the Duke sailed to South Africa to serve with the Imperial Yeomanry in the Boer War. The gossip in London, according to *Town Topics*, was that the Duke of Marlborough's decision to go to war had much to do with his marital problems and that Alva had rushed over from New York to sort things out. 'London persistently gossips that there has been trouble between the Duke and Duchess of Marlborough; that this was the reason for Mrs O. H. P. Belmont's hurried visit and for the Duke's determination to risk his life in the Transvaal,' it related. *Town Topics* was not, of course, so simple-minded as to believe such stories itself. 'The cabled reports of the Duke's departure, last Saturday, state that he left his "bachelor flat" and that this seems to give some colour to the stories; but a "bachelor flat" does not necessitate a bachelor life.'[16] At the same time, Colonel Mann was at a loss to see how Alva's presence might help since she was known to be a forceful advocate of the Boer cause. It is true, however, that Alva and Harold were at Blenheim in January 1900, and years later *Town Topics* reported more gossip that Consuelo had been ordered by Alva to travel to Southampton for a public gesture of farewell to her husband.[17] Though she would later evince little enthusiasm for the Boer War herself, saying that no-one in England was very proud of the campaign, Consuelo did her duty before the Duke's departure and attended a farewell dinner in Woodstock for those from Oxfordshire who had volunteered for active service with the Imperial Yeomanry and presented a wrist watch to each man.

The Duke departed for South Africa on the *Kinfauns Castle* on Saturday 20 January. Regardless of *Town Topics'* speculation, the Duke, like many of the 800 others on board, had volunteered for active service amid the upsurge of patriotism that followed unexpected reverses in South Africa in December 1899; over 3,000 British soldiers had been killed in 'Black Week' alone, administering a profound shock to the aristocratic certainties of the Victorian Age.

On arrival in South Africa, the Duke, who was a lieutenant in the Imperial Yeomanry, joined the commander-in-chief, Lord Roberts, and subsequently became his assistant military secretary on 16 April. This led to criticism in *The Times*, for it was felt that the appointment should have gone to a military man. Taken together with objections in the radical press at the unduly aristocratic nature of Lord Roberts' staff (it included the Duke of Norfolk and the Duke of Westminster), the Commander-in-Chief decided to leave the Duke of Marlborough out of the march on Johannesburg and Pretoria. This distressed him greatly but thanks to the intervention of Winston Churchill – then working as a war correspondent for the *Morning Post* – he was reassigned to the headquarters staff of General Sir Ian Hamilton alongside Cousin Winston. They thus saw some sharp military action in each other's company. They rode into Pretoria together as it capitulated, and liberated a prisoner-of-war camp from which Churchill had earlier made a daring and somewhat controversial escape:

> Marlborough and I cantered into the town. We knew that the officer prisoners had been removed from the State Model Schools, and we asked our way to the new cage where it was hoped they were still confined. We feared they had been carried off – perhaps in the very last train. But as we rounded a corner, there was the prison camp, a long tin building surrounded by a dense wire entanglement. I raised my hat and cheered. The cry was instantly answered from within. What followed resembled the end of an Adelphi melodrama. We

> were only two, and before us stood the armed Boer guard with their rifles at the 'ready'. Marlborough, resplendent in the red tabs of the staff, called on the Commandant to surrender forthwith, adding by a happy thought that he would give a receipt for the rifles. The prisoners rushed out of the house into the yard, some in uniform, some in flannels, hatless or coatless, but all violently excited. The sentries threw down their rifles, the gates were flung open, and while the rest of the guard (they numbered 52 in all) stood uncertain what to do, the long-penned-up officers surrounded them and seized their weapons. Someone produced a Union Jack, the Transvaal emblem was torn down, and amidst wild cheers from our captive friends the first British flag was hoisted over Pretoria. Time: 8.47, June 5. Tableau![18]

The Duke was later mentioned in dispatches for the part he played in the Boer War, and returned home in late July 1900, when it seemed (erroneously) that the campaigns of Roberts and Hamilton had brought Boer resistance to an end. He sailed back in the company of Lady Sarah Wilson who had been in Mafeking reporting for the *Daily Mail*, and had been taken prisoner by the Boers and released after she wandered behind enemy lines. When she read about Lady Sarah's experiences, even Queen Victoria was heard to comment that she thought Lady Sarah Wilson was quite capable of looking after herself.

In addition to the Duke, Lady Sarah Wilson and Winston Churchill, Lady Randolph Churchill was now in South Africa, in part to see a young subaltern, George Cornwallis-West, whom she would later marry. This left Consuelo on her own in England without house parties to host. The only visitor of note to sign the Blenheim visitor's book in the five months the Duke was away was the French painter, Paul Helleu, who came to stay on 11 May 1900. Helleu, a friend of John Singer Sergent, was an established society portraitist whose sketches of society women had become extremely fashionable. He favoured a technique known as dry-point etching which allowed him to sell limited editions. 'Helleu was a nervous, sensitive man with a capacity for intense suffering that artistic temperaments are prone to,' wrote Consuelo. 'He thought himself something of a Don Juan, and with his black beard, his mobile lips and sad eyes he had the requisite looks, but he was too sensitive for the role.'[19] In spite of the fact that some of his warmest and most touching drawings are of his wife, Helleu was thought to have affairs with some of his beautiful and fashionable clients, though he did not have the salacious reputation of his friend and rival Boldini.

Helleu's daughter believes that he and Consuelo probably had an affair between 1900 and 1901. While he was staying at Blenheim he is thought to have done two pastels, five dry-point etchings and several drawings. Some of these are delightfully tender and intimate, particularly a drawing of Consuelo dozing on a sofa at Blenheim, a dog asleep on her lap. Consuelo went to Paris herself on 9 June 1900, and other drawings have backgrounds that match photographs of Helleu's studio. It was either in his studio or at Blenheim that he produced his most famous dry-point etching of Consuelo in a high black feathery hat with a long feathered peak and a fine fur wrap around her neck – textures lending themselves to the velvety lines produced by the diamond stylus of the dry-point etching technique. In the more relaxed atmosphere of Paris, there would have been no objections to going alone to Helleu's studio for a sitting, or having lunch together in a café afterwards. For a time, it was, at the very least, a great friendship. Helleu was a well-known wit, a yachtsman and a cultured man who was a friend of Proust and other artists of the *belle époque*. He offered companionship when she needed it as well as limitless admiration of Consuelo's blossoming beauty. Their friendship later soured, when he persisted in making copies of the dry-point etching of Consuelo in her feathered hat into the 1920s without her permission – to raise money – but during the first period of their friendship she celebrated it with gifts, notably a slim cigarette case which he could carry in a jacket pocket.

By the beginning of 1901 the gossip mongers were back in business reporting further trouble between the Marlboroughs. It was suggested that the Duke was bitterly disappointed at being passed over for the job of Lord Lieutenant of Ireland. He felt his chances had been ruined by the American press and was holding Consuelo responsible: 'Sunny . . . is not the best tempered of men, and the young Duchess has to bear a lot of ill-natured remarks from her lord and master about things American generally and about American journalists especially.'[20] According to *Town Topics*, select items from American newspapers suggesting that Consuelo was already considering herself the 'Vicereine' and looking forward to having her hand kissed were shown to 'Old Judy' by the Earl and Countess of Dudley, who were anxious to secure the appointment for themselves. The Queen was said to be furious at the presumption on display, and all further discussion about the Duke of Marlborough's appointment came to an abrupt end.

There may be little truth in this, since the Queen's strength was already ebbing away around this time and she died on 22 January 1901. In her memoirs, however, Consuelo alludes to Lady Dudley's manoeuvring at Queen Victoria's funeral: '[She] seemed preoccupied, interrupted all my conversations and never left me a moment alone with those who wished to speak to me. I discovered later that she was anxious to have her husband appointed Lord Lieutenant of Ireland and, knowing that Marlborough was spoken of as a candidate, she feared that I was pleading his cause with the political personages present.'[21] The Earl of Dudley subsequently held the post until 1905 leaving Sunny excluded from a post that had been held by his grandfather and later by his cousin, Ivor Guest.

The Duke may well have chosen to blame both the US press and his wife for this and let his vituperation show. 'It is hard lines on the Duchess of Marlborough that she should be placed in this awkward position for she is in no manner to blame for it . . . Strawberry leaves are, no doubt most becoming to the American girl, but was it worthwhile to pay so big a price for a coronet of them, only to be abused in vile language by her husband in public?' asked *Town Topics*.[22] It is clear, however, that Consuelo was unenthusiastic about a move to Ireland for she later wrote that it was a 'relief' when Lord Dudley became Lord Lieutenant and her husband was given the Garter and appointed Under-Secretary of the Colonies.

This may have been held against her by her husband, for it would have damaged his chances of being appointed had she made her feelings known.

Whatever the cause, the rift between the Duke and Duchess early in 1901 was sufficiently serious for *The New York Times* to remark in April that the Marlboroughs were now travelling together from Paris to London and that 'the fact is regarded as indicating that if there was any ground for the rumour of discord between the two the unpleasantness has been smoothed over'.[23] The amount of time the Marlboroughs had been spending apart was causing concern, the newspaper noted, hinting strongly that they did not sleep under the same roof in Paris, and were only reunited when it was time to leave. 'The Duke of Marlborough, after travelling for a month in the South of Spain, came to Paris a week ago and stopped at the Hôtel Bristol, on the Place Vendôme. While the Duke was in Spain the Duchess was in Paris. The last three weeks prior to her departure for London she spent at her father's mansion in the Avenue des Champs Elysées. After the Duke returned from Spain he visited his father-in-law, and there saw the Duchess. This morning the Duchess drove to the Hôtel Bristol, whence she and the Duke proceeded to the railway station in company.'[24] It was against this background of rumour and gossip that Gladys Deacon came to stay at Blenheim in the late summer of 1901.

Gladys Deacon first appeared in London society in the autumn of 1897, just after Consuelo gave birth to her first son, Blandford. Gladys met the Duke of Marlborough while Consuelo was recovering at Blenheim and charmed him. On discovering that Gladys was not only American but an old Newport neighbour of his wife, the Duke invited her to Blenheim to provide some companionship for Consuelo, which meant that Gladys first arrived at Blenheim when the Marlboroughs were at their happiest.

Strikingly beautiful, Gladys (pronounced 'Glaydus') was intelligent and cultivated, a perfect antidote to those English sporting types who regarded culture with dark suspicion. She was, according to Consuelo, 'a beautiful girl endowed with a brilliant intellect. Possessed of exceptional powers of conversation, she could enlarge on any subject in an interesting and amusing manner. I was soon subjugated by the charm of her friendship and we began a friendship which only ended years later.'[25] It is very likely that Consuelo knew something of the Deacon family background before she met Gladys, because it was highly scandalous. Although Gladys Deacon was American by birth and had lived in Newport, she was not an heiress, although her mother was very well off and she counted at least one captain of the revolution among her distinguished forbears. Born in Paris, and European in upbringing, Gladys was one of four daughters of Edward and Florence Parker Deacon, who mixed with the *gratin* of *belle-époque* society. Edward Parker Deacon was unstable and violent, or so it was alleged by Mrs Deacon. He once beat his wife when he disliked the way she had done her hair, and again when he found her crying after their only son had died.

Gladys's upbringing was dogged by her parents' marital difficulties, for if Edward Parker Deacon was unstable, her mother was adulterous. When Mrs Deacon took one Emile Abeille as her lover Mr Deacon followed them both to Cannes and – in the face of considerable provocation and many lies it must be said – proceeded to shoot Emile Abeille dead in the Hôtel Splendide before giving himself up to the police. Naturally the entire affair gave rise to a murder trial that was publicised internationally (and also attracted the interest of Henry James). It ended with Edward Parker Deacon being sent to prison for unlawfully wounding Abeille, though the court accepted that he had not intended to kill him. Mrs Deacon's adulterous behaviour counted against her however, and when Mr Deacon was released from prison he gained custody of his three eldest children, including twelve-year-old Gladys. He removed her from her mother to live in America, precipitating a long and bitter custody struggle.

Between the ages of twelve and fifteen, Gladys was educated in America, first in Newport and later at a boarding school in Massachusetts. By 1895, when she was fourteen, she was already showing signs of becoming cultured and clever. She also, according to her biographer, Hugo Vickers, developed the beginnings of an obsession. In October 1895, she read about Consuelo's engagement to the Duke of Marlborough in the newspapers and wrote to her mother: 'O dear me if I was only a little older I might "catch" him yet! But Hélas! I am too young though mature in the arts of woman's witchcraft and what is the use of one without the other? And I will have to give up all chance to ever get Marlborough.'[26] After Gladys left school in 1896, her mother successfully regained custody and took her back to Paris. She never saw her father again, for by 1897 his behaviour was becoming more and more eccentric. After he fell into an uncontrollable fit he was admitted to the McLean hospital in Boston where he later died of pneumonia.

Though she would become a regular visitor to Blenheim, Gladys spent several months in Germany after her first visit, completing her education. She learnt to speak and read in German and Italian, and took up Latin. Though Bernard Berenson said rather sourly of her later: 'She did not really know how to "learn", but she would retain everything which would be useful to her in making an effect,'[27] Gladys continued to study with tutors for several more years emerging with seven languages, exceptionally wide general knowledge, a great interest in mythology and a life-long love of art, literature and poetry. Since she was a very great beauty – a perfectly moulded face with large grey eyes and a 'rose-leaf' complexion – and an outstanding conversationalist, Gladys soon began to attract attention. She was taken up by the poet, nobleman and aesthete, Comte Robert de Montesquiou, a friend of her mother's; she was drawn by Boldini, and met Bernard Berenson, the great connoisseur and art critic in St Moritz in 1899.

Although Bernard Berenson was already in love with Mary Costelloe and would marry her the following year, he became infatuated with Gladys, later telling his wife that he had even thought about marrying Gladys instead. Gladys's infatuation with Berenson ('You are not a person to me, you are an *état d'esprit et d'âme*'[28]) survived a three-month trip to California and his marriage to Mary, however, and in March 1901 she and her mother went to stay with the Berensons at their villa, I Tatti, near Florence, where mother and daughter forged a friendship with them both. After Gladys left, Berenson slipped into lethargic gloom. 'Even Gladys has faded out of his grasp,' wrote his wife, who was fascinated and disconcerted by Gladys in turn.[29]

Leaving the newly married Bernard Berenson suitably subjugated at I Tatti, Gladys went to Blenheim for several weeks in August 1901. She immediately caused emotional mayhem (and attended the first great Unionist rally). She began by enthralling the Duke's cousin, Ivor Guest. By 10 August 1901, the unhappy Duke of Marlborough was also helpless in the face of her beauty, intelligence, intensity and capricious charm. He removed himself to Harrogate for a health cure, and immediately wrote her the first of many letters. As in some of his later letters he started by apologising for his own dullness: 'This is the first letter that I have had the pleasure of writing to you. I hope that you will not suppose that its peculiar dullness and stupidity is a correct specimen of me?? – but truly this place allows one little opportunity of writing anything of interest.' But he changed to French to express his true feelings, quoting Rochefoucauld and expressing the sentiment that absence fanned the flames of grand passion, while diminishing mediocre sentiments, as the wind extinguished candles but fanned the flames of fire. He lingered on the hope of fire 'and the possibility of its burning' but remained pessimistic about his chances of success.[30]

He was right to be despondent because two conquests were not enough for Gladys Deacon that August. The Duke returned to Blenheim from Harrogate to host a visit from the Crown Prince of Germany, arranged by the Kaiser himself. The house party was small – the Duke's relation, Viscount Churchill and his wife Verena, Gladys and a selection of German courtiers including the German ambassador, Count Metternich, Count Mensdorff, Count Eulenburg and Colonel von Pritzelwitz. The Kaiser must have imagined that, supervised by such a committee, the Crown Prince could come to little harm, but he reckoned without Miss Gladys Deacon. Much to Count Metternich's anxiety, the Crown Prince proceeded to fall under her spell like everyone else, assisted, according to a later magazine serialisation of the romance, by beautiful sunsets and games of tennis. Consuelo felt sorry for Count Metternich who had a harrowing week 'anxiously preening a stiff neck in vain endeavour to follow a flirtation his Prince was happily engaged in pursuing'.[31]

During the visit (and unobserved by Count Metternich) the Prince gave Gladys the ring his mother had given him at his first communion, while Gladys presented him with a bracelet. When they all went on a sightseeing expedition to Oxford, the Prince, who had never driven a coach before, let alone a coach and four, insisted on driving. Consuelo sat beside him ready to grab the reins at any point – a sensible precaution since the Prince constantly turned round to gaze at Gladys with a lovesick expression, to the great anxiety of the other passengers. The Marlborough groom had the presence of mind to clear the road by repeatedly sounding the horn which ensured that they all reached Oxford without any damage other than that inflicted on Count Metternich's mental health. When the Prince returned to Germany, his father the Kaiser noticed that the ring had disappeared and demanded it back in a fury. Consuelo eventually arranged for its return through Colonel Pritzelwitz, who exchanged it for Gladys's bracelet. 'So ended a foolish and completely futile conquest,' wrote Consuelo much later.

Shortly after the Prince's return to Germany, Consuelo and Gladys decided that this was just the moment to take a sightseeing trip to Berlin and Dresden. There followed one of the happiest interludes of this period in Consuelo's life. In spite of the fact that their arrival in Germany made the imperial court extremely jumpy, and that they were assigned a most dour and irritating aide-de-camp 'impervious to woman's charm' to keep them away from the Crown Prince, they explored San Souci and the galleries, and chatted about life and art from dawn till dusk. They took steamers down the Elbe and went to the opera in the evenings. Consuelo had never experienced this kind of friendship, and by the time their short holiday ended, and Gladys returned home to Paris, she was almost as entranced by Gladys as her husband had been earlier, exchanging locks of hair, referring to herself as 'your old Coon' and signing herself 'your loving Consuelo'.

Letters from Consuelo to Gladys in this period give intriguing glimpses of Consuelo's state of mind. Although Consuelo was older, she clearly regarded Gladys as the teacher in the relationship, and teased her by calling her Pallas Athene, saying: 'Do not bury yourself in study – you know quite enough already. I, who know nothing real of a good deal of life must still learn.'[32] She asked Gladys not to accuse her of moods, writing: 'I know only too well what it is to have them; but now that I am busy they do not assail me so often', adding that 'I have no-one to disburden myself to at all.'[33] She entreated Gladys to be 'the Dresden Gladys' again, and here and there she dropped hints about the state of her emotional life, writing: 'I am very good – like you – and although it irritates me at times – on the whole it is comforting?' and: 'Don't laugh at me! I think I should have tried to have been a Vestal Virgin and forever nourished the fire of life and rejoiced in enforced virtue, recognised as such! Or – like Cleopatra – I hate the middle course.'[34] In another letter Consuelo wrote ironically: 'Vive l'amour! But then goes on to say 'But I am not sure – I think I prefer la paix?'[35] In October 1901, Consuelo added a wistful PS, writing: 'I wish we were at Dresden! Please say you do too.'[36]

After Gladys returned home to Paris, the Duke of Marlborough took his turn to call on her, returning with presents for his sons. Sunny was also penning notes of his own to Gladys, arranging another meeting in Paris, where he would be travelling with Consuelo, and signing off 'am looking forward to Sunday with much pleasure and some feelings of excitement. Bless you. Fair friend.'[37] On the surface, however, the Marlboroughs patched up their difficulties in 1902, finding distraction in the great ceremonial set pieces which the Duke particularly enjoyed, and starting work on Sunderland House. Gladys Deacon, meanwhile, returned to the Berensons who took her on a trip round Italy. Mary Berenson was both deeply attracted and appalled by Gladys and wrote some highly perceptive descriptions of her at this period in her diary: 'So beautiful, so graceful, so changeful in a hundred moods, so brilliant that it is enough to turn anybody's head. Part of her mysteriousness comes from her being, as it were, sexless. She has never changed physically from a child to a woman, and her doctor said she probably never will . . . Brought up by a mamma who thinks of nothing but Dress & Sex, her mind plays around all the problems of sex in a most alarming manner with an audacity and outspokenness that make your hair stand on end. She is positively impish.'[38] Gladys held views about sex that were far in advance of her time. Even in old age she told Hugo Vickers: 'One is changed by ecstasy. Physical love is the creator of all things. It is a recreation, and when I say a recreation I don't mean a recreation, but a re-creation!', though she regarded marriage as 'something very different'.[39]

It is not difficult to see that if such views were expressed to either of the Marlboroughs they would have had a somewhat unsettling effect. Mary Berenson was worried by persistent rumours about relations between Gladys and the Marlboroughs, and Gladys confirmed her fears by telling her in confidence that Consuelo was 'nearly broken hearted because the Duke would make such wild love to her'.[40] But Mary Berenson wrote that Gladys often told lies, and Consuelo does not appear to have been at all 'broken-hearted' as she escorted her friend round a brilliantly successful London season, before playing her role in the postponed coronation on 9 August. Not long afterwards, Consuelo made a trip back to the United States, on her own. It was the first after her marriage. She stayed with her mother in Newport, where her girlish demeanour and great elegance was the subject of much comment.

The Marlborough marriage came under new pressure that year, however. In the wake of the visit to Russia in 1902, Consuelo caught a bad cold which left her slightly deaf. Instead of clearing up in the months that followed, the deafness became worse to the extent of becoming a serious handicap in English society where people spoke in subdued tones, particularly when addressing someone they regarded as exalted, or if they had something particularly interesting or scandalous to impart. 'This was not only tiring but also exasperating, not to say humiliating, and was causing me a great deal of anxiety, for to be cut off from human intercourse at so early an age appeared to me catastrophic.'[41] Her anxiety was justified for in spite of seeing specialist after specialist, and a number of painful operations, it gradually became clear that nothing could be done to reverse the problem. Consuelo's deafness soon came to irritate Sunny intensely, and there was little sympathy from the wider world. Queen Alexandra also suffered from deafness and in 1904 *Town Topics* suggested unkindly that it had become a fashionable affliction: 'Mrs Paget is, dear soul, deaf! She is nothing if not smart, and not to be as deaf as Queen Alexandra and the Duchess of Marlborough is to be out of the running.'[42]

Consuelo's deafness would prove to be one more source of stress in a strained marriage, affecting relationships with those around her until hearing aids were developed to the point where she could hear almost normally again. 'There were so many hiatuses that had to be filled in, so many half-tones that had to be divined, so much intensive guesswork towards significance, that even when the purport had been grasped the end barely justified the trouble. And later, when loved voices grew dim and I saw the flicker of annoyance that follows an unappreciated jest, I became withdrawn and lived in a world peopled with characters of my own choosing, at once apprehensive and apprehending. Solitude can become fiercely possessive ... It became a solace for me to remain what Lord Curzon called a "black swan", aloof in the soundless shadowed waters where I could choose the mirrored scene.'[43]

In November 1902, Consuelo travelled to Austria for an extended course of treatment from a specialist, accompanied by her two young sons. Count Mensdorff had provided her with letters of introduction to Austrian society and she spent many pleasant evenings at the opera, at Viennese operettas, and at suppers afterwards in the company of its younger cosmopolitan circles. She was presented to the Emperor but she did not warm to the Viennese aristocracy. Though they were polished, well-educated and well-bred, she remarked scathingly that it was 'a pity they could express their thoughts in so many different languages when they had so few thoughts to express'.[44] This did not stop her flirting with at least one of its scions, as Daisy, Princess of Pless noticed when she met her in Vienna:

> . . . She is looking very well and I think fascinatingly pretty, with her funny little turned-up nose and big brown eyes. She had made friends with Z before I arrived. And her first words to me were: 'Of course, Daisy, you know Prince Z? I have been saying to myself that I knew you would come and take him away from me.' I replied: 'You dear little silly, you can't take away in five days a man who has been a loyal faithful friend to me for more than five years.' As a matter of fact I had sent him ahead of me from Pardubitz to Vienna to take care of her; she leaves tomorrow for England, and from there goes to India for the Durbar.[45]

Unlike Princess Daisy, when Gladys Deacon learnt that the Marlboroughs were going to India for the Delhi Durbar, she was disapproving. She wrote to the Duke: 'Sterne says that 3 travel – Those who do so from Imbecility of mind or from Incapability of body or from Inevitable necessity. The first he says, comprise all those who travel by land or by water labouring with pride, curiosity, vanity or spleen. As for the 3rd class they are all the peregrine martyrs who live by the traffic & importation of sugar-plums, we'll say, & boots. Does this explain the Durbar? You, of course, belong to the first & second, most obviously. Don't you now?'[46]

In the Marlboroughs' absence, Gladys spent the winter in Paris with her mother and sisters. She became ill, however, causing her mother so much concern that she was eventually sent to a *maison de santé* for a complete rest. It was in this highly strung and nervous state that Gladys took a step which would come to haunt her. According to her biographer, Hugo Vickers, Gladys had long been fascinated by her own face. 'Mabel Dodge Luhan wrote that "she was content to lie for hours alone on her bed, happy in loving her own beauty, contemplating it". As she gazed, she worried about the slight hollow between the forehead and the nose, which, to her mind, denied her a true classical beauty. Her bizarre love of the classics, combined with a reckless vanity, set her to consider how she might achieve the perfection of the Hellenic profile she had so often admired in museums and galleries.'[47] Gladys's next step was to visit Rome where she explored the Museo delle Terme and measured the distance between the eyes and nose of a statue there. Returning to Paris, 'she had paraffin wax injected into the bridge of her nose to build it up and form a straight line from the forehead down to the tip. At first the endeavour had a measure of success, but in the long term it was a disaster.'[48]

When Mary Berenson visited Gladys in her sanatorium in April she found her greatly changed. Gladys lied to her about what she had done, although her nose was still swollen. Mary Berenson was more alarmed by her mental state, her inability to concentrate and her habit of saying half a dozen contradictory things in the course of one visit. By 8 April, however, Gladys discharged herself from the sanatorium. Another suitor, Lord Francis Hope, was pressing his case, and the Duke of Marlborough, recently returned from the Delhi Durbar, was due also to arrive in Paris for a few days. His appearance worried both Gladys's mother, Mrs Baldwin, and Mary Berenson. Mrs Baldwin feared that in such a highly strung state Gladys was plotting to elope with him to spite Consuelo of whom Gladys now seemed to have become extremely jealous on account of her apparent hold over the Duke.

'What a hateful silly muddle it is,' wrote Mary to Bernard Berenson, 'and poor Gladys with her excited brain drifting about in it with no one to guide her.'[49] Gladys continued to be elated at the idea of seeing the Duke. It is unclear what transpired during his visit to Paris in 1903 but before long she was being courted by the recently widowed Duke of Norfolk and the prospect of upstaging Consuelo by becoming the wife of the 'Premier Duke of Great Britain' appears to have taken over from any desire to elope with her husband. Gladys returned to England in the summer of 1903 for a further period of recuperation, enjoying another successful season alongside her younger sister Audrey, followed by another stay at Blenheim. At the end of the year Mrs Baldwin decided she had tired of Paris, and moved her household to Rome where they occupied the Palazzo Borghese.

The Duke of Marlborough, meanwhile, suddenly found himself occupied by politics. He became Under-Secretary to the Colonies in 1903 after Arthur Balfour succeeded Lord Salisbury, and his duties were more onerous than those as Paymaster General. During his period as Under-Secretary to the Colonies, the Colonial Secretary Alfred Lyttleton formed a good opinion of the Duke's capabilities, writing that he was 'proud of my lieutenant' when he persuaded the House of Lords to accept Chinese labour for the Transvaal on behalf of the government. Balfour wrote in similarly glowing terms after a meeting in Bolton where Sunny was instrumental in setting up the Imperial Cotton Growing Association alongside Joseph Chamberlain. Later, Balfour would write to the Duke that he could not have wished for 'a better or more loyal colleague', that in the Lords he had 'never made a mistake' and 'showed not merely great mastery of the subject but much tact and address'.[50]

The Marlboroughs took possession of their great new London home, Sunderland House, at the end of 1903, and for a brief period it was used for political entertaining, the purpose for which it was built. The house was designed by Achille Duchêne 'in eighteenth-century style'. It was as much Consuelo's creation as the Duke's, with up-to-date plumbing, adequate bathrooms and an American emphasis on bas-reliefs in the long gallery. Not everyone liked it when it was finished. Daisy, Princess of Pless could not stand the dining room which was 'not high enough and the windows have always to be covered over with stuff as they are right on the ground and the people in a dirty little back street can look in'.[51] Consuelo did not enjoy the role of political hostess either, later recalling: 'Long lists of important officials and guests had to be memorised, for Colonials were proverbially touchy and had Mr Smith from New Zealand been presented to Lady Snooks as coming from Australia, or vice versa, the result would have been disastrous. These receptions were by no means a pleasure.' She found debates in the House of Commons rather more interesting, particularly when the members were arguing about Home Rule and both sides demonstrated furious eloquence. Political disagreements often extended to the ladies watching from above. 'There were times when, sitting in the Speaker's Gallery between Lady Londonderry, whose sympathies were for Ulster, and Margot Asquith, an ardent Home Ruler, I found it difficult to keep an impassive neutrality. It was a relief when Mrs Lowther, the Speaker's wife, imposed the silence it was the rule to observe.'[52]

Official receptions apart, the Marlboroughs appeared together in public less and less often from 1904. Increasingly, Consuelo was beginning to forge an independent interest in philanthropy beyond the confines of the Blenheim estate. Her first experience of this had come before Sunny left to fight in the Boer War, when she assisted Lady Randolph Churchill and others to equip a hospital ship to send to Cape Town. Wealthy American wives in London had never defined themselves as a group in this way. Equipping a hospital ship was no mean task and it was Consuelo's first taste of an ambitious philanthropic undertaking which involved raising substantial funds, garnering expertise and wielding influence. It may have been little more than a taste – she still was young and inexperienced in the ways of charitable committees in 1899, and Jennie Randolph Churchill's sister, Clara, described her in a letter as 'about the most useless member of all'.[53] Nonetheless, it gave Consuelo a glimpse of what could be done by deploying American energy and English aristocratic connections.

In 1902, when she was asked to open a bazaar in aid of West Ham Hospital (later Queen Mary's Hospital) in London's East End, Consuelo decided to put a cautious toe in philanthropic waters herself. She was so intrigued by its work that she proposed herself as patroness, lending her financial support, name, social network and even Blenheim in the years that followed (she eventually became president of the hospital in 1916). Thereafter there is clear evidence that her philanthropic activity increased as her marriage become more miserable and in 1904 an American journalist, Kate Masterson, reported to *Town Topics*: 'The Duchess is very busy these days; she is always presiding at things. The latest to her credit is the Industrial Exhibition at the St James-the-Less Building, Sewardstone Road. But she looks awfully well, despite her English clothes.' The Duke meanwhile, was in a 'perpetual grouch just as he had been in New York, but it goes with the sturdy British atmosphere, and must be the right thing.'[54]

* * *

In the spring of 1904, Gladys Deacon suddenly faced tragedy. In April, not long after the family moved to Rome, her favourite sister Audrey developed the heart condition endocarditis. It was clear she would not have long to live. By May she was given morphine to help her rest, and Gladys came to sit with her before she died.

Gladys was devastated by Audrey's death, writing to Consuelo that: 'the pain in me has burned away all but what is the nucleus of immortality in us . . . If this living be of importance, why should she have been withdrawn?'[55] At the end of June, after visiting Audrey's grave in Florence with her mother, and an evening with Bernard Berenson that left him feeling profoundly uneasy, she joined the Marlboroughs in Paris. A couple of weeks later, when the Marlboroughs were back in London, they both wrote to her in Italy, from the same house, on the same day, apparently unaware that the other was making contact. Consuelo's letter was filled with love and affection, asserting that: 'I love you & feel for you and you want love and sympathy now. Women can give to each other what no man can give us – for their love is interested . . . [while ours is] in opposition to self interest. Surely then it must be real and stronger than self. Gladys dear I want you too, I long for your clever and deep thoughts for all that makes you so attractive and dear.'[56]

The Duke's letter, however, was one of real pain, suggesting that the encounter with the grief-stricken Gladys had moved him profoundly, and awakened all his old feelings for her. It implied that there had been intimacy in the past, but that she had now rejected (by letter) any suggestion that he should come to visit her in Italy. 'You are perfectly right to express yourself with frankness to me. I am glad that you have done so, for it is silly to live under a false impression,' he wrote. He expressed great concern at rumours from the 'social houses' that she was very ill, saying that 'everything that concerns you is of the same acute importance to me as if I was living near you', and emphasising that it was this concern that prompted him to 'trespass on her privacy'. He had only done so in 'tender recollection of much kindness that you have shown me in times gone by'. He had 'experienced such difficulty in resigning myself to the position of one who no longer may claim to be included in the inner circle of your instincts, thoughts and emotions'.

The Duke begged her not to place him on the same level as some of her other admirers saying: 'I shall always retain the deepest and truest gratitude to you for kindness which I believe no man has ever received at a woman's hands.' He would travel to Italy alone in any case. 'I will not attempt however to come to see you, the most I truly intend to do is the hope that passing through Florence I may see your mother to ascertain from her the extent of your illness. I shall then wander on to the Italian lakes which I have always wished to visit, and I will come home early in October. The solitude amidst the beautiful surroundings will do me good . . . I love Italy and I love the sunshine and under certain circumstances I love solitude – so I daresay I shall enjoy myself . . . I do not feel inclined in this letter to write to you about the silly things of the world in London. I hate the life here so it is distasteful even to write about it.'

Sunny then declared that he had spent a day at Blenheim re-reading her letters: 'The task was both a sad and pleasant one and I treasure these documents as a remembrance of some of the happiest and brightest days of my life and in recollection of one who possessed a heart kinder and more sensitive than any I have known.'[57] Later in the day, like lovers the world over, the Duke panicked about the letter he had written earlier in the day and sent another: 'I have many misgivings that the letter I sent to you this afternoon may be construed by you as unfeeling and unkind,' he wrote, consumed with insecurity. He asked her to make allowances for his distressed frame of mind since he was 'dead with fatigue' and 'exhausted with 12 months stay in this city of filth and heat . . . My temper and wellbeing are disturbed and I am prone to expressions of irritation.' But at its end he could not prevent another howl of pain: the anxiety caused by their long separation was almost too much to bear, and sometimes he wished they 'had never met'.[58]

The Duke appears to have kept his word and stayed away from Gladys, who was being pursued by other, more available suitors.

One was Roffredo Caetani, Principe di Bassiano and son of the Duke of Sermoneta; another was the Duke of Camastra. In October 1905, Consuelo made a further visit to the US, this time for an operation on her throat which it was hoped might alleviate her deafness. *Town Topics* reported that Alva was beside herself with delight, and showed her daughter off to all and sundry, driving her 'up and down Fifth Avenue past the clubs, and round and round the Waldorf Astoria'.[59] For the first time a Marlborough divorce was openly discussed in the American press, though *Town Topics* dismissed such talk as 'the same story that was in circulation in London and Paris last season, and it caused the same flutter'. A year later, the same gossip columns maintained that during 1905, Consuelo had a 'violent quarrel with the dashing young Anglo-American, Miss Gladys Deacon, whom she openly accused of flirting with the Duke'.[60]

It was in this atmosphere that the Duke commissioned John Singer Sargent to paint the family portrait that hangs in the red drawing-room at Blenheim, intending it to be a companion piece to the painting by Sir Joshua Reynolds of the 4th Duke and his family which hangs opposite. Sargent was somewhat taken aback to find that in Reynolds's painting there were eight people and three dogs, and questioned whether he could achieve what the 9th Duke desired. 'When, however, he realised that there was no choice he began to study a composition that would place us advantageously in architectural surroundings.' Consuelo wrote. 'We were therefore depicted standing in the hall with columns on either side and over our heads the Blenheim Standard, as the French Royal Standard captured at Blenheim had become known. I was placed on a step higher than Marlborough so that the difference in our height – for I was taller than he – should be accounted for. He naturally wore Garter robes.'[61]

For Consuelo, Sargent chose a black dress whose wide sleeves were lined with deep rose satin, modelled on the one worn by Mrs Killigrew in a portrait by Van Dyck on the north wall of the red drawing-room at Blenheim. Blandford wore a costume of white and gold and Sargent put Ivor in blue velvet, playing with a Blenheim spaniel at Consuelo's side. Sargent, whom Consuelo came to like very much as she took her sons for sittings at his studio in Tite Street, told Sir Edgar Vincent that he planned to emphasise the Spanish Infanta in her appearance, and for this reason he refused to let Consuelo wear her famous pearls, much to the horror of one of her sisters-in-law, 'who remarked that I should not appear in public without them'.[62]

In spite of all its pointers to a magnificent family history – the Blenheim Standard, the Duke's Order of the Garter – the painting conveys a deep sense of unease. There is no suggestion here of the relaxed informality of, for example, Sargent's paintings of the nouveau-riche Wertheimer family. The painting is dominated by Consuelo. The sense of movement in her draperies, and the dash of colour in her sleeve give her life and energy, though her face is sad. Blandford, the 'link in the chain', looks graver and more inert, held in place by Consuelo as if he could burst into mischief at any moment. Oblivious to the rest of the group, Ivor plays at Consuelo's side. There is great tension between the dynamism of the rest of his family and the seated figure of the Duke, positioned as a counterweight to his wife, sons and dogs. A wide space between the faces of the Duke and Consuelo underlines the Duke's lonely position. He seems disconnected from the family group, inscrutable behind an expression of melancholy amid emblems of a glorious Marlborough past.

There would be one more portrait of Consuelo before the end of 1906, by Boldini whom she met through Paul Helleu. She agreed to sit for him 'provided his behaviour remained exemplary', since his reputation with women was poor. He restrained himself to sighing 'Ah, la Divina, la Divina!' – he said this to a number of his female clients – and the portrait he produced was one of his best, showing Consuelo at her most vibrant. Towards the end, it gave Boldini problems. 'He had difficulty in getting my left arm on which my weight rests in proper position, and at one time I resembled a Hindu goddess, with no less than three arms protruding at different angles.'[63] The Marlboroughs decided to buy it. They had it enlarged to include Ivor and it was hung in the dining room of Sunderland

House. In almost every other respect, 1906 was a difficult year, starting and ending badly.

The year began with a general election that resulted in a landslide victory for the Liberal Party, putting an end to the Duke's career in government. At first glance, the reason for the Liberal Party's success appeared to have been its celebrated programme of social reform which represented an early attempt to grapple with the problems of industrialisation suffered by those whom Churchill described as 'the left-out millions'. However, the Liberal programme of social reform only evolved after the party took power, driven forward by Lloyd George from 1908. In 1906, Liberal success had as much to do with the unpopular policies and problems of the Conservative Party whose standing was severely damaged by the failures and cruelties of the Boer War, and associated revelations about the poor physical condition of young British soldiers. The need to reinforce the bonds of Empire after the divisive effect of the Boer War prompted Conservative politician Joseph Chamberlain to argue for tariff reform – a preferential system of duties for goods from the Empire. This split the Conservative Party down the middle and co-incidentally caused Winston Churchill to join the Liberals who supported free trade.

The Conservatives suffered a further blow when Lord Salisbury resigned in July 1902. He was replaced by his nephew and the Marlboroughs' friend, Arthur Balfour, who may have been, in Consuelo's words 'gifted with a breadth of comprehension I have never seen equalled', but whom historians blame for a series of misjudgements which gave the impression that the aristocratic Conservatives were becoming detached from the problems of their middle- and working-class supporters. These included the pro-Anglican church school Education Act of 1902 which deeply alienated the Nonconformist vote and induced the uncommitted to defect to the Liberals; a solution to divisions over tariff reform widely perceived to be inimical to working-class interests; support for the right of mine owners in the Transvaal to import low paid Chinese workers (proposed by the Duke in the House of Lords) which alienated working-class supporters; and its refusal to reverse a judgement in the Taff Vale case of 1901 which had severely restricted the ability of workers to strike. Conservative difficulties were adroitly exploited by the Liberals who forged an electoral pact with the emerging socialist Labour Representation Committee (LRC) in 1903 to avoid splitting the progressive vote.

The return of the Liberals to power with a huge majority marked a change in atmosphere and attitude in Edwardian politics. Even if the Liberal programme of social reform only evolved after 1906, it marked the end of an era for the Conservatives (though they would successfully reinvent themselves) and the beginning of the end of aristocratic government. As early as 1892–5, the Liberal Chancellor William Harcourt had made a distinction between 'unproductive' and 'productive' wealth, introducing death duties on large incomes in an early signal that the Liberals were now prepared to impose taxation on the rich, particularly the landed rich – a principle that would be extended to pay for social welfare reform in Lloyd George's People's Budget of 1909. After the Liberal victory, there were far fewer aristocrats in Parliament to argue their case. In 1906, the number of MPs from landed backgrounds fell to below one-fifth of the total. Leading members of the squirearchy such as Henry Chaplin lost their seats (after thirty-seven years in his case). 'Even Balfour, normally detached and unflappable, admitted that "the election of 1906 inaugurates a new era",' writes David Cannadine. Sir Henry Lucy went further as he contemplated a House where almost half the members were sitting for the first time, noting that 'its tone and character were "revolutionary"'.[64]

Apart from one brief sighting by Daisy, Princess of Pless on 21 January 1906 when they joined a house party at Eaton, the Marlboroughs seem to have spent the first part of 1906 doing their best to avoid each other. This may well have had something to do with Charles Stewart Henry Vane-Tempest-Stewart – Viscount Castlereagh and eldest son of the Marquess of Londonderry – a cousin of the Duke's and Unionist MP for Maidstone. Although he had married Edith 'Edie' Chaplin seven years earlier and they had two young children, he already had a reputation as a philanderer. He was many things that Sunny was not: tall, extremely good-looking, sensitive and sympathetic. Consuelo was beautiful, lonely, miserable and sorely in need of rescue. As Anita Leslie points out in *Edwardians in Love*, it was not always possible to observe the rule that extra-marital affairs should be discreet and restrained. With so many marriages arranged within a small aristocratic circle for reasons other than romantic love, it sometimes happened that an adulterous couple lost their heads, fell in love and threw caution to the winds. This seems to have been what happened: Consuelo and Lord Castlereagh either went off to Paris together, or were proposing to go when their plans were discovered.

Though rumour had it that the Duke wired Consuelo and told her not to return, Castlereagh's mother, the formidable Lady Londonderry, was determined to save her son from a scandal that would ruin his political career just as it started, and destroy his marriage to Edie into the bargain. It was said that she even involved the King and Queen who applied pressure of their own – this would have had little effect on Consuelo who already found royal circles dull and irritating – but it might, if it is true, have had some effect on Lord Castlereagh. The offence was, of course, that they took so little trouble to disguise their relationship. Such pressure was applied that the affair ended in the early summer and they both agreed to put a stop to it.

Consuelo continued to like Theresa Londonderry, in spite of her role in putting an end to the relationship with her son, and even came to appreciate her 'shrewd worldly wisdom' which 'proved a wholesome antidote to any sentimental tendencies on my part'.[65] Charley's wife, Edie, on the other hand, was made thoroughly miserable by the affair, one of the few of her husband's dalliances which she believed was a genuine threat to their marriage. 'I did experience a real deep shock in the early summer and it showed me with a clearness and abruptness how foundations which one imagines to be firmly built on rocks seem as nothing when a great wave – of what shall I say? – feeling anything you like, comes along,'[66] she wrote to him afterwards. She took an uncharacteristically long time to forgive him, even though he was deeply penitent after it was over and wrote begging for forgiveness. In the end she did, writing that it was wrong to blame other people and that she believed that they would both have to change. 'After this happened, I set out to alter the old conditions of things to the new. It is not in my nature to do otherwise. God alone knows what I expected in the summer when I felt that everything was slipping and sliding away from me.'[67]

The Castlereaghs were reconciled on a trip to Spain for the wedding of Prince Alfonso XIII to Princess Ena on 31 May 1906. The atmosphere between the Marlboroughs in the same month, on the other hand, was reported to be dreadful. It cannot have helped that Lord Castlereagh was elected to a parliamentary committee on South African and Colonial affairs while the Duke's political career languished. Writer Pearl Craigie wrote from Blenheim where she was part of a large party that included Winston Churchill that: 'I could not lead the life of these houses. I'd sooner die in an attic with an ideal. There is no affection in the atmosphere: the poor Duke looks ill and heartbroken.'[68] Sightings of the Marlboroughs that summer suggest that they remained anything but reconciled. Whenever Daisy, Princess of Pless saw Consuelo, she was alone. In July the Duke was away taking another health cure while Consuelo went with Daisy to the opera accompanied by Mr A. E. W. Mason, a Liberal MP and author of *The Four Feathers*, and the young good-looking Duke of Alba. The Duke was still away when they both joined at a house party at the Desboroughs at Taplow Court.

The break finally came in the middle of October 1906, when the Marlboroughs were both back at Blenheim. It was later said by the American press that it had come during a dinner at Blenheim when Consuelo had mentioned – in front of guests – that she was intending to go to Paris to buy her winter wardrobe, whereupon the Duke lost his temper, and shouted that she should go to Paris and stay there. If the story of this outburst is true, it may have been caused by a certain sensitivity about Paris as a destination; he may have suspected she was planning to see Lord Castlereagh again. The Duke later wrote to Churchill that he had asked Consuelo to leave on the grounds that he refused to be 'complicit' in her affairs any longer, though the impression was given in the American press that Consuelo departed because she could not stand another insult from her husband.

William K. Vanderbilt immediately came from Paris to join Consuelo while Winston Churchill wrote to his mother from Blenheim on 13 October: 'Sunny has definitely separated from Consuelo, who is in London at Sunderland House. Her father returns to Paris on Monday. I have suggested to her that you would be v. willing to go and stay with her for a while, as I cannot bear to think of her being all alone during these dark days. If she should send for you, I hope you will put aside other things and go to her. I know how you always are a prop to lean on in bad times. We are v. miserable here. It is a miserable business.'[69]

His mother, who had recently married George Cornwallis-West, had already heard the news. She was at Sunderland House by 16 October and their letters crossed:

> At the last moment when I was in the train – I gave up Floors [staying at Floors Castle with the Duke and Duchess of Roxburghe] and came here as I was wanted – you will be I know awfully sorry to hear that Sunny and Consuelo have separated. It is a terrible thing, and I can't tell you how painful it all is – Mr Vanderbilt is here and in a few days the legal separation will have been signed – and then it is finished forever – poor Consuelo is utterly miserable and dignified and quite calm . . . as far as I can I avoid everyone for fear of being asked questions – I feel as sorry for Sunny as I do for her – and I am obliged to say that he is justified in taking the course he has – it does not make it any easier that she has brought the whole thing on herself – how the women who have had 20 lovers and are kept by rich Jews et autres will be virtuously shocked.[70]

Both Winston and his mother did their best to help both sides in the immediate aftermath of the separation. Winston sent presents to the boys at Blenheim, while Jennie took Consuelo off to Salisbury Hall near St Albans – the home she shared with her new (much younger) husband George Cornwallis-West. George wrote to Winston – who was simultaneously trying to sort out a loan to cover his stepfather's losses – 'Poor little Consuelo is here I do pity her with all my heart, what a tragedy, the whole thing reminds me of Hogarth's series of satyrs [sic] "Marriage à la mode." Take my advice and if ever you do marry, do it from motives of affection and none other. No riches in the world can compensate for anything else.'[71]

Malicious gossip whirled round London, some of it put about by the Duke's supporters. They were led by Lady Sarah Wilson who spread tittle-tattle via Minnie Paget, well known to be London's biggest gossip. 'Sarah has behaved like a perfect beast,' wrote George Cornwallis-West to Churchill, 'and the sooner she is told the better, I heard every little detail, unsavoury and otherwise, from a man today who was told by Minnie Paget who got it all from Sarah. I naturally did not discuss the case, but merely denied any knowledge of such details. If Sarah thinks she is championing Sunny's case by casting mud (some of which will undoubtedly come back to him) she is much mistaken. Surely the obvious line for all your family to take is to decline all discussion on the matter, let alone volunteering disgusting gossip with the most renowned of gossip mongers.'[72]

Unsurprisingly, it proved impossible to dam the flow of talk. Alice Vanderbilt saw William K. Vanderbilt's wife, Anne, in Paris and wrote to her daughter Gertrude afterwards:

> It is very sad about Consuelo, is it not? Anne told me they were to be separated but did not specify whether there is to be a divorce, but as you will probably see Anne you will hear all about it. She made no charge against M. in talking to me except to say he was impossible and that he had insulted C. in every possible way and that for two years there had been trouble. Of course, the English will point to the example of her Mother and Father (i.e. on their divorce) which is unfortunate as that does give M. a leg to stand on, but they certainly

> cannot put anything to C's charge, although I hear the Churchills are furious and are going to be as unpleasant as possible . . . It's excessively sad, and as C. will always be an object of observation under any circumstances, the outlook is unpleasant.[73]

The problem was that malicious and inaccurate gossip about Consuelo's extra-marital relationships endangered not just her social position, but the extent to which she might be able to see her children. Jennie's letter to Winston suggested that she, at any rate, did not think that Consuelo was one of those 'women who have had 20 lovers and are kept by rich Jews et autres', but the Churchill camp were now implying something different. The Duke was certainly making 'charges' wrote Alice Vanderbilt to Gertrude, 'implicating 3 some say 6, but the real reason seems to be that she is physically repulsive to him and that he cannot bear to be near her'.[74] One reason for treating the Duke's 'charges' with caution can be found in another furious letter from George Cornwallis-West to Winston Churchill. He had discovered that Sarah Wilson had been going round suggesting that he, George, had been one of Consuelo's lovers. 'She actually had the impudence to tell Consuelo that she [Sarah] was not certain whether I was her [Consuelo's] lover, and the only reason that she had any doubt about it was because I had borrowed money from Sunny. I don't want to start a row between Jennie and the Churchill family, but I am sorely tempted to go and tell Sarah what I think of her, and if I did she certainly would not forget it in a hurry. I'll come and look you up some time tomorrow. What a liar that woman is.'[75]

In spite of George and Jennie Cornwallis-West's friendly support, however, they had an upset with Consuelo on 2 November that ended in a quarrel at Sunderland House. Jennie felt strongly that discretion in the matter of affairs was essential. She may have let slip her view that Consuelo had, at least in part, 'brought the whole thing on herself' (presumably because of her lack of discretion in going to Paris with Lord Castlereagh) and that Sunny was justified in taking a stand. Consuelo may have seen such even-handedness as betrayal. Afterwards, Jennie Cornwallis-West wrote to her that her conduct was 'inexplicable' and that she was '*deeply* wounded'. 'I make every allowance for the frame of mind you must be in during such a terrible crisis in your life – hurt that you should turn on *me* who have not only been a true friend to you, and had you been a sister could not have shown you more loyalty or affection . . . I do not regret it – God knows you are not in a position to alienate a friend – therefore I will still call myself one.'[76] But the breach was serious, for even by the end of 1907 (when they had, on the face of it, made up the quarrel) Jennie Cornwallis-West refused an invitation to spend Christmas with friends, the Ridleys, because Consuelo would be there; Winston's brother, Jack, felt compelled to refuse the invitation too. From then onwards, the Cornwallis-Wests seem to have spent more time supporting the Duke. There was a quarrel between Churchill and Sarah Wilson too, which the Duke wrote telling him to forget because it had been caused by Consuelo being 'mischievous and malicious'.[77]

Winston Churchill now did his best to bring about a separation agreement between the Marlboroughs which would simultaneously do the least damage to Sunny and protect Consuelo's reputation – essential if she were to see her children and live the life of a separated woman in Edwardian society. Her reputation would be irrevocably damaged if her infidelity with Lord Castlereagh came out in court. This may have been one reason why William K. Vanderbilt was adamant there could be no question of divorce, though it is also tempting to suspect that he did not wish his daughter to detach herself entirely from the house of Marlborough, whatever she might feel about the matter. Even Alice Vanderbilt, no friend of divorce, found his attitude peculiar, writing to Gertrude: 'Her Father will not listen to there being any divorce; queer is it not after his own experience!' Alice also thought it was wrong of William K. to go back to Paris in view of the publicity the matter was inevitably attracting. 'How like a man to get out of the way. Why could not W. K. have stayed in London for at least a few days after the publicity? He got well out the way *before* the papers got the story!'[78]

One reason that William K. may have 'got out of the way' was to avoid Alva who was steaming across the Atlantic to London, ready for a fight. In spite of accusations made by *Town Topics* in the past – that she was all too ready to take advantage of her position as 'the American dowager' – she seems to have been much more ready than her ex-husband to contemplate divorce for Consuelo. The problem was that the terms of Edwardian divorce law gave Sunny the upper hand here – in 1906 a husband could simply accuse his wife of infidelity while a wife had to prove both infidelity and physical cruelty, or desertion and non-support. However, the more sensible of the Duke's friends and supporters felt it was more important to stop him from taking a vindictive and unfair step which would damage his own standing, and make his children's lives a misery. Even if it were true that he was in the stronger position legally, his feelings for Gladys Deacon were well known, and it was the view of the American press, at least, that the Duchess's lawyers had not a shadow of doubt that as a result of the evidence against the Duke, the court would give the custody of the children for at least half the year to their mother.

Hugh Cecil, a supporter of Sunny's, wrote two long and thoughtful letters about the Duke's position to Winston Churchill who was emerging as chief intermediary. He was concerned that the Duke appeared to be prepared to disgrace Consuelo publicly while Alva was threatening legal action to stop him. In the first he argued that: 'I am satisfied after hearing much talk that Sunny is in danger of falling between two stools. What I said to you is evidently true: what he is doing pleases neither the Christians nor the fast set. The Christians feel that whatever his wife may have done at any rate he is blame as himself unfaithful: the fast set do not like a fuss about such a matter and the implied rebuke at their own lives.'[79] Saying that Sunny's position was that 'his wife is unfit to live with him because she went wrong before he did and because the standard for women in these things is higher than for men', really would not impress anyone very much, he added.

In the next letter Hugh Cecil discussed Sunny's view that the timing of their respective affairs somehow made a difference to their relative degrees of guilt. Here, it is unclear exactly what he meant. Consuelo may have had an affair – possibly with Helleu –

as early as 1900 – and a relationship between the Duke and Gladys Deacon may have taken place after that, or in 1903 after he returned from the Delhi Durbar. It is also possible that the Duke and Gladys Deacon finally became lovers in 1906 at the time of Consuelo's affair with Lord Castlereagh; but that too is pure conjecture. In the event, Gladys's name was not introduced into any legal discussion, although it came up frequently in reports of the separation in the American press, along with vague references to 'actresses'. Whatever the truth, Hugh Cecil was clear about two things. Sunny had married Consuelo without loving her, which he should not have done; and Sunny had also been unfaithful. However badly Consuelo had behaved 'he is not and has not been absolved now by her misconduct and that therefore his unfaithfulness is a thing of which he ought to be seriously ashamed. The doctrine that the obligations of the two sexes are quite different is not now believed even by the world to anything like the degree that it used to be'. Cecil went on to say that even if Sunny believed it was his duty to clear himself of all complicity in Consuelo's affair, 'he may not rightly cast stones'; the children's welfare mattered above all else, and that it was 'really very wrong indeed for Sunny out of vindictiveness – for now there can be no question of dishonourable complaisance – to insist on a separation which will ruin their lives'.[80] Hugh Cecil felt strongly that in view of this, the Marlboroughs should attempt a reconciliation.

His views were studiously ignored for both sides were now committed to a legal separation. Far from being wrapped up in a few days, as William K. and Jennie Cornwallis-West had thought, negotiations continued for weeks, punctuated by threats of court action by Alva who tried hard to insert a 'social clause' into the agreement to stop Sunny and his supporters from making the kind of vindictive allegations which George Cornwallis-West had so vigorously repudiated. This in turn was met by threats from Sunny to the effect that if 'the Hag' persisted in talking about going to court, Consuelo could expect no protection 'formal or informal' from him ever again.[81]

By Christmas, a settlement had almost been reached: neither

party would allege infidelity publicly and the boys would spend six months of the year with their father, and six months with their mother. There were concerns on the part of Sunny's cousin and best man, Ivor Guest, that this was tantamount to an admission of Sunny's guilt and he was also worried about the Duke's physical and psychological state: 'He seems quite worn out physically and mentally with the struggle and his judgement I think suffers in consequence', and that: 'He allows himself to be talked over by his opponents (ie Lady Lord Londonderry [sic]).'[82]

Just before Christmas, both Consuelo and the Duke of Marlborough thanked Winston for his help in settling matters thus far. He replied to Consuelo with a charming note in which he said he would always cherish the memory of their friendship. Then, on 4 January 1907 Churchill wrote an uncharacteristically angry letter to his cousin, which he never sent: 'As I fully expected, everything is back again on the war basis . . . Of course, I cannot save you from yourself. If you cannot fight and will not make peace you must just be hunted down and butchered. When I think how near we were to a satisfactory settlement it makes me heartsick to see you cast away your last chance of a decent life by folly and weakness.'[83]

The issue once again seemed to be Sunny's persistent allegations against Consuelo, which Winston Churchill felt were unreasonable and unfair. 'All you were asked to do is give up the pleasure of blackguarding your wife. Rather than surrender that, you will immerse yourself in such shame and public hatred that no one will ever be able to help you any more . . .' Once again, Alva was poised to take legal action to fight defamation of her daughter's character. 'If you are not equal to the task of settling this social difficulty, why won't you entrust it to Cecil or to Ivor? Let them talk to Mrs B. There is still time. But the days are slipping by: and at any moment a step may be taken that will be fatal.' Winston's exasperation with his cousin finally got the better of him. 'Why on earth can't you face the situation like a man? Do your best to help Consuelo to have a fair chance in life, under the new conditions, and forget for a moment your petty pride, your shoddy consistency . . . and the damned fools to whom you listen . . . From the bottom of my heart

I feel for you in your distraction; but if you muddle this business any longer compassion is all you will ever get in the world, and that only from a few dumbfounded friends.'[84]

Although the Duke never received this letter because Winston wisely withheld it, there was clearly a *froideur*. The next letter to Winston came from the Blenheim estate office asking him when his three ponies might be leaving. But at the end of January the Duke wrote a warm-hearted letter in response to one from Winston who was concerned by his silence, and they made up the quarrel. By now, terms had finally been agreed and ratified by William K., although there were still rumblings about whether or not Consuelo should be allowed to visit Blenheim. The Duke had delayed making contact with Winston, he explained, because he knew that Mrs Belmont felt that Winston was her intermediary and 'relies on your influence to work on me'. She had also tried to manipulate Ivor Guest and Hugh Cecil, but as soon as she realised that the Duke had broken off contact with them she agreed to the terms of the settlement very quickly. 'My dear you tried to bring pressure on me to do what you wished but not what I wanted,' wrote the Duke. 'You must forgive me if . . . I had to pretend to be a stranger from you. It was the only way I could triumph over that old Hag. She is now utterly deflated.'[85] *Town Topics* concurred with this latter description, observing with ill-concealed glee that on Oliver Belmont's Long Island estate even the rare black and white pigs were squealing in sympathy, the Jersey calf brayed sonorously and 'the stepfather of the Duchess looked the picture of dead hope'.[86]

Newly separated, Consuelo accepted an invitation from her father to escape publicity's glare by joining him on a *Valiant* cruise in the Mediterranean, accompanied by her two sons. Between the beginning of January and 18 February 1907, she found herself back on board, drifting once more in the company of her Vanderbilt family from France to Italy, Greece, and Tunis. Of the two, however, it was the Duke who appeared more miserable. It was agreed even by those who liked him that he had not married Consuelo for love; but he seemed to suffer more when they separated, afflicted again by the melancholia that was never far away. In April 1907, he went with Daisy, Princess of Pless and Jennie Cornwallis-West on a motoring expedition from Beaulieu-sur-mer, on the Riviera, to Avignon. But everything seemed to go wrong. The Duke's car ran out of petrol and he arrived in Avignon hours late; and after dinner when he pushed a table out of the way, it collapsed. 'Down the whole thing went,' wrote Daisy, 'Dessert, wine, butter, olives, dates, plates: the corner of the room into which everything fell looked like a pig-sty . . . The Duke was miserable; by the way he looked at the debris one might have thought he was peering at his own life, which at the present moment is in much the same state.'[87]

PART THREE

8

Philanthropy, politics and power

DURING THE SEPARATION PROCEEDINGS those close to Consuelo were worried about whether London society would accept her as an independent duchess, particularly when an indignant King Edward VII and Queen Alexandra made it clear that both Marlboroughs were *persona non grata* for the heinous offence of bringing their marital difficulties to public attention. Shared custody of the children was regarded as a concession in Consuelo's favour, however, and Winston Churchill's valiant efforts to protect her reputation by keeping the case from the courts proved successful.

London society rallied round both Duke and Duchess, refusing for the most part to take sides and accepting invitations from both. Although some of the Churchills remained hostile, the rift with Jennie Cornwallis-West never became widespread knowledge, and Consuelo was helped by the continuing support of the powerful Lady Londonderry, well aware of Charley's record as a philanderer, and anxious to sweep the whole affair under the carpet. Consuelo's popularity and widespread sympathy for her position also helped. In America, the public concurred with President Roosevelt's private opinion that 'the Duke of Marlborough [was] a cad',[1] and the press collectively decided that the Duchess was the victim of the story. Articles appeared denouncing heartless mothers and international marriages, and many people she had never met wrote to wish Consuelo well.

Nonetheless, Consuelo's position as a separated woman in society was anomalous and difficult. She was a celebrated beauty and was only twenty-nine when the Marlboroughs parted. Great self-discipline was required, for the slightest hint of another scandal would have played into the Duke's hands had he been minded to act vindictively and made her vulnerable to losing her children. As it was, she only had the care of her sons for six months each year, and for much of the time they were away at school.

After the separation, Winifred Fortescue, who met Consuelo as a star-struck seventeen-year-old when her father was Rector of Woodstock, came to London from Oxfordshire to train as an actress. The move, which was designed to ease the strain on her family's finances, had come about at Consuelo's suggestion, and Winifred lived in lodgings close by in Shepherd's Market, partly so that Consuelo could keep an eye on her. She was a frequent visitor to Sunderland House and knew Consuelo felt isolated. 'Her loneliness now tore my heart,' Winifred Fortescue wrote in her autobiography. 'She had everything in the world except the things that matter most. After she had kissed me farewell I hated to hear those little heels of hers clicking away from me across the marble floors into the dim desolation of that great French palace, fragrant with lilies and incense and lovely with antique brocades and priceless porcelain.'[2] The Duchess needed an absorbing interest that would enhance rather than damage her reputation in the difficult period after the separation and she soon found one.

Consuelo had become involved in philanthropy beyond the confines of Blenheim and Woodstock long before separating from the Duke, and these activities gained momentum in the year before her marriage collapsed. From January 1906 she was to be seen accompanying Daisy of Pless to open a club for young men in Dulwich and returning the vote of thanks; selling needlework by women in Church Army homes; and presenting prizes to the London School of Medicine for Women – an event presided over by Mrs Garrett Anderson and Lady Frances Balfour, sister-in-law of Arthur Balfour, and leading campaigner for both suffrage and women's education. At the prize-giving Consuelo remarked that the hospital and its supporters 'had a justifiable source of pride in knowing that women were independent and able to contribute as freely as men to the general utility of the world' and that 'in the medical profession the services rendered by women were invaluable and scope for their activity was still limitless',[3] a strikingly feminist note for 1906 and one to which she would frequently return.

Consuelo now turned to serious philanthropy where she would focus more consistently on social problems she had already encountered, but where marriage to Sunny and his particular strand of conservatism had left limited room for independent manoeuvre. Her first ambitious undertaking after the separation was at the behest of the charismatic Prebendary Carlile, founder of the Church Army. She had already opened a labour depot run by the Church Army in 1905 and she invited Prebendary Carlile to Blenheim in June, where he signed 'Church Army' beside his name in the visitor's book to make it clear that he was present in an official capacity rather than succumbing to the lures of society.

The Church Army had been formed in 1882 as an attempt to reconnect the Church of England with its working-class base by sending groups of working-class men and women to work as evangelists in the poorest districts of Westminster. Its image as an evangelistic movement was burnished by the Reverend Carlile himself who regularly led marching bands into the slums wearing his surplice and playing the trombone (causing at least one indigent – of whom George Bernard Shaw might have been proud – to enter a protest at 'the insolent condescension of well-fed persons who intrude themselves in this way upon the sufferings of the very poor'[4]). Soon after its foundation, the Church Army also initiated a host of projects designed to tackle problems it found in the slums, such as homelessness, hunger, and alcoholism.

As his trombone-enhanced performances suggest, Prebendary Carlile had a certain flair for publicity. 'I have always found that money spent in advertising comes back in the collection,'[5] he once said. (He was the first to introduce the cinematograph into a parish church and his sermons often featured sporting analogies, causing one old lady to remark: 'Since my old man has been coming to this show church, he don't go no more to the music hall.'[6]) Prebendary Carlile also had a talent for harnessing the power of celebrity.

Although he castigated extravagance and luxury – or perhaps because he did – he was extremely good at winning the support of members of the English aristocracy with a conscience. By 1907 he knew Consuelo well enough to enlist her help with a project that he knew would appeal to her in her own difficult position. It was for prisoners' wives – powerless women rendered helpless by marriage, 'punished for the guilt of others' as Consuelo wrote later, which was 'essentially unfair'.[7]

The project she directed for Prebendary Carlile was designed to empower prisoners' wives by giving them financial independence. It was part of Carlile's genius that he not only appealed to Consuelo for her money and celebrity value, but to her executive competence, developed by running large households for over a decade. The Church Army project involved her in buying and equipping a centre in two adjoining houses in Endsleigh Street, providing them with laundries and sewing-rooms, and establishing a crèche where children could be looked after while their mothers earned a living wage. It was a venture of much greater complexity than any of her previous philanthropic undertakings, and it proved that she had a talent for organisation. When the centre opened at the end of May 1907, *The Times* made it clear that she had been in charge.

The centre also provided a swift education in reality for an inexperienced duchess drawn into the day-to-day running of such an establishment for the first time. Part of its purpose was to help first offenders into work when they emerged from prison and at least one felon talked Consuelo into giving him some tools which he promptly used to go burgling again. This was bad enough, but Consuelo was startled to be told off in no uncertain terms by his wife – who thought that she should be allowed continued use of the facilities – that the Duchess had been as good as accessory to the crime by naively acceding to his wishes. As part of her work for the Church Army, Consuelo was asked to provide religious leadership. 'At Prebendary Carlile's request, I closed our day's work with prayer, and I can still feel the emotional tension with which those sorrow-laden souls filled our simple service, "We like the Duchess to read to us," they said, "but she always makes us cry."

For me there was comfort in feeling that for once their tears were not bitter.'[8]

It is easy to scoff at such class-bound sentimentality, to write off such engagement with the poor as emotional therapy for an unhappy aristocrat, or to dismiss this kind of involvement as motivated by Consuelo's need to re-establish her social position after a much-discussed marital breakdown. But even though there may be a grain of truth in such assertions, it would be wrong to stigmatise *serious* philanthropy in this way. In recent years, historians have re-examined nineteenth- and early-twentieth-century female philanthropy and come to see it as an important factor both in an extension of female power in general and the campaign for the vote in particular. 'Women have traditionally used these activities to wield power in societies intent upon rendering them powerless,' writes Kathleen D. McCarthy. 'Unlike men, who enjoyed a host of political, commercial, and social options in their pursuit of meaningful careers, women most often turned to non-profit institutions and reform associations as their primary points of access to public roles. In this process, they forged parallel power structures to those used by men, creating a growing array of opportunities for their sisters and themselves.'[9] This was partly because philanthropic work was not always as easy as the Lady Bountiful image suggests: to work, often unwelcome, in inner-city slums required courage and resilience. More important, philanthropy has also been reinterpreted by historians as an important social force, closely linked with both the history of social welfare provision and the emergence of feminism, in that it provided women with a route into the public sphere.[10]

In the first instance, women were able to use philanthropy as an entry point into public life without agitating conservative elements because it was seen as an extension of their role as wives and mothers and, in the case of rich or aristocratic women, as part of their traditional role. Motivation for taking on philanthropic duties ranged from traditional expectations, an escape from boredom, a sense of duty, to religious impulses – the historian Frank Prochaska suggests that Christianity was often a crucial factor in propelling women into this kind of work, and to this extent Consuelo's involvement with the Church Army was characteristic.

Whatever their reasons for taking up philanthropy, however, some women then discovered that they rapidly had to develop professional skills such as public speaking in order to be effective. This in turn changed the way that women's capabilities were perceived. Such changes came about only gradually, touched off by women determined to get their message across and willing to test convention by addressing charity meetings, social science congresses, and trades union gatherings. 'By such actions they broke down the prejudice against women speakers and made it easier for the less forthright to express themselves in public without fear of obloquy. By enlarging the scope of women's activities, charitable work also modified the way in which people interpreted the possibilities inherent in the female character.'[11]

As important, late-nineteenth-century women involved in serious charitable work found it took them to the heart of topical issues. Philanthropy was central to the Victorian approach to the problems of industrialisation. It was regarded as crucial to the fabric of society, a reliable and wholesome remedy for social ills and the principal conduit for the redistribution of wealth. This meant that in entering philanthropic roles women were part of a key expression of Victorian social values and found themselves at the centre of Victorian – and Edwardian – debates about social policy. One consequence of this – and this was soon true of Consuelo – was that it was often a short step from executive involvement in philanthropy to involvement in issues that many liberals and progressives increasingly regarded as political – such as the care of the elderly, child welfare, family poverty and exploitation of female workers. It was then a very short step to women becoming intensely frustrated at being unable to influence social policy in these areas because they were denied the vote.

Between 1907 and the end of 1908, partly as a result of emancipation from the Duke's conservatism, and partly as a result of her growing experience as a philanthropist, Consuelo's interest in social welfare problems, the position of women and politics developed rapidly. Her project in Endsleigh Street opened her eyes to the problems faced by women who lacked the training to do work of their own. In an article written in 1908, Consuelo argued strongly in favour of the advantages for women of lobbying for change in alliance with men through the trades union movement and was critical of working-class women who refused to engage: 'It still remains a problem how to induce women to fight for their rights,' she wrote. 'The writer herself has visited Clubs in the East End of London, composed of girls working in factories, where they earned a miserable pittance and worked from eight to twelve hours a day. When she tried to impress on them that these were not fit conditions for their acceptance, and that their Club should become a co-operative union to resist unjust terms, instead of being merely a social centre, they smiled hopelessly as if at some wild but yet pleasant fancy and returned to the dreary monotony of things as they were and would to them remain.'[12]

By April 1908 Consuelo was taking such a keen interest in working-class problems that she was forced to defend herself against accusations of having become a socialist, telling reporters at the dockside in New York: 'I am not a Socialist . . . Of course I do not mean by that that I do not want to do all that I can for those who are unfortunate . . . but that does not make me a Socialist. I have never expressed such views, and I cannot understand where they get such ideas about me.'[13] Given her support at that time for women fighting for their rights through the trades union movement, the mistake was perhaps understandable, but Consuelo was not and never would be a socialist. Away from her staunchly Tory husband, she was now freely endorsing 'New Liberal' thinking, a political position expressed cogently by cousin Winston Churchill in 1906, two years after he crossed the floor to join the Liberal Party: 'No man can be a collectivist alone or an individualist alone . . . No view of society can be complete which does not comprise within its scope both collective organisation and individual incentive. The very growing complications of civilisation create for us new services which have to be undertaken by the state.'[14]

Consuelo would have concurred with this wholeheartedly. As

her exposure to social welfare problems increased, she soon came to think that the state had an important role in welfare reform, and that some problems of industrialisation were too great to be solved by the individual alone. Consuelo agreed with T. H. Green, a key writer on New Liberalism, that the state had a role in regulating individual behaviour particularly in relation to child health. Her own experience, even at Blenheim, probably would have left her unsurprised at the findings of Seebohm Rowntree and Charles Booth that about 30 per cent of Britain's town dwellers lived in poverty, and that among agricultural workers it was even worse. She strongly supported the Liberal government's programme of social reforms, led by David Lloyd George and Winston Churchill – especially the children's acts which provided free school meals and medical inspections, the introduction of the old age pension and the National Insurance Act of 1911. At the same time she shared New Liberal views that individual effort and initiative were important, that the distinction between deserving and undeserving poor should remain, and that wherever possible, social welfare services should be paid for by voluntary effort rather than the state.

The philosophy of New Liberalism had emerged from England's new professional middle classes. Consuelo was unusual in possessing great wealth and finding the key tenets of New Liberalism appealing, though she was certainly not unique. By 1908 she was arguing publicly that it was the task of plutocrats to distribute their own wealth in a socially useful manner. Far from being a socialist, however, she believed that a widespread sense of social responsibility on the part of the rich, in conjunction with a political programme of social welfare reform, was the best way of drawing the sting of socialism which, like many of her wealth and class, she regarded as an evil. These were certainly not the views of a radical hothead, but they had moved on some way from those of the Consuelo who had gently teased Millicent Sutherland during the trip to Russia in 1902 by calling her the 'Democratic Duchess'. Such opinions also set Consuelo at a great distance from the opulent fantasy of Alva's Gilded Age palaces and the Duke's belief that poverty was a question of individual responsibility and that the poor should know their place.

Consuelo's serious interest in philanthropy both enabled her to continue participating in public life and stood her in good stead in English society. *The New York Times* printed rumours of a 'boycott' by society after the separation, but there is no evidence of this at all. It also printed a front-page story that she had 'returned' to society by November 1908.[15] In her memoirs, Consuelo thought that the breakthrough in attitudes towards her had come about a year after the separation (around 1908) when the *beau monde* turned out in force for a glittering reception at Sunderland House to hear the violinist and composer Fritz Kreisler. Fragments from a diary kept by Consuelo in 1908 talk of her being placed beside the new prime minister, Herbert Asquith at dinner, and of six parties in one day and three balls on one night.[16]

Once again it was the Duke who appears to have suffered more from feelings of exclusion. Understandably he felt the pain of expulsion from royal circles more keenly than Consuelo, and at the end of 1908 he was the one whom friends changed plans to support. Daisy of Pless wanted her brother George and sister-in-law Jennie Cornwallis-West to stay on with her in Furstenstein for Christmas, but they felt they had to go back to England 'as "Sonny" Marlborough is a bit low, Jennie says'. One reason for his depression was that 'his mother and sister go to Consuelo's London house and help her at her charity meetings and so on. Jennie says: "Of course they do as she has all the money",' an explanation which Princess Daisy trenchantly observed was nonsense.[17] Consuelo did not, of course, have 'all the money' since the Duke's part of the Vanderbilt marriage settlement was unaffected by the separation.

One testimony to Consuelo's charisma as a philanthropist came years later from the actress Dame Anna Neagle who was brought up in a working-class area of East London. 'I was six when I saw the then Duchess of Marlborough . . . When I saw her she would have been in her thirties. She came to have tea in our house when we lived near the London Docks, at West Ham. My mother was the honorary secretary of the local branch of the British Sailors' Society, and I imagine the Duchess was giving her support to some bazaar or sale of work for this. All I remember is her loveliness: not just her face, but in everything about her. The exquisite sapphire blue of the long velvet gown she wore; the touches of fur at her wrists and throat; the pale serene beauty of her face, the charm of her smile. She left some sort of spell over my childish mind so that, when she had gone, I carefully wrote her name on the underside of the chair in which she had sat.'[18]

Consuelo also began to attract international attention as a result of her philanthropic work, particularly in America. In the spring of 1908 she made another trip back to the States. As usual, her arrival attracted attention, and this time Mrs Cornelius Vanderbilt (wife of her cousin Cornelius and later *the* Mrs Vanderbilt) gave a well-publicised dinner in her honour at 677 Fifth Avenue, with much of the New York *gratin* in attendance. A week later, after some arm-twisting by Colonel George Harvey, editor of the distinguished American magazine the *North American Review*, she made her American public speaking debut at a dinner in honour of Mrs Humphrey Ward to raise funds for children's playgrounds in America. Consuelo took advantage of this platform – and the attention she was getting – to articulate a new conviction.

Rich American women were too idle, she said. They should follow the example of their English counterparts and make a useful contribution to society. Life as an English duchess had given her a chance to observe the extent to which, in England 'the influence of women has permeated every field of human endeavour, political and philanthropic'. In England, said Consuelo, donating money was not enough. 'Personal direction is exacted and freely given, and it is this expression of human sympathy that knits closely together the widely varying elements of the community.' America was different because there was no tradition of such activity. 'But does not the mere privilege of citizenship in a Republic such as this involve personal responsibility and place it on a far higher plane?', Consuelo asked. 'Is it not possible for the women citizens of this great Republic to recognise that personal obligation on its ethical basis and to turn it to account in practical works?'[19] Alva, who was present at this dinner with Oliver Belmont and several hundred of New York's finest, can only have been delighted by what her daughter had to say. It was nothing less than a great public vindication of her original decision to marry Consuelo off to an English duke.

Consuelo sat down to considerable applause, although Mrs Humphrey Ward – soon to emerge as the pre-eminent English anti-suffragist – rose to her feet and remarked somewhat tartly that as far as playgrounds were concerned, America was well ahead of England in all respects. This was largely lost on the New York press. The *New York Evening Journal* marvelled at the change which had come over 'the shy little American girl, timid, fearful, unable to realize the immense power of the millions of dollars which were hers'.[20] Colonel George Harvey promptly invited Consuelo to expand what she had just said into a series of articles for the *North American Review*. This was a great honour and a deeply flattering endorsement of her new role as a figure in public life: the *Review* was a highly distinguished magazine and the leading arbiter of opinion in the United States; its contributors had included no fewer than ten presidents by the time George Harvey acquired it in 1899.

Consuelo developed her April speech into three articles entitled 'The Position of Women' during 1908. *The Times* printed extracts from advance sheets on 30 December and the three articles then appeared in consecutive issues of the *North American Review* in January, February and March 1909. *The Times*'s extracts made it clear that Consuelo was explicitly in favour of women having the vote as one remedy to social welfare problems and anyone reading the articles as they appeared in the *Review* would have been left in no doubt at all that this was her view.

The theme of all three articles was the exclusion of women from public life, the 'gradual narrowing and restricting of her sphere to the present day, when woman is at length attempting to re-establish the balance of primitive rights as well as to gain the economic and political equality civilisation brings in its wake'.[21] Much of the first and second articles was taken up with a long historico-anthropological analysis that would not pass muster now. Nonetheless it was a bold attempt to grapple with the historical background to the exclusion of women from public life in the absence of any serious contemporary study of the question. It showed too, quite clearly, how feminist Consuelo's thinking had become by 1908, and how exercised she was by the question of female exclusion from economic and political influence.

In these articles Consuelo argued that women had power in primitive societies which was usurped when men ceased to be hunter-gatherers and took the female agricultural role for themselves. But women had not done enough to resist these and other forms of economic domination, she argued, nor to resist the pernicious ideal of blind obedience to men that had traditionally been imposed on them: 'It is in my opinion the necessity to adjust herself to man, to be judged by his individual standard and to conform her whole personality to his ways of thinking, that has robbed woman of the power, strength and influence she could have exerted as a united and independent majority.'[22] This, wrote Consuelo, was a moral code unfit for anyone other than a slave.

Although she would later dissociate herself from militant suffragette tactics, Consuelo wrote these articles at a time (1908) when suffragette militancy was still emerging in England and was not the divisive issue between campaigners that it soon became. In these articles Consuelo expressed an understanding of female militancy represented by the Pankhursts and their organisation, the Women's Social and Political Union (WSPU), and was supportive of their attempts to reclaim political power for women. 'It is because womanly measures have failed to open the gates that they have resorted to more masculine ones,' she wrote. 'Not because they enjoy going to prison or making themselves objectionable, but because they know that no great reform has ever been brought about without public agitation of a more or less aggressive character on the part of those directly concerned.'[23] As a general principle Consuelo believed that women were most effective in resisting exclusion at times of great upheaval. Once powerful organisations such as the Church achieved stability, however, a woman's public life 'became more cramped, and narrowed to one of pure domesticity, and her influence in affairs outside the domain of the home was *nil*'.[24]

The third article, entitled 'Expanding Activities and Opportunities in America and England', was the most explicitly political. Having analysed the manner in which women had historically been squeezed out of public life, Consuelo proceeded to suggest a way in which married women in America could reintegrate themselves. Here she returned once again to the same analysis that Alva had used to justify marrying Consuelo off to the Duke of Marlborough. Rich American females were too ready to lead a life of frivolity into which 'many brilliant women are apt to degenerate into from lack of opportunity and purpose'.[25] They should copy their English sisters by setting it aside and taking a role in public life. The presence of women would de-contaminate American public life from scandal. If they did not take up the challenge and socialism took hold, argued Consuelo, the wealthy would only have themselves to blame. Although 'the responsibilities of great wealth' were often 'puzzling, discouraging and strenuous', such work could be of 'intense interest' to the rich women of America, 'where men leave so much of the distribution of wealth in their hands'.[26] This was 'advancing socialism in its most favourable aspect', Consuelo thought, an arrangement by which every member of the community had a 'given task and an appointed place in the working of the great state machinery'.[27] She ended by saying that this kind of public work was an important dress rehearsal for the time when the campaign for female suffrage was ultimately successful.

Consuelo repeated a slimmed-down version of these views several times throughout 1908 and 1909. 'American Duchess Condemns Idleness' screeched a headline in *The New York Times* in 1909 when she opened a flower show in the East End of London. 'She said she was a great believer in the sound judgement of the working man,' the newspaper reported breathlessly. 'She was a great believer in work as the best discipline and she wished everyone rich as well as poor was obliged to work a certain number of hours every day.'[28]

It all gave Alva great satisfaction when she discussed Consuelo's marriage with Sara Bard Field in 1917. 'Once when [Consuelo] had come back to America to make me a visit she confessed that she was appalled by the emptiness of the lives of rich American women . . .', Alva said. 'Their lives were vapid and meaningless, starved and bored. "In spite of all that has happened," said my daughter, "I am glad I married an Englishman." Looking about me at the ineffectual living of the leisure woman in American, I echoed her gladness.'[29] What Alva overlooked, however, was that it was only *after* Consuelo separated from the Duke that she was able to translate the role of Duchess of Marlborough – and her share of the Vanderbilt wedding settlement – into a position of independent influence.

Alva's pleasure at Consuelo's acceptance of the benefits conferred by becoming an English duchess was deeply ironic. After her marriage to Oliver Belmont in January 1896 she became the embodiment of precisely the aimless life led by American society women criticised by Consuelo. Her elder son, Willie K. Jr[30] had left Harvard to marry Tessie Oelrich's younger sister Virginia 'Birdie' Fair in 1899 (they intended to spend their honeymoon at Idle Hour as Consuelo had, but it burnt down on their wedding night). Harold lived with the Belmonts throughout his teens, though like his older brother he was educated at boarding school and was often away.

Alva's marriage to Oliver Belmont had started in controversy when she antagonised her Belmont in-laws by refusing to part with the trust fund set up for her by William K. Vanderbilt at the time of their divorce in 1895. The outcome, however, was that the Belmonts became a spectacularly rich couple with little to do, moving between a house in New York, three estates in Newport (Belcourt, Gray Crags and Marble House) and a house at Hempstead on Long Island where they spent the greater part of each year. This was Brookholt, commissioned by the Belmonts from Richard Howland Hunt, son of Richard Morris Hunt in 1897, an interesting triangular collaboration which resulted in a cross between an eighteenth century French chateau and an English Palladian house with exceptionally large stables designed in the French manner.

The lifestyle of the leaders of Newport society, meanwhile, reached breathtaking levels of vapidity in the years before 1914.

Many of the worst offenders were the Belmonts themselves and their immediate circle. Mrs Astor's power had slowly waned in the late 1890s and ebbed away entirely in the years between 1905 and 1908 when her memory began to fail. 'A legend sprang up that discreet servants still went through the routine of announcing guests and pretending to pass dishes and pour wines at a long empty table, at the end of which the witless old lady – dressed in a Worth gown and festooned with chains of diamonds – continued to converse left and right with imaginary guests long dead,'[31] wrote Louis Auchincloss. (The legend of Mrs Astor's last years would inspire Edith Wharton's short story 'After Holbein'.)

In the vacuum left by Mrs Astor's slow fade from the real social world to an imaginary one, there was a sharp tussle for Newport supremacy between Alva and Mrs Ogden Mills – for whoever won Newport became de facto queen of New York's elite. 'Between Mrs Belmont and Mrs Ogden Mills there existed a perpetual state of feud which extended itself to their respective courts. In temperament they were diametrically opposed . . . Mrs Belmont, warm-hearted, impulsive, aggressive . . . Mrs Mills . . . cold, sarcastic and aristocratic,' wrote Alva's friend and fellow socialite Elizabeth Drexel Lehr. There were difficulties with both camps. Mrs Ogden Mills 'made a cult of rudeness' and was so extremely exclusive that she slashed the magic figure from 'the Four Hundred' to the 'Two Hundred' and maintained there were only twenty families in New York who counted. This turned out to be taking exclusivity too far. Alva, on the other hand, was incapable of diplomacy. Here, the general view was that it was 'impossible to have a Queen of "the Four Hundred" who could sign only declarations of war'.[32] In the end, there was no outright winner because, according to Elizabeth Drexel Lehr, Mrs Odgen Mills had too few friends, and Alva too many enemies.

Before Mrs Astor's death in 1908, however, the press finally pronounced in favour of a triumvirate. Mrs Ogden Mills was beached by her own exclusivity, and the crown was awarded jointly to Alva, her great friend Tessie Oelrichs and Mrs Stuyvesant – or 'Mamie' – Fish. The triumvirate was aided in its social endeavours by Harry Lehr, their pet cotillion leader and Ward McAllister's natural successor. (McAllister had died, unmourned by most of those he had assisted including Mrs Astor, in 1895.) Lehr's arrival signalled the end of the solemn certainties of Mrs Astor's world for he had the nerve to tell her, among other things, that her diamonds made her look like a chandelier. Mrs Astor – after a pause – decided she was amused, and took him up briefly before her decline. Lehr then migrated to the great triumvirate, and to Mamie Fish in particular, whose amanuensis and court jester he became.

Knowing Harry Lehr as they did, it seemed extraordinarily naive (at best) of Alva, Tessie Oelrichs and Mamie Fish to have encouraged Lehr to propose to Elizabeth Drexel, a young heiress from Philadelphia. Elizabeth's wealth enabled Harry Lehr to live comfortably as a gentleman of leisure, but she endured years of unhappiness when he rejected her savagely on their wedding night, saying that he hated women, had lied about loving her, and would only tolerate a *mariage blanc*. 'He delighted in women's clothes and was never so happy as when he was helping some woman or other of his acquaintance to choose dresses ... In his diary he wrote "... Oh, if only I could wear ladies' clothes; all silks and dainty petticoats and laces, how I should love to choose them,"'[33] according to his wife. Few people (in her version of the story) ever suspected how vicious he was to her in private; but her mother hated the idea of divorce and Elizabeth remained married to him until he died. Her unhappiness, however, caused her to view Newport society with an unusually detached eye while moving at its centre, resulting in two memoirs written several years later which are among the best of the period.

The parties Harry Lehr organised with the triumvirate hovered uneasily between magical fantasy and teasing cruelty, complicated by a collective streak of ambivalence on the part of its new leaders towards Newport society in the first place. Mamie Fish in particular saw herself as a social anarchist. 'Make yourself at home, and believe me, there's no one who wishes you there more heartily than I,'[34] she is said to have told one caller. To another, searching for her niece before leaving a party, Mrs Fish is said to have enquired

The Gold Room at Marble House, Newport, Rhode Island.

Consuelo's bedroom at Marble House, much as it would have looked in 1892–5.

Left A photograph of a portrait of Alva by Benjamin Charles Porter, commissioned while she was still married to William Kissam Vanderbilt. She destroyed the painting after she took up the suffrage cause.

Below Consuelo styled as an English duchess by Carolus-Duran, commissioned by Alva in 1894.

The 9th Duke and Duchess of Marlborough with Their Sons the Marquess of Blandford and Lord Ivor Spencer-Churchill by John Singer Sargent, 1905.

The Duchess of Marlborough and Her Son, Lord Ivor Spencer-Churchill
by Giovanni Boldini, 1906.

The water terraces at Blenheim designed by Achille Duchêne, built by the 9th Duke of Marlborough during the 1920s.

Opposite Consuelo at home at Casa Alva, Palm Beach, May 1950.

Consuelo and Jacques Balsan on Fisher Island, Maine, 1952.

silkily whether she had looked under her (male) secretary. On the magical side of the balance sheet, Mrs Oelrichs gave a *bal blanc* where everyone came dressed in white, white swans swam in her fountains and a white flotilla of illusory, illuminated boats shimmered on the ocean just beyond her garden balustrade to create a mirage of a house beside a harbour. On the debit side, Mrs Stuyvesant Fish almost sacrificed her position as a leader of society when she conspired with Harry Lehr to give a dinner in honour of 'Prince del Drago'. Newport society flocked to meet him, only to discover that the 'Prince' was a chimpanzee.

Part of the problem was boredom. The summer season became dull, the social competition tedious. 'Newport again! Another season; the same background of dinners and balls, the same splendours. The same set, the same faces, here and there a few lines on them carefully powdered out – no-one could afford to get old, to slip out of things,' wrote Elizabeth Drexel Lehr. There was an infamous 'Dog's Dinner' where owners fed foie gras to their dogs, while the poor starved; and, in just as poor taste, a 'Servants' Ball'. 'The door of "The Rocks" was opened by Henry Clews, attired as a valet and holding a duster in one hand and a kitchen pail in the other.* Behind him was Mrs Oelrichs with a large mop, industriously polishing the floor. Oliver Belmont, a little feather dusting brush stuck into his cap, was acting as cloakroom attendant, taking charge of coats. The funniest part of the evening was the dinner, cooked by the guests.'[35] The servants, who presumably had to clear up the mess, were given the night off in a late outbreak of sensitivity but they could have been forgiven for thinking that this was a grim sort of fun.

Town Topics kept up a campaign against the Belmonts from the time of their marriage onwards. The magazine's preferred line was that Oliver and Alva were a seedy couple who did little to earn the respect that they felt they deserved, who more or less wrote their own publicity material in the *New York World* and the *New York Herald*, and who had high hopes for the benefits conferred by their relationship with Consuelo. It particularly liked to publicise

* His wife was said to budget $50,000 a year for sartorial mistakes.

moments when it looked as if these had been dashed. In June 1897, for example, a correspondent wrote that 'the home-coming of Mr and Mrs Oliver Belmont before the Jubilee celebrations together with the absence of their accounts of their entertainment at Blenheim, has naturally given rise to the notion that the worthy couple were just a little disappointed in their reception by the Marlboroughs, and that the rift in the lute, which, it was predicted, would start as soon as the young Duchess had become at all acclimatized, is already visible'.[36] Any suggestion of a rift should be treated with caution, for Alva was back at Consuelo's side for the birth of Blandford in September 1897. However, *Town Topics*' implication that the Marlboroughs were wary of providing an entrée for the Belmonts into English society may contain more than a germ of truth, as may its allegation that the Duke found William K. Vanderbilt the more congenial in-law. The Blenheim Palace visitor's book certainly suggests that he was a far more frequent visitor than Alva, though it is equally possible that he simply signed it more often.

It should also be noted that Colonel Mann of *Town Topics* bore particular animus towards the Belmonts, because, to their credit, they consistently refused to surrender to the Colonel's attempts at extortion. In 1906, Oliver took the stand as a witness in a court case which exposed the Colonel's blackmail racket and, according to Andy Logan, 'testified that in 1899 Colonel Mann had come to him and asked him for $5,000 in return for some shares of *Town Topics* stock, that when he refused Mann had written a letter, which Belmont produced in court, soliciting a straight loan of $2,000, and that after Belmont had again refused (and also turned down an invitation to appear in another magazine, *Fads and Fancies*), some fifty abusive items about him appeared in the magazine'.[37]

This did not mean that all the abuse was untrue, however. The Colonel went to some lengths to make sure that at least some of what he wrote about society's high fliers was accurate because this gave him power. Oliver Belmont's stand against Colonel Mann was in marked contrast to that of William K. Vanderbilt, who turned out to have been a leading 'immune' from the mid-1890s onwards – a privilege for which he paid out more than $25,000 and which may be a reason why his affair with Consuelo, Duchess of Manchester, was treated gently by the magazine.

Throughout the years of her marriage to Oliver Belmont, *Town Topics* constantly poked fun at Alva, who was not an 'immune', reminding everyone that she, the Mother of the Duchess, flaunted cables from Consuelo whenever she could – especially at the opera when she thought people were watching – and 'exhibiting her daughter' at an exhibition while the Duchess was in New York in 1908 (*Town Topics* went on to say 'had Alva worn a yashmak the two of them together would have been beautiful'[38]). The magazine reserved particular scorn for the time Alva auctioned off Consuelo's old child's donkey cart in aid of the Nassau Hospital at Minneola. This attracted huge attention from 'thousands of free, virtuous and intelligent American citizens . . . thoroughly imbued with the democratic spirit',[39] just as Alva must have known it would. The magazine's correspondent was particularly incensed by the way Mr and Mrs Belmont were then unable to stop themselves from announcing to the assembled company that Consuelo would soon be coming to stay.

Despite *Town Topics'* mockery, and a later change of perspective, Alva looked back at these years with nostalgia. 'It was an ideal life,' she said later; 'thoroughly refined but full of gaiety and fun.'[40] By the standards of other periods in her life, however, her achievements during her marriage to Oliver Belmont were slight. Given her strong feelings on the subject of parental involvement in choosing a child's spouse, it is not surprising that she was said to have orchestrated and encouraged the engagement of her son, Willie K. Jr, to Virginia Fair. After his marriage to 'Birdie' in 1899 at the age of twenty, Willie K. Jr pioneered the introduction of the automobile to Newport, a move in which he was publicly backed by Alva and Oliver, who drove a car of their own, decorated with flowers and two large butterflies, at the first automobile parade down Bellevue Avenue. Newport finally objected to the noise and dust of this new invention, repelled, perhaps, by the 'hilarity' of new diversions such as automobile obstacle courses involving nurses with dolls in perambulators.

As a result, Willie K. Jr moved his automobile experiments to Long Island. Alva and Oliver continued to lend their public support as he encouraged American manufacturers to develop high-performance engines by starting the Vanderbilt Cup Races in 1904, and developed the Long Island Parkway in 1908 where cars could race and drivers with a taste for speed could test new engines 'without keeping one eye on the constabulary'.[41] The Belmonts shocked society by introducing the first shower, at Belcourt; and on 18 July 1907 it was reported that Alva had taken control of the Newport Bridge Club. 'It was a great day for Mrs Belmont,' wrote *Town Topics*, 'and she bustled about with all the importance of a shipwright about to launch a battleship.'[42]

Otherwise, Alva's battles were restricted to taking up the cudgels on behalf of those whom she felt were suffering unfairly from the playground ethics of Newport society. One example was the manner in which she championed the cause of Mr and Mrs William Leeds, the 'Tinplate King' and his wife. For two years their fate in Newport hung in the balance. They made all the right moves, arriving quietly, and renting large houses while their owners were away, but acceptance seemed to elude them. One socialite who had let them her house was roundly castigated by the ultra-exclusive Edith Wetmore who is alleged to have said: 'How can you lease to those horrible, vulgar people? Why, the whole house ought to be disinfected after them!' In her memoirs, Elizabeth Drexel Lehr relates how Alva saved them: 'The Leeds found a champion in Mrs Oliver Belmont, that valiant warrior to whom opposition was as the breath of life. Nothing made her happier than the knowledge that she was pitting herself against the rest of the world. She loved to see herself as a pioneer, to make others bend to her will, to have them follow her in the end, meek, sheep-like . . .'[43]

In spite of the empty tenor of her lifestyle, Alva's marriage to Oliver Belmont was far happier than her years with William K., despite all their achievements. As his resistance to Colonel Mann suggests, Oliver Belmont was a powerful as well as an eccentric character, well able to stand up to his wife's assertive nature. He was also more complex than his fondness for stuffed horses, his insistence on leading a life of leisure, his architectural priorities and his social antics suggest. In 1899, three years after he married Alva, he entered politics endorsing Populist policies and the candidature of William Jennings Bryan, who had often attacked August Belmont, another Democrat. He published a newspaper, the *Verdict*, which broadly endorsed the Populist programme, and after a series of manoeuvres with the party's political machine attended the 1900 Democratic National Convention as a congressional candidate, serving one brief term.

The Populist Democrat programme that Oliver Belmont endorsed was remarkably close in spirit to the reforms proposed by the Liberal Party in England – a surprising platform for such a wealthy man. Their measures included the introduction of income tax, inheritance tax, a church tax, unions for workers and a strong anti-imperialist line – the latter almost certainly influencing Alva's avowed support for the Boers during the South African War. As ever, Oliver's interest in politics soon fizzled out, but it is reasonable to suppose that Alva took an interest in his activities and would apply some of the lessons learnt in the years ahead.[44]

Because they were well matched, Alva was prepared to give ground to Oliver. Her changes to Belcourt, for example, did little to alter the guiding principle of its design (which was Oliver's) – at least in his lifetime. (On one occasion a tourist guide was heard to bellow through his megaphone: 'Here you see before you the new home of a lady who is much before the public eye . . . A society lady who has just been through the divorce courts. She used to dwell in marble halls with Mr Vanderbilt. Now she lives over the stables with Mr Belmont.'[45]) She also accepted Oliver's manservant, Azar, and even promoted him to major-domo on her arrival. Azar was a fixture at every Belmont event, welcoming guests between two English footmen in court liveries with powdered hair, and standing beside Oliver's chair at dinner 'tall and handsome in his picturesque zouave jacket and embroidered fez'.[46] Once again, the splendours of pre-war Newport and Edwardian England ran closely in parallel.

Even *Town Topics* had to concede reluctantly that the marriage was happy, writing later that 'when everything was running smoothly there is little doubt that she looked upon . . . [Oliver] as the grandest man in the world' – before going on to say 'the problem was that bliss seldom reigned in the Belmont domicile'. According to Colonel Mann and his correspondents, dishes flew, tempers frayed and one of Oliver's tactics when he was annoyed with Alva was to sing: 'Would someone kindly tell me, for I would like to know, why I got a lemon in the garden of love, where only peaches grow',[47] and to request this ditty wherever he found an orchestra. Overall, however, Alva was not simply being sentimental when she characterised the difference between her two marriages thus: 'We were interested not primarily in what each could get out of the other but what each could put into it and from the blending get the most satisfying results.'[48] One clue to the scale of Oliver's affection for Alva was his present to her for her forty-ninth birthday. It was a life-size statue of Joan of Arc sculpted by Prosper d'Epinay (it had been exhibited at the Salon des Artistes Français in 1902); a solid marble love token which suggests that however tempestuous the relationship may sometimes have been, he understood his wife's fiery and determined nature and admired her for it.

There is every reason to suppose that if Oliver's life had run its expected course, Alva would have disappeared from view, relegated to memoirs of the Gilded Age as one of its more colourful and powerful female figures like her friend Mrs Stuyvesant Fish, with a well-deserved footnote in histories of American architecture and a question mark over her behaviour regarding her daughter's choice of husband. This was not to be. Her second marriage only lasted twelve years. Oliver became ill at Brookholt on 1 June 1908. At first his doctors thought he had developed an ailment of the liver, and it took another three days before he underwent surgery for appendicitis. In the days before antibiotics this was a fatal delay, for he developed peritonitis and septic poisoning. In spite of a weak heart and considerable excess weight, Oliver rallied briefly on 6 June. Alva was at his bedside when he died at Brookholt on 10 June 1908.

* * *

Neither of Oliver Belmont's brothers came to his funeral, though one of his nephews did attend – a gesture which touched Alva so deeply that she left his sons and daughters a legacy in her will. Oliver's death also marked the beginning of an uneasy rapprochement between Alva and the Belmonts after they offered her burial space for Oliver in the family plot in Newport – an offer she turned down with great care, for she and Oliver had already bought a burial plot at Woodlawn Cemetery.[49] Alva was undoubtedly devastated by Oliver's sudden death. William Gilmour, still superintendent of Marble House in 1908, in spite of coming close to resignation many times, reported in his notebooks that Alva 'looked very badly indeed' when she appeared in Newport eight days later on 18 June. Her grief appears to have touched him, for in a rare display of emotion he also commented: 'I found her very nice however.'[50] Understandably reluctant to stay in Newport for the summer season, Alva told Mr Gilmour that she planned to sail for Europe and return in September. Accompanied by Azar (who would stay with her for many more years) and by her daughter-in-law, 'Birdie' Vanderbilt, Alva departed on the *Mauritania* for Europe on 24 June, where, she later wrote bleakly: 'I was ill for a long time.'[51] *Town Topics*' only mention of Oliver Belmont's death was to question whether there could truly be a Newport season without 'Alva the magnificent domineering the bridge players'.[52]

By 3 September the magazine was speculating as to how Alva would now occupy herself. 'Not since her divorce has there been so much interest exhibited in Mrs Alva Belmont . . . There is a keen desire to know if Mrs Belmont will retain all her properties . . . and finally what disposition she will make of Mrs Belmont herself.'[53] Alva remained in Europe until late September when she returned to take a health cure at Virginia Hot Springs, where, it was rumoured, she intervened to shore up the now floundering marriage between her son Willie K. Jr and 'Birdie' (once again it was reported that she had been assisted by the pliable Mrs William Jay in piling on the pressure.) There were other rumours too. Alva was planning to sell Belcourt and Brookholt; and she had plans to open a colony for a group of social intimates at Virginia Hot Springs, although *Town Topics* felt obliged to point out that because Alva had so many enemies, such a group would be very small.[54]

None of these rumours materialised. Nonetheless, the twelve months between June 1908 and June 1909 undoubtedly marked the greatest upheaval in Alva's life. She suddenly found herself widowed, bereft of one of the few men she had ever loved. Both marriages she had arranged for her children had either collapsed publicly or were on the point of doing so, for Willie K. Jr and his wife finally separated in 1909, embarking on a period of separation that lasted until 'Birdie', a Roman Catholic, agreed to file for divorce in 1927.

The social world that Alva had dominated for so long was already the focus of widespread criticism and without Oliver it is easy to see how it lost much of its charm. Although she would continue to build, restore and alter houses well into old age, this now became part of her life rather than an all-absorbing interest. *Town Topics'* barbed remarks about her small circle were partly true, for Alva was now more isolated than she had once been. Her sister, Julia, who had married the Comte de Fontenilliat, died in 1905, while her eldest sister, Armide Vogel Smith, died in 1907. Jenny Tiffany lived in France; and Alva was on uneasy terms with two important and extended New York families, the Vanderbilts and the Belmonts. On top of this there was the conventional view of widowhood itself – a retreat to the shadows, another kind of exclusion from the theatre of life, another kind of purdah where pale rays of 'sunshine' would be few and far between.

In accounts of her conversion to the suffrage cause, a move that would rescue her from the half-life of a rich widow, the number of months Alva spent in Europe immediately after Oliver Belmont's death has generally been overlooked. So is the fact that in the summer of 1908, when she travelled to Europe, both Alva and Consuelo were living alone for the first time in either of their lives. Though this was almost certainly irrelevant to Alva in the first stages of grief, Consuelo's interest and involvement in social welfare problems, her range of activities and successful consolidation of her position in society cannot have been lost on Alva as the weeks went by; nor can the work Consuelo was undertaking for her articles for the *North American Review* during the second part of 1908. Consuelo had made her speech criticising idle American females at the dinner for Mrs Humphrey Ward in April 1908; Oliver died in June; and throughout that summer and early autumn, Consuelo was drafting and rewriting 'The Position of Women' so that *The Times* could print advance extracts by December 1908.

Alva remained in Europe throughout the summer of 1908 and was back there again in February and March 1909, which coincided with the publication of Consuelo's articles. There is no record of exactly how much time they spent together during the months Alva was in Europe, but it is reasonable to suppose that it amounted to a period of several weeks, and that they would have discussed Consuelo's work. Alva may well have contributed her own ideas while Consuelo was drafting the articles, but at this stage Consuelo was well ahead of Alva in her thinking. It is likely that it was Consuelo who persuaded her mother that if she did not wish to lead a half-life as a widow, she would have to take a step outside the 'great gilt cage'. At the very least, Consuelo's example made Alva consider the proposition that the public life of the English aristocratic woman might now apply to her too.

Put in the context of Consuelo's views about the position of women, the exclusion of women from power, her support for the English suffrage campaign and her involvement in social welfare issues affecting women, Alva's subsequent decision to throw her energies into the fight for the woman's vote in March 1909 seems less like an overnight conversion. Soon after her return to New York in March 1909, Alva accepted an invitation to a lecture on woman's suffrage by Ida Husted Harper in the home of Consuelo's bridesmaid, Katherine Mackay (nee Duer), the only American society woman of any standing already taking an interest in the cause. Ida Husted Harper was the press chairman of the National American Woman Suffrage Association, the leading suffrage organisation in America. On 18 March 1909, its president, Anna Howard Shaw, was invited to dinner by Alva and they talked about female suffrage till one in the morning. This caused Anna Howard Shaw to miss her train and spend another night in a hotel, but the expense was worth it, she told her board because 'I got her for a life member of the National Association before leaving'.[55]

Later Alva wrote that she had taken time over her decision to embrace the suffrage cause and that it had not been taken lightly. 'When it dawned on me that there was some serious humanitarian niche in life which I, with my opportunities, might fill, it was not a matter which I could decide in a day,'[56] she wrote. She and Consuelo both shared the view that happiness could only be achieved through 'self-subjugation' and that this was the best way to deal with profound misery. 'Don't think of yourself, it is the greatest medicine of all,' Alva once wrote to Harry Lehr when he was feeling depressed.[57] 'It is this I saw years ago, when the light of my life burnt out with him,'[58] she wrote to Sara Bard Field years later. Alva also agreed with Consuelo about the dangers of female idleness. 'It is a mistake to believe that any woman, no matter what her financial condition of life, can lead an idle existence. It is merely a question of the worthiness of her activities,'[59] she wrote.

Alva also suggested that she had given serious thought to becoming involved in some great philanthropic undertaking. But between Oliver's death in June 1908 and the spring of 1909 when she joined the National American, she gradually came to the conclusion that her daughter and her philanthropist colleagues were approaching social problems from the wrong direction. '[Some] of my personal acquaintances, with money at their command, had gone into the slums, and I assure you that many of them did ameliorate the hardships of those among whom they worked, until here and there the eddy widened to a whirlpool,' she wrote. Alva believed that improvements achieved by philanthropists were only ever temporary and superficial. 'In the end, after much deliberation, I was forced to the conclusion that the uplifting of the slums could, at best, be but ephemeral unless the very conditions which created the slums were overcome,'[60] she wrote in 1911. These conditions, she argued, had evolved entirely at the behest of men and were a direct result of 'their strength and directness' without the compensating balance of female 'delicacies of perception and intuitive forms of reasoning which the masculine mind lacks'.[61] As a result (and here her argument did coincide with Consuelo's), any woman concerned with issues of social welfare was deprived of influence over the matters which most directly concerned her as wife, mother and philanthropist. The difference was that Alva argued that the female vote was needed *before* the fundamental causes of poverty could be tackled.

In 1909, however, both mother and daughter's views on the suffrage issue appeared to be identical and watchful commentators assumed that Consuelo was behind Alva's 'conversion'. 'Everybody says' remarked *Town Topics*, 'that Mrs Belmont's daughter the Duchess of Marlborough (we all know who her daughter is, but it tickles Mrs Belmont to see it in cold type) is responsible for instilling these suffragist ideas into her head. Consuelo is wrapped up in similar movements in England.'[62] The *New York Mail and Express* put it slightly differently. 'The influence of her daughter, the Duchess of Marlborough, is understood to have contributed to this marvellous change in her mother from the aristocrat to the socialist, for the daughter's chief interest in London is now centred on the great social problems of the British metropolis. She is continually ministering to the wants of the poor and helping by every means in her power to uplift the lower classes'.[63]

Common cause between mother and daughter, however, masked the beginnings of a profound disagreement about tactics and priorities. While Consuelo saw female involvement in philanthropy as a dress rehearsal for the time when women would have the vote and become full citizens, Alva argued that good citizenship was impossible without the vote and that a fundamental right was being withheld, a difference in approach that would eventually cause considerable tension both with Consuelo and the leaders of the American woman's suffrage movement.

Alva's decision to fight for women's suffrage rather than apply herself to philanthropy is open to a number of different interpretations. One is simply a matter of temperament. Throughout her life, she had been drawn to a fight, and she relished a big battle as a way of distracting herself from grief. There were also those in society who maintained that her thirst for publicity was so great that she simply embraced the suffrage cause to keep herself in the public eye once her life was changed by widowhood, a suspicion later entertained by some of those within the suffrage movement itself. Such evidence as there is, however, suggests that Alva took a considerable time to make up her mind before she became seriously involved, and that she spoke the truth when she said she had been an instinctive feminist all her life.

Harry Lehr's comment that: 'The dear Old Warrior has got something to fight for at last'[64] was only part of the story. In her later attempts at autobiography she may have endeavoured to make her feminism seem more credible, but there is evidence of Alva's feminist instincts well before 1909, including Oliver's gift of the Joan of Arc statue, and *Town Topics* accurately reporting in 1906 that Alva wanted a career for Consuelo when she married her off to the Duke of Marlborough. Elizabeth Drexel Lehr commented that when Alva took up the women's suffrage movement as a cause, 'none of her friends were very much surprised. They remembered that Alva had always been a fighter, that she had always championed her own sex, taken the woman's part in any discussion.'[65]

The American woman's suffrage campaign had, by the time Alva joined it in March 1909, stagnated to the point of torpor. The energy of the first phase of the movement, led by Elizabeth Cady Stanton and Susan B. Anthony, which started with the Seneca Falls Convention in 1848, had fizzled out amid bitter disagreements about tactics. One camp favoured securing voting rights for women by way of a federal amendment. By 1908 all serious attempts to push this through had come to a standstill. The second approach was to tackle the issue of female franchise state by state. Wyoming and Utah enacted women suffrage provisions in 1869 and 1870, followed by Colorado in 1893. The most recent state to follow suit had been Idaho and there had been no further progress in the previous twelve years.

The National American Woman Suffrage Association (NAWSA),

or 'National American', was the only campaigning organisation still active, but it was bedevilled by weak leadership and an amateur approach to political campaigning. Its headquarters were in Warren, Ohio because its treasurer happened to live there. Some key supporters of the National American, particularly Elizabeth Cady Stanton's daughter Harriet Stanton Blatch, were already deeply frustrated by the time Alva joined. Anna Howard Shaw, president of the National American, now sought to make the best use of her formidable new recruit. She was hoping for a substantial injection of funds and in an effort to encourage Alva's interest she asked her to act as a delegate to an International Woman Suffrage Alliance Convention in London in April 1909.

Alva was exasperated to find, however, that the general tenor of the conference was placidity and conservatism. Fired up by her new interest, she decided to investigate the working methods of the highly active English suffrage campaign while she was in London. What she found in April 1909 was a British movement broadly divided into two camps. The larger camp was represented by the National Union of Women's Suffrage Societies (NUWSS), led by Millicent Garrett Fawcett, and was committed to securing votes for women through constitutional methods – an approach similar to that of the National American and supported by Consuelo. A far smaller camp numerically, but one with a high public profile, was represented by a splinter group led by the Pankhursts, whose Woman's Social and Political Union (WSPU) had split from the main organisation in 1903, frustrated by slow progress and impatience with the failure of the constitutional approach. This group had been nicknamed 'the suffragettes' by a journalist in 1906, a name which came to be synonymous with the militant suffragism of the WSPU.

After 1903 the Pankhurst's WSPU evolved a series of highly publicised tactics which began with 'mild militancy' – heckling Liberal politicians, hiding in rooftops and disrupting political meetings and holding rallies. By 1909, however, WSPU tactics were characterised by an undertow of violence and the English campaign was far more divided than in 1908. Widespread criticism of the

Pankhursts simply piqued Alva's interest. 'I was curious to hear these women whom all the public – men, press and "society" abused,' she said, so she took a box for a WSPU rally at the Royal Albert Hall in April 1909 to observe them in action for herself. 'Already the strong hand of a "Liberal" government was endeavouring to suppress these impatient and unreasonable women' and she had trouble finding anyone to go with her. 'I invited my daughter, the Duchess of Marlborough, to accompany me. Though a suffragist, she refused. Some friends were importuned. My invitation did not prove popular. Just as I was resigned to going alone, I found two women brave enough to accompany me.'[66]

In her box at the Albert Hall that evening, Alva was mesmerised by what she saw. 'It was as if millions of atoms were being hurled towards a solution. It was as if a hitherto unheard cry was becoming articulate.' For the first time, she understood that the fight for the vote was far more than a campaign for full citizenship – it was a fight for 'the eternal world-old demand for justice and liberty'. 'I longed to throw myself into this turbulent tide and to feel myself strengthened by the substance of the whole' she wrote later. 'What a revelation! I could not believe my eyes. Such electric fervour I had never seen nor felt in all my experience . . . The speakers, most of whom had served prison terms for asking questions in public meetings, or attempting to hold street meetings and various other "heinous crimes" took the platform amidst tumultuous applause . . . I was exalted.' What astonished her most was 'the astounding grasp of things political which the speakers showed. Their minds were vigorous, alert, imaginative . . . they already perceived what we in America only realized seven years later – that the battle must be fought in the political arena.'[67]

It was another of the defining moments in Alva's life. She returned to the States with a clear vision of what needed to be done. Just as she had once seen the Vanderbilts as Medicis, and Consuelo as an English duchess after the visit to Government House in Calcutta, she was now gripped by a new vision for the American women's suffrage campaign, one which drew on the lessons she had brought home from the Pankhursts but which also tallied with her experience as a leader of society. Since the finances of the National American were in a parlous state, Anna Howard Shaw and her colleagues had little alternative but to listen.

The first step, Alva argued, was to raise the visibility of the American suffrage campaign, an observation that came directly from her experience as society leader. What the organisation needed was 'commanding headquarters', a large, high-profile building that would be visible to press and public – a political equivalent to 660 Fifth Avenue. The National American should move from Warren, Ohio where no-one could see it, and establish itself in New York city like every other political party. Here, it should stop behaving like a parlour sect with meetings in the Martha Washington Hotel where tea was served to the same inward-looking group of fifteen women every time. Her suggestion was debated and agreed (much to the fury of the Ohio-based treasurer who liked working from home) and on 19 July 1909, Alva leased the seventeenth floor of a building on Fifth Avenue near 42nd Street, for which she would pay the rent for two years. As important, she undertook to pay the costs of a press bureau run by Ida Husted Harper, including salaries for three workers. The move, and her financial support for the Press Bureau had an impact almost immediately – sales of National American literature increased from around $1,200 in the Warren, Ohio years to around $13,000 in the first year in the New York headquarters.[68]

Alva had also noticed the extent to which the English campaign was a broad social coalition, successfully drawing on support from all classes including the aristocracy. Feeling that the social elite was one branch of society that she could deliver to the American campaign, Alva now set to work. 'Newport was the ideal place from which to begin. The social and summer-publicity centre of the nation. So in the summer of 1909 we undertook to introduce suffrage for the American women to the men and women of wealth who have the leisure to think of current problems.' For much of her marriage to Oliver, Alva had mothballed Marble House, using it only for its superior laundry facilities and allowing her son Willie K. Jr and his wife to occupy it during the Newport season when it suited them. In the summer of 1909 Newport society was agog at the news that Alva would be opening up Marble House once again to raise funds for the suffrage movement. '"Shades of Allah! Suffrage in Newport! Not a person will attend! A ticklish time for Mrs Belmont! Will she fail?" And so on ran the comments of the press and the male wise-acres in their comfortable gentleman's clubs,'[69] wrote Alva later.

Mrs Belmont did not fail. In 1909, she organised two highly successful suffrage meetings at Marble House towards the end of the Newport season, at the end of August, though the first received rather more press attention than the second partly because of its celebrity guests. The first meeting took place on Tuesday 24 August 1909. Alva successfully prevailed on her old friends Mrs Stuyvesant Fish, Tessie Oelrichs and Harry Lehr to attend, knowing that their presence would ensure that the rest of the *ton* followed like lambs. She also secured the presence of the venerable Julia Ward Howe, year-round Newport resident, author of 'The Battle Hymn of the Republic' and a longstanding champion of women's rights – the only survivor of the pioneering suffragist group that met at Seneca Falls in 1848.

The suffrage flag with its four stars for the four suffrage states flew over Marble House, while Elizabeth Lehr and Miss Lily Oelrichs acted as guides to the house which could be toured for $5. Strange juxtapositions were immediately leapt upon by the delighted press: 'Society Meets Suffrage at Mrs Belmont's Marble Houses in Newport' shrieked one headline, 'Harry Lehr and Mrs Howe under the same roof!'[70] The newspaper also maintained that visiting suffragists were so overawed by the 'Louis XIV magnificence' that they almost forgot about the cause as they wandered round and gazed in wonder.

It was rumoured by one newspaper that most of the tickets had been bought by architecture students, but queues of suffragists and society ladies formed early as well as a 'mixed crowd' of artists and other summer residents anxious to have a glimpse of such a famous establishment. Those less intensely interested in architectural history or who knew the house already confined themselves to the garden and were charged $1, where they listened to the orchestra from the Newport Casino. A tent was erected for the speeches at the rear of the garden in the shade of plane trees. Alva (still nervous about public speaking) opened proceedings with a very short speech introducing Julia Ward Howe who received a five-minute standing ovation, before delivering a speech which not everyone could hear on account of her frailty. The main address was given on the suffrage campaign by Dr Anna Howard Shaw. She spoke for an hour and finished by saying: 'The aid socially, financially and educationally of such women as Mrs Belmont has been a positive gain to the cause. But we as women suffragists are appealing to all classes. Women of all classes are joining our cause in England.' And by way of co-opting support from the absent Consuelo added: 'The Duchess of Marlborough holds the same view as her American mother here' which was not strictly true but offered reassurance to worried society persons.[71]

The second meeting took place on 28 August. In spite of *Town Topics'* dismissive description of it as a 'complete frost', this drew a crowd even larger than the five hundred or so who had appeared at the first meeting. Many came from out of town, and some from as far away as Philadelphia to hear an address from Professor Charles Zueblin, Professor of Sociology at the University of Chicago, in the presence of the Governor of Rhode Island.

What was actually said on both occasions, however, was generally considered to be rather less important than Mrs Belmont's audacity in holding suffrage meetings in Newport in the first place. One thing was clear: As *Town Topics* put it: 'Mrs Ollie Belmont is thoroughly launched on the suffragette sea and is one of the most militant mariners of the fleet'[72] and this in itself was enough to focus national attention on 'the cause'. 'It was admitted that she has caused gossip on all sides not in Newport alone but throughout the country. That this gossip will have an excellent effect no one who is interested in the great question has any doubt,'[73] wrote the *World*. 'Newport has capitulated to an idea,' wrote the *Evening Sun*. 'Woman suffrage has usurped as topic of talk the place of piffle and pink teas.'[74] Even the London *Spectator* remarked shortly afterwards that though the American woman of means 'appears to have no function in life except the dismal one of providing herself with perpetual amusement' there was now 'a type of woman of wealth who needs to get out of herself and her circle of luxury and income spending by identifying herself and her fortune with some popular cause; and the woman's suffrage movement, judging by developments within a year, bids fair to enlist not a few women who will be able to finance the propaganda in a generous way'.[75]

This did not mean that Newport society was universally delighted by the rallies. 'Mrs Belmont has become almost scandalously democratic for she hobnobs with people these days who in the past she would never have condescended to look upon,'[76] said *Town Topics*. 'The avenue was lined for a mile with automobiles . . . the sorry nag of the station livery team standing next to the $15,000 touring car of the millionaire,'[77] said the *World*. This was Newport society's objection. 'Why let loose a horde of fanatics on the stronghold of your friends? Said everyone,'[78] wrote Elizabeth Lehr. Harold Vanderbilt (who was now twenty-five) was also said to be outraged. Although Alva wanted him to introduce the speakers, he 'resisted violently', according to Elizabeth Lehr, and hid indoors. The other person who was deeply unhappy was Azar. Chimpanzees were one thing, suffragists quite another. 'Poor Azar's consternation was boundless as hundreds of women from New York, Boston, Chicago – every part of the country, swarmed into the house that had earned the reputation of being one of the most exclusive in Newport, and wandered in the garden in groups of three and four. Women in shirtwaists, their jackets hanging over their arms, women carrying umbrellas and paper bags. Man-hating College women with screwed-back hair and thin-lipped determined faces; old countrywomen red-cheeked and homely, giggling shop girls. Azar had never seen such guests . . . It was too much for him!'[79]

While most of Newport society refused to be converted, the majority were still too afraid of Alva to do much more than grumble. Most Newport summer residents did not wish to cross her since there was an ever-present danger that she would stop inviting them to parties. As one newspaper remarked: 'The advocates of woman's suffrage have made a shrewd move in associating "votes for women" with visiting lists,'[80] and she was helped by the support of her friends. Mrs Stuyvesant Fish would only remark to reporters that she 'loved novelty' while Harry Lehr confined public criticism to the hideous yellow of the suffragist buttons. In private, however, he could not resist a prank. After the rally at Marble House was over, Alva discovered that her suffrage banner bearing four white stars had been stolen. A week or so later, she was dining at the Elisha Dyers' in Newport when an unexpected guest was announced in the form of Carrie Chapman Catt, a leading figure in the National American – already on record as resenting Alva's interference in the movement. Everyone turned round in amazement as 'a lady of noble proportions, majestically draped in blue with a flowing court train whose four white stars on a blue ground' walked into the room. 'Mrs Belmont's eyes fell on it. She gave a cry of astonishment: "Why it's Harry Lehr! And *you* were the culprit who stole my banner!"'[81]

Harry Lehr, it seems, could not pass an opportunity to dress up as a woman.

9
Old tricks

REGARDLESS OF THE SUCCESS of both Marble House meetings, Alva's involvement in the American suffrage campaign soon caused problems. For the first time in her life Alva felt compelled to bow to the leadership of others and she found this very difficult. She firmly believed that a united front was vital. She never sought to displace or undermine the president of the National American, Anna Howard Shaw, and supported her for several years in the face of fierce criticism. This did not mean, however, that Alva instantly set aside a lifetime's habit of vehement certainty about what needed to be done. She resolved some of these tensions, in the short term at least, by setting up her own campaigning group, the Political Equality Association (PEA) on 16 October 1909, working under the National American umbrella and sharing premises in the New York headquarters. But the arrangement was uneasy from the outset, and there was open speculation as to how long it could possibly last.

Alva's opening moves were looked on warmly by Anna Howard Shaw. Exposure to the English suffrage campaign had convinced Alva that the American struggle must broaden its social base. In August 1909 she had courted the aristocracy of Newport. In November an opportunity to enlist working-class support suddenly presented itself when female garment or 'shirtwaist' workers in New York's Lower East Side went on strike against appalling pay and conditions. Many of the strikers were refugees from Russia and their difficulties in asserting their rights through the organisations such as the Ladies' Shirtwaist Makers' Union were compounded by poor English and the arrogance of male trades union leaders. Garment manufacturers responded to the strike by having many of the women arrested for picketing in spite of the fact that it was within the law. Police treatment of some of them was tantamount to sexual assault: younger women had their clothes torn, were dragged through the streets to the police station and subjected to 'gross indignities' and 'improper proposals'.

Such abuse of police power at the behest of garment manufacturers soon became as great an issue as the original causes of the strike. Alva swung all her energy and influence behind the shirtwaist strikers in a manner that had male trades union leaders and most of New York society gasping, and garment manufacturers begging her to see matters from their point of view. Alva paid no attention. In November 1909 she led a parade of strikers down the Bowery; she rented a 6,000-seat arena for a rally at the Hippodrome, where she arranged for the strikers to be addressed by Anna Howard Shaw; and she raised money by inviting women from the Shirtwaist Makers' Union to the exclusive new female Colony Club to tell their stories, and to extract contributions. To the delight of New York's press, who tracked her every move, Alva sat up in the all-night courts glaring at the magistrate while he heard the strikers' cases, interrupting him in such an intimidating manner that they were all acquitted, bar one whose fine she paid. The experience confirmed Alva's views about political priorities too, for she saw at first hand how unfairly working-class women were treated by the male judicial system. 'We must have radical changes,' she wrote. 'And I cannot believe that they can be brought about until women are recognized on an equal footing with men.'[1]

In early 1910 Alva took another step which caused uproar and demonstrated just how far she had stepped out of her society fortress in the space of a few months. Alva embraced the idea that white suffragist groups should affiliate with 'coloured' female suffrage, and encouraged black women to join her PEA, albeit in a separate grouping. This brought calumny down upon her head from all sides, with some calling her 'an evil influence'.[2] The Negro Men and Women League of the PEA was formed on 23 February

1910, greatly discomfiting many of the National American workers with whom the PEA shared offices. 'With characteristic feminine disregard for logic, the social leader . . . would confer the ballot on Negro women at a period in American history when public opinion in the North has begun to doubt the wisdom of the Fifteenth Amendment,' expostulated one newspaper.[3] This was just the kind of reaction that Anna Howard Shaw had hoped for when she enlisted Alva. She wanted her money, certainly, but it is clear that she also wanted Alva's social power and her celebrity value as Mother of the Duchess of Marlborough. Because Alva had taken care to keep her public profile and relationship with Consuelo well burnished during the years of her marriage to Oliver Belmont, almost any move she made now ensured a barrage of press attention. By October 1909 the *World* was calling her the 'unofficial leader of the woman's suffrage cause'.[4]

While Anna Howard Shaw was pleased about this, some of her colleagues in the National American were less than thrilled. 'Seemingly, there is a feeling that Mrs Belmont is getting more attention than is her due,' said the *Washington Times*. 'We would not for a moment suggest that it is a feeling akin to jealousy . . . [but] it all goes to show that if women do get the ballot, there will be interesting times for the woman who gets rather more limelight than her sisters.'[5] Behind Alva's back, some within the National American dismissed her interest as the passing fad of a rich dilettante while others saw her as a compulsive self-publicist who had simply found a new platform. Harriet Taylor Upton, an ally of Carrie Chapman Catt, thought that 'last year Mrs Mackay was very generally noticed in the New York papers, and Mrs Belmont wanted to do the same thing'.[6] Such conservative reaction missed a serious point. Extrapolating from her great society campaigns, Alva was certain that the American fight for the vote must now embrace publicity. 'The American woman has been brought up to shun publicity, but we must forget our personal inclinations for the sake of a great cause,' she said in 1910. 'To be successful in any phase of politics one must give one's life more or less to the public, and that is the lesson the American suffragist must learn.'[7]

When Anna Howard Shaw publicly congratulated Alva on her extraordinary impact on press attention (the press office counted more than 3,000 articles between April 1909 and 1910, whereas they had previously considered themselves lucky to get a mention), the response of Carrie Chapman Catt, ex-president and key figure in the National American, revealed the depth of conservative resistance. 'I know you think the number of articles in the press is an evidence of progress,' she wrote. 'I think we should make more progress if we had considerably less.'[8] When Alva hired a lobbyist in Albany to campaign for New York State suffrage, Catt was even more furious, fulminating that: 'If the legislators wish to pass it they will do so because of the wish of their women constituents, without the intervention of a paid worker.' The use of a lobbyist in Albany was regarded as so unladylike that it caused Katherine Mackay to resign from the National American, a split which was then blamed on Anna Howard Shaw by those plotting against her leadership.[9]

This kind of reaction caused Alva great frustration. Realising that she would make little progress through the National American leadership, she took matters into her own hands through the PEA. When the National American proved deaf to her pleas that campaigners should descend from the seventeenth floor of the New York building and connect with the masses, Alva started PEA 'settlements', or offices, all over the city including Harlem and the Bronx. ('When men want successful recruiting, they do not hide their stations on the 17th floor,' she wrote later.[10]) She argued that suffrage campaigners needed to create 'centres of interest' and a 'demonstrative form of agitation'. To show what she meant she bought a separate building for a PEA assembly district branch at 140 East 34th Street, putting an experienced suffragist, Mary Donnelly, in charge.

Away from the inhibitions of the National American, this became the focal point for early experiments in 'demonstrative agitation', much to the horror of male neighbours on both sides who put up cast-iron railings against suffrage contamination. On Valentine's Day hundreds of heart-shaped balloons with a suffrage message attached were released into New York's skies; at Christmas a Santa Clausette 'arrayed in bright red modified hobble skirt, red coat and red hood, all trimmed with white fur'[11] handed out hundreds of free gifts to local children. Alva's most original idea, however, was to open a self-service lunchroom which drew in all kinds of new supporters who were bombarded with propaganda over their mashed potatoes.

Another underlying cause of tension with the National American leadership ran much deeper, however. From the moment that she witnessed the Pankhurst rally at the Royal Albert Hall in London in 1909, it was clear that Alva was a natural militant while her colleagues in the National American emphatically were not. The first two articles in a clippings book which she started in 1909 go straight to this fault-line. On 19 June 1909 Alva was reported in the *Washington Post* as declaring: 'We women of America are not one whit less determined than our martyred English cousins to enforce our right to the ballot . . . If we cannot gain our end quietly and peaceably we will gain it otherwise'; while in a second cutting Carrie Chapman Catt was reported in the *World* as denying that she was in any way a militant or even intended to hold a parade.[12] 'Mrs Belmont's militant announcements in which she intimated that the suffrage fight must be won even if bloodshed be necessary has not been approved of by the more conservative members of the association . . .' said the *New York Evening Journal* and quoted Ida Husted Harper: 'When Mrs Belmont makes an announcement, she gives her personal views.'[13]

Although at this stage Alva was aligning herself with the militancy of the British WSPU, she was not yet in favour of the destruction of property or attacks on politicians which she thought inappropriate to the American situation. But she deeply admired the English militants' embrace of visibility and their flair for publicity; and she felt strongly that ruling out non-constitutional tactics put the women's campaign at an immediate disadvantage. She was also deeply irritated by the National American's insistence that the issue of votes for women was an ethical one, best resolved by not antagonising men. 'Nowhere could we find this noble example set by [men] of not "antagonizing" their enemies,' she wrote later.[14]

Until 1912 Alva remained publicly ambivalent on the question of militant action in the interests of unity. When asked by a reporter from the *New York Sun* whether she would object to being called a *suffragette* she replied: 'I should not, if I could claim the title, but it is usually applied to the radicals whereas I am a conservative.'[15] Set against this, she was always publicly supportive of the Pankhursts themselves. She particularly admired Mrs Pankhurst whom she described as a 'strong, defiant, erect and liberty loving leader of the women of the world . . . a flaming torch which lighted my way'.[16] During Emmeline Pankhurst's visit to the USA on a speaking tour in October 1909, Alva struggled to persuade the National American to honour her with a reception in New York, though on this occasion she won the battle and paid for the flowers. 'I took every minute I could spare to share the company and inspiration of this remarkable woman-leader,' she wrote. 'I agreed with Mrs Pankhurst when she said to me: "With your conservative suffragists in America suffrage is still an ethical principle. With us it is a political principle. You talk. We act." '[17]

In June 1910 Alva made a trip to England to stay with Consuelo at Sunderland House, ostensibly for a rest after a very busy year. She took lessons in public speaking, and, in an episode which might have appealed to P. G. Wodehouse, practised by giving an after-dinner speech on suffrage to Consuelo's guests who included Mr and Mrs Higgins, Lord Rocksavage, the Hon Miles Ponsonby, Alva's old friend Mrs Leeds, and Consuelo's mother-in-law Lady Blandford.[18] Alva's presence in London that summer meant that she could observe the passage of the Conciliation Bill through Parliament, and she attended debates in the House of Commons with Consuelo, Anna Howard Shaw, and Millicent Fawcett, leader of the English NUWSS.

Had this bill succeeded, it would have granted a limited form of women's suffrage in England and could reasonably have been hailed as a victory by both the constitutional and militant camps. There were high hopes for its success; but Alva was there to see it defeated by one of the more disgraceful pieces of political chicanery of the early twentieth century. It passed its second reading with a huge majority in the House of Commons but Prime Minister Asquith, a committed anti-suffragist, was determined to block it in spite of this result. As bad, David Lloyd George who had always publicly spoken (and voted) in its favour, realised that the proposed extension of the franchise would hand an electoral advantage to the Conservative Unionists. Between them, they succeeded in having the bill referred to a committee of the whole house, denying it parliamentary time to complete its passage and thus 'torpedoing' it.

Although this would eventually cost the Liberal government dear, the reaction at the time was outrage – and a sharp division between the two camps of the English suffrage movement. The constitutionalist NUWSS – and the Liberal government – blamed the failure of the bill on an upsurge of militancy by the suffragettes as they realised the bill was going to fail. The WSPU argued that the bill's failure demonstrated the extent to which 'constitutional' non-militant methods were simply no use, and embarked on a campaign of militancy which would become steadily more violent. At this fork in the road, Anna Howard Shaw took the NUWSS position and Alva sided with the Pankhursts. Unable to contain her disappointment, she gave an interview to the *Daily Mirror* at Sunderland House in which she was quoted as saying that she approved of everything the militants had done; chivalry was humbug; women should own property jointly with men; women should have equal rights over children; and that they should be compensated for housework which was only regarded as degrading because it was free.[19] Both Alva and Anna Howard Shaw joined a protest rally of over 500,000 in Hyde Park where their presence under the United States' suffrage banner made headline news in the American press.

Convinced now that the American campaign had to become more 'political' and less 'preachment and propaganda', Alva returned to America deeply frustrated by the refusal of Shaw and Catt to change tactics in spite of what had just happened in England. For the time being, however, she bit her tongue in the interests of unity and turned away briefly from suffrage politics to another initiative of her own: an agricultural school for young women – or 'farmerettes' as they were quickly dubbed by gentlemen of the press. Once again, this project could not be accused of lacking ambition. The idea, which drew heavily on an anti-urban, arts-and-crafts strain in contemporary thought, was to take a handful of young women factory workers and give them basic agricultural training on Alva's estate farm at Brookholt. It was envisaged that some might then wish to become landscape gardeners while others would band together to buy small farms sold to them by Mrs Belmont.

The very notion sent sub-editors across America into paroxysms of delight, particularly when they saw the first illustrations of the initial intake – sixteen fresh-faced girls clad in broad-brimmed hats and blue bloomer suits specially designed for them by Alva. 'City Girls Make Dainty Farmers in Mrs Belmont's Newest Eden' shouted the *New York City Mail* on 13 May 1911, while the *Los Angeles Record* admired the 'Wan City Girls Turned Into Healthy Happy Maud Mullers on Adamless Farm'.[20]

At the start, all went well. The PEA lunchroom was the lucky recipient of the first crop of the agricultural school's 'vegetablettes', as the *New York City Tribune* called them, which then became a propaganda tool unique in the history of suffrage campaigning. 'Everybody who visited the headquarters was presented with a suffrage radish with "Votes for Women" stamped around its middle in yellow letters,'[21] wrote a *Tribune* reporter who happened to be there for his lunch. This demonstrated quick thinking on the part of the lunchroom organisers for the radishes were 'perfectly aldermanic' in size. 'Some of them measured ten inches round, and if they were a trifle pithy inside the suffragettes who ate them didn't mind. They said suffrage vegetables ought to have plenty of pith to match suffrage arguments.' (Some suffrage rhubarb which came with the consignment was also unusually large but was made into pies for that day's lunch.)

Aldermanic radishes, however, turned out to be the least of the agricultural school's problems. By June, press reports were suggesting that some farmerettes found the Adamless life unsatisfactory. 'The suffragettes point out with pride that a farm run by women can be beautifully conducted, but after that what?'[22] Some of the girls were from good homes and were horrified at having to do their own housework. Others were resentful that promised tennis parties never materialised. According to one account, the superintendent marched out, leaving the girls to fend for themselves.

It is perhaps not surprising that towards the end of 1911 Alva, now in her late fifties, was beginning to feel thoroughly overextended. She had been extraordinarily active, by anyone's standards, and she had given herself little time to come to terms with Oliver's death. On top of the breakdown of discipline among the 'farmerettes', Mary Donnelly upset her deeply by excluding eight black people from the PEA lunchroom. The historian Peter Geidel also suggests that her various initiatives may have been more of a drain on her financial resources than she cared to admit. In quick succession she foreclosed on a mortgage to Hempstead Hospital run by women; closed all the PEA settlements scattered across New York; and shut down Brookholt Agricultural College almost as quickly as she had opened it, much to the fury of some students who claimed that they were only just finding their feet.

In the face of this rationalisation, and the continued refusal of Carrie Chapman Catt and Anna Howard Shaw to retract their outright condemnation of militancy, Alva decided to concentrate her activities in one PEA building, in a manner she could directly control. Increasingly aggravated by the internal politics of the National American, by Catt's attempts to undermine Shaw's presidency and by the organisation's timidity, Alva did not offer to renew the lease for their joint headquarters in 1911. ('You see she is a spoiled rich woman,' wrote Harriet Taylor Upton to Catt when she heard the news.[23]) She selected a new building at 15 East 41st Street and although she offered the National American space in her new PEA headquarters, it was not accepted.

Though there was no question at this stage of a split with the National American, Alva was now finally able to concentrate much of her vision in one highly visible theatre of activity and to give free reign to her idiosyncratic views. In a widely reported speech at the opening ceremony of the new PEA building, she declared that women's suffrage was merely the first step towards a complete re-alignment of gender relations, which required nothing less than a full-scale remodelling of the male brain. 'Our work . . . is to remodel man's past ideas of what is safe for women. To make him know us as we are, not as we were; to ask him to believe we are just as good, though perhaps more advanced, certainly more desirous of assuming responsibilities now existing in the business and political world, where once we found them in the home only.'[24]

The new PEA headquarters became a hive of activity. It ran its own orchestra,* a musical school for anyone over sixteen, a PEA women's chorus, monthly National American meetings, meetings of the PEA 'Lawyers' League', free dance classes, an Artist's League for sculptors and painters and a new and improved lunchroom. Once again Alva played off all social classes against her celebrity persona. Once again there was a striking parallel between the new headquarters and her 'society' houses, for the building was magnificently furnished with a trellised garden, a hall of white marble and a liveried servant who opened the door to visitors. Here, the lower social orders were given a chance to brush with a social celebrity in an ersatz society setting. Alva played the role to perfection, arriving each morning in a large motor car, attended by a footman who descended from the outside box, held open the door and stood to attention while she entered the building. She then conducted a detailed tour of inspection, tasting the food and ordering the waitresses to behave like ladies. At the same time, she corralled her society friends onto the premises to endorse the cause by eating in the self-service lunchroom. This was regarded as a hilarious novelty, for few of them had ever helped themselves to food before let alone carried a tray.

What really grabbed the headlines, however, was a maverick Alva initiative on the first floor. This was the 'Department of Hygiene' which dispensed health advice to women and sold the PEA's own brand of 'Victory' beauty products including 'Satin Skin

* Advised by Walter Damrosch, director of music at Consuelo's wedding.

Cream', 'Glycerine Jelly with Rose', 'Camphor Ice' and 'Violette Water', and its own brand of toothpaste. Like Emmeline Pankhurst, Alva was firmly of the view that women were much more effective if they looked attractive, and felt strongly that the suffrage movement should harness the power of the emerging cosmetics industry spearheaded by women such as Elizabeth Arden. Beauty, she insisted, was an instrument of female power. The product which caused the greatest stir, however, was the PEA 'Victory' laxative. Alva sometimes sold these 'liver pills' herself and a visit by one lady journalist suggests that in addition to her other qualities, Alva had some of the characteristics of a snake-oil salesperson.

'I have taken them every night since last February and wouldn't be without them,' Alva is alleged to have said, reporting that they were also taken by other PEA workers including Miss Donnelly and remarking that 'no doctor will prescribe them for you, they're so good', before ratcheting up the sales pitch with: 'You know, when a person gets to be over thirty, and has to have a clear brain when they have to do lots of thinking, as I do, they need something for the liver.' But when the intimidated lady journalist enquired how many pills she ought to take each night, Alva suddenly had a lapse of memory and was forced to ask Miss Donnelly.[25]

It was all too much for Mary Donnelly, who resigned a month later. 'Not that I don't believe in suffragists using these creams and hair restorers,' she said. 'Every woman ought to try to be as beautiful as she can. I do myself. But I'm for suffrage first, last and all the time, and when it comes to suffrage being swamped in face creams I'm done.'[26]

One of the complexities of Alva's position during these years was that she derived much of her power within the suffrage movement not simply from her wealth but from her position as a society figure and Mother of a Duchess. One visitor to the new headquarters, noticed that the driver of the Fifth Avenue bus told passengers where to alight by shouting: 'This is the Newest Hall of Fame! It is the Victory home of the suffragettes and was built by its commander-in-chief, Mrs O. H. P. Belmont, Mother of our own Duchess of Marlborough.'[27] Doris Stevens, a young fellow suffragist who came to know Alva well after 1914, maintained that: 'She was always in conflict between living sumptuously and selfishly and being amused by a court jester and her marked identification with the purposeful life of Joan d'Arc. It is my belief she never reconciled these conflicts.'[28]

This assessment overlooked the fact, however, that it was in the interests of the suffrage movement that Alva maintained her status as a social celebrity and it was sometimes a difficult balancing act. During the Newport summer season of 1912, for example, she simultaneously opened a local suffrage office, introduced her beautiful suffragist protégée Inez Milholland into society, and enlisted the support of the social elite with a large open-air party on Easton Beach (causing much hilarity in *Town Topics* as its spies watched Mrs Fish and Mrs Oelrichs pick their way across the boardwalk). In her desire to win over rich people to the suffrage cause she was careful to avoid criticising them, telling *The New York Times* in a much misquoted article: 'I know of no profession, art or trade that women are working in to-day as taxing in mental resources as being a leader of society.' She then added that she was no longer sure the result was worth the effort 'because humanity at large is so little helped, and I would not consider now the putting forth of such energy in that way as intelligent service'. But she flattered her old friends – and herself – by declaring: 'But there is no doubt of the essential quality of sheer brain power being a *sine qua non* to the social leader.'[29]

At the beginning of 1912, Alva's relationship with the National American snagged on another disagreement. The National American campaign had continued to focus on winning women's suffrage state-by-state across the USA, and had made some progress. Washington had been won in 1910, California in 1911, and by 1912 women had secured voting rights in a total of ten states. By 1912, however, Alva had come to believe that state-by-state progress was too slow and that if the campaign continued in this vein it would make very little headway. In her opinion, the debate needed to move on to a national level and the National American should try to engage the support of at least one of the national political parties. When the

National American showed limited enthusiasm for this tactic, Alva took matters into her own hands once again.

She had discovered in recent years that newspapers welcomed her articles, and that journalism was an excellent means of circumventing her dislike of public speaking. She now exploited this interest – and her own fame – to take the suffrage debate on to a national level by lending her name to a series of articles from the 'Department of Hygiene' in the *World* which ran from 1911 till the middle of 1913 and writing a weekly column in the *Chicago Tribune* from April to November 1912. The latter series drew directly on her own experience, and many familiar themes emerged: the 'vices and immoral life' of the promiscuous husband, whose wives were 'paid legitimate prostitutes'; the trivial and wasteful lives imposed by rigid convention on society women; and the greater value placed on a boy than a girl. The contrast with her withdrawal from the problems of everyday life in her great gilt palaces at 660 Fifth Avenue and Marble House could hardly have been clearer and she now explicitly rejected her former flight from reality. 'Beware the Fairy Tales of Life', she wrote in the *Chicago Tribune*. 'It will be only after the castle is entered that the walls may crumble. It is hard to undo what is done. It is far better to face a situation with knowledge than to acquire it a day too late.'[30]

Her utopian streak remained intact, however, though it had a very different focus. She was essentially a single-issue politician gripped by a conviction that all would be well if only women could have the vote. Although she became an advocate of higher taxation her political radicalism should not be overstated. When she bailed out Max Eastman's socialist magazine *The Masses*, for example, it was because he supported women's suffrage. Indeed, until 1913 she conceded many debates in the interests of suffrage unity. She even agreed to take part in a suffrage demonstration march up Fifth Avenue organised by the more radical Harriet Stanton Blatch – a surrender partly because she thought Blatch was divisive but mainly because Alva never went anywhere on foot if she could possibly avoid it. So it was a great concession to take part in a demonstration march. Elizabeth Lehr watched from the Hotel

St Regis and wrote that Alva 'looked as serene and unselfconscious as though she had been in her own drawing-room', although Consuelo wrote later that her mother hated parading in this manner: 'I did not realise what such a conspicuous public act must have cost her until she later confessed.'[31] Before long, however, Alva would become so disenchanted with the National American that festering tensions would erupt into the open.

During this time Consuelo's own views on the issue of militancy and female suffrage also crystallised. She may have defended militancy while drafting 'The Position of Women' during 1908, but that was the last time on record. In October 1908, the Pankhursts attempted to rush the House of Commons, and were arrested. In common with many other supporters of the English women's suffrage campaign, it seems likely that Consuelo thought this was a militant step too far. Thereafter she aligned her views with the non-militant wing of the English suffrage movement represented by the NUWSS and Millicent Fawcett. In April 1909, when Alva invited her to her box at the WSPU rally in the Albert Hall, it is clear that Consuelo had no wish to endorse the Pankhursts because she already felt their actions were damaging the suffrage cause. The figure to whom Consuelo said her views came closest were those of her friend Lady Frances Balfour, Liberal daughter of the Whig Duke and Duchess of Argyll, sister-in-law of the Conservative Prime Minister Arthur Balfour and a leading supporter of the NUWSS and of women's education. 'She held my views on women's suffrage,' wrote Consuelo,' 'believing in the more conservative approach rather than in the distressing exhibitions of martyrdom which were shocking society'.[32]

Unlike Alva, who already saw votes for women in terms of human rights, Consuelo continued to regard female suffrage as an issue of citizenship and a privilege for which women should prepare and educate themselves. It is therefore unsurprising that Consuelo, like Lady Frances Balfour, was drawn to the question of higher education for women. Her own teenage intellectual interests had been short-circuited by Alva's insistence that a grand European marriage was a faster route to power, though she had apparently proved herself capable of reading for an Oxford degree. She now attached herself to the campaign to develop Bedford College for women (part of London University, the first in Britain to grant degrees to women) which needed to move from cramped conditions in Baker Street in London to a new site in the middle of Regent's Park.

Consuelo held the first of many fundraising meetings for Bedford College at Sunderland House on 30 June 1910. As soon as she became honorary treasurer of the appeal, the campaign started to attract support. Consuelo was not shy of publicity when she felt she was acting in a good cause. Like Alva (but with rather more success), she set about enlisting society's help in a manner that led the principal, Dame Margaret Tuke, to comment later that 'the activities of Bedford College between that year and 1914 were so similar to those practiced by other public beggars as to be easily imagined'.[33] Consuelo turned out to be extremely good at fundraising and in her history of Bedford College, Dame Margaret was unstinting in praise of her efforts: 'It is not possible to say how much the College owes to her able support and influence. She interested her friends . . . She gave wise advice. She held meetings in Sunderland House. She succeeded in keeping the appeal in the public eye so that the name of the College became known to the outside world, which had never before shown much interest in such causes.'[34] Consuelo became president of the college in 1913 and the new building in Regent's Park was finally opened by Queen Mary on 4 July 1914. As far as Dame Margaret was concerned, Consuelo's efforts on behalf of Bedford had blown it firmly 'into port'.

Consuelo's early involvement with the problems of prisoners' wives continued, and broadened to include other initiatives designed to help less well-off women earn an independent livelihood. One of these was a campaign to build hostels for single women in large cities where they could live respectably and cheaply, following the exposure of dangerous and degrading conditions in common lodging-houses by Beatrice Webb and Mrs Mary Higgs. 'The number of women workers is increasing every year,' wrote Consuelo in an article for *The Nineteenth Century and After*. 'The obligation to provide housing accommodation for their needs is bound to become more and more urgent . . . so that the self-respecting woman worker shall be at least as fairly treated as the self-supporting man.'[35] In spite of this feminist slant, Consuelo was successful in persuading George Curzon, a leading exponent of anti-suffragism and 'separate spheres', to finance and open such a hostel in 1913, in memory of his wife Mary. Another friend, Sir Edgar Vincent, was similarly helpful in giving her the use of a house at Esher Place to provide holidays for single self-supporting women.

More generally, Consuelo's philanthropic projects can be best understood as a series of overlapping concerns with the welfare of women and children at the centre. (Her keen interest in maternal and child health inspired sections of the press to nickname her 'the Baby Duchess'.) Her involvement with organisations such as the National Physical Recreation Society was typically Edwardian in that her interest was also triggered by contemporary debates about National Efficiency, a wide movement that emerged around the turn of the century and embraced Fabian socialists such as Sydney and Beatrice Webb as well as the collectivist conservatives like Milner and Chamberlain. Broadly speaking, all factions within this movement were reacting to the perceived decline of industry and agriculture during the previous twenty years and the corresponding enfeeblement of Britain as a world power, a process that had seemed to contribute to the disasters of the South African War and subsequent revelations about the degeneracy of the British population in the Report of the Inter-Departmental Committee on Physical Deterioration in 1904. The evidence showed that it was among the professional and upper middle classes that this decline was greatest; this in turn gave extra impetus to eugenic worries that the 'better' elements of the British population might be swamped by the lower classes.

Apprehension about the physical deterioration of the 'national stock' stimulated a variety of remedies such as the improvement of secondary education, initiatives such as those of the National

Physical Recreation Society, and a bolder application of state power to create a healthier population and more efficient workforce. To this extent, Consuelo's interest in social welfare issues was not simply driven by feminism and New Liberal ideas about a larger role for the state but by National Efficiency as well. Her interest in the latter movement also accounted for the occasional eugenic flourish in her writing: 'White Slave traffic, the degeneracy of the race, and the high rate of infant mortality can be traced to the absence of moral supervision and the impossibility to acquire clean and healthy habits of life in the lodging-houses,'[36] she wrote apropos of the need for hostels.

This pattern of overlapping concerns applied equally to her campaigns for improved maternal welfare. Once again this was not simply a feminist issue but was related both to National Efficiency issues and the need for social reform. Consuelo was not alone in this. 'Scrutiny was directed at the newly urgent problem of the ability of the nation's working-class mothers, who reared Britain's foot soldiers, to bring up healthy children. Motherhood was no longer a natural attribute of women but a *problem* and an achievement,'[37] writes Ellen Ross. In May 1913, Consuelo gave a speech in which she said that 'mothers of the present generation left school with little knowledge of domestic science or hygiene and that in consequence they were incapable of cooking a wholesome meal or giving proper attention to bringing up their children,'[38] and then announced that she was setting up the Marlborough School of Mothercraft to train health visitors who could address such problems. Such views on the education of mothers did not make her universally popular. An angry left-wing newspaper retorted that: 'Schools for mothers are all very well, but when "slumming" Duchesses talk about the failure of elementary education to turn out better workmen and more competent mothers, they simply disclose the failure of upper-class education to turn out persons who can think.'[39]

Such interests led Consuelo to become involved in a new initiative proposed by one Mrs Fitzstephen O'Sullivan, that a political party should be formed with the express purpose of getting women elected to municipal councils. Mrs O'Sullivan felt this was necessary in response to a growing perception that women were not taking advantage of their right to vote or stand as candidates in municipal elections which they had won in 1894; that women's interests were thus not properly represented even at local government level; and that this lack of interest gave opponents of suffrage valuable ammunition since women were not making use of the right to vote they already possessed. Mrs O'Sullivan (rather than Consuelo, as she claims) formed the Women's Municipal Party in March 1913, and announced its existence in the *Daily Mail* on 26 March.[40] This prompted a favourable response from leading newspapers, all asserting that women did not value sufficiently the vote they already had. After several conversations with Mrs O'Sullivan and some careful thought, Consuelo agreed to become its chairman.

In an Alva-like move, Consuelo took over responsibility for financing the party and its offices at 32 Victoria Street and started to enlist the support of friends and like-minded philanthropic associates, just as she had done for Bedford College. By September members included Lady Frances Balfour. Throughout the summer and autumn of 1913 the Women's Municipal Party continued its work, putting up candidates (who were non-partisan and simply dedicated to pursuing women's interests at local government level) or supporting candidates who endorsed the party's agenda, at local, borough and parish council elections. The party was officially inaugurated on Thursday 5 March 1914.

At first glance, the party seemed explicitly feminist in its aims. 'The Women's Municipal Party is of the opinion that a definite policy should be formed to invite women in *one* party which shall stand for their interests and represent their needs. Our candidates will hold our policy on all questions affecting the interests of women and children but on other questions they will vote according to their own political views,' wrote Consuelo to birth-control campaigner Marie Stopes in October 1913 in an attempt to persuade her to join up.[41] When she spoke at its launch, however, Consuelo was just as keen to emphasise its role in local welfare reform, drawing comments of surprise from *The New York Times*. 'A speech of more than a half hour's duration on so complicated a subject as local government is not delivered every day by a Duchess, but the Duchess of Marlborough performed the feat with remarkable facility at a meeting of women at Bedford College to-day,' it declared. 'She showed herself thoroughly conversant with her subject, and moreover, evinced an extensive knowledge of the housing conditions of the poor in the East End, of reforms that should be adopted, and a hundred and one other details of municipal affairs.'[42] Indeed, though always in favour of women having the vote, Consuelo went to some lengths to draw a line between the Women's Municipal Party and the national suffrage campaign, stressing that the WMP was a lever for greater control over social welfare issues by experts close to local problems – who were often women – rather than feminism by the back door.

By November 1913, Consuelo was explicitly rejecting suggestions that her many philanthropic activities were principally driven by feminism. She hosted a large conference at Sunderland House organised by two important trades union officials, Margaret Laurence and Gertrude Tuckwell, in conjunction with another organisation, the rather alarmingly titled National Anti-Sweating League. The League had been formed in 1906 to obtain a legal minimum wage for workers in the sweated trades and had been instrumental in persuading the Liberal government to pass the Trade Board Act in 1909, which resulted in minimum rates for ill-paid women such as chain makers, lace finishers and matchbox makers.

Consuelo's Sunderland House conference focussed on the need for similar protection for women in other sweated trades. Individual women were asked to step forward and tell their stories of living on less than 15 shillings a week. One newspaper asked its readers to imagine the horror of various dignitaries who imagined they had been invited for 'trite moral reflections' and 'gratuitous champagne' confronted by 'twelve poor but respectable old women, who had each spent from 20 to 50 years of a long life bearing the yoke of industrial slavery in its cruellest form.'[43] The chance to see the interior of Sunderland House clearly acted as a magnet to some of the 500 guests, and to this extent there was a parallel with Alva's manipulation of nosey social climbers at her Marble House rallies. One reporter from *Pittsburgh Times* said that the sight of Consuelo on the platform at Sunderland House at the conference jointly hosted with the Anti-Sweating League reminded him of Alva at the head of a suffrage parade. In reality, the resemblance was superficial. When militant suffragettes interrupted a speech by a Liberal politician during the conference, Consuelo pacified them tactfully. Years later, she took pleasure from an article which remarked that her conference had 'probably done more to advance the cause of female suffrage in Britain than the violent combined efforts of the militant suffragettes'.[44] And when an interested visitor said to her that afternoon: 'This is the age of feminism' she replied: 'Not so much feminism as humanism.'[45]

One explanation for Consuelo's circumspection may be that she wished to distance herself both from the Pankhursts' violently destructive militant suffrage campaign in England during 1913, and from Alva, who had finally exploded into public support of it. Consuelo may have played a key role in launching her mother 'on the suffragette sea' in 1908, but by 1913 Alva had sailed miles past her when it came to suffrage radicalism.

The event which provoked Alva into an open declaration of support for a very unpopular position was the arrest of her heroine Emmeline Pankhurst – on very flimsy evidence – for firebombing Lloyd George's country house, and Mrs Pankhurst's subsequent sentence to three years of penal servitude. Alva was due to attend the Congress of the International Woman Suffrage Alliance in Budapest as a National American delegate, accompanied by Consuelo, in July 1913. On the eve of her departure to Europe in April, she announced that, in view of the British government's treatment of Mrs Pankhurst, she would only stay in England for as long as it took to arrange a passage to France and that she refused to spend any money there because of the Liberal government's cruel treatment of the suffragettes. 'It is all right for women to fight to win man's freedom. It is only when they fight to win woman's freedom that they are called hysterical, viragoes, criminally insane etc. Men fight for what they want. There is lots of talk about arbitration, but show me the man who would arbitrate for forty-five years!'[46] Her views aroused deep hostility in newspapers across America. 'Mrs Belmont is mixing in the suffragette demonstrations in London and if she doesn't get into jail and have some food squirted down her throat many thousands of her countrymen will be disappointed,'[47] said the *Kansas City Journal*.

When Alva arrived in London on 21 April 1913, she did not stay at Sunderland House as she normally did, knowing full well that her much publicised support for the Pankhursts would embarrass Consuelo who had just launched the Woman's Municipal Party (the official explanation was that 'spring cleaning' was in progress). The press had a field day. Journalists noted that Alva took a British train to Paddington, tipped a British porter and took a British taxi to the very British Ritz. 'With the customary brutality of the English toward women, the management made her settle,'[48] said one. 'What if our dukes should "girlcott" America?',[49] enquired another. None of this improved Alva's mood. 'How I hate and loathe England,' she told journalists. 'With what brutality you treat your women . . . We are disgusted with England. The whole world is ashamed of you . . . Mrs Pankhurst is the greatest woman of the age. You will probably kill her, just as Joan of Arc was killed.' She went on to say that she thought that the Pankhursts were remarkably restrained. 'Do I believe in burning down houses? Certainly, after what women have had to put up with. Yes, I would go even further than leaving bombs about, if women are to be killed and tortured.'[50]

Consuelo arrived in the middle of the press conference where Alva made these remarks and greeted her mother effusively to avoid talk of a rift. Staying at the Ritz, however, was as far as Alva was prepared to go in sparing Consuelo's blushes. Shortly afterwards, Alva insisted on making a visit of support to WSPU headquarters, only to find police raiding the building. They demanded that Alva and other visitors should keep away from the premises, arrested staff working on the WSPU newspaper the *Suffragette*, and took away papers and documents. The raid was probably designed to intimidate rich supporters from abroad, though Alva knew perfectly well that her position as Mother of the Duchess of Marlborough offered her considerable protection from arrest. Before she left England she was reported to have had a brisk passage of arms with her relation by marriage Winston Churchill, no friend of the suffrage movement since he was held responsible for particularly rough policing during one suffragette demonstration. At the end of the argument Churchill reportedly said to Alva: 'At least you'll admit that man has a great deal more will than woman?' to which Alva is said to have replied: 'Not at all; he's only got more won't.'[51]

Alva left England for a rented house near Deauville on the Normandy coast on 1 May and then travelled on to the International Woman Suffrage Alliance Congress in Budapest where Consuelo met her in mid-June. If Alva had hoped to press militancy on the congress (an international meeting of many different women's societies) she was to be disappointed. Although the delegates gave her a warm reception, it was the non-militant suffrage view which prevailed, with leaders of the National American particularly anxious to avoid upsetting American men in the run up to another New York State referendum in 1915. While Alva was disgusted by the timidity and strategic ineptitude of the congress, Consuelo was rather pointedly impressed by Anna Howard Shaw who was recommending a more cautious approach. English militant suffragettes were not formally admitted to the conference at all, and only had a chance to speak at the end, when they denounced their exclusion and proposed a resolution to honour Emily Wilding Davison, killed when she threw herself under the King's horse at the Derby on 4 June. 'We did manage to get in a few cheers for the militants, to the great disconcert [sic] of the more conservative delegates,'[52] wrote Alva later. It was noted that when the militants spoke, she stood up and waved her handkerchief wildly in the air.

After the conference was over, Alva returned to the rented house in Deauville, where she took further steps to support the WSPU. While the British government did its best to disrupt and intercept funds coming into WSPU accounts, Alva fought back by using her own bank account to channel money to Emmeline Pankhurst from England to France. She also invited the Pankhursts to stay – they were living in Paris to avoid arrest that summer – so that Consuelo met both Emmeline and Christabel in Alva's Normandy house. Even if she disapproved of Mrs Pankhurst's tactics, Consuelo admired her courage: 'a fine undaunted woman, her delicate body held the flame that animates crusaders, the spirit that willingly endures suffering and pain for an ideal'.[53] On the other hand, Consuelo thought Christabel Pankhurst was ridiculous: 'To hear Christabel Pankhurst orate against the male sex, as if their presence in this world were altogether superfluous, made one wonder how far prejudice could contaminate a brilliant intellect,'[54] she would write later.

She may also have been irritated by the extent to which Alva became quite infatuated with thirty-three-year-old Christabel, who spent several weeks of that summer under her roof. Alva later called the summer of 1913 'one of the most interesting periods of my life.' It was almost as if she had found a surrogate militant daughter in Christabel in whom she saw all her own independence of spirit. While she stayed with Alva, Christabel and a colleague worked on editing the *Suffragette* which was then smuggled by courier back into England. There was a brief respite on Thursdays when the paper was 'put to bed' for the week, whereupon Alva and Christabel enjoyed 'brisk walks at the seashore – rides through the country or moonlit walks in the garden overlooking the sea'. Alva was amazed by this 'slip of a girl' discussing the 'great tragedy of womankind . . . with the moon shining on her brave, heroic young face she would discuss her ambition to so improve the status of woman that she could never again fall into the unspeakable slavery of body and soul she was now in'.[55]

Christabel thoroughly enjoyed the luxurious surroundings provided by Alva for guests she admired. Set up in an office with a 'rose-coloured paper and draperies', and a secretary of her own, she wrote and dictated editorials for the *Suffragette* looking out on to a large and beautiful garden, and discussed ideas for her book on venereal disease, *Plain Facts About A Great Evil*, with Alva, who seems to have had insights of her own to contribute. 'I shall always be glad that we had that long time together in the Summer,' Christabel wrote afterwards. 'It helped me very much indeed to be able to talk over the book with you and to hear all the things that you had to tell me.'[56] For her part, Alva felt that 'this precious July and August at Deauville passed too quickly',[57] and would later keep a photograph of Christabel on her desk in New York.

Alva's hospitality to Christabel caused problems for Consuelo, however. On 13 September 1913, a *New York Times* correspondent wrote that he had 'the authority of the Duchess of Marlborough herself to deny the report that she is a sympathizer with the militant movement. The report originated from the fact that the Duchess was at Deauville while her mother entertained there Mrs Pankhurst and Miss Christabel Pankhurst but reports that she has in any way identified herself with the militant campaign are entirely without foundation.'[58]

That autumn, when she returned to the States, Alva sponsored a US tour by Mrs Pankhurst whose arrival in the country on 18 October 1913 put the whole question of militancy back on the American agenda, particularly when she was held by the authorities on Ellis Island and threatened with deportation. Anna Howard Shaw not only did little to help, but refused to speak on her behalf. Tensions over Mrs Pankhurst's tour brought Alva to the point of making a move which would redefine her suffrage activity until the end of her life. Early in 1913, yet another party formed under the umbrella of the National American. As far as Alva was concerned, this group offered a much more exciting strategy, despite having given itself a dull name: the Congressional Union for Woman Suffrage (CU). It was led by the redoubtable Alice Paul who argued that the American campaign should switch to campaigning for woman suffrage by way of a federal amendment; and should therefore be based in Washington where it could lobby more effectively. Moreover, Alice Paul had learnt her politics from the WSPU in England where she had been imprisoned and force fed for WSPU activities, and become something of a cause célèbre as early as 1909.

In November 1913, at the annual National Suffrage Convention, Alva heard Alice Paul's 'brilliant' report on some work she had already done for the National American in Washington. Alva immediately proposed a resolution moving the headquarters of the National American to Washington, but was defeated. The National American, sensing a growing rift, took steps to ensure that it was less dependent on Alva's money, successfully enlisting financial support from other well-off women. Alice Paul, meanwhile, decided to approach Alva for support, for she too realised that a split between the CU and the conservative suffragism of the National American was now almost inevitable.

Some of Alice Paul's radical colleagues in the CU were very wary indeed about the idea of enlisting Alva. They cautioned against soliciting help from a society woman whose reputation for misplaced aggression had preceded her. 'She really enjoys eternally fussing about details, and would use up all your energy and then would not be satisfied' wrote one member; Alva lacked 'constructive ability' and was 'an impossible member of the committee'; she was 'in the habit of running things absolutely', was 'very rash with newspapers' and was 'constantly making statements which you wish she wouldn't make'.[59] Another influential member of the CU committee, Mary Beard, supported an approach to Alva, however, arguing that she would be a valuable ally if treated properly rather than 'merely a money bag'.[60] Alice Paul finally visited Alva in person on 18 January 1914 and impressed her greatly. Alva then wrote the Congressional Union her first cheque, for $5,000, and joined its executive committee, a condition of her financial support. When reports of her defection to the radical CU appeared in the press, she confined herself to anodyne comments that the National American needed better leadership.

Behind the scenes she was far less polite: 'I was tired of having to pull along with me a heavy mass of suffrage conservatism . . . I was happy to find at last in America a fearless group unafraid to fight for its principles.'[61] The decision to commit to a fearless rebel group gave Alva a surge of energy. Once again she resuscitated society techniques. She searched for a prominent house in

Washington to act as new headquarters for the CU and decided to raise the new group's profile by giving a ball. Washington society women queued up to join the ball committee ('pathetic' was how one CU worker described it). Meanwhile her new political colleagues were obliged to get used to the quirks of their important new backer and those who had been wary of Alva felt soon felt their reservations were justified. While plans were being made for the ball, she threw a tantrum over a bill for a prize cup, feeling that she had already contributed enough. Alice Paul quickly learnt that Alva had to be approached with care for her subventions. There was also disagreement over an important question of tactics, that is: whether the political party in power should be held responsible for failing to give women the vote. When reminded quietly by a CU secretary that the organisation did not share her views on this issue, Alva flew into a towering – and familiar – rage. '[She] said that if she had to be dictated to she would do nothing; that when she was requested to take any action, no matter what the nature of it might be, she must be left to decide just how it shall be done. She knew what was required better than anyone else, and her method must be pursued or none at all,'[62] an alarmed Caroline Reilly wrote in confidence to Lucy Burns. It was a letter that could almost have been dictated by Consuelo.

That summer, Alva returned to Newport to repeat a familiar manoeuvre. Just as she had held two suffrage meetings at Marble House when she first joined the woman's suffrage campaign in 1909, she now decided to use her Newport power base to help the more radical CU, both by opening summer headquarters and holding a CU meeting there later in the season. As it happened, Alva had already started plans for a conference at Marble House in July 1914. This was for distinguished American women involved in social welfare work (a 'conference for social workers'[63] was how Alice Paul described it). She had planned this for the benefit of Consuelo, who had agreed to come to America from England after a long absence. That, at least, was the official line. An alternative interpretation is that in arranging for Consuelo to meet eminent professional American women at Marble House at the height of the summer season in Newport in 1914, Alva was up to her old tricks once again.

On the one hand, there is no doubt that Alva was genuinely proud of Consuelo's work and wished to introduce her (and parade her) to like-minded distinguished women; on the other, she knew perfectly well that if she wanted to guarantee a good turn-out and maximum publicity for her own cause, the Duchess of Marlborough was an incomparable draw. Consuelo's name was certainly most effective in recruiting distinguished speakers. '[My daughter] is intensely interested in the achievements of women in executive positions the world over,'[64] Alva wrote to Ella Flagg Young, superintendent of Chicago schools, who immediately agreed to attend.

Once she was sure that the chance to meet such women would definitely bring Consuelo to Newport, Alva set about using her asset to maximum advantage. Consuelo's presence offered an outstanding opportunity to shore up Alva's social position which was always close to being damaged by her much-publicised suffrage activities and her absences from Newport (she had missed the 1913 season completely while entertaining the Pankhursts in France). Alva now let it be known that she would be giving a Chinese ball in late July to celebrate a spectacular new Chinese tea house she had just built in the grounds; and that the ball would be given in honour of Consuelo. Those who wished to be invited, and to meet Consuelo, had little difficulty in working out that they stood a better chance if they signed up for the July rally for distinguished women too.

Initially, Alva seems to have had some idea of combining the meeting of 'great women' with a rally for the CU, but even she appears to have been hesitant about manipulating prominent American women and Newport's social elite into such an overt association with a rebel suffrage group. Instead she decided to open CU offices in Newport for the summer season and create an implicit relationship between the rally, Consuelo and support for suffrage militancy in the minds of the press and public. In practice, organising both a rally (under the auspices of the Political Equality Association) and a ball at Marble House was almost too much for Alva, with the result that she had little time to help with finding the CU offices in Newport as she had promised. This fell to Doris Stevens, a young CU organiser and activist, who would become another of Alva's young suffragist protégées, and who ended up by feeling highly ambivalent about her. After a frantic search, Doris Stevens settled on premises near the Newport Casino before embarking on a ceaseless round of public speaking and fundraising to pay for the entire enterprise including her own salary.

From the middle of June 1914, it was clear that Consuelo's presence at the conference for social workers would turn it into an event that Newport society could only ignore at its peril. Her arrival in the US was trailed by a slew of newspapers, who frequently did Alva's work for her by printing headlines such as: 'Duchess of Marlborough Expected To Make Many Converts to Suffrage Here', and calling Consuelo an 'avowed militant'.[65] The Duchess's impending arrival was tracked from Boston to Philadelphia, but there was a difference of opinion about what Consuelo would be doing when she arrived. Alva's press agent announced that the Duchess was on her way to Newport to support the cause and to address the meeting. Consuelo had given no such undertaking and had certainly not agreed to address the assembly on the subject of 'Battling for the Ballot in England',[66] as reported in one newspaper. Indeed, in the one interview that Consuelo gave on board ship to the *New York Herald* on 27 June, she said that she would not be speaking at all; but even the assassination of Archduke Franz Ferdinand in Sarajevo on 28 June did nothing to damp down intense press interest in her imminent appearance.

As soon as she arrived in the States, Consuelo found herself trapped in a dilemma indelibly associated with Marble House. She was immediately caught between not wishing to embarrass her mother publicly and great reluctance to submit to Alva's will – in this case, by endorsing militancy. Sensing a difference of opinion, the newspapers pounced. When asked by *The New York Times* what she thought of the 'militant suffragists' she replied: 'I hardly dare express an opinion on the matter; it is rather dangerous,' before going on to say: 'But I may say that I am a suffragist, though not a militant.' When asked whether she disapproved of English militancy, Consuelo reached for a slightly different formulation. 'I won't say that . . . It is for history to decide whether English militants are justified in their methods or not. It is impossible to compare English and American suffragists, the conditions are so different. The men of England are not so open-minded as those of the United States. They are more stubborn.'[67]

The most accurate summary of Consuelo's views probably appeared in *The New York Times* on 6 July when she was reported as saying: 'Militancy will never win the ballot in England . . . If the methods of the Pankhursts and their followers are persisted in I cannot say what will be the result. The hostile papers, however, have greatly enlarged on the horrors of militancy. They have failed to give the conservative suffragists credit for victories won, and have used the acts of the militants to disparage the work of the so-called constitutional suffrage organisations.'[68]

Misrepresentation of her views was compounded by misreporting throughout her stay. One interviewer suggested that Consuelo believed that rich women needed the vote to protect themselves from spendthrift sons-in-law – a piece of nonsense that drew a vigorous letter of protest to the editor from Alva afterwards. Consuelo was also alleged to have said: 'I cannot say what will happen if the Pankhursts and their kind continue their present tactics. But the Englishmen who cried "Let Them Die!" deserve death themselves for voicing so inhuman a thought,'[69] – a gloss on her views that she strenuously denied the following day in *The New York Times*. Consuelo's more immediate problem was that she had learnt that her mother was desperate that she should make some kind of speech in order to secure maximum publicity and ensure the success of the event. It is quite possible that Alva repeated a familiar tactic by telling the press that Consuelo would be speaking before she arrived and then making Consuelo feel she would humiliate her mother if she refused. In the end Consuelo surrendered, but only with considerable reluctance.

When the rally finally took place on 8 July, it was a triumph of adroit manoeuvring by Alva. Although this rally was purportedly a meeting of women involved in social welfare issues, the whole town of Newport gave itself over to the suffrage cause for the day. Suffrage colours 'flew from Ruth Law's aeroplane 1,000 feet in the air, fluttered in front of the Casino during the gayest period of the day, decorated shop windows along Bellevue Avenue, stood out before the deep green of many a fashionable front yard and even found vantage ground on some of the hundred or more yachts here for the regatta of the Eastern Yacht Club and the trials of the cup defenders',[70] wrote the *New York Herald*. The crowds started to assemble outside Marble House about 1 p.m. although the gates were only opened at 2.30. As they entered, a military band from Fort Adams played and young PEA workers distributed suffrage literature, pencils and fans – favours, in fact, but ones that could not have been anticipated in 1895. Once again, visitors could pay $5 to see the house, though the price of admission to the grounds and the new Chinese Tea House had gone up to $2.

Many observers later commented on the mix of women at the 1914 rally and expressed amazement at the polite manner in which they listened to each other. On the platform there were women from politics (Helen Ring Robinson, a Colorado State senator), the justice system (Chicago Juvenile Court judge Mary M. Bartelme), education (Ella Flagg Young), philanthropy (Maud Billington Booth of the Volunteers of America), reform (Florence Kelly of the National Consumer's League); and in a great coup for Alva, a representative of labour in Rose Schneiderman of the National Women's Trades Union League. Consuelo was presented by Alva as 'the worthy daughter of her mother', which may not have entirely pleased Consuelo but certainly reinforced the link between Mrs Belmont and the Duchess of Marlborough in the unlikely event that anyone was ignorant of the connection.

The 500 visitors included many representatives of high society – the ever-loyal Mamie Fish and Tessie Oelrichs, but Harrimans, Kernochans, Burdens, Warrens, Pells, Howlands and Perrys too. They sat politely through speeches defending the rights of Colorado miners, a particularly emotive subject in the wake of the Ludlow massacre of miners and their families; an attack on New York's judges for spending $18 million on a marble courthouse while the poorest women in New York wrapped their legs in newspapers to keep warm; and a blistering attack on rich industrialists from the socialist Rose Schneiderman who denounced 'an industrial system that throws women on the scrap heap when they are worn out'.[71]

The crowd reserved its warmest support for Consuelo who negotiated her way through a difficult situation by speaking first of her work with prisoners' wives, and then about the campaign for a decent municipal standard for lodgings for women. While the greater part of Consuelo's speech focussed on these initiatives, she finished with a concession to her mother by explicitly connecting these ventures with the fight for the vote. 'It is in order to obtain reforms such as these that women are asking for the vote,' she said. 'Those of us who are engaged in any form of social service realize that without legislation, individual and voluntary work can accomplish but little and must necessarily be sporadic. I therefore close by wishing those of you who have labored strenuously in many fields bearing the torch of civilization and progress before us, a speedy and a successful ending to the great work of women's enfranchisement which you have undertaken.'[72]

Afterwards, society women helped to sell tea and cakes, while the distinguished women on the platform passed warm comments about each other for the benefit of reporters. Senator Helen Ring Robinson told journalists that Consuelo impressed her as 'a woman of unusual quality of mind and heart' who struck 'a high spiritual note'.[73] Doris Stevens was impressed too, writing to Alice Paul that Mrs Belmont had been a superb chairman and that 'even the Dutchess [sic] was charming and made a most finished speech'.[74] This did not mean that coverage was universally polite. The *Ohio Star* was pleased to see that no-one on the platform had been experimenting with the hunger-strike business. The *Boston American* thought that the entrance charge was steep. The *New York Call* was irate at the way Rose Schneiderman was upstaged by Consuelo: 'Never mind Rose of the brave heart . . . How can you expect to be mentioned by America's plutocracy-truckling press on the same page as a dainty duchess, who is sweetly ready to do everything for the poor, except get off their backs?'[75]

There was something in this, for as Alva well knew, the American press still suspended its collective critical faculties when confronted with Consuelo. Her presence as 'the scintillating star of suffragettes' was reported the length and breadth of America, from California, to Louisiana to the states of the Mid-West. 'While the meeting would undoubtedly have been memorable in any case, it will probably be longest remembered by the women who attended because of the slim tall figure of the Duchess with her uplifted hand, as if she were fastening some mythical glove . . . she still has the power that draws – the desire to meet and talk with her – that she had when she married the Duke,'[76] gushed one reporter.

Almost every press photograph featured either the Duchess or the Chinese Tea House in accounts of the rally, a point not lost on *Town Topics* a few days later when it reported on 'a regular orgy of equal franchise'. 'One must give Mrs Belmont all the credit due for knowing how to stage a play. I would not say that she built the now famous Chinese pagoda and imported the Duchess of Marlborough as special features in her fight for the ballot, but as accessories to last week's conclave they were undoubtedly invaluable.' As Alva's guests left the rally, they had showered her with congratulations. In the two-faced world of Newport society, however, all was not as it seemed. 'I regret to say that many of them were tempered by the feelings of one gracious dame who, after, almost weeping in the excess of her admiration, remarked as she waited for her motor to take her away: "Well what do you expect? Alva Smith simply must be doing something to keep her before the public".'[77]

As far as Alva was concerned, however, there was more work to be done with her duchess daughter. The day after the rally, with much of the press corps still in Newport, she turned her attention to the opening ceremony of the CU's summer headquarters at 128 Bellevue Avenue. Until this point Consuelo had managed to steer a careful path between embarrassing her mother and her great reluctance to be drawn on the issue of militancy. That afternoon she was outmanoeuvred. Doris Stevens had asked her to attend, and as soon as Consuelo promised to come she immediately lined up a Pathé camera 'to get her as she comes into headquarters'.[78] In fairness to Doris Stevens, both she and Consuelo seem to have been hoodwinked by Alva, for it later transpired that Doris Stevens was quite unaware of Consuelo's resistance to militancy, while Consuelo was unclear about the CU's support for the Pankhursts. In front of a horde of reporters and photographers, and probably out of politeness, Consuelo contributed $50 – and appeared to be joining up.

This endorsement was misinterpreted by a wildly excited Doris Stevens who immediately wrote to Alice Paul that they had secured Consuelo as a CU member. This gave Alice Paul a bright idea: Consuelo might like to lead a delegation to President Wilson while she was still in America. She wrote to Alva saying that it would be 'a wonderful way of again focusing the eyes of the country upon the President's refusal to help'.[79] She enclosed a copy of this letter to Doris, adding: 'I hope and pray that we can carry through this plan of having the Duchess go to the President. Will you not present it to her in as favourable a light as possible, if the opportunity offers?'[80]

A few days later, Alice Paul received a short note from Alva explaining that it would be impossible for Consuelo to take part in the delegation since she was leaving for England on 1 August. Undeterred, Alice Paul came up with an alternative plan. 'We were indeed disappointed to learn from your letter, received today, that the Duchess of Marlborough could not take part in the deputation to the President. If you think the plan of having another deputation a good one and if your daughter would be at all interested in it, would it be possible for her to come sometime during the latter part of July?'[81] Alva, who may have thought that she could bring Consuelo round to the idea given enough time, was finally forced to explain the true position to Doris Stevens. Consuelo had never spoken on a suffrage platform in her life and she had been most reluctant to speak at the Marble House rally at all. 'Her mother by the way says she is not at all in sympathy with the militant movement,' wrote Doris Stevens to Alice Paul. 'Obviously my judgement . . . was in error. Her mother of course is way ahead of her in suffrage spirit.'[82]

Before she returned to England, Consuelo was obliged to assist Alva with one more task, reminiscent of the summer of 1895. This was to help Alva bolster her social position by association with her daughter's aristocratic standing. Newport's summer residents fell over each other to entertain the Duchess of Marlborough in much the same way as they had once competed to entertain the 9th Duke. She even had a dance named after her by two professional dancers from New York (the 'Consuelo' was eventually performed for her during a small party at Marble House). Her sister-in-law, Mrs William K. Vanderbilt, gave a dinner at the Clam Bake Club, and Mrs Stuyvesant Fish announced that she would be giving a ball in her honour.

The high point of the 1914 Newport summer season, however, was Alva's own Chinese ball, held on 24 July to celebrate the opening of her new Chinese tea house, though even this social event enjoyed its share of controversy. This time, it was the men of Newport society who rebelled, not at the suffrage associations with Marble House, but at having to dress up as Chinamen. The problem was a) the Chinese pigtail and b) the effort of changing. News of the revolt even reached the *Chicago Record*: 'Too much of a change,' they said, 'to get one's self up like an oriental after an arduous day, beginning in tennis flannels and running the gamut of bathing suits, afternoon dress and evening clothes of any description,'[83] though when Newport's women responded with cries of 'Don't be a crow!' they surrendered as one and duly appeared at Marble House, pigtails and all. Suffrage colours faded out of the shop windows on Bellevue Avenue and Chinese embroideries faded in. 'The only wonder to me is that the streets are not hung with yellow flags emblazoned with red dragons and that squads of [Chinese] are not selling chop suey and chow mein at the entrance to the Casino,'[84] said *Town Topics*, whose correspondent spent an enjoyable afternoon following Mrs Stuyvesant Fish up and down Bellevue Avenue as she tried to find something Chinese to wear. Her companions screeched 'Mamie, you do look a fright' at every turn, so it was 'a trying as well as a trying-on day'.

Alva's Chinese ball on 25 July was preceded by a Chinese dinner given by Mrs Stuyvesant Fish for 100 guests that almost upstaged it. Her house was decorated to represent a summer temple in Peking and it was said that some of the *objets* had been looted during the Boxer rebellion. In the main hall of the house the Emperor Keon Lung, who reigned in 1700, was seated on a throne and the guests saluted him as they passed. A standard bearer, a giant Manchu general, gave a greeting in Chinese to each guest on arrival. ('The house itself was ugly. It needed these things,'[85] said Alva.)

Feeling that her Chinese Tea House was the main attraction of the ball that followed, Alva confined herself to illuminating it and the rest of the garden with small electric lights and Chinese lanterns. Characteristically, Alva came dressed as the Dowager-Empress of China. Consuelo, rather daringly, wore a tunic and trousers as 'Lady Chang', consort of a Ming dynasty emperor, an outfit said to have been sent from China. Half way through the night, the Chinese lanterns in the house flickered out and the ball was plunged into darkness. One woman guest immediately detected an anti-suffrage plot, but most guests thought it was a deliberate ploy, intended to throw the illuminations in the garden into bolder relief – until Alva erupted into one of her rages. 'She could not be seen, but she could be heard, for she lost her temper completely and all the veneer of good breeding dropped from her . . . She poured her ire on everybody from the scullions in the kitchen to the flunkies in their black smalls, like a veritable suffragette fury,'[86] according to *Town Topics*. Terrified engineers from the local lighting company quickly appeared; the lanterns flickered on again; and Alva's secretary handed out reports that the ball was a triumph to waiting reporters, which is exactly how it was reported the following day.

'Again and again whenever she has wished, she has done something in the social world that has distinction and originality in sufficient degree to make people of the social world fall over each other in their attempt to get within invitations from her,'[87] wrote the

Holyoke Transcript. 'In the bright lexicon of Mrs Belmont,' remarked another newspaper, 'There is no such word as fail.'[88] But even Alva could not hold back the onward march of European history. The stuttering lanterns at her Chinese ball illuminated the death throes of the Gilded Age. A few days later Mamie Fish cancelled the ball she had planned in honour of Consuelo – for war had been declared in Europe.

10

Love, philanthropy and suffrage

AS SOON AS she heard that war had been declared, Consuelo left Newport to return to England. In New York, she discovered that the German liner on which she had booked a passage had been forbidden to sail, and returned to Marble House while she waited for another crossing. This, according to *Town Topics*, made at least one heart register an extra joyous beat.

> The organ was in the possession of one Boris Yonine, secretary of the Russian Embassy. There was reason a-plenty for the elation for if the eyes of Newport, mine included, may be taken as capable and veracious recorders of the things that are, Consuelo was leaving America with a very large slice of the genial Yonine's tender sentiments stowed away among her other belongings. For reasons very well known to everybody – that the Duchess is already blessed(?) with a lord and master, although they do not abide under the same roof – the affair could not extend to its very limits. But that the swarthy Muscovite was deeply smitten was as plain as the nose on your face. Wherever the Duchess went there also went Yonine. He carried her fan and her parasol and her wraps with as much faithfulness as he will probably carry a musket for his Imperial Master, and gazed at her with a fervour that distinctly said 'Where speech fails read it in my eyes.' Oh I assure you it was quite a touching little passionette – a poem by Blanche Wagstaff – and since the disturbing war clouds in Europe have compelled the Duchess's return to Newport I am sure that Mr Yonine, who accompanied her on the return, will say that it is a

> shockingly ill wind that blows nobody good. What are empires compared with tender sentiments?[1]

Nothing more is known of poor Yonine. One can only assume that he was left broken-hearted on the New York dockside as Consuelo's liner became a dot on the horizon. But the story shows that in 1914, at the age of thirty-seven, Consuelo had lost none of her power to charm and that it would be a mistake to see her life exclusively in terms of philanthropy and suffrage. The self-confidence that came with her achievements in public life attracted many admirers, in addition to those who simply assumed that a young, attractive duchess possessed of a great fortune and living on her own in a large London house was fair game.

Her position was complicated, however, by frequent rumours of reconciliation with the Duke. These appeared in the press almost annually between 1907 and 1912, often accompanied by suggestions that a reunion would be warmly welcomed by Consuelo's father. William K. seems to have continued to hope that the separation was temporary, if only out of concern for Consuelo's difficult position. His readiness to host some of her great receptions at Sunderland House suggests that he was sensitive to her anomalous standing and he may well have thought – somewhat unimaginatively – that this was best addressed by a rapprochement with her husband.

There is little sign, however, that any kind of reconciliation was ever likely if only because their differences became more pronounced with the passage of time. The Duke seemed increasingly beached by the tide of history, speaking out mainly on agricultural matters when the pressing political agenda was urban. Though he held another huge Unionist rally at Blenheim in 1912, he was not a wholly unreconstructed Tory for he did his best to grapple with some pressing issues of the day such as reform of the House of Lords. But his solutions – like his remarks about Consuelo at the time of the separation – often seemed designed to please no-one. 'My dear Sunny,' wrote Lord Lansdowne on 9 May 1911, after he had received the draft of a memorandum which the Duke wished to publish on the future of the House of Lords. 'I am as much convinced as you that a fundamental scheme of reconstruction is called for; but I am not convinced that it *must* be on your lines. I should, if I were you, think twice before I confidently recommend a scheme which assumes that 500 peers have been created, which gives us a House of Lords consisting entirely of hereditary peers, and which confers upon the Prime Minister the right of dissolving the Second Chamber. Is it necessary for you to publish anything? If not, might it not be wiser to remain silent? Yours aff. L.'[2]

Consuelo, on the other hand, felt thoroughly at home in the new Liberal zeitgeist. She drew on the support of a wide circle of aristocratic and American friends in her philanthropic ventures; in turn these reinforced friendships and opened up new vistas. Her interest in social welfare problems brought her to the attention of Beatrice and Sidney Webb. Asked to dinner by the Webbs, she found herself sitting next to George Bernard Shaw who became a friend. In the years after the separation she built up literary acquaintances by giving a series of dinners on Friday evenings at Sunderland House, and although she was never regarded as an eminent *saloniste*, guests flocked to these evenings, where the company was stimulating and the food delicious. Her early supporter, Lord Rosebery, acted as host to one such dinner, given in honour of H. G. Wells (Lord Lovat put it about afterwards that she had invited another H. G. Wells, but he was wrong) and attended by Bernard Shaw, J. M. Barrie and John Galsworthy – whose 'Forsyte Saga' was taking London by storm. W. B. Yeats was more often to be found at dinner with her godmother, Consuelo Manchester, and later she wrote that it was difficult to know who had been the most brilliant conversationalist – Yeats, Shaw or the 'outrageously handsome' George Wyndham. 'Looking back,' wrote Consuelo later, 'I cannot imagine a more gifted company in any country. No matter what the subject, the talk was never heavy, for there was always a flash of British humour to enliven it, and with Maurice Baring, Harry Cust and Evan Charteris to add fuel to the fire we were often privileged to assist at a marvellous display of pyrotechnics.'[3]

In her memoirs Consuelo remarked that 'niches give me

claustrophobia'[4] and the manner in which she moved between different cliques in Edwardian society suggests that she preferred to resist classification. She was close to the group of aristocratic and self-consciously intellectual Edwardians known as the Souls who favoured intense conversation, tennis and long walks over gambling and racing and whose relationships were characterised by a complex web of intimacy and extra-marital relationships. Consuelo was a regular visitor to one Soul stronghold, Taplow Court, the Desboroughs' home near Marlow, where there were 'shady walks in the woods and always an agreeable cavalier as escort'. When she invited Souls to her literary dinners, she liked to add 'a judicious mixture of a more frivolous element' to discourage excessive solemnity. She was also a popular guest at great weekend house parties that harked back to the eighteenth century, like those of the Duke and Duchess of Portland at Welbeck Abbey. But much as she enjoyed a few days 'enshrined in a hyper-aristocratic niche where sorrow or want or fear were unknown',[5] she was always ready to leave after a few days.

Sunderland House played host to many leading figures of the international *belle époque* too. When her memoirs were published, one man who lived in Curzon Street as a child wrote to tell her that he would steal out of bed to catch a glimpse of her great receptions through the windows of the salon facing onto West Chapel Street, enthralled and captivated.[6] Had he found his way inside, he would have encountered an array of Astors, Vanderbilts, Keyserlings and Lichnowskys, Russian and French aristocrats and young attachés, as well as ambassadors of the Great Powers like Count Mensdorff, so enchanted by Consuelo's flair and talent as a linguist that they would ask her to act as hostess at their own parties.

Feeling the need to escape from London at weekends, Consuelo took over the lease of Crowhurst Place in Surrey from the architect George Crawley in 1910. Crowhurst, which she breezily described as 'a small house', was a fifteenth-century manor house belonging to the Gainsford family, which Crawley had already begun to convert and restore until forced to stop by financial problems and his wife's ill-health. Consuelo retained George Crawley to complete the restoration, adding an extra wing, filling the house with old furniture, and creating a beautiful garden, which Mrs Crawley disliked because 'civilisation and shrubbery have diminished the drowsy air of a vanished day and . . . have spoiled much of the charm'.[7]

Hers was a lone voice of dissent however. The restoration has been described by one connoisseur of the genre as a masterpiece[8] and Consuelo herself loved it, remembering Crowhurst as a pastoral idyll, a weekend escape from city problems and the pressures of committees. Crowhurst was also an antidote to the baroque splendours of Blenheim. Here there were flagged floors, furniture polished to a 'honeyed sheen', rafters, owls, an oriel window, swans in a moat, and a herb garden with a splashing fountain. Like Alva's attempt to start a school for young women farmers at Brookholt, there was a discernible anti-industrial, arts-and-crafts note to Consuelo's affection for the house, the first on which she really had a chance to impress her own personality, an impulse which extended to having her monograph stamped on the drainpipes. The house was too small for large house parties, but there were many visitors, including three different Asquiths – the former Prime Minister among them after 1916 – on three separate weekends.

It is quite likely that there were admirers too, away from the watchful, prying eyes of London society. In 1913, Consuelo had dazzled another young diplomat, the writer Paul Morand, on his first posting as an attaché to the French embassy in London. That year he tried his hand at a series of written impressions of Consuelo devising a process which he described as 'layers of superimposed replicas'. The 'likeness' that emerged has to be treated with great caution, for one critic believes it is probably a composite of several women in Edwardian society including Margot Asquith. Here and there, however, his insights ring true. His Consuelo was also difficult to place, and Morand was annoyed with himself for being unable to form a clear judgement. 'She sums up all others, and yet is herself,' he wrote. Morand paints a picture of a charming, warm-hearted impulsive woman quite changed from the shy teenager who made her debut in London society in 1896. His Consuelo calls the pompous by their first names, wires a new friend once an hour for a week in the first flush of new friendship, sends flowers, rushes to the railway platform as one's train is leaving, tends to sick ambassadors, arrives late for her own dinner parties 'dazzling from the day', laughs easily and takes over the conversation like a 'wonderful firework'.[9]

The real Consuelo hated these impressions so strenuously that she begged Paul Morand not to publish them and he delayed for a number of years. She was clearly afraid that the world might take him too literally. She might have tolerated unhappily his portrait of a giddy female who was an enthusiastic friend of the successful, but cared less for the obscure; she could probably have shrugged aside his trivialisation of her own political activity ('she sees herself as mistress of the world, and wants to make the liberal candidate lose in the following week's by-elections'[10]), though his failure to note the practical, committed streak of his muse would have been irritating. The difficulty was that Morand implied that their intimacy was one of the boudoir; he quoted one princess who described her as 'immoral' and hinted strongly that Consuelo had at least one lover in the background.[11]

Friendships and admirers were one thing. Affairs were quite another. If Consuelo took lovers after the separation with the Duke she was careful to cover her traces. Many men attended her dinners and soirees and some of them, like Harry Cust, were celebrated philanderers; but it can only be a matter of conjecture whether any of them were more than close friends, a judgement made more difficult by Consuelo's friendships with individual Souls. While the Souls took a relaxed view of extra-marital relationships it is also true that this was a circle which prided itself on a new tone in male-female friendship, where intensely intimate platonic relationships were common, and flirtatious letters *de rigueur*. Consuelo is sometimes said to have been a lover of one Soul, Sir Edgar Vincent, who lent her a house in Esher as a holiday home for women workers, but one letter to Sir Edgar calling him a 'perfect eugenic specimen' is not enough on which to build a case, especially since she also sent her warm regards to his wife.[12]

Another Soul with whom she has sometimes been linked is her

old friend George Curzon whose first wife Mary Leiter died in 1906 and who did not start an affair with his second, Grace Duggan, until 1915. Here, anyone wishing to play detective encounters the problem of Curzon himself, for his romantic life was a constant source of speculation. He was extremely attractive to women, had many affairs and was rumoured at one stage to be involved with at least three women simultaneously. Curzon and Consuelo were deeply attached to one another in spite of their political differences, and stayed at each other's houses. Consuelo often went to Hackwood to join Curzon's great Whitsun house parties. Like Sir Edgar Vincent, Curzon supported Consuelo in her philanthropic ventures, opening the Mary Curzon Lodging House for Women in 1913. However, according to Curzon's biographer, David Gilmour, there is no evidence that their relationship was anything more than another *amitié amoureuse*, though Curzon would certainly have burnt any adulterous correspondence. It seems equally plausible that Consuelo could have decided that romantic involvement with such a strong-willed character with pronounced anti-suffrage views was not for her. She observed that his first wife Mary had to subjugate her character to him almost entirely, though even Consuelo was startled when she was once his house guest to discover that Curzon had changed all the books that his second wife Grace had picked out and put beside Consuelo's bed.

Consuelo's position as a separated woman in society was bound to attract gossip, some of it stoked by jealousy. The affectionate and flirtatious charm which Morand implied made her so many friends may also have misled the susceptible. It certainly confused one nameless gentleman who so misread the signals that he resigned from his clubs, convinced that they were about to elope. The most compelling reason for thinking that Consuelo enjoyed having admirers at her feet but generally kept them at arm's length is that she would have been highly concerned about destabilising her difficult relationship with the Duke, particularly in the years just after the separation. Her reputation had already been threatened by the affair with Lord Castlereagh and she had worked hard to salvage it. She remained vulnerable to salacious gossip which could all too easily have given an angry Duke grounds for challenging favourable custody arrangements while her sons were young, something she was anxious to avoid at all costs.

There is also evidence that the Duke actively tried to stop her from having love affairs, the only area of her life in which he felt he still had any power. In 1913, the year that Morand wrote his 'likeness', whispers of romantic dalliance grew louder. Consuelo was only thirty-six, but by now, Blandford and Ivor were sixteen and fifteen respectively and her position in London society was secure. A measure of Consuelo's difficulties in pursuing any romance at all is revealed by a letter in 1913 from the Duke to Gladys Deacon from Beaulieu-sur-Mer on the French Riviera in 1913 (theirs had been an on-off affair until 1911, when they started to correspond regularly again and entered a phase which would end with their marriage in 1921). In it the Duke wrote that Consuelo was enjoying an affair with one of his first cousins, Reginald Fellowes, second son of the de Ramsays, who were furious about it. 'I ought really to have him watched,' he wrote to Gladys. 'I must do so when I get back.'[13] He told Gladys that the de Ramsays wanted their son to marry a girl who was 'cut and dry for him, anxious to ally herself with his dusky loins', but that Consuelo was preventing him.[14] He wrote to Gladys again on Easter Sunday saying that he had met 'the buck' and cut him dead and was highly satisfied that Consuelo had now been forced to stay in London. 'I have spoilt this plan for the moment,'[15] he told Gladys.

For his part, the Duke seems to have become more embittered towards Consuelo as the years passed. He certainly resented the fact that she always seemed to have the upper hand in public relations, for her status as both serious philanthropist and victim of ducal greed charmed the popular press. In a letter to Gladys from Blenheim on 13 September 1912, the Duke enclosed a press cutting from the *Daily Chronicle* which remarked that: 'No London hostess takes greater practical interest in outdoor sports than her Grace of Marlborough, who is a fine horsewoman and a fairly good shot. Golf likewise claims her as an enthusiast.' An infuriated Duke annotated the cutting for Gladys's benefit: 'Philanthropist, Beauty, the used wife, Patriotic Yank, what else!!!'.[16] He also appeared to have enjoyed moments when Consuelo was mocked for her liberal leanings by others, sending Gladys a cutting from *Town Topics* in 1913, after the magazine had heard that Consuelo was speaking publicly on the need for women's lodging houses. 'Really Consuelo,' jeered *Town Topics*, 'even for a democratic Duchess – this is a bit tough.'[17]

It did little to help that Consuelo allied herself politically with David Lloyd George who launched a blistering series of attacks on aristocratic government in general and dukes in particular from 1909 onwards. On one occasion at Mile End in 1910 Lloyd George got the ear of the crowd by demanding: 'Since when have the British aristocracy started despising American dollars?' When a wag in the crowd yelled 'Marlborough!' Lloyd George replied: 'I see you understand me. Many a noble house tottering to its fall has had its foundations underpinned, has had its walls buttressed by a pile of American dollars.'[18] The Duke responded to this in print by saying that it was 'cowardly to attack lords through their ladies', but Consuelo did not exactly ease the tension by befriending Mrs Lloyd George, and attending committee meetings at 10 Downing Street where the Prime Minister's wife took the chair.

A. L. Rowse's assessment that the Duke suffered from depression, and the misanthropy and sense of persecution that often accompany it, is borne out by some of his letters to Gladys. He was certain, for example, that Consuelo was manoeuvring against him in 1915, when the ceremonial post of Lord Lieutenant of Oxfordshire fell vacant. Consuelo, according to the Duke, was not only supporting the candidacy of Lord Lansdowne but had become a central figure in the campaign. 'Wd you believe that little *L* is my opponent on the C. ticket. Is it possible, you will say!!!!'[19] he wrote to Gladys.

It is difficult to understand why Consuelo would deliberately have taken such a course of action: she was preoccupied with her philanthropy and war work, and it was not in her interests to annoy the Duke. He, however, left no-one in any doubt about his feelings on the matter, storming round to see Lansdowne and losing his temper. Lord Lansdowne wrote to Winston Churchill that he had

Above Consuelo at a bazaar at Stratford Town Hall in aid of West Ham Hospital.

Right Menu card for the dinner given by the Playground Association of America, 31 March 1908, at which Consuelo made her US speaking debut.

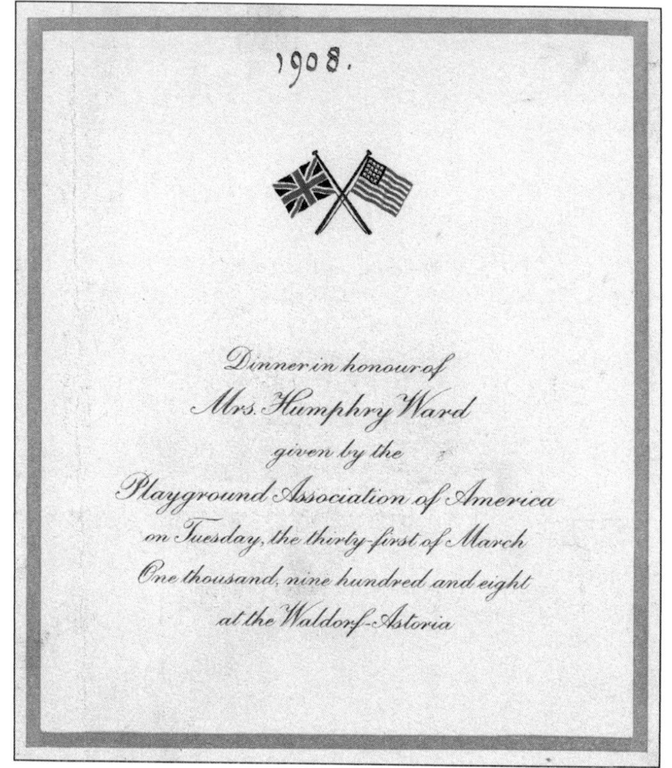

1908.

Dinner in honour of

Mrs. Humphry Ward

given by the

Playground Association of America

on Tuesday, the thirty-first of March

One thousand, nine hundred and eight

at the Waldorf-Astoria

Consuelo surrounded by knights at Blenheim House, a boys' club in Dulwich, 1904.

In descending order: the Marquess of Blandford, Lord Ivor Churchill and a friend in Wengen, 1912.

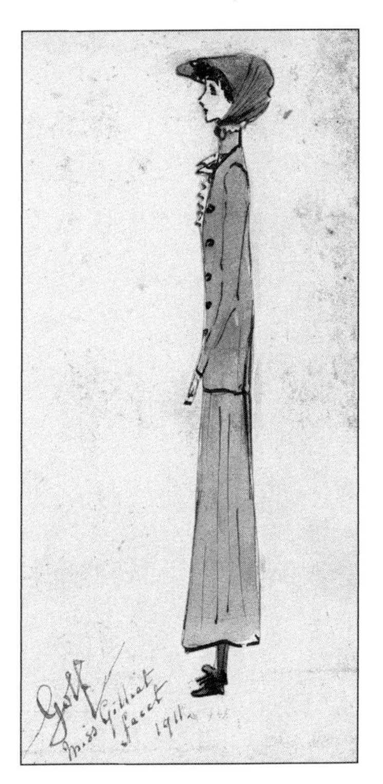

Drawing of Consuelo: 'Golf Miss Gilliat Facet, 1911.'

Above Consuelo and Alva at Marble House during the suffrage rally in July 1914.

Below Society and suffrage: a *thé dansant* to raise funds for the Congressional Union during the Newport summer season, 1914.

Above Congressional Union headquarters, Newport 1914. Alva stands at the back on the left; Doris Stevens is on the edge of the group at the far right. The banner reads: 'We demand an amendment to the United States Constitution enfranchising women.'

Right Gladys Deacon in 1921, the year of her marriage to the Duke of Marlborough.

Left Consuelo in the garden at Crowhurst, *c.*1919.

Below Consuelo in a pensive mood at Crowhurst, *c.*1919.

Top left Mrs J. B. Battelli of Ohio, Alva and Miss Lucy Hill with banner, at the dedication of the new headquarters of the National Woman's Party in Washington, 21 May, 1922.

Above Alice Paul greets Alva in a rather striking hat outside the National Woman's Party Headquarters, 1922.

Left Alva *c.*1927, as President of the National Woman's Party.

Opposite Consuelo in the garden of Old Fields, Oyster Bay, 1952.

Four generations at Blenheim *c.*1958. From left to right: Robin Muir, Lady Elizabeth Spencer-Churchill, Hugo Waterhouse, Ed Russell, Consuelo, the 10th Duke of Marlborough, Lady Caroline Waterhouse, Lady Rosemary Muir. Children in the cart from left to right: Robert Spencer-Churchill, Alexander Muir, Michael Waterhouse, Elizabeth Waterhouse.

an open mind on the subject of his nephew's wish to become Lord Lieutenant, but that his 'violent and abusive'[20] language had not helped him. Winston Churchill, F. E. Smith and others made representations to Asquith but the Duke had to endure a painful wait during which time he wrote to Gladys: 'You have no conception of the vileness of the intrigue against me and I am overwhelmed with despair ... I am in one of those moral convulsions of anger and despair which shake my whole being.'[21]

Under the circumstances it is perhaps not surprising that the Duke's thoughts should now turn to divorce, for Gladys Deacon too was placed in a difficult position and he was anxious to marry her before she tired of waiting. One interpretation of his behaviour towards Consuelo over both the matter of affairs and the very long delay in seeking a divorce, suggested by one member of the family, is that he did not wish to see her remarry until she was past child-bearing age, partly out of spite and partly because he feared that the Vanderbilt settlement might be dispersed in directions other than Blenheim. There are signs that Consuelo, too, was reluctant to give up her position as Duchess of Marlborough – which gave her such influence – without an alternative life in sight. The delay can also be attributed to a need to wait for changes to English divorce law, without which they would both face a complicated charade and a publicity barrage which could only rebound unfairly on their sons. There had been discussions since 1909 of an amendment which would make desertion for more than three years grounds for divorce. The outbreak of the war delayed the bill, however, leaving the Duke and Gladys in marital limbo, and consigning Consuelo to a lonely future.

When Consuelo returned to England from America in August 1914, she expected to find her adopted country tense and sombre in the face of war mobilisation, and was surprised by the general mood of sang-froid. The outbreak of war meant the dissolution of an international web of friendships, including a sad goodbye to Count Mensdorff, the Austro-Hungarian ambassador; but she was soon caught up in a flurry of war work, which Janus-like, looked backwards and forwards. Some of it reflected the passing of an aristocratic age, such as an employment agency for domestic servants put out of work by the closure of great houses and a highly successful scheme where women were asked to give a jewel to save a child's life, donating 'cameos, chatelaines, diamond brooches, bracelets and rings, stomachers in their old-fashioned settings and diamond tiaras and necklaces',[22] and other accessories of Edwardian opulence that they would rarely need again. Other projects pointed to the future rather than the past. Consuelo quickly became involved with the Women's Emergency Corps, one of the main mechanisms by which women took over the work of enlisted men, and became chairman of the American Women's War Relief Fund which ran a military hospital in Devon.

The rapid influx of women into every part of the wartime economy changed the nature of many of Consuelo's pre-war interests. She turned her attention from Bedford College to raising money for the Medical School for Women at the Royal Free Hospital in London, the only hospital in England where women were allowed to practice, setting up an annexe with eighteen beds as a lying-in hospital, staffed by women doctors and students. In the same year, she was the first woman to give the Priestley Lecture and used the opportunity to collect her thoughts on infant mortality, a process which took her 'three weeks to write and an hour to read'[23] and caused something of a sensation. Consuelo departed from a conventional script in attributing almost half the number of infant deaths in England to syphilis, causing several distinguished ladies in the audience to walk out.

Unlike many great London houses, Sunderland House proved unsuitable for temporary conversion to a nursing home, though it remained a useful place to have meetings. Its cellars doubled as a shelter during Zeppelin raids (the same man who had gazed at her parties as a child thanked her for the hospitality of the Sunderland House basement during the war). In spite of this, it became clear that it was somehow regarded as unpatriotic to live alone in such a large house when the country was at war; it also became uncomfortable and difficult to run, for young women servants preferred to work in the munitions factories. 'When the tenth housemaid in two weeks gave notice I asked her to tell me frankly what was wrong. "Well", she said, "I thought I had come to a private house, but I find it's the Town Hall, and I'm sick of washing that there marble floor after those meetings and refreshments." '[24] Towards the end of the war, Consuelo put the house at the disposal of the British government and decamped with Ivor to a smaller house in Devonshire Place.

The strain and tension of war did little to improve Consuelo's relations with the Duke, though there were brief moments when they found themselves in accord. The Duke became chairman of the Women's Land Service Corps, a movement which Consuelo also supported. In his inaugural speech, however, he could not help opine that if he could make do with the services of boys, old men and women as land workers, so could everybody else. As Consuelo's support of the liberal-left progressive position on a range of social welfare issues crystallised, the Duke wrote: 'Poor Demos, how he is struggling and dragging us all down with him. When will he learn the truth?' In contrast with his wife, he often felt his existence had little point and was prone to attacks of self-pity. 'I do not see what use I am in the world. I am so hopelessly tied down to an impossible existence . . . I am full of gloom for I see nothing ahead. Nothing.'[25]

The Marlboroughs, in common with thousands of other parents of sons, had to cope with acute anxiety over the Marquess of Blandford who had now left Eton for a short course of officer training at Sandhurst before joining the army. Even this was the cause of tension between his parents. In a letter to Gladys, the Duke raged that he had 'arranged everything with the WO [War Office], having taken B[landford] before the General concerned and having settled all formalities – after all these uncertainties B[landford] informs me that he prefers the 1st Life Guards.' Consuelo, it transpired, also preferred the Life Guards for Blandford and had taken steps to arrange it. 'The trouble is that the boy will keep on saying that he is in a second rate Regiment and his brother will keep on saying the same thing. B was perfectly contented a week ago. Only when he gets with his mother do these changed views appear. It is disturbing as it shows a weakness of character.'[26] In a further letter the Duke wrote that he was 'going to London tomorrow to see if I can undo the mischief that C has wrought re Blandford. I go with a heavy heart, the thing is so revolting.'[27] His visit, however, achieved little for Blandford went straight into the 1st Life Guards as a second lieutenant.

The war had one unexpected benefit for it rapidly improved the chances of English women being given the right to vote. As soon as war was declared the exhausted Pankhursts wisely declared 'a patriotic truce'; the vast influx of women into men's roles rendered anti-suffragist arguments about separate spheres of activity unsustainable; and concern about his ability as a war leader led to the replacement of Asquith by Lloyd George in 1916, removing an implacable opponent of women's suffrage from the head of government. It is worth noting that many historians now consider the role of the constitutionalist mainstream movement led by Millicent Fawcett – supported by Consuelo – more important to the eventual success of the campaign than the Pankhursts' militancy which, it is argued, may have ultimately delayed an extension of the franchise. In America, conversely, the war in Europe caused difficulties for suffragists. America would not enter the war until 1917, but it dominated the political agenda and made it more difficult for suffragists to be heard.

The meeting of the Congressional Union of Woman Suffrage at Marble House at the end of August 1914 went ahead as planned, and to Alva's great pleasure, it now adopted the principle over which she had lost her temper with Caroline Reilly. This was the Pankhurstian approach of holding the party in power responsible for failure to deliver votes to women, regardless of support for the cause by individual members. In America this meant that the CU now targeted both the Democratic Party and President Wilson, who claimed to have an open mind on the suffrage question but consistently refused to take action. Alva, according to Doris Stevens, was the one who led the way in persuading the CU that the best way to apply pressure on male politicians was to make women's suffrage an election issue. The key to this was that women in the nine 'suffrage states' who should be urged to withhold their votes from the Democrats in the forthcoming congressional elections until the party agreed to support a federal amendment. 'We knew this policy would be called militant and in a sense it was,' wrote Doris Stevens. 'It was strong, positive and energetic.'[28]

Apart from financing the CU, the adoption of these two political manoeuvres represented Alva's principal political contribution to the radical group. She was always prepared to trust the judgement of women she regarded as stronger than herself, and though she met very few, Alice Paul, the CU's leader, was one of them. Indeed, by financing the energetic CU led by Alice Paul, Alva suddenly found herself with less to do. For the time being she continued to run the Political Equality Association in New York, dividing her energies between the CU's national campaign and the New York State campaign for another referendum in 1915, for both approaches to suffrage were still in play.

In 1915 Alva finally sold Belcourt to Oliver Belmont's brother Perry, and bought a piece of land on the north shore of Long Island where she orchestrated the construction of another extraordinary house, Beacon Towers, at Sands Point (the ruins can be seen to this day). Designed by Richard Howland Hunt, it took as a starting point Alva's fascination with France, this time with the chateaux of Burgundy and the paintings of the *Très Riches Heures du Duc de Berry*. Indoors, there were great murals of Alva's militant heroine, Joan of Arc. Outdoors, the architect grafted a strikingly modernist note. Walls were surfaced in white stucco and the windows lacked surrounds in the style of Corbusier, in a manner 'appropriate to new structural materials in a new age'.[29] In photographs taken from the shore, Beacon Towers seemed to float weightless and fantastic, as if Alva's dreams of living as a princess in a castle had now meshed with ambitions of modernity. The house was sold to William Randolph Hearst in 1927, and demolished in 1943. Had it survived it would perhaps have been Alva's greatest architectural achievement.

Back in the political world, any hopes of a rapprochement between the Congressional Union and the National American were finally dashed as the result of action taken by Alice Paul in the wake of the *Lusitania* crisis. On 7 May 1915 the British Cunard liner was sunk off the Irish coast by a German submarine. Of the 1,195 passengers who died, 128 were US citizens and included Alva's nephew by her first marriage, Alfred Gwynne Vanderbilt, a son of Cornelius II who had avoided the sinking of the *Titanic* three years earlier when he changed his passage at the last moment. Although the German embassy published advertisements in the morning papers warning American passengers not to sail, they appeared far too late and the sinking of the *Lusitania* increased pressure on President Wilson to take America into the war against Germany. Alice Paul felt this would be a good moment to ratchet up the pressure on the President and allowed two members of the CU, Florence Harmon from Alva's PEA and Mabel Schofield, an English suffragette, to 'besiege' him by attempting to deliver a letter to him at a banquet. When that failed, they pursued him in a taxi to the pier, eventually giving the letter to his stenographer.

Criticism of the CU and Alva for bothering the President at such a moment was widespread. Anna Howard Shaw was among those who denounced 'English militants' who 'were employed by Mrs Belmont to make the outrageous attack on the President'.[30] Alva damped down criticism with some well-judged remarks and then proceeded to the San Francisco Voters Convention where she publicly waved off members of the CU on a transcontinental journey to Washington with a petition to which they added more than a million signatures as they travelled from coast to coast. This garnered many headlines, again deflecting press coverage away from the war, and succeeding in generating real political discussion between the CU and two senate committees when the petition was delivered in Washington. When the New York State suffrage campaign was defeated in another referendum in 1915, however, Carrie Chapman Catt and Shaw blamed the 'militant' action by the CU and Alva for its failure. Catt took over as leader from Anna Howard Shaw at the end of 1915, and remained implacably hostile to Alva, though she slowly began to adopt tactics that Alva had advocated for years.

Defeat in the New York referendum in 1915 left all New York's suffragists uncertain of how best to proceed. By state law, another referendum on woman's suffrage could not be held until 1917. The immediate problem was keeping the suffrage issue alive. Alva stepped in with a bright new idea, the most idiosyncratic of all her schemes for uniting suffrage with society. She announced that she would put on a suffragette operetta, to be sung by New York debutantes, assisted by suffrage workers. The music – and some of the lyrics – would be written by Elsa Maxwell, whom Alva had befriended that year when she arrived in America from England bearing a letter of introduction from Consuelo. Elsa Maxwell was, in her own words 'a short, fat, homely piano player from Keokuk, Iowa, with no money or background [who] decided to become a legend and did just that'.[31]

Alva took to her on sight, not always an immediate reaction, since Elsa Maxwell's appearance was as singular as everything else about her. 'Prince Christian of Hesse once saw me swimming off Eden Roc, and mistook me for a rubber mattress,'[32] she once wrote. She would eventually succeed in becoming an international society celebrity and a famous giver of parties. Doris Stevens was deeply disturbed by Maxwell's lesbian relationship with Dickie Gordon and later wrote: 'Many people hate her passionately and others are indifferent, but wherever she is functioning she is always a subject of speculation.'[33] But Alva tended to see anyone who was always a subject of speculation as a person after her own heart.

If Alva and Elsa Maxwell had set out to create an antidote to gloomy news from Europe with their production *Melinda and Her Sisters*, they certainly succeeded. *The New York Times, Tribune, Telegraph, Sun, Journal, World* and even *Musical America* tracked the operetta to the night of its premiere. The leading role of Melinda was taken by a professional actress, Miss Marie Douro; the rest of the cast comprised an exciting mix of debutantes and suffragists.

Marie Douro, in an early exercise in method acting, became a suffragist overnight for the part. 'I did not know that suffrage meant all that,' said Miss Douro to *The New York Times* when she read the libretto. 'It appeals to the very highest and best there is in me'[34] – though not, it would appear, to her vanity, for there was rather a tussle over her preference for classical sandals rather than the flat-heeled sensible shoe so characteristic of the suffragette. Some of the debutantes were astonished that just before the first performance they would be required to rehearse *for a whole day*. Alva, in her self-styled role as theatrical impresario, announced that she expected *Melinda* to be produced all over the country and that it would make at least a million dollars for the cause. As it turned out, *Melinda and Her Sisters* had one performance at the Waldorf Astoria on Friday 18 February 1916. Boxes cost a staggering $150 each, but the parents of the debutantes queued up to buy them and it was soon completely sold out.

The debutantes' parents did not go unrewarded. The plot, devised by Alva, concerned the eight daughters of Mrs Pepper of Oshkosh, all talented and beautiful except Melinda, a suffragist. When Mrs Pepper has a birthday party she omits to invite Melinda on the grounds that her embarrassing suffrage politics will ruin her sisters' social chances. The seven sisters dance with their friends – one of them performs the dance of the faun from Diaghilev's ballet *L'Après midi d'un Faune* – but Melinda appears uninvited, with a suffrage parade in tow and an assorted collection of 'laborers, factory girls and salesladies'. In a twist of the plot which is never quite explained Mrs Pepper sees the error of her ways, converts to the cause of female suffrage, forgives Melinda and they all live happily ever after. This plot summary does not, however, do justice to some of Alva's lines. Her heroine comes off worst with persistent calls for women to unite. The wicked characters do rather better. 'It doesn't take money to get a villa at Newport, it takes brains,' says Mrs Grundy. 'It takes brains to make money,' replies Mrs Malaprop. 'Any fool can make money, it takes a clever person to spend it,' replies Mrs Grundy. 'Society is the key to the Higher Life – Publicity,'[35] remarks another character elsewhere. But as the reviewer from *Theater Magazine* pointed out, 'each present thought the quip was intended for his or her neighbour and accordingly laughed with unsuspecting heartiness'.[36]

Any impresario tempted to revive *Melinda and Her Sisters* would be repelled by its racist undertow, prevalent throughout the American suffrage movement at the time. Alva may have been a pioneer in including black women in the PEA but that did not stop her taking the opportunity to mount one kind of argument also to be found in the writings of Carrie Chapman Catt: '. . . all the blacks, the negros, they also are allowed their vote . . . and imbeciles,' says Melinda to Mr Dooless. 'By denying women the political right to vote and by allowing old black Joe that same right, you place old black Joe mentally and economically in a position superior to that of the late Mrs Dooless, your capable and very good wife.'[37] Sadly, none of this seemed unusual comment to the audience and the press was captivated. Never had the ballroom of the Waldorf Astoria witnessed 'a more triumphal and brilliant affair'. 'It proves once more that Mrs Belmont is full of original ideas,'[38] said the *New York Telegraph* on 19 February 1916. The *New York Sun* attributed some of *Melinda*'s success to 'Marie Dressler's dress, and Miss Pam Day's undress'.[39] But other newspapers pointed out that anyone who picks out the cream of society's debutantes, casts them as love-letters, sporting sisters, and castle dancers, and throws in a serendipitous Zeppelin raid over the Astoria for good measure was not only guaranteed success but positively deserved it.

At a national level, however, the most pressing political issue was the Presidential election in 1916. 'It was Mrs Belmont's persistent desire that a women's party be formed which would stand as a wedge between men's political parties and force them to accept equal suffrage,'[40] wrote Doris Stevens later. The CU formed itself into a 'one-plank' political party in Chicago in 1916 with the sole aim of persuading women voters in the suffrage states to use their votes to punish any political party that did not support a federal amendment. Alva's instinctive flair for marketing led her to press an important change on the CU – its name. She was convinced that 'Congressional Union' was meaningless to most people, and urged the organisation to take the opportunity to rebrand itself as the Woman's Party. In an interview with Amerlia Fry many years later, Alice Paul said the name change came about 'because of this effort of Mrs Belmont to have something called the Woman's Party. Then we thought, "The name seems to be generally accepted. People like it . . ." I remember Mrs Belmont was so pleased that we did it . . . and so full of interest, that she got up and pledged on the platform some tremendous sum of money.'[41]

Perhaps unsurprisingly, Alva suffered from high blood pressure, though the weak heart with which she threatened Consuelo in 1895 had held up amazingly well in the intervening years. In the summer of 1916 she was advised by doctors to take a rest. She chartered a yacht, the *Seminole*, to sail up the coast to Newport, and begged Doris Stevens to accompany her. Alva told Doris they both needed a break, but she had another motive. According to Doris Stevens, Alva 'was not certain of some of the guests she had invited and that since she was taking Mr X along for the whole trip she must not be left with gaps on the boat when no other guests would be aboard. She said people might talk.'[42] Mr X was J. Ralph Bloomer, a real estate agent thirty-years younger than Alva, whose name occasionally appeared in *Town Topics*, attached to a list of other young bachelors at minor social events. Alva probably met Ralph Bloomer on one of the occasions when she was trying to sell Brookholt, and he may also have advised her on the sale of Belcourt to Perry Belmont the previous year.

Somewhere between 1916 and 1917, Alva, who was now sixty-three and increasingly deprived of close friendships by death and suffrage, fell in love with the much younger Ralph Bloomer. She was painfully aware that this love would never be requited, and that what she most wanted was truly unattainable. Her feeling for him was so strong that it worried Doris Stevens, who feared not only for Alva but for the cause, thinking that Ralph might lure Alva away from women's suffrage. This did not happen, though he was still in evidence at Marble House a year later in 1917 and Alva was still miserable. 'She talks to me freely of him' Sara Bard Field wrote to Charles Erskine Scott Wood, and 'searches the whole world for contemporary women who have married much younger men, drags out the illustration of Lady Churchill and George Cornwallis-West and when I say "Yes, but that didn't last," she responds "No, of course not. It was on a fleshly basis only but when she did divorce him he turned right around and married another woman older than Lady Churchill, Mrs Pat Campbell." You can see she is doing what we all do – rationalising her impulse – or trying to.'[43] When she finally met him, Sara Bard Field was surprised to find that she liked Mrs Belmont's 'young beau' more than she expected. 'I think he has some possibility,' she wrote to Wood. 'He is a clean-looking, clean-living chap, neither drinks or smokes (athletic influence while in Harvard) and with a wholesome love of the outdoor life and sports and especially flowers. He is sick of the insipid society women and perhaps he does find a lure in Mrs B's vigorous mentality and uncompromising selfness.'[44]

It was a forlorn hope. Alva's unrequited feelings for Ralph Bloomer in the summer of 1916 can be felt in the wistful undertone of a log she wrote for the benefit of those who joined the *Seminole* during the month-long trip. Apart from two days when he went ashore to visit relations (a visit which Alva revealingly describes as his 'holiday'), he kept her company the entire time. As in the old days of the great yachting trips, other guests came and went. They included Doris Stevens and the Democrat Collector of the Port of New York, Dudley Field Malone, who was a leading male supporter of what had just become the Woman's Party and was passionately in love with Doris Stevens. 'I have passed through the market; she is still at the threshold, and yet it would seem we are both vague, unable to arrive at any conclusion,'[45] wrote Alva.

The trip was punctuated by unseasonable storms. According to the log, it was Alva and Ralph who sorted out difficulties, taking decisions with the yacht's captain about how best to proceed. One night, a raging storm made it impossible for Alva to close her porthole, leaving her 'lamenting my want of power' for everyone was asleep. And then: 'Oh joy! Ralph, who says he sleeps lightly – and I believe he does – comes to the rescue. I am to disappear and he is to close the storm out. Again, I envy masculine strength.'[46] But this was as far as things went, leaving Alva quite unable to sleep and awake until dawn. 'Those in harmony should never go,' she wrote elsewhere in the log. 'In conjugating the verb "to be" only the present tense should be used.'[47]

It may have been her feelings for Ralph Bloomer that caused a striking sense of disorientation when Alva returned to Newport that summer. Arriving by an unusual route and in very different company she certainly saw its 'imaginary importance'[48] afresh. She did not open up Marble House that season, much to the annoyance of her friend Tessie Oelrichs, but stayed on the yacht, taking her guests to visit old friends and relations, including the Perry Belmonts at Belcourt. 'This is not the place or hour to give the impressions of what Belcourt did for me,' she wrote. 'I will simply say, to show what one can do, I made myself believe I was not myself and all went well.'[49]

Newport was as charming as ever and she was surrounded by friends, but to her, 'Life seemed careless, bright and the people like so many butterflies.'[50] After Tessie Oelrichs had given them all dinner Alva found Rosecliff 'breathing of luxury, of what wealth can do, charming every sense, conducive to idleness, dulling perhaps, alas! our standard of – shall I say, duty?'[51] Even when she watched society athlete Eleonara Sears play tennis, Alva felt uneasy at the sight of four rich women exhausting themselves in a game, while over the fence 'two men plough and dig, and with the sweat of their brows eke out a meagre existence for themselves and perhaps others. What a dark farce it all is!'[52]

Alva soon found herself grappling with more serious problems, however. The 1916 Presidential election was dominated by the issue of whether America should enter the war, and members of the Woman's Party – far more of them than in 1914 – campaigned to the point of exhaustion for the Republican candidate, Hughes, who had agreed to support a federal amendment for woman suffrage as part of his manifesto. In spite of their efforts, President Wilson was narrowly re-elected in the autumn of 1916. He continued to parry demands for votes for women, even when visited by a deputation on the death of Alva's beautiful protégée Inez Milholland, who had campaigned despite ill health in 1916, and then collapsed on a suffrage platform gasping: 'Mr President, how long must women wait for liberty?', and died weeks later. Convinced that more confrontational action was now needed where politics had failed, Alice Paul thought up a new way of generating publicity for the cause. In January 1917 she became the first person in history to lead a picket to the White House. 'We could wait no longer,' wrote Doris Stevens. 'Volunteers signed up for sentinel duty and the fight was on.'[53]

The 'militant' wing of the American suffrage campaign, represented by the Woman's Party, never came close to the violent and destructive acts of Mrs Pankhurst and the suffragettes in England. The White House pickets, which began in January 1917 were, by any standard, an extremely tame affair. Typically, a dozen women would stand in a line outside the White House, eight of them carrying purple, white and gold banners reading: 'Mr President, what will you do for woman suffrage? How long must women wait for liberty?' – Milholland's last public words. This caused an enormous stir and triggered headlines describing the women as 'unsexed', 'undesirable' and 'dangerous'. It all pleased Alva hugely – so much, in fact, that she immediately sent Alice Paul a cheque for $5,000.

The picketing continued for three months. President Wilson was mildly amused and would sometimes tip his hat to the suffrage ladies as he drove to the White House. On 6 April 1917, however, America joined the war against Germany and the suffrage movement as a whole came under pressure to act as Englishwomen had and declare a patriotic truce. The National American immediately offered its services to the government, but Alice Paul was determined to go on pressing for suffrage. The argument was summed up by a picket banner which read: 'Tell the President that he cannot fight against liberty at home while he tells us to fight for liberty abroad.' Alice Paul was also emboldened by events in England, where the new franchise bill was making its way through Parliament in wartime with few problems and would eventually lead to the passing of the Representation of the People Bill in February 1918, allowing most women over thirty to have the vote.

Unlike many members of the Woman's Party who were political radicals, Alva supported America's entry into the war. This may have been one reason that, most uncharacteristically, she did not attend the ceremony during which the Woman's Party became the National Woman's Party (as it is still known) in Washington on 2 March 1917. If she disagreed with Alice Paul's wartime picketing strategy in private, however, she supported her stand in public, and continued to offer financial support. This was not as straightforward as it had once been. Income tax of 7 per cent had been introduced in 1913, had risen to 15 per cent by September 1916 and would briefly rise to a top rate of 75 per cent before the end of the war; and war inflation would drive up the cost of building Beacon Towers; at the same time financing the National Woman's Party (NWP) became much more difficult after 1917 because no-one wished to be associated with a 'treasonable' group.

The summer of 1917 was a challenging time for the NWP. In June the administration began to clamp down on the picketers who were briefly imprisoned on a bogus charge of obstructing the traffic outside the White House. In July matters deteriorated sharply. A picket, which included Doris Stevens, was arrested and sent to the Occoquan Workhouse where they were stripped naked in front of each other, forced to wear prison clothes and denied the most basic sanitary items including lavatory paper. They were pardoned by the President after a few days of vociferous protest. For the rest of July and the first part of August 1917, the pickets resumed and for the time being, the arrests stopped.

In the middle of August events in Washington suddenly moved into a different and much more violent phase. As the President continued to disregard the case for democracy at home, the NWP pickets took to taunting him with banners calling him 'Kaiser' and 'Autocrat'. In the build up to military action that now gripped the country, soldiers and sailors roamed the streets of Washington tense with pre-war nerves and ready for a fight. In mid-August the 'treasonable' banners of these women provoked a furious reaction from one male mob who attacked the female demonstrators, knocked them to the ground and tore at their clothes. On the first occasion the police stood by and let them get on with it. During the second, on 16 August, they led the attack themselves. At least fifty women were arrested and sent to the workhouse, where black women prisoners were ordered to attack them by the warders; conditions were so filthy and degrading that a hunger strike was the only practical response.

When Lucy Burns led a campaign inside the workhouse for treatment as political prisoners, she was sent to the district jail where treatment was so shocking that Dudley Field Malone resigned his government job in protest. In October, Alice Paul was also sent to the district jail for picketing where she led a hunger strike to which the response was force-feeding. When she refused to call off the hunger strike the authorities responded by attempting to have her certified, sending her to a psychiatric ward with other criminally insane inmates, and keeping her awake by shining electric lights in her eyes every few minutes. On 17 November, more pickets were arrested, including a frail seventy-three-year-old, Mrs Mary Nolan, who was sent to the workhouse and subjected to what was later known as the 'Night of Terror'. On the orders of the warden, the suffragists were attacked by a group of men. Women picketers were manhandled, knocked to the ground, and denied water and all medical attention for their injuries, even when one of them had a heart attack. There was a huge protest outside the prison, and the prisoners were eventually released in late November after serving a writ of habeas corpus. The NWP was eventually vindicated when a federal court ruled that every arrest had been illegal under the First Amendment, and that the NWP picketers had been held in the workhouse illegally. Alva rushed to their defence, denouncing the soldiers and sailors who had attacked them. 'It is grievous to hear that sailors, wearing the uniform of our nation, supposed at this very moment to be preparing for the fight for democracy, find nothing better to do than to knock down women likewise seeking democracy,'[54] she raged.

During the lull between the first arrests in July and mob attacks in August, Alva went to Newport for the summer season, where she was joined by Sara Bard Field who had agreed to help her write her memoirs. Sara Bard Field was a poet, a radical, a feminist and a campaigner for social reform – an area in which she first became interested while married to a Baptist minister by whom she had two children. She had been a paid organiser for the campaign that won suffrage for Oregon in 1912 and sprang to prominence as one of the leaders of the great transcontinental suffrage petition in 1915. By the standards of the day, her private life was highly unconventional. She divorced her husband in 1914 around the time she met and fell in love with the poet and writer Charles Erskine Scott Wood. In an interview with Amerlia R. Fry in the 1970s, Sara Bard Field thought she had first come to Alva's attention at a meeting in New York in 1916: 'I evidently spoke with a kind of fervour that Mrs Belmont approved of and had herself. I would call it, perhaps, in her case vehemence. She couldn't do anything by halves, it was always all the way or nothing.'[55] When Doris Stevens suggested she would be a suitable person to ghost Alva's memoirs, it was Sara Bard Field who hesitated and Alva who was determined. 'She seems gluttenous [sic] to get at my spirit and have it here, just as she buys gowns and jewels,'[56] wrote Field to Wood.

Sara Bard Field accepted the job because she needed the money. It was agreed that she would travel to Newport, where she stayed for almost a month, lodging at a boarding house on Coggeshall Avenue. A routine began to emerge: Alva would call for her after shopping in the morning or would send a car; after lunch at Marble House, Alva would talk and reminisce for two or three hours. 'She is not at all embarrassed with me. I take copious notes. I interrupt with a hundred questions. If she is alighting some point of vital interest I urge on her memory by suggestion and interrogation,'[57] Field wrote to Wood. Later in the day, Sara Bard Field would type up her notes and begin to divide the material into chapters 'dashed off while in the spirit and color of her own recitals'.[58] Sometimes Alva's phrases and observations were transcribed verbatim. At other times Sara Bard Field clarified and rewrote what Alva had said. 'She is often so vague I have to define all her ideas myself and I hope to god she will stand for some of the things I have interpolated. Of course this is *her* life not mine and I have no right to put my views into her mouth unless she subscribes to them and makes them her own.'[59] As a socialist, Sara Bard Field was particularly fascinated by what Alva said about Joseph Choate's attempts to dissuade her from divorce on the grounds that the rich must not break ranks; and by Alva's observations on marriage in general and Consuelo's marriage in particular. 'Well the chapter on Consuela's [sic] marriage was some job! It made a discussion of the old lady's reasons for furthering such a marriage necessary and she certainly does put it over in fine shape,' she wrote to Wood on 24 August 1917.

Sara Bard Field hated the luxury with which she was surrounded at Marble House. Indeed she hated the whole of Newport in 1917. It was filled with 'soldiers and sailors about to kill and be killed', and the lifestyle of the very rich revolted her. The soldiers and sailors stank in the summer heat, but attempts by Alva to engage society women and generals' wives in a campaign to give each serviceman a change of underwear met with no success. 'These mindless creatures of inherited wealth or of private riches live their lives of selfish indulgence and unvirtuous sloth,'[60] Sara Bard Field wrote to Wood. 'No wonder the Goths sacked Rome when Rome had been reduced to such as these.'[61] Alva's disgust at the insipidity and passivity of Newport's society women seems initially to have misled Sara Bard Field who was naive enough to think she might be able to convert her employer to socialism. 'I have sworn to the gods to try to help this dominant, fearless old lady to see the light – more light than she now sees,'[62] she wrote in one letter. This turned out to be rather more difficult than she imagined and she was soon forced to conclude that 'once she made her mind up about anything there was no power on earth that could convince her about anything'.[63]

At one point, Harold Vanderbilt appeared. Like his brother, Willie K. Jr, he had joined the US navy after war was declared and was now a first lieutenant in charge of nearby Block Island. Sara Bard Field was surprised to find she liked him, and that he shared her views about the pointlessness of war. She was intrigued by Alva's relationship with her sons. Alva flushed with pleasure when Harold turned up unexpectedly, and was proud that he seemed to enjoy talking to an intelligent suffragist so much. At the same time, Field wrote to Wood, Alva seemed determined not to let her affection show. 'She is a strangely loving mother who had denied herself the luxury of showing it till her boys should respect her views.' Later that day, Sara Bard Field had the greatest trouble turning down an invitation from Alva to come to dinner and get to know Harold better. Field had not felt like it, she told Wood, because it was her 'sick time (aren't you relieved – I am)', but it had taken all the force of character she possessed to stand up to a furious Alva even on such a minor matter: 'What! I return the invitation of a Vanderbilt – sought as the most desirable eligible in all the land . . . Everyone does as he asks. That was all implied in her displeased replies to my refusal.'[64]

There were moments when Sara Bard Field was touched by Alva's vulnerability and loneliness. At other times, she was horrified by Alva's materialism. When Sara spoke of her love for Wood, Alva responded by saying angrily that she was a fool, and that she ought to let Alva find her a rich husband. 'If he [Wood] loved you as he says he does, he ought to hope and pray you would find a man who could give you an honourable place as his wife . . . It's all very well to talk about the beauty of free love but you pay too heavy a penalty for it – the woman, I mean.'[65] Later, Sara Bard Field was treated to an insight into Alva's relationship with Newport society.

> I remember one day when I was still working in her library . . . she came back from a bridge party and she was in a very, well, angry state of mind. She threw her gloves down on the table and said, 'I *loathe* these people!' I said, 'Well, why do you go to them, Mrs Belmont? If you're not happy?' And she said, 'I'm determined to keep my position among them.' And I said, 'Well, why do you loathe them?' She said, 'They care for nothing but money.' 'Well,' I said, 'I would say that it was strange that you feel that way, because of course you have a great deal of money.' And she looked at me fiercely and she said, 'Surely I have. Money is power.' I can't tell you with what angry sense of arrival at a conclusion that was just

> impossible for her to escape from, even though she had met women who thought there were things other than power.[66]

The aspect of Alva's personality which most repelled Sara Bard Field was the sharp discrepancy between her support for the feminist movement and her treatment of women she regarded as 'inferior'. 'I remember one night when I was still there and had refused to go out to some elaborate party, her personal maid came into my room, knocking on the door, of course, first, and flung herself down in a chair weeping, and I said, "What is the matter my dear?" She said, "Oh, she struck me so hard with her hairbrush because I didn't get her hair up in a hurry" . . . I had a feeling of the chasm between her attitudes towards the individual and her attitude toward women in general, when they were fulfilling her desire, which was always, it seemed to me, based on some kind of revenge.'[67] This story resonates with another in the archives of the Preservation Society of Newport. Elsie Powers told an archivist in 1997 that her parents John and Ruth Townsend worked as a married couple at Marble House in 1914, until her mother became pregnant with Elsie. When she was told about the pregnancy, Alva was furious, telling the Townsends that she did not want a baby in the house, and that when it was born they must give it up for adoption. The Townsends chose to leave her service.[68]

In the end, Alva seemed to lose interest in publishing the memoir she started in 1917 and it was never revised or completed. Sara Bard Field later remembered that she departed from Newport that summer 'extremely depressed' about the realities of wealth, feeling that it 'was not only insignificant but inimical to people's development and growth'. But though she never warmed to Alva she felt that, in retrospect, Alva had been one of the very few to use great wealth constructively. 'Without resigning her social position and giving a great deal of her life to the social activities, she really had a desire, probably from her will to power, to perform the work she did for women, and it is not to be despised by any means.' At the same time it should not be confused with a 'great self-giving for its own sake'. 'Perhaps,' she continued, 'I have no right to look into motives, only into results, and the results were good. She was a force throughout all the effort for national suffrage.' Sara Bard Field believed that deep down, however, Alva harboured a secret dream. 'She believed ardently that some day we would have a woman president, 'said Field. 'I think she had dreams of being president one day.'[69]

Consuelo first met Louis-Jacques Balsan, or Jacques as he was always known, on the night of her Paris debut at the party given by the Duc de Gramont in 1894, when she was seventeen. He is said to have gone home and told his mother that he had just met the girl he wished to marry, but he correctly surmised that his suit stood no chance whatsoever with Alva who was then toying with the prospects of a German princeling for Consuelo.

Jacques Balsan's background was affluent *haut bourgeois* rather than aristocratic. The Balsan fortune was derived from a great textile factory in Châteauroux in the Indre, where the wool of the local sheep provided such fine cloth that Louis XV had created the Manufacture Royal du Chateau du Parc in the town in the eighteenth century. The enterprise was eventually bought by the Balsan family in the mid-nineteenth century, and flourished as it cornered an international market in cloth for military uniforms. In an age of intense nationalism, large armies and constant warfare, this became a highly lucrative business. As the enterprise flourished, the Balsans built large houses, one of which incorporated an earlier chateau, in the park near the factory, and the family came to dominate the town of Châteauroux, building houses for its workers, schools and a hospital. In Jacques' generation, the sons were still expected to work for the business – but they were also able to enjoy life as wealthy young bachelors of the *belle époque*.

Jacques' glamorous younger brother Etienne was a lover of Coco Chanel, whose career he helped to launch. Jacques himself was a friend of Count Boni de Castellane, a French rake who married Anna Gould, another American heiress of fabulous riches, in the same year that Consuelo married the Duke of Marlborough. As a young man Jacques moved in international high-society circles, staying at Blenheim at least once, in 1898, but the Balsan family did not allow him to dawdle. Much of his early twenties were spent travelling round the world on behalf of the Balsan enterprise, finding new sources of wool for the ever-expanding business – journeys which took him to South America, Russia, Australia, China, the Far East and included a period when he was said to have lived with a tribe in the Philippines. After 1900 day-to-day management of the Balsan business passed by agreement to Jacques' brother, Robert, his cousins Henri and Jean, and a brother-in-law, Roger de la Selle, leaving Jacques time to pursue a long-standing passion for aviation.

Jacques already held 'Balloon Pilot Certificate No 90' and had his own balloon, *St Louis*, in which he set a record in 1900 by remaining airborne for thirty-five hours and nine minutes. He then set an altitude record a month later by taking the *St Louis* to a height of 8,558 metres – about five miles. A month after that he flew the balloon another record distance of 850 miles from Vincennes, near Paris, to Opoezno in Russia. This enthusiasm for flight soon transferred itself to an interest in the experiments conducted by the Wright brothers on heavier-than-air craft in Biarritz in 1910. Jacques bought his own plane and became 'Pilot No 18' to hold a licence in France.

Jacques was not simply an aviation pioneer. He was one of the first to grasp that this new invention, the aeroplane, would play a critical part in twentieth-century warfare. He volunteered for two tours of duty during the French campaign in Morocco (from March to June 1913 and for six months in the first half of 1914), taking part in early attempts at aerial bombardment. 'Très courageux, plein de sangfroid' said his *note de service* afterwards, 'il a rendu de très grands services a l'aviation du Maroc par sa grande connaissance des choses techniques de l'aviation, par son sens de l'adaptation de ce moyen au but pour suivi'.[70]

This experience stood him in good stead when war broke out in August 1914. He became a captain in the new French air force, making long and dangerous flights to reconnoitre German troop movements on the eve of the Battle of the Marne and contributing information about General von Kluck's advance on combined French and British forces which eventually led to a crushing German defeat. As the war progressed, Jacques was promoted to colonel and commanded a group of Scout planes. Before one particularly dangerous mission he sent Consuelo a postcard wishing her well – he feared he might not return and told her later that he wished to greet her before he died.

It may have been Consuelo's father, William K. Vanderbilt, who finally brought them together. Living mainly in France after 1903, William K. was vocal in his opposition to America's position of neutrality up to 1917 and became a generous supporter of the American war effort in Europe. When some Americans in Paris formed a squadron of volunteer air pilots known as the Lafayette Escadrille, he gave it generous financial support. The credit for founding the squadron is given by aviation historians to three rich Americans – William Thaw, Norman Prince and William K.'s doctor, Edmund L. Gros. For some time, the reaction of the French authorities to a volunteer American squad was negative. As one American airman said: 'There was really no place of need for volunteer aviators. Hundreds of young Frenchmen were clamouring for admittance to this new and romantic branch of the service,'[71] but after the French air force sustained serious combat losses, the French authorities relaxed their attitude. Edmund L. Gros secured financial help from William K., who began by donating $20,000 and continued to bankroll the Lafayette Escadrille until the end of the war, helping to pay the passage of American volunteers who wished to join. (In her memoirs, Consuelo wrote that Jacques was also a founder of the squadron, but its historian does not support this and, in a memorandum of his war service drawn up for the French government at a later date, Jacques makes no such claim either. It is more likely that as a distinguished French military aviator his championship of the idea was important in convincing the French authorities to allow young American volunteers to fly. He certainly tried to revive the idea at the beginning of the Second World War, which may account for the mistake.)

When Jacques was sent on a mission by the French military

authorities to England in 1917, William K. Vanderbilt may well have encouraged him to call on Consuelo. Forty-nine years old to Consuelo's forty, Jacques was unmarried – he had contracted a civil marriage to Marie Odile Destors in 1903, possibly to legitimise a child who died, and they divorced in 1906. He was soon a guest at one weekend house party at Crowhurst, when Consuelo had invited Asquith to stay, and courtship soon began in earnest. (When he was home on leave, the twenty-one-year-old Marquess of Blandford spent much of his time in London theatres pursuing musical comedy actresses. He once said: '. . . I used to sit in a box every night, and one night, my mother, who disapproved, was sitting in the next box being courted by Jacques Balsan.')[72] Although they would be separated again during the last few months of the war, Jacques and Consuelo were an established couple by the time it ended, appearing together at a party in Paris given by Elsie de Wolf to celebrate the Armistice.

By now Consuelo's sons were old enough to look after themselves (though the Marquess of Blandford's taste for musical comedy actresses – especially one Betty Barnes – worried both his parents, with the Duke attributing his tastes to his 'common American blood'[73]). Consuelo had other commitments however. In 1917 a vacancy came up for the North Southwark seat on London County Council and it was pointed out by her fellow members of the Women's Municipal Party that its president ought really to stand. Unlike her mother, Consuelo had no dreams of high political office or even of entering politics. She did, however, have a strong sense of obligation. She agreed to stand as a Progressive, part of a radical ginger group committed to pushing the Liberal Party in a more socialist direction. Her candidacy was supported by E. A. Strauss, the Liberal MP whose constituency included North Southwark. (Coincidentally, the Duke also went back into the national wartime government as Joint-Parliamentary Secretary to the Department of Agriculture, representing it in the House of Lords.)

Since all elections were suspended for the course of the war, Consuelo's candidacy was assessed by a selection panel. She was introduced to its members by Mr Strauss and met about sixty of her potential constituents, to whom she made a speech lasting over an hour. Consuelo was then sent outside the door to await the committee's decision. It was almost an hour before they announced their decision, making Consuelo rather despondent for although she had no particular wish to stand 'there was now a question of pride involved in the issue'.[74] Eventually Mr Strauss emerged to say that she had indeed been selected and that the delay had been caused by the number of people who had leapt to their feet 'to explain that they could not see how a Duchess could be a Progressive nor how she could understand a working-man's point of view, but since hearing your speech they were convinced of your sincerity'.[75]

Women had been empowered to sit on London's County Council since 1907, and its remit covered the whole of inner and outer London from Hackney in the north to Norwood in the south. Its wartime work suited Consuelo for debates – which she would have found difficult because of her deafness – were rare. Instead, important decisions were taken by committees on health, education and housing and included planning the new garden suburbs. It was hard work nonetheless, for hearing problems meant much extra preparation. 'I have often been asked how I could sit on committees and do the work I did, handicapped as I was. I had some help from an instrument I wore under my hat, but it was mainly an effort of concentration. In preparing a subject on which I was to speak I invariably had answers ready for the questions I thought I might be asked.'[76]

As Consuelo entered local government, Alva continued to do battle on behalf of American woman's suffrage. New York State finally agreed to a referendum on the issue in 1917, and Carrie Chapman Catt was correctly credited with its success for she ran a highly professional political campaign. This in itself might not have irritated Alva as much as it did had not Catt and Anna Howard Shaw continued to attack the National Woman's Party, to the extent of supporting beatings, arrests and the administration's attempt to have Alice Paul certified. Alva was eventually unable to stand the way Catt edited her out of the New York victory story any longer. 'I brought the National Suffrage Association to New York from Warren, Ohio ... Dr Shaw was practically unknown here till I brought her to New York ... Dr Shaw and Mrs Catt have forgotten who was the person responsible for this victory. But I don't care if they have. I shall go down in history,'[77] she told the *World*.

A Senate vote on the suffrage amendment failed by two votes on 1 October 1918, and a further attempt was postponed as President Wilson went to the Versailles peace conference. Alva, realising that all important political activity would henceforth be located in Washington, donated the New York PEA headquarters to the Salvation Army for demobilisation work, and closed the PEA down. This made the *New York Tribune* positively nostalgic: 'Mrs OHP Belmont's suffrage shop ... famous for the votes-for-women cold creams and lip salves, which did so much to make women suffrage and beauty synonymous in New York City, and for the votes-for-women beef stew and apple pie, which made many a convert to the cause before the suffrage orators got in a word, passed into history yesterday,'[78] it wrote mournfully.

As the prospect of a federal amendment becoming a reality edged ever closer, NWP members impatient with slow progress took to burning President Wilson's speeches as 'watch fires for freedom' outside the White House and sporadic attacks by mobs started up once again. Alva had private reservations about civil disobedience when the cause was now so close to victory but, as ever, she supported them publicly. 'Whenever men fight for their political freedom they are willing to kill thousands and thousands of people, guilty or innocent alike. Women are unwilling to sacrifice human life, but we will fight for our liberty with every weapon at our command.'[79] The NWP launched a 'Prison Special' of members who toured round the US giving talks on their treatment at the President's hands. Wilson, concerned to defuse a situation which could have ended by handing a victory to his political rivals, finally came out strongly in favour of an amendment in May 1919. The House approved it on 21 May; the Senate on 4 June. It then went out to the states for ratification, a process which took over a year with the decisive vote coming from Tennessee on 18 August 1920.

Unsurprisingly, both camps in the suffrage movement claimed credit for the victory. Carrie Chapman Catt's 'Winning Plan' which combined a state-by-state approach with lobbying for a federal amendment from 1916 certainly had more widespread support than the confrontational attitude of the NWP. But even the US Vice-President, Thomas R. Marshall, later conceded that the pressure put on the administration by Alice Paul and the National Woman's Party had an important effect. 'The amendment,' he wrote later 'really was submitted to the people in self-defense to get rid of these women in order that some business might be transacted.'[80]

By the end of the war Consuelo had temporarily lost interest in society and had become wholly immersed in her philanthropic and political activities: she watched the great victory parade of the Allied armies through the capital from the London County Council offices rather than from the window of a great private house. In 1919 she stood by her commitment to the Woman's Municipal Party, contested her London County Council seat for North Southwark in the peacetime elections and won on the Progressive ticket once again. She could often be seen escorted in her campaign by children chanting: 'Vote, vote, vote for Mrs Marlborough!' though the *Daily Telegraph* wrote that the Duchess had more motor cars at her disposal than some of the other 'less Liberal' candidates.

She would soon revert to some aspects of her old life, however. In spite of their romance earlier that year, Jacques was not much in evidence during June and July 1919, when Grand Duke Dmitry Pavlovich Romanov made an appearance. 'An exceptionally handsome man . . . he had fine features, and the stealthy walk of a wild animal,' wrote Consuelo.[81] He was exceedingly glamorous. The grandson of Tsar Alexander II, Dmitry had been close to his first cousin Tsar Nicholas II until he took part in the conspiracy to murder Rasputin and put an end to the so-called holy man's baleful influence over the Tsarina Alexandra. In December 1916, the grand duke had been present when Rasputin was poisoned, shot and drowned at Prince Felix Yussopov's house. Ironically, Dmitry survived the Russian Revolution because he was exiled by the Tsar to Persia as punishment for the murder. He arrived in England in December 1918.

Consuelo and Dmitry were introduced by the Curzons. On 28 May 1919 he wrote in his diaries that he had talked to her properly for the first time, had found her 'very kind and sympathetic' and added, 'it happened imperceptibly between us.' Dmitry was studying political economy with Professor Hugh Dalton at the London School of Economics and in spite of the age difference (Dmitry was twenty-seven while Consuelo was forty-two) their relationship quickly became intense. Six days later he arrived at Crowhurst, which he thought was 'one of the most beautiful places I've ever seen'. The other guests included Consuelo's mother-in-law Albertha Blandford and they paid a visit to Winston Churchill and his family. 'I slept well but very little – ???' he wrote that night, suggesting that he already had expectations of a physical relationship with his hostess. A week later he returned and met Ivor, whom he described as 'awfully nice . . . He has broad views and is not at all like a narrow-minded Englishman.' On 16 June Dmitry was back at Crowhurst again. 'Went to bed early, but didn't sleep much!!'[82] he wrote, implying that he and Consuelo were now having an affair or – possibly – that he was reduced to sleeplessness by overcharged emotions. He continued to see Consuelo throughout July, though usually in the company of Ivor and other friends. His diaries suggest that after the end of June they rarely met alone and that Consuelo encouraged Ivor to keep him company, but Lady Cunard also asked Consuelo to bring Dmitry to lunch, suggesting that she knew of their friendship.

Before long, however, warm feelings suddenly cooled. 'I was much less happy at Crowhurst than before,' he wrote on 21 July. They played tennis and bridge but he went to bed early, writing that Consuelo was 'apparently exasperated with me and (*in English*) therefore she's all the time snearing (sic)'. He cut short his visit and returned to London. Either Consuelo now regretted an impulsive affair or had decided that her kindness had been misconstrued and that Dmitry's smouldering attentions were inappropriate and tiresome. She may also have suspected his motives, since he was known to be short of money. Thereafter, his references to her were much less frequent and a great deal ruder. She was 'stupid', 'inelegant' and a dinner party at her house on 23 October was 'boring'.[83] The episode may have highlighted the drawbacks, as well as the advantages, of an independent life for by September, Jacques had reappeared and was pressing the case for marriage.

There were other changes too. Local politics became increasingly frustrating in the face of mounting opposition from the Conservative Municipal Reform Party. Consuelo fell asleep at one committee meeting and regarded the creation of a children's playground in North Southwark as her only success. Blandford's enthusiasm for musical comedy actresses was set aside in favour of marriage to the daughter of Viscount Chelsea, Mary Cadogan on 17 February 1920, of whom Consuelo warmly approved. When William K. Vanderbilt died on 22 July 1920 after months of illness, she was at his bedside. This brought a particularly poignant sense of the passage of time as she returned with his body to New York and attended his funeral at 660 Fifth Avenue. In spite of the mood of change, however, Consuelo did not rush to embrace a new identity. It was clear that marriage to Jacques would mean not only a change of nationality but a very different life in France. She had a strong sense of obligation to the organisations with which she was connected and regarded her commitment to them all as her work, her professional occupation.

When they knew that Consuelo would finally be leaving England, many of those who had worked with her presented her with an illustrated book containing notes of appreciation from seventeen charitable organisations where she had played a role, an idea suggested by Miss Aldrich-Blake of the Royal Free Hospital and London School of Medicine for Women and warmly endorsed by Dame Margaret Tuke of Bedford College. There were entries from the American Red Cross, the Children's Jewel Fund and many others. 'Her name will live in the hearts of many for whom she did so much to benefit in this country,'[84] read the address in the book, but Consuelo knew it was the name of the Duchess of Marlborough that would linger, rather than that of Madame Balsan.

Later, George Curzon, who understood Consuelo as well as anyone, asked her if it had been worth the sacrifice, 'giving up being the beautiful Duchess of Marlborough and all that it meant'.[85] She turned aside the question with a jest, amused that snobbish George could find the idea of parting with a grand title so difficult. But Curzon asked the right question, even though he would have been baffled by the answer, for the sacrifice involved stepping back into a private life. 'Leaving my work was a wrench and saying farewell to my fellow workers saddened me,'[86] Consuelo wrote later. But she did not regret saying farewell to the inwardness and snobbery of Edwardian society even if she sometimes missed the pageantry and power that came with her title. She consoled herself with the thought that during her time in England she had least tried to achieve something. 'The best that ordinary mortals can hope for is the result which will probably come from sustained work directed by as full reflection as is possible.' This might be adversely affected by circumstances beyond one's control she wrote. 'But if we have striven to think and to do work based on thought, then we have at least the sense of having striven with such faculties as we have possessed . . . and that is in itself a course of happiness, going beyond the possession of any definite gain.'[87]

It had been one of the ironies of her story that, after she left the Duke, the position of Duchess of Marlborough helped Consuelo to evolve an independent life, which in turn brought self-respect, confidence, autonomy and freedom, and quasi-professional roles far beyond anything that Alva had envisaged when she first arranged the marriage. Turning her back on such positions of influence was undoubtedly a sacrifice, and Consuelo had to think long and hard before she finally decided to trade it for 'the promise of happiness that had now come my way'.[88] But she was forty-three: Lord Ivor Churchill had come of age; she and Sunny had finally taken the first steps towards securing a divorce; and, in the end,

Blandford's happiness with his new wife Mary Cadogan gave her the reassurance and confidence to risk it.

Although attitudes to divorce were less censorious than they had been in 1906, the Duke and Consuelo still had to go through an elaborate charade which made their previous reluctance to proceed more understandable. The problem was that, under English law, a legal separation freed a husband from the obligations of marriage so that he could not then be divorced for desertion with adultery, the most straightforward route. This meant they had to appear to live together again, the Duke had to leave her, and Consuelo had to request him to return. He was then required to refuse to do this (in writing), whereupon she was required to go to court and seek an injunction granting 'restitution of conjugal rights'. When he failed to comply, she was then free to seek a divorce, provided she had also amassed the evidence to prove infidelity. The Duke of Marlborough had visited Crowhurst the previous December, accompanied by his sister, Lilian Grenfell, and left again. The necessary correspondence, largely drafted by the lawyers who gave it some curious rhetorical flourishes, had been concluded. On 28 February, the Duke spent the night from 10.30 to 8.30 the following morning at Room 193 at Claridges with an unnamed woman.

On 22 March 1920, Consuelo was obliged to go to court to enter the plea for restitution of her conjugal rights, a step which naturally attracted a great deal of attention. She stood in the well of the court, rather than the witness box, on account of her deafness, while Sir Edward Carson took her through the facts. 'Only as she described receiving the letter telling her that her husband had left her for the second time did her dull, lifeless tones betray any emotion,'[89] reported *The New York Times*, completely carried away. The decree was granted and Consuelo then managed to give the battery of photographers waiting by a large black car outside the slip and left in a taxi waiting by a side door. In April, Consuelo resigned her membership of London County Council on grounds of 'ill-health' paying the statutory fine of a sovereign on 28 April. Later that summer, Consuelo moved out of Sunderland House and Crowhurst, going to a house in Paris that her father had given her shortly before he died.

On 10 November the divorce petition was finally entered. '*Thank heavens it is all over*,' the Duke wrote to Gladys. 'The last blow that woman could strike over a period of some 20 years has now fallen – Dear me what a wrecking existence she wd have imposed on anyone with whom she was associated.'[90] The decree finally became absolute on 13 May 1921, with all sides believing that the matter was now settled.

The Duke was the first to re-marry. The wedding took place in Paris, with a civil ceremony at the British consulate on Saturday 25 June, followed on the Sunday by a religious ceremony at the Paris home of Gladys's cousin, Eugene Higgins. Gladys had a sudden attack of nervousness: 'I feel again the thrill of terror which ran through me when I read it in the D. Mail. I loved him but was fearful of the marriage,'[91] she wrote later. The Duke was astonished when she hesitated. 'Haven't you had time to make up your mind?'[92] he asked. The engagement was announced on 1 June and for four weeks the wedding was the subject of much press interest. This time, however, the Duke was pleased about the excitement it was generating, even though the old Deacon family dramas were resurrected once more. His immediate relatives were charming to him about it all, even Lady Sarah Wilson who said that she thought Gladys was 'very beautiful', though in Paris, Marthe Bibesco was upset about 'the conquest of an historic castle by one American over another, the latter very lovable and much loved'.[93]

Consuelo and Jacques married quietly on the morning of 4 July 1921 in London. They too were married first in a civil ceremony, at the registry office in Henrietta Street, Covent Garden, and then in a religious service at the Chapel Royal, Savoy, just off the Strand. Consuelo was determined both to avoid the publicity surrounding the Duke's remarriage and to ensure that her second wedding was as unlike her first as possible. This time everything was done to avoid the attentions of the press. 'In shedding the lustre of the coronet it was my hope to avoid publicity,'[94] she wrote in her memoirs. The witnesses were Consuelo's cousin, General Cornelius Vanderbilt (present at this wedding even if he had been excluded from the first), and the American ambassador, George Harvey – the former editor of the *North American Review* who had helped to propel Consuelo into public life by persuading her into her American public-speaking debut and commissioning her articles.

Consuelo wore a dress of deep grey satin charmeuse and was accompanied to the registry office by Ivor, with the Marquess and Marchioness of Blandford in attendance. After the service at the Chapel Royal the wedding party went back to the Blandfords' house at 1 Portman Square. Alva was absent from the ceremony which, twenty-six years later, saw her daughter surrender the power that came with being Duchess of Marlborough. Instead she busied herself with the Newport season, entertaining her grandson William K. Vanderbilt III, knowing that her appearance in England would mean the end of the secret. There was, therefore, no opportunity for tears as Colonel and Madame Louis-Jacques Balsan left England for France, appropriately enough, by aeroplane.

PART FOUR

II

A story re-told

THE SUMPTUOUS AND ELEGANT HOUSE in Paris that William K. Vanderbilt bought for Consuelo shortly before he died in 1920 was an idyllic setting for the newly married Balsans. Standing near the Eiffel Tower at the corner of Rue du Général-Lambert and Avenue Charles Floquet, it had been built by the French architect Sergent a few years earlier and overlooked the gardens of the Champ-de-Mars near the Ecole Militaire. Consuelo had always loved France and the French, drawn by the unabashed sensuality of Gallic life, writing that here 'there was no false shame in their frank acceptance of life's joys'.[1] Now she was part of the *va et vient* of Paris in her own right. 'In the mornings I awoke to the gay gallop of cavalry officers as they rode past our windows on the bridle-path that encircles the gardens . . . At midday workmen ate their luncheon on the benches that lined the paths . . . Men and women sat reading their papers under the trees . . . In time I came to know the papers they read and the children who answered their call.'[2]

Jacques enjoyed creating houses as much as Consuelo and building a collection for the Paris house together was a new experience for them both, helped by the trust fund of $2,500,000 set up by William K. for Consuelo before she remarried, and an additional $450,000 of United States Liberty Loan Bonds transferred in December 1919, shortly before his death.[3] The Balsans were drawn to eighteenth-century French art and furniture in the grand manner, but they were set apart by the refinement of their taste. 'We used to wander up and down the quays and streets with eyes glued to shop windows for a find . . . Later, but still casually, we would draw near to our find, and the inevitable bargaining would begin,' Consuelo wrote in her memoirs. This was not bargain hunting as understood by most people, however. 'I will never forget the excitement with which we first came upon Renoir's "La Baigneuse",'[4] she commented blithely at one point. A great Boucher tapestry in the dining room came from the estate of William K. and the Boldini portrait of Consuelo with Ivor came from England and hung in the hall. Elsewhere, Aubusson and Savonnerie carpets, Gobelin tapestries, busts by Clodion and Houdon, "Berthe Morisot" by Manet, drawings by Hubert Robert, portraits by Nattier, paintings by Ingres, Boucher and Fragonard, chinoiserie screens, walnut bergères, *brûle-parfums*, marquetry-inlaid cellarets, and gilt bronze gueridons in the style of Louis Quinze, were lovingly collected and arranged by the Balsans themselves.[5]

Once the house was ready, Jacques and Consuelo became stars in the Parisian social firmament. They often invited friends from elsewhere to meet their French acquaintances, knowing that foreigners were rarely asked to Parisian homes. Consuelo particularly enjoyed parties at the British embassy where she was amused to observe the protocol headaches caused by the separation of French society into the old aristocracy of the *ancien regime*, the *noblesse d'Empire* created by the Bonapartes, and the diplomatic corps. The most memorable Parisian parties were those given by remnants of the *ancien régime* in its own style, in houses that had been spared from looting and had changed little since the 1770s, though she stressed in her memoirs that she and Jacques also consorted with the bourgeoisie. 'It was not only aristocratic parties that pleased us. A luncheon in a bourgeois home stands out with equal charm,'[6] wrote the American democrat, without a trace of irony, as she recalled the wife of the Minister of Public Health asking in an anxious whisper whether it would be a breach of etiquette to check the chef's progress in the kitchen.

Meanwhile, work began on a house at Eze on the Riviera, just above Eze-sur-Mer, where Alva had recently bought the Villa Isola. Consuelo had been enchanted by the region when she stayed with her mother just before her divorce from the Duke and soon afterwards she and Jacques set about trying to buy the land in the hills above Alva's home. Jacques had to bargain with around fifty peasant owners for small plots, a process impossible for anyone but a Frenchman. 'I marvelled at my husband's persistence as he bantered and bartered with them, for they were cunning and cautious, and, although anxious to sell, were loath to part with their land. Just as we thought a bargain had been made we would find one parcel had not been included, and inevitably it was the choicest bit . . . and negotiations would start again.'[7] Lou Sueil (meaning 'the hearth'), was modelled on the Convent of Le Thoronet in Provence, and was designed with assistance from Achille Duchêne. Cloisters opened on to an inner garden on a rocky promontory looking out over the Mediterranean. Rooms were panelled as a background for the furniture from Crowhurst and the Balsans spent 'joyous weeks' arranging Chippendale chairs and Ispahan rugs when they finally turned up from England in vans.

Like the Riviera houses of the *belle époque*, Lou Sueil was designed exclusively for winter occupation. (The idea of going to the Riviera in the summer only started later in the 1920s with the new craze for a suntan and it was not until 1931 that the hoteliers of the Côte d'Azur agreed to stay open all year.) Lou Sueil soon delighted scores of visitors from every part of Consuelo's life, from her brother Willie K. Jr and Rose Warburton (whom he would finally marry in September 1927), to new Riviera neighbours such as Roderick Cameron. 'Madame Balsan had immense flair for making anything she lived in attractive,'[8] wrote Cameron, though he thought that living at Blenheim had influenced her taste more than she was prepared to concede. 'One could see at once, as one entered Lou Sueil, that it was a loved house,'[9] wrote Gloria Vanderbilt (married to Cousin Reggie). 'My Darling I am writing to you in bed in this marvellous scented nest – It is really almost too beautiful & too comfortable. One simply wallows,'[10] wrote Clementine Churchill to Winston in 1925.

Winston Churchill was one of the Balsans' first visitors, coming to stay in January 1922. He loved the 'paintatious' light and landscape of the south of France, but he was equally happy in the pacier atmosphere of Riviera houses belonging to Maxine Elliott, Daisy Fellowes or Lord Rothermere. Clementine Churchill disliked these establishments, and according to her daughter, Mary Soames, was always much happier staying with the Balsans. 'Both [Consuelo] and her husband were persons of culture and distinction, and their friends reflected their tastes and characters; with them Clementine found herself in an atmosphere in which she was at ease.'[11] At the end of 1924, Winston Churchill re-embraced the Conservative Party, becoming Chancellor of the Exchequer; and it became the custom for Clementine to spend February with the Balsans on her own.

Clementine Churchill found Jacques and Consuelo's love for each other deeply touching: 'Nothing could be kinder & more charming than Consuelo and Balsan – They love each other very much & it is a most peaceful & restful atmosphere . . .'[12] A year later, on another visit – when Consuelo was forty-eight – she wrote: 'Consuelo looks younger and more ethereal every year. Her hair is more silvery but on the other hand her cheeks are pinker & and her eyes brighter. Her Jacques surrounds her with *petits soins*.'[13] Jacques was shorter than Consuelo, like the Duke of Marlborough, but there the similarity ended. 'It is difficult to appraise a character so completely in harmony with one's sympathies,' Consuelo wrote later, before going on to say that she loved him for 'the charm of his personality . . . the keenness of his varied interests . . . the subtle intelligence of his understanding . . . the wit of his conversation, and above all, the profound goodness and kindness of his nature'.[14] Although Jacques spoke good English he preferred to speak in French, which could sometimes make him appear remote to the more linguistically challenged English and Americans of their acquaintance. But even those who did not know him well found him warm, charming, hospitable and full of preternatural energy as he ran up and down Lou Sueil's tumbling terraces.

Clementine Churchill was not alone in finding Lou Sueil agreeable. Elizabeth Lehr was staying with the Balsans at Eze when she heard that Harry Lehr had died. Elsa Maxwell came to stay and (with Consuelo's permission) asked Charlie Chaplin for lunch. Consuelo noted his socialist tendencies and the 'melancholy undertone' of a great clown, but was reduced to helpless laughter by his account of hunting – for the first time in his life – with the Duke of Westminster a few weeks before. Prince Serge Obolensky, a Russian émigré who had known Consuelo since he was an undergraduate at Oxford, found himself playing tennis with Arthur Balfour, who wrote: 'He could hold a crowd spellbound about philosophy, history, foreign affairs, literature, art, economics, with a relaxed and measured air that was like the art of a great actor. We certainly got an earful during our visit.' Obolensky did not rate Balfour as a tennis partner either. 'He was sparing and deliberate, and only played the balls that came near him – I did the running.'[15] Perhaps the Balsans were wise to arrange for Balfour to be partnered by the great tennis player Suzanne Lenglen whose repartee, according to Consuelo, was often 'as good as her volleys'.[16]

Margot Asquith was another link with pre-war England, arriving to stay at Lou Sueil just after Asquith's death. Always a law unto herself, she was hyperactive in grief. Consuelo was hard pressed to stop her from wearing a red scarf with her black dress when visitors were expected, knowing that French guests would be shocked. She also managed to scandalise English visitors who arrived to find her executing a high kick over Jacques' head in time to music. 'With admirable courage she fought the sharpness of her sorrow, delighting us with the spontaneous and irrepressible gaiety that was her greatest charm,'[17] Consuelo wrote in her defence (though Clementine Churchill wrote to Winston suggesting that Consuelo found Margot a far from easy guest).

Perhaps the most touching visitor from Consuelo's English days was George Curzon who visited the Balsans a few weeks before he died. He enjoyed his time at Lou Sueil far more than he expected and was so enchanted by Eze that he even toyed with buying some property nearby. 'Consuelo says he was so changed & sad & *humble*,' wrote Clementine to Churchill. 'He is more unhappy than King Lear, for of his 3 daughters not one of them is a Cordelia ... Consuelo says he was a charming guest & entertained them with witty anecdotes ... When Curzon left for home he shed tears ... said he had not had so happy a holiday since he was a young man.'[18] This was not surprising since he and Consuelo were close and she had been a superb hostess, laying on everything he needed in order to write to the extent of having black-out curtains made so that he could sleep without being disturbed by light. In the face of such hospitality it is not surprising that the Balsans sometimes had difficulty ejecting visitors. 'Strong measures were required with such delinquents,'[19] wrote Consuelo darkly.

Even without house guests, delinquent or otherwise, there were plenty interesting neighbours between Cannes and Menton. The Balsans built Lou Sueil in the decade when the Côte d'Azur was at its most bewitching, causing all who knew it then to lament the passing of paradise later. The French politician André Tardieu lived nearby, as did the American composer Sam Barlow. Their neighbour Dr Serge Voronoff, whose experiments in transplanting primate glands on to male patients to slow down the ageing process were creating intense interest and controversy, would sit and tell risqué stories on the terrace with Jacques. If one tired of looking out over the cypress, eucalyptus and mimosa there was opera in Nice or ballet in Monte Carlo where Diaghilev spent the winter with the Ballet Russes (his dancers came to parties at Lou Sueil). Consuelo heard Horowitz play for the first time on the Riviera, and his engagement to the daughter of Toscanini, who lived nearby, was the neighbourhood romance. Consuelo counted at least seven nationalities at one of Lou Sueil's almost daily luncheon parties. Signatures in the visitor's book included 'HRH The Duke of Connaught', and a Japanese prince who took ten minutes to sign his name while everyone looked on in embarrassed silence.

Within two or three years the garden at Lou Sueil attracted as much admiration as the house. The Balsans created a series of terraces and gardens of studied naturalism that reached its peak in late spring before they departed for a second helping of springtime in northern France. 'Drifts of spring bulbs; carpets of hyacinths and bluebells mixed in with the grass under the olives,'[20] wrote Roderick Cameron. 'The garden is a dream – carpets of purple gold & cream flowers on the emerald green grass,'[21] Clementine Churchill told Winston. There was a 'seasoned order of tulips, peonies and daffodils,' wrote Consuelo. 'Almond trees bloomed first, in pink and white showers; then came the *prunus* and the Judas trees with their bronze and scarlet foliage.' In reality, the casual aplomb of Lou Sueil's garden required 'endless toil'. 'In September, after the first rains that softened the soil, we scattered thousands of bulbs;'[22] a small army of gardeners followed close behind on their knees with trowels – and dug the same bulbs up again each May to protect them from the summer heat when the proprietors had gone.

This was an era of Riviera gardening when enthusiasts went plant-hunting by Bentley and Rolls-Royce. The Balsans loved 'motoring over those beautiful mountains into some hidden valley in search of Alpine flowers' in the company of gardening friends such as Lawrence 'Johnnie' Johnston (whose garden at Hidcote is still one of the great gardens of England), Henry May and Norah Lindsay. One of the other outstanding gardeners of the Riviera with whom they also exchanged notes was Edith Wharton, who had bought the Château Sainte-Claire near Hyères after the First World War. She had already created a beautiful garden at the Mount at Lenox, and was regarded as an expert on garden design after the publication of *Italian Villas and their Gardens* in 1904, a year before *The House of Mirth*. After 1919 she set about making another famous garden at the Château Sainte-Clare.

In an introduction to a book by Harriet Martineau in 1924, Edith Wharton wrote of the northern gardener's pleasure in encountering the myriad possibilities of the south when walking round Mediterranean gardens. One of these was the garden of Lou Sueil. In Wharton's papers a list of gardening questions to be asked 'chez Madame Balsan' still survives. What was the 'mélange de graines anglaises et italiennes' that the Balsans used for their lawns and what was the address of the *grainier*? What was the name of the white azalea that grew so well beneath their trees? She wanted to know about the 'oeillets [carnations]' too (ce sont de beaucoup les plus beaux que j'ae vue); and what was the 'nom de petit iris bleu ciel plante sous les arbres, pas mauve mais de la couleur d'une jacinthe bleue'?[23]

Consuelo had known Edith Wharton slightly for many years.

She visited Blenheim at least once, and was photographed with Consuelo and Lady Randolph Churchill on the steps in 1904. Wharton may have been Lady Randolph Churchill's guest that day, for her New York set was older than Consuelo's and she knew both Consuelo Yznaga and the Rutherfurds well. She may have had a crush on Winthrop Rutherfurd herself, for she wrote in her autobiography that the Rutherfurd boys were outstandingly handsome and 'the prototypes of my first novels'.[24] Alva appears to have asked Edith Wharton and her husband to lunch in Newport while the Duke was staying in 1895; but Consuelo was certain that Edith Wharton and Alva cordially disliked each other.

Consuelo found her difficult too, though gardening was a common bond. 'Edith Wharton came often to Eze and we delighted in visiting her at Hyeres on the Riviera or at the Pavillon Colombe near Paris where she had created lovely gardens. She liked to show them, being proud both of her taste and her horticultural knowledge, for she had the faculty of remembering the name of every plant, however rare.' Remarking that touring gardens with an amateur gardening enthusiast could be tedious, Consuelo remembered that this was never the case with Mrs Wharton who 'was a passionate lover of flowers and would dissect their mutations with the same ruthless precision she practised in analysing the characters she portrayed'. This was the problem with her novels, thought Consuelo, who felt that her writing 'lacked the glow of humanity' and she found Mrs Wharton's 'hard, ambitious types of American womanhood' particularly unpleasant. Like many others, Consuelo found Wharton herself forbidding, 'reserved and frigid' even, and wondered whether 'the warmth of her nature had found its only blossoming in her garden'. Consuelo may also have had the uncomfortable feeling in Wharton's presence that she was being closely observed – she wrote that Wharton lacked spontaneity and 'gave the impression of intellectually controlling her emotional contacts'.[25] She could have been right, for Consuelo is said to have been the model for Nan St George, the sensitive and dreamy heroine of Wharton's final novel *The Buccaneers.*

* * *

After the Nineteenth Amendment giving women the vote was approved in America in 1919, Alva started showing signs of disorientation and she only played a minor role in the campaign leading up to its ratification in 1920. Like others who had expended so much energy on the suffrage battle, she felt ready to focus on her private life again after the struggle was over. Even Alice Paul seems to have toyed with the idea of taking another law degree and several leading members of the National Woman's Party now got married, or, as the *Washington Star* preferred to put it: 'Iron Jawed Angels Succomb [sic] to Matrimony . . . Majority of Once Prominent Suffragist Pickets, Charged with Being "Unsexed" and "Home Destroyers", Now Happily Married'.[26] Alva felt tired too, hardly surprising for she was sixty-eight and suffering from high blood pressure; her doctors now instructed her to rest.

Alva's decision to purchase the villa at Eze-sur-Mer in 1921 had been the first step in a slow process of re-aligning her life to Europe from America, where she still had three substantial homes: Marble House, Beacon Towers and a house in New York. As Consuelo settled into married life with Jacques, Alva continued to cross the Atlantic on post-suffrage business and often summoned her suffragist protégée, Doris Stevens, to be with her. Doris was eight years younger than Consuelo and had been a full-time organiser for the National Woman's Party since 1914. Over the years Alva's admiration for Doris had deepened and it had become an informal but established practice for Alice Paul to send Doris Stevens to Alva when she needed help. As Doris Stevens later wrote: 'Because of her alleged oft-expressed fondness for me and the trust she placed in my work, more demands were made upon me than upon any other of my colleagues. She always wanted me to execute her plans.'[27]

In spite of her affection for Doris, Alva's association with her eventually became complex and difficult, as Doris would later testify. Doris's testimony has to be treated with great care for it was written during litigation after Alva's death but it seems that her relationship with Alva slowly became a reverse image of Alva's with Consuelo: Alva began by being impressed by Doris's independent spirit, but was subsequently unable to stop herself from trying to dominate her. Part of the problem was that the financial basis of their arrangement was never clarified. Sometimes Doris Stevens worked for Alva as a representative of the National Woman's Party, though her small salary was paid for indirectly by Alva. At other times, she acted as Alva's personal companion but never received a regular salary. Unlike Alva, and indeed Alice Paul, Doris Stevens had very limited resources of her own. Alva knew this, appreciated that payments made to Doris Stevens by the NWP were pitiful in relation to the work she did and would sometimes give her a generous cheque or a large present. On other occasions Doris would be asked to act as Alva's companion, staying with her in smart hotels with her expenses paid but receiving nothing in the way of salary. According to Doris, Alva promised to smooth out financial inconsistencies by treating her generously in her will. In her lifetime, this promise was enough to give Alva considerable power over the impoverished Doris. Conversely, Alva exploited her financial power to ensure that Doris kept her company.

The relationship was more complex than this however. Doris Stevens was often described as beautiful and dynamic. She was also young. Another beautiful, young and dynamic woman, Clare Boothe (later Clare Boothe Luce) was also briefly taken up by Alva and prevailed upon to work in the NWP headquarters in Washington in the early 1920s. Clare Boothe only lasted in headquarters for ten days (though she maintained a relationship with the party itself for longer than that) because she detected an undertow of lesbianism that she found disconcerting.[28] There is no suggestion that she is referring to Alva here, and her reaction may have been confused by the presence of Elsa Maxwell when she first met Alva. It would be too easy in a post-Freudian age to characterise Alva's enthusiasm for younger women as sexual, but it is probably wide of the mark. As Sara Bard Field put it, Alva was 'gluttenous [sic]' for qualities she perceived in others and felt she could exploit. Outliving many of her friends, her children busy with their lives in Europe and America, she enjoyed the company of intelligent independent and attractive feminists more than the company of her tired narrow-minded circle in Newport, even if she had to pay for it.

Enthusiasm for these qualities in the young was not in itself problematic. It posed no particular difficulty that Christabel Pankhurst and Doris Stevens partly represented that militant suffragist daughter that Alva never had. The problems only started when Alva forgot that Doris Stevens was neither daughter nor servant and tried to control her life. Doris exasperated Alva intensely in the winter of 1918–19, for example, by refusing to marry a 'rich but very commonplace widower' in Florida, whom they met when they were campaigning together. This was a man, according to Doris Stevens, 'whom she was determined to have me marry in order that I might have enough money to help her continue the work. She very carefully explained to me how much he had, secured an invitation from him for me to be his house guest with her in Florida, which I declined, and explained that a marriage to him would help relieve her of the burden of having to carry on what she considered a heavy financial load.'[29] Clare Booth Luce had a similarly alarming experience which may be another reason for her precipitate departure from NWP headquarters. 'The thing that scared me,' she later told Elsa Maxwell, 'was Mrs Belmont's wild desire that I should marry her son, Willie K. – who, though charming, appeared rather an elderly gentleman to me at the time.'[30]

The role of Alva's companion was not to be envied. Consuelo and Alva may have drawn close in 1920 but Alva seems to have saved her temper for Doris, summoning her to Paris for a most trying two months in November and December. 'I stayed with her in the Hotel, accompanied her on shopping expeditions, sightseeing expeditions, theatre with her, provided dinner and recreational companions when she desired, read aloud to her the whole of Wells' "Outline of History" as a sample etc. During all this time she was under a doctor's care taking a cure, subject to the most rigid diet, which kept her in an almost constantly irascible state of temper. I was a nervous wreck at the end of two months of hearing her quarrelling with the servants, waiters in the hotel, trades people and finally with me.'[31] Harold, according to Doris Stevens, stayed with his mother for three days during this time and then escaped.

Long stays with Alva did allow Doris to see her vulnerable side, however. 'I was on demand as companion in her 51st Street home, often spending the night with her, when she said she was lonely for companionship.' It became clear that Alva was much more fearful than she first appeared. 'She was often afraid to go alone with her chauffeur at the wheel, to a friend's house for example, for dinner. She would ask permission to bring me along. She was afraid, she said, she might be robbed of her jewels.'[32] As Alva grew older her fear of situations she could not control grew worse. One of Doris's tasks was to pilot Alva through crowds which 'she feared with a deadly fear; she always thought even a small crowd was equivalent to a mob which might attack her'.[33] Alva always disliked staying in houses where she thought someone might have died, and according to Elizabeth Lehr, she also developed a terrible fear of cellars or anything subterranean. This phobia became so bad that on her way to the villa at Eze, Alva would always leave the train at a station north of Lyon, drive through the town and catch it again a few miles down the road to avoid the Lyon tunnel.[34]

Between 1920 and 1923, however, Alva hit on a new idea that for three years became the focus of her restless energy. Across America the end of the suffrage campaign brought the movement to a crossroads. Many women simply saw winning the vote as the end of a long battle. Carrie Chapman Catt stepped down from the National American which re-created itself as the League of Women Voters to provide civic education for women. The National Woman's Party, on the other hand, decided not to disband for it collectively regarded the passing of the Nineteenth Amendment as the beginning of a new battle, though here there was disagreement about the nature of the battlefield. Alice Paul – who had given up ideas of doing another law degree – thought the NWP should now focus on the issue of equal rights for women which, it transpired, were rather more equal in some American states than others. Alva had an entirely different vision. She was determined that the NWP should do nothing less than turn itself into a third American political party, of which she would be president. This new third force in American politics would form a wedge between the Democrats and Republicans, representing the interests of newly enfranchised women and holding the balance of power.

The idea of becoming a third political force in American politics was ambitious; but the implications of her next suggestion were even more extraordinary. After involving herself in the suffrage campaign for over ten years, Alva announced that until the NWP could reconstitute itself as a third American party, it should be telling newly enfranchised women *not* to vote. 'The vote is only the beginning of what we need – it is only our tool, our new broom, our pick, our shovel, our axe,' she wrote. 'But we have no intention of seeing these excellent utensils put in a museum.'[35]

There was a certain ruthless logic to her position. Men had been in charge for hundreds of years and made a mess of things. Existing political parties had been evolved by men entirely in their own interests. Women should not waste their precious new vote on bolstering support for decaying, archaic, male institutions which would simply continue to work against women's best interests. 'Personally, I think it would be far better for women to stay out of all parties and away from all elections, if they can find no other medium of expression than the existing decrepit man-dominated parties,' she wrote in an article called 'Women as Dictators' in September 1922, which appeared in the *Ladies' Home Journal*, sandwiched between advertisements for 'Fuller Brushes Sold by 3,500 Courteous Men' and 'Let's Make the First of Our Fall Hats'. 'The day is not far off when the Woman's Party, of which I am president, will be strong enough to impose any measure it may choose,'[36] Alva declared.

When it came to outlining what the National Woman's Party would actually do when it became a third force in American politics, its president-in-waiting was rather vague: 'I hope that one of the first acts that women will try to pass will be an act providing that the people who govern them shall be normal.'[37] Not that she was against older people becoming involved in politics, she hastened to add. 'But fitness is not determined by years. Man has taken the position that he never grows old. When women grow old they are neglected.'[38] More generally, women would bring new standards into public life with immediate benefit for future generations. 'Men have paid great attention to the breeding of horses, dogs, cats and pigs. They have paid no attention to the breeding of the human race. Women will do their utmost to change this state of affairs. When we put up our standards, men will have to put up theirs. They will have to change ... There is nothing so stimulating as opposition.'[39]

Needless to say, Alva's separatist manifesto attracted immediate attention in the press. 'As long as the voting women of America are content to remain in their husbands' and brothers' political groups, this condition will continue. That is why we will urge the women to affiliate with us,'[40] she told the *New York Journal* in October 1922. *The New York Times* paid her the compliment of analysing the views she set out in 'Women as Dictators' carefully and at some length: 'Mrs Belmont sees more barriers than are visible to some of us: but she is perfectly clear, comprehensive, logical and determined. Women had better stay out of politics if they are simply to imitate men in politics. Better have no vote, she says, than "scramble for jobs instead of fighting for principles".' The problem, the paper continued, was that this was a counsel of perfection: 'Much might be said for the view that men have made a mess of the world, and that government by women or even by children couldn't be worse and might be more intelligent.' *The New York Times* was also prepared to concede that women had had a raw deal from male politicians. 'It will be admitted by politicians in their frank private moments and is visible to clear-sighted women that, so far as politics is concerned, the cheese is for men while to the woman is graciously left the rind.' But all this was going to change now, according to the paper, and its correspondent remained unconvinced that a third political party led by Mrs Belmont was really the best way forward. Indeed, the prospects for poor old men were quite terrifying if her vision ever became reality. But there was a glimmer of hope. It lay in 'the most fortunate peculiarity of the Woman's Party which, like the Ancient and Honorable Artillery Company of Boston, consists entirely of officers. It is all leaders and no led.'[41]

Behind the scenes, Alice Paul privately agreed with this. She did not believe that the National Woman's Party had the resources to become a third force in American politics and as fast as Alva promoted the idea Alice Paul tactfully downplayed it, trying to refocus attention on an equal rights campaign where she thought the NWP could make an impact. In this she had the support of most of her colleagues. In May 1923 one life member of the NWP, Mary Winsor went as far as describing talk of a woman's party as 'extraordinary nonsense'.[42] Part of the problem was that Alva had, quite simply, become too extreme even for her own radical colleagues and was years ahead of her time. Much later, in the 1970s, Alice Paul told an interviewer that in retrospect, they should have taken the idea of becoming a third party in American politics much more seriously.

In 1922, Alva became president of the NWP after she gave it money for new Washington headquarters which she continued to hope would serve as a women's parliament. Alice Paul, meanwhile, accepted the post of vice-president and proposed that the focus of the new fight should be an equal rights amendment (ERA). With remarkable tact and diplomacy, Alice Paul slowly persuaded Alva to drop the idea of a third party and throw her weight behind the amendment, making her one of the very few people who ever persuaded Alva to change her mind. Alva gradually came to see that lack of enthusiasm for her grand plan made it unworkable and it was in this spirit that she led the first ever ERA deputation to the White House and presented it to President Coolidge in November 1923.[43]

Although Alva supported the campaign for an equal rights amendment she was concerned about its exclusively American focus. Between 1924 and 1926 she slowly found a way of resolving her disappointment over the future of the National Woman's Party by becoming more involved in the issue of international women's rights. Spending less and less time in America after 1923, she began to argue that women must be ready for the time when the United States joined the League of Nations. Feminists round the world should get themselves organised, she thought, so that when international conferences were set up to draw up codes of law, they could ensure that women had equal status and in 1926 she formed the International Feminist Committee (which became known as the International Advisory Committee of the NWP) to monitor international agreements and make sure that women's interests were protected.

The problem was that her old enemy, Carrie Chapman Catt, had the same idea and old battle lines were suddenly redrawn. Much of the new bitterness was focused less on international rights than Alice Paul's equal rights amendment, but the two soon became closely intertwined. Championship of ERA involved dismantling protective laws for women as the weaker sex. This was fiercely opposed by moderate suffragism in its new incarnation, the League of Women Voters backed by Carrie Chapman Catt. This battle would tear American feminism apart for nearly forty years and would only finally be resolved by the passage of the Civil Rights Act in 1964 which extended protective legislation to men, but in 1926 the schism was already beginning to show and it was not long before the brand new battle extended to the arena of international women's rights as well.

Although she was now seventy-three, the first part of 1926 represented a return to form for Alva. Her appetite for belligerence showed no sign of slackening and after a long silence she started writing articles again. In Paris she invited French suffrage groups to the house she had bought at 9 Rue Monsieur, frustrated by their inability to collaborate. In June 1926 Alva decided that her NWP international rights committee should affiliate with the larger and more powerful International Woman Suffrage Association, the announcement being made at a conference in Paris. She dispatched her delegation from New York with a great send-off dinner, only to learn that their request for affiliation had been rejected – and that the person behind this was none other than the old foe, Carrie Chapman Catt. By the time she arrived in Paris herself, Alva discovered that Catt had been working behind the scenes to exclude the National Woman's Party on the grounds of their 'aggressiveness' and their championship of the equal rights amendment.

Alva erupted at Carrie Chapman Catt in the press: 'I worked with Mrs Catt for years. I took her out of obscurity. During those years, I always believed in unity, and that we should accept each others views as best we could but I always objected to her idea of what was wise. I did not leave Mrs Catt until it seemed the better for woman suffrage. This is the first time since 1909 I have said anything against Mrs Catt. I wish to make it clear however, that Mrs Catt has not failed to oppose me on every occasion. In the present cause her hand is evident.'[44] Alva then said she would take the money she had pledged to the International Women's Suffrage Association elsewhere. None of it had any effect on the Association. Although she kept up a brave front, and was heralded by *The New York Times* as one of the three main leaders of the international suffrage campaign in the press that summer, Alva was deeply irritated by Carrie Chapman Catt's victory for she realised she had been wholly outmanoeuvred.

Alice Paul was already increasingly concerned by Alva's commitment to the NWP, fearing that at seventy-three, with houses in France, she might be losing her appetite for American politics and could withdraw her financial support. Sensing Alva's discouragement after the upset with Catt, Alice Paul urged NWP members to write to their president and 'really work on reviving her interest'.[45] Alice Paul even organised a banner-waving delegation to meet Alva when she arrived back in New York in October 1926. Jane Norman Smith was part of the dockside reception committee and after it was over she wrote to Alice Paul in some dismay. Alva had not been herself at all, she said. In fact, she looked haggard. 'When she got off, she simply greeted us and then said she would "see you tomorrow; I have relatives waiting for me" and then turned aside to greet her two sons!'[46] Four days later, Jane Norman Smith met Alva again. Her demeanour was just as alarming. She was suffering from high blood pressure and was extremely irritable. To the relief of the NWP's members, however, it soon emerged that Alva's black mood had nothing to do with them.

In September 1926, Clementine Churchill was staying with the Balsans at Grantully Castle in Scotland when Consuelo gave her some startling news. Clementine was so taken aback by what Consuelo had to say that she conveyed it to Winston in a note marked 'Private' and told him to burn it 'becos of les domestiques'. 'Consuelo tells me that her marriage to Sunny has been annulled by the Pope!' she wrote. 'I was so staggered that I lost the chance of cross-examining her about it. It reminds me of the opening lines of the hymn: "God moves in a mysterious way his wonders to perform".'[47]

Clementine's astonishment was understandable for she clearly assumed that the only grounds for annulment of a marriage by the Vatican were non-consummation of the marriage. Consuelo and the Duke not only had two sons but had recently become grandparents: the Blandfords had produced two daughters, Sarah and Caroline, in 1921 and 1923, and the Marchioness had given birth to another link in the chain, John George Vanderbilt Henry – the future 11th Duke – a few months beforehand in April. It was, therefore, a difficult concept for anyone to grasp. But Clementine regained her equilibrium sufficiently to probe Consuelo just a little further to discover there were other grounds entirely. 'After I had recovered from my embarrassment I asked her on what grounds she had suggested to His Holiness that He should operate the miracle – And she replied "Coercion, being under age at the time".'[48]

Clementine's attempt to keep the news secret was soon overtaken by events. Within a matter of weeks a furore erupted over the Marlborough annulment which lasted almost as long as newspaper coverage of the original wedding preparations. At one point bitterness became so intense that the Duke felt obliged to deny that it had ever been his idea, issuing a statement through his solicitors that the application for the annulment had been made by Madame Balsan and not by him. Clementine Churchill's note to Churchill bears this out. 'She says Sunny is enchanted as rumour has it that he is to be received into the Church of Rome & will then be able to marry Gladys properly – I suppose Jacques' family suggested it as they are strict Catholics & consider him to be living in sin with Consuelo.'[49] This summed up the position neatly. Requesting an annulment six years after a legal divorce and five years after remarriage may seem odd, but in 1926 it suited both the Duke and Consuelo to go through the process for very different reasons.

The Duke's reasons were religious. Although he had started life as a 'total pagan' he had slowly been drawn to the Roman Catholic faith and had begun instruction three years earlier with the Jesuit Father Martindale, a fashionable priest responsible for a number of high-profile conversions during the 1920s. Because the Duke had been married before, Father Martindale could not agree to receive him formally into the Church, but the Duke had taken to attending Mass at St Aloysius Roman Catholic Church in Oxford as a non-communicant. The Duke's move towards the Church of Rome was partly driven by fury with the Anglican Bishop of Oxford (or the 'Burgundy Burge' as Gladys called him), after he barred the Duke from attending the Diocesan Conference in 1922 on account of his divorce and re-marriage to which the leaders of the Anglican Church were still deeply opposed. The Duke had been accepted as a communicant in local churches throughout Oxfordshire and in London; he had never wished to attend the conference in the first place (he was an ex-officio member as Lord Lieutenant of Oxfordshire); and he was embarrassed by unpleasant publicity for weeks afterwards. But both Father Martindale and the Duke's friend, Shane Leslie, believed that beneath his complex and difficult exterior the Duke had a profoundly religious temperament. Without an annulment to wipe out the 'stain' of divorce, becoming a full member of the Catholic Church remained impossible.[50]

Consuelo's reasons for seeking the annulment related entirely to Jacques. 'Sanctified as our marriage had been in an Episcopal Church, it was to me completely valid; and I would have been content to ignore the ultra-religious views that prevented the orthodox Catholics from recognising it,'[51] she wrote later. The difficulty was that orthodox Catholics who felt strongly about this were not simply confined to a few grand French families, but included Jacques' immediate relations such as his aunt, the widowed Madame Charles Balsan, who was head of the family and 'ensured the traditional discipline of the Catholic Church'.[52] Jacques' own divorce from his earlier civil marriage posed no problem; but his second marriage to the divorced Consuelo could only be recognised if she remarried Jacques according to the rites of the Catholic Church. This in turn was only possible if she voided her first marriage through an annulment. In the first months of marriage the attitude of Jacques' family did not disturb either of them, but by 1926 Consuelo was anxious to take steps to heal a family rupture of which she was inadvertently the cause. It then transpired that the only way of doing this was to prove that one party in the Marlborough marriage had been coerced into marrying. Under Catholic ecclesiastical law this was enough to declare it invalid and have it annulled.

'It pained me to approach my mother for her consent, but on learning that the proceedings were entirely private we agreed to take the necessary steps,'[53] wrote Consuelo. Though there was pressure for an annulment, it was not critical to anyone's happiness. Had those involved been able to predict the storm of publicity ahead it seems likely that they would have hesitated and possibly desisted. As it was, the assumption that the annulment proceedings would remain private was perfectly reasonable. Witness statements were heard in private by a tribunal in the Roman Catholic diocese of Southwark in London. These statements were so confidential that any priest who divulged their contents risked excommunication. The tribunal's findings would then be forwarded for consideration by the Rota, the Catholic court in Rome, for final approval. In the event that this was granted, a summary of the judgement would be published in *Acta Apostolicae Sedis* in Latin, apart from quotations from witness statements which appeared in French. Since the press were not in the habit of scrutinising *Acta Apostolicae Sedis*, and most of them would not have understood the Latin text even if they had, everyone could proceed with a reasonable degree of confidence, provided the news did not leak in any other way. As it turned out, this possibility should have been given more thought for though the annulment proceedings themselves would remain private, the steps that the Duke and Consuelo had to take afterwards would not.

One reason for this unaccountable failure for the participants to think things through may have been timing. The tribunal hearings took place in July 1926. Alva, who was most likely to be affected and by far the most political of those involved, had spent much of June dealing with the setback caused by Carrie Chapman Catt's vindictive exclusion of the National Woman's Party from the International Women's Suffrage Alliance. Immediately after the tribunal hearings her attention was diverted once again because her sister Jenny Tiffany, who had been one of the witnesses, died suddenly. The Duke then travelled to Rome to be received in audience by the Pope, a move which Consuelo later blamed for the story breaking. One likely cause of Alva's ill-temper when she arrived back in New York in late October 1926, therefore, was that she knew it was only a matter of time before the dam of publicity burst.

When the story appeared in *The New York Times* on 13 November it started a bitter controversy that would rage for several weeks. The Bishop of New York, Bishop Manning, led the attack. His first position was one of 'amazement', saying that he could hardly credit the authenticity of the despatch from abroad: 'It seems incredible to me that the Roman Catholic Church which takes so strong a position against divorce, should show such discrimination in favour of the Duke of Marlborough . . . It would be a serious thing indeed, and most dangerous in its implications, if the Roman Catholic Church should claim the right to annul a marriage such as this, which was entered into in entire good faith, which resulted in the birth of two children and which was accepted as binding by both parties to it for many years.'[54] Bishop Manning's suggestion that dukes were somehow favoured by the Catholic Church was quickly rebutted by an anonymous English prelate declaring: 'Nothing more can be done for the Duke than a tramp.'[55] Indeed, throughout the whole affair the Vatican repeatedly pointed out that the social importance of the persons involved had nothing to do with the case and that Vanderbilt wealth had no influence. The 'Sacred Rota Tribunal' had been in existence since 1326, and had famously refused to annul the marriage of Henry VIII to Catherine of Aragon; more recently Count Boni de Castellane had made no fewer than three attempts to have his marriage to Anna Gould annulled and had failed every time.

The following day the news was even more sensational. It was the former Duchess who had brought the suit, on grounds of coercion, and *The New York Times* had it on good authority that the tribunal had accepted Madame Balsan's plea that undue pressure had been brought to bear when she was not yet of age by her mother and other relatives. The Duke had not contested the plea either, and so the judgement had been promulgated accordingly. Madame Balsan had married a French Catholic and as one priest put it, she doubtless wished 'to regularize his position with the Church which had become irregular by his marriage to a divorced woman'.[56] It was pointed out that this was an ecclesiastical action with no civil effect and that the legitimacy of the Marquess of Blandford and Lord Ivor Spencer-Churchill was not affected. This, it was hoped, would lay the matter to rest.

However, earlier that year Alva had picked a fight with Bishop Manning and had made him look foolish. She was already on record attacking the Church for its treatment of women. She took particular exception to the church's habit of accepting female support and money while denying women the right to be priests, and had taken a radical swipe at the ecclesiastical hierarchy for 'dictatorial discrimination' in her article 'Women as Dictators' in 1922.[57]

In April 1926 a bureaucratic blunder by Bishop Manning's office gave Alva the perfect opportunity to show everybody what she meant by 'dictatorial discrimination'. Years beforehand Alva had founded the Trinity Sea Side Home for Sick Children near Islip on Long Island, run by Protestant nuns. It had been her first foray into public life and for this reason perhaps she was particularly attached to it. Alva had long been aggrieved that after her divorce from William K. Vanderbilt, Bishop Manning had forced her to stand down as president of the children's home and then list her subsequent donations to the organisation as 'anonymous' because she was a divorcee. In April 1926, Alva received a routine solicitation from the Bishop's office requesting funds to complete the building of the Cathedral of St John the Divine. Alva, who always enjoyed an epistolatory spat, saw her chance and decided to retaliate.

Although Doris Stevens tried to persuade her to tone down her response, Alva wrote a furious public reply, whilst castigating Doris for being 'too mild'.[58] 'Will you allow me to remind you, dear sir, that it is only a few years back you would not permit my name to appear as the President of the Trinity Sea Side Home for Sick Children . . . What I fail to understand is only this change on your part, dear Bishop. My status remains the same. I am still a divorced woman.' Her response appeared on the front of page of *The New York Times* under the headline 'Mrs Belmont Twits Bishop About Gift'.[59] The only reaction was an embarrassed admission from Bishop Manning that his recent request for funds had been an office error, and the press could not resist noting with glee that Alva had made him look a fool. Alva's 'stinging precision' had driven Bishop Manning to cover, wrote the *Chicago Tribune*. 'She has become famous for the vigor of her speeches, and for the swiftness and pungency of her retorts . . . The present Bishop of New York knows, or ought to know, something about this.'[60] It all gave Alva great satisfaction at the time, but it was a piece of teasing that she would live to regret. As Elsa Maxwell later remarked, offending Bishop Manning was tantamount to social suicide.[61]

Confronted with the annulment of the marriage of Mrs Belmont's daughter, Bishop Manning now behaved like a man whose time had come. On 15 November, he graduated from being 'puzzled' to saying that the matter was 'amazing and incredible' and that this was one of those moments when one was required to stand 'four square like the tower of your church'.[62] As far as he could see, this annulment was nothing less than a direct attack on the sacredness of the marriage tie itself – a sacred tie of which the Roman Catholic Church had always been a doughty defender. The problem, in his view, was that Mrs Belmont and those like her wished to sweep away not just the male-dominated hierarchy of the Church but the sacred institution of marriage itself with their attitude that divorce represented liberation. It was extraordinary that the Catholic Church should support this, lamented Bishop Manning, and in any case what right did it have to annul an Episcopalian marriage?

This last point became the focal point of a battle fought in an atmosphere of unending confusion over the religious and civil nature of a Vatican annulment and a general feeling that it was one more sign that society was going to hell in a handcart. Some, disregarding the fact that the parties were already divorced, saw the annulment as divorce by the back door. Canon Carnegie of St Margaret's, Westminster, opined that 'if such action could stand, no one's marriage would be safe'.[63] 'Everybody's prayers,' said the Rev Dr S. Edward at the Bedford Presbyterian Church, Nostrand Avenue and Dean Street, Brooklyn, 'should be offered for the confused young people of today'.[64] One Episcopalian magazine suggested that if there was no distinction in the mind of the Rota between a duke and a tramp, the Church should produce the tramp; and that since there had been no marriage the Duke might like to consider returning his American millions.[65] The *Church Times* thought it might herald revolution. 'It is disconcerting that this annulment should be announced at a moment when it is reported that marriage is entirely abolished in Russia,'[66] it said.

This provoked a sharply personal counterblast from the *Tablet*. 'The Vanderbilt nullity has been pronounced on grounds which every man, and especially every woman, should respect, namely that the Catholic Church rejects the heathen practices of marriage by capture and of marriage by purchase, however plausible their renaissance may be dissembled by professing Christians of vast wealth and high position.' The 'vulgar' barter of coronet for dollars had taken place in Bishop Manning's diocese complete with '20,000 sprays of lilies of the valley, eight miles of roses, chrysanthemums equal in bulk to a haystack'[67] and the Anglican clergy should have paid more attention.

On 25 November, Bishop Manning came charging back in again with a statement read from the pulpit before the sermon at the Thanksgiving Day service in the Cathedral of St John the Divine. He called the annulment 'an unwarrantable intrusion and an impertinence, a discredit to the Christian Church and an injury to religion'.[68] His second point, however, came in response to accusations that Anglican clergy had turned a blind eye to the Marlborough marriage. 'Many who were present at the marriage, and were associated closely with the Marlboroughs at the time, have informed me they saw no sign that the bride was acting under compulsion but quite the contrary,'[69] he said. Dr Brown had made enquiries at the time and was satisfied that gossip about coercion was untrue. It was a 'preposterous' claim, said the Bishop. And finally he turned his attack from Alva to Consuelo. It was a scandal that a woman of middle age should be able to make such comments about the behaviour of her parents when at least one of the parents was no longer alive to defend himself against the claim.

In an attempt to extinguish this firestorm the Vatican decided to make public the text of the Rota's reasons for granting the annulment. When this appeared it stoked the flames of the controversy higher for it presented a version of the story that would not have been out of place in a Gothic novel by Mrs Radcliffe, its effect greatly heightened by the translation of the text from English into Latin and French and back into English again. This highly coloured version of events was then reflected back in the American press.

Consuelo was a girl of 'youth, beauty and a great fortune' and 'endowed with every womanly grace' who had 'plighted her troth to Rutherfurd, whom she violently loved, and her mother, who strongly opposed the match, brought overpowering forces into play to wrest the heart of her daughter from that man and prevent her marrying him'. Alva, meanwhile, 'driven by her desire for a title of nobility, substituted another man for the one whom Consuelo loved passionately'. Consuelo, as the appellant, was quoted as putting her case in similar style: 'My mother tore me from the influence of my sweetheart. She made me leave the country. She intercepted all letters my sweetheart wrote and all of mine to him. She caused continuous scenes. She said I must obey. There was a terrible scene in which she told me that if I succeeded in escaping she would shoot my sweetheart and she would, therefore, be imprisoned and hanged and I would be responsible . . .' It seems a little unlikely that Consuelo would really have called Winthrop Rutherfurd 'my sweetheart' three times in as many sentences, but that was certainly the phrase favoured by the translator quoted in *The New York Times*.

Corroborative witness statements had to convince the tribunal that there had really been coercion, and 'in this connection the appellant party must produce proofs which shall cause moral certitude that intimidation was actually used'. Even allowing for the effect of translation and a desire to make the charge of coercion stick, most witnesses appear to have stuck carefully to their own version of events and did their best to avoid saying that there had been outright intimidation in the case of Consuelo's engagement to the Duke. The Duke of Marlborough said that Consuelo told him 'that her mother had insisted on her marrying me, that her mother was strongly opposed to her marrying Rutherford [sic] and that she had used every form of pressure just short of physical violence to reach her end'. Aunt Jenny Tiffany said: 'My sister was continually causing scenes and tried to soften her daughter by saying she was suffering from heart disease and that she would die if she continued to cross her.' Mrs Jay confirmed that Consuelo had disliked the Duke for his arrogance. Even Alva managed to avoid saying that she had actually tortured Consuelo to the altar, though she did say she 'ordered' her. 'I forced my daughter to marry the Duke. I have always had absolute power over my daughter, my children having been entrusted to me entirely after my divorce. I alone had charge of their education. When I issued an order nobody discussed it. I, therefore, did not beg, but ordered her to marry the Duke . . . I told Consuelo he was the husband I had chosen for her. She was very much upset . . . I considered myself justified in overriding her opposition, which I considered merely the whim of a young inexperienced girl.'

The only other person who came close to alleging outright coercion was Mrs Lucie Jay. When asked by one of the judges: 'Do you believe that the means used to force the marriage were moral persuasion or actual coercion?' She replied: 'No persuasion at all. Coercion absolutely. This I am aware of and this I know.' The tribunal concluded that Consuelo was 'gentle and mild of character and accustomed to obedience', and in no way fitted to resist her mother's 'ruthless, stubborn temper, intolerant of opposition, imperious in the extreme and ready always to bend everybody to her will. The seventeen-year-old girl's fear that her mother would die if crossed in her wishes was very real to her, it added, especially as it was confirmed by a doctor.' 'Despite all this,' the tribunal continued, 'Consuelo did not easily fall in with her mother's wishes. It is on record that she broke down and wept when first told she must marry the Duke and did so again the next day when her engagement to him was announced in the newspapers. She had no-one to whom she could turn, not even her father, both because her father and mother were separated by divorce and because all the witnesses unanimously declared that she would have fallen under the influence of the obstinacy and imperiousness of her mother.' Alva testified that she had put a guard on Consuelo's door on the morning of the wedding. Consuelo's statements about her state of mind were further supported by the Duke who said: 'She came very late to the wedding and appeared much troubled.' The tribunal stressed that it had made 'very special inquiries about all the witnesses who had testified and received information that they all were worthy of the utmost confidence', and that it was impressed by the precise nature of the depositions. Under canon law, the Rota simply had to be convinced that Consuelo had felt 'deferential fear' for Alva and that she was frightened that her failure to enter into the marriage would result in her mother's intense hostility. Confronted with ample testimony that Alva was indeed 'an imperious mother unused to any arguments against what she said and bending everything to her way of doing things', and, taken together with the further evidence of marital unhappiness evinced by a legal separation and divorce, the Rota upheld the original decision of the Southwark Diocesan Court.[70]

As controversy seethed about them, all the main characters in the drama remained steadfastly silent. Journalists attempted to speak to all the witnesses but no-one would be drawn. The Duke was said to be secluded in his 'castle' although the 'hamlet of Woodstock' was 'getting a thrill'.[71] Mrs William Jay was tracked down at the Plaza but refused to make any comment; Winthrop Rutherfurd was eventually cornered in his apartment at 274 Park Avenue and would only say that he had known Miss Vanderbilt at the time and had admired her; and Alva, who was forced to decamp from staying at the Colony Club in New York to the home of her son Willie K. Jr, confined herself to remarking: 'This is merely one of those adjustments that come into the lives of people.'[72] The only person to crack was Consuelo herself who appears to have had an outburst at a reporter from the *People*, which promptly claimed it had a scoop.

According to the newspaper, the Balsans' house in Paris was 'besieged' by journalists camped outside, and they had received hundreds of telegrams and letters 'asking for statements and offering blank cheques in return'. These were being returned unopened, or dealt with by secretarial staff and 'the former Duchess refuses to look at them'. The reporter from the *People* asserted that Consuelo agreed to speak to him because he knew her family well. He reported that she was furious at the suggestion that she had been forced into a loveless match by her socially ambitious parents and quoted her as saying:

> I say, once for all, that the suggestion of undue pressure is the foulest slander that could have been uttered against my father and mother, both of whom thought only of my happiness and would never have been party to a match had they not thought they were studying my welfare . . . I may have been a little romantic and consequently over-enthusiastic at the time. To that extent, perhaps, I was easily persuaded in my own heart when the glamour of a first love was on me, that it was for my happiness that I was taking the step; but I want you to be clear that the step was mine and that I alone was responsible for it . . . It is only because my parents have been attacked that I speak out now.[73]

But the report should be treated with caution, not least because it contained key factual inaccuracies: the paper referred to Consuelo as Madame Joseph Balsan and incorrectly reported that Alva was on her way to Paris to discuss the original marriage settlement in relation to the Duke and Consuelo's children, a wholly spurious piece of news, but one of which the man from the *People* seemed particularly proud. The paper's so-called 'scoop' shared the front page with other 'leading' stories, namely: 'Rain Foils Diabolical Indian Plotters' and 'Killed by 39th Lion: Champion Hunter's

Shocking Death'. Consuelo would not be the first person to find that she had apparently 'given an interview' to an importunate reporter while showing him the door, only to find an impromptu remark embroidered beyond recognition in next day's papers. What is clear is that Consuelo's ham-fisted attempt to play down the extent of the coercion to the *People* was driven by distress at the portrayal of her parents as unfeeling and socially ambitious tyrants, the last thing she had ever intended. Unfortunately, the 'interview' was immediately re-printed almost word for word in *The New York Times*, giving Consuelo's so-called remarks more credibility.[74] Taken together with Bishop Manning's assertion, based on hearsay, that she had been a happy bride, this helped to create the persistent impression that the annulment testimony had been a fix.

The affair rumbled on into early December. *The New York Times* published an editorial on 27 November stressing that whatever the truth of the matter, 'the voiding of the Marlborough marriage is binding only upon the conscience of Catholic communicants', with no legal effect, though it pointed out that Father Martindale in London 'had a just sense of what has happened when he wrote that, though the ecclesiastical authorities had acted strictly in accordance with canon law and precedent, the effect could not fail to be, at least temporarily, unfortunate for the Church'.[75] The Vatican responded to suggestions that Vanderbilt wealth played a role by pointing out that the annulment only cost the ex-Duchess $240 (of which $40 went to defray the expenses of the court in Southwark and $200 to defray the cost of the tribunal in Rome) and that anyone who could not afford this did not have to pay at all. It then confined itself to suggesting that Bishop Manning had an imperfect understanding of the principles regarding the annulment of marriages.[76] One Mr Charles C. Marshall quoted the letters exchanged by Consuelo and the Duke at the time of the divorce in 1921 when she begged for restitution of her conjugal rights and asked where the evidence of coercion stood now.[77] Bishop Dunn's reply to this was so arcane that everyone suddenly became bored of the whole argument and the storm finally died down.

* * *

So, was Consuelo coerced into marrying the Duke of Marlborough, or were the annulment proceedings based on a lie, as Bishop Manning suggested? Some parts of Alva's testimony were clearly inaccurate. The Duke did not go off to Canada as soon as the engagement was announced; the Marlboroughs did not separate in 1905, but in 1906. Other parts of the testimony was downright misleading. The impression was given that Consuelo was only seventeen at the time of her engagement when in fact she was a year older (Alva was never strong on dates of birth). In other places there is a non-coercive explanation for what took place: the footman may have been placed outside Consuelo's door on the morning of her wedding to guard against kidnappers; her tears may simply have been caused by nerves. On the whole, however, it appears that the witnesses in the proceedings told the truth about what happened. The Duke would not have agreed to a conspiracy to deceive the Catholic Church for he wished to join it. Appeals for annulments on grounds of coercion often failed, a serious matter when it prevented Catholics from remarrying. In this instance, there was simply not enough at stake to prompt perjury.

An equally credible explanation is that the parties agreed to proceed because they felt they had a case if coercion was defined as 'deferential fear'. It was then up to the Rota to decide whether it was strong enough to justify annulling the marriage. The discussions of the tribunal in Southwark and the Rota remained secret, but the judgement in Consuelo's favour may have been a closer call than it later appeared. Had the Rota's members been familiar with Alva's behaviour at other times, however, they might have felt even more confident that they had taken the right decision. Even in 1917, Alva was still maintaining to Sara Bard Field that Consuelo had no right to choose a husband, that this decision was a major parental responsibility, and that it would have been a sin not to exercise it with due care and attention. Even after a happier life with Oliver Belmont, she continued to take a wholly instrumental view of marriage, urging Sara Bard Field to let Alva find her a rich husband, becoming intensely irritated with Doris Stevens for refusing to marry a rich Florida widower as part of a Belmontesque economy drive, and attempting to convince Clare Boothe Luce she should marry her son, Willie K. Servants, suffragists, secretaries, and correspondents for *Town Topics* bear ample witness to her imperiousness, her frightening temper and her utter determination to get her own way. She was controllable by very few people, had a long track record of overbearing behaviour towards those over whom she had any authority and little instinct for negotiation when she could have a fight instead.

Bishop Manning's allegation that Consuelo did not seem unhappy does not undermine the judgement of coercion defined as deferential fear either. One consistent theme in Consuelo's life was that she never wished to cause her mother embarrassment. It seems likely that Consuelo was intimidated by her mother to the extent that she dared not risk disrupting Alva's carefully laid plans and tried not to let unhappiness show in public.

One person who would have been fascinated by the coverage of the annulment proceedings was Edith Wharton. She began thinking about what would be her final novel, *The Buccaneers*, in 1928 – two years after the annulment proceedings. Its story concerns the marriage of a young American heiress to a kind but dull English duke and focuses on a group of young American women and their mothers who are shunned by New York society in the 1870s, but whose success in English society prises society doors open back in America. The youngest of the Americans, Nan St George, enters into an unhappy marriage with the Duke of Tintagel, but is eventually helped to escape by her English governess, Miss Testvalley, into a new life of 'deep and abiding love' with Guy Thwarte. Miss Testvalley, however, sacrifices her own chances of romantic happiness by helping her charge, an act of expiation for having moved Nan towards the Duke in the first place.

Edith Wharton undoubtedly based some of the characters in this novel on people she knew, besides Consuelo. The character of Conchita Closson, one of Nan's friends, is a thinly disguised (and highly sympathetic) portrait of Consuelo Yznaga, Duchess of

Manchester. The ambitious mothers, however, bear no resemblance to Alva, and seem to be more closely modelled on Mrs Paran Stevens, mother of Minnie Paget, who was responsible for breaking off an engagement between the young Edith Jones and her son Harry. It may have been as a result of the annulment proceedings that Edith Wharton became aware of the fascinating but shadowy figure of Miss Harper, Consuelo's governess. One commentator thinks it likely that Edith Wharton then transposed her into the figure of her own governess Anna Bahlmann, who later became her literary agent and secretary and supported Edith Wharton emotionally through her divorce in 1913. It is also possible that Edith Wharton's glimpses of the Balsans at Lou Sueil suggested that 'deep and abiding' love was possible and helped her towards the ending she proposed in a synopsis of the novel.

During the writing of *The Buccaneers* Wharton suffered a stroke which caused her to lose the sight in one eye; she worked on with increasing difficulty and had another stroke two years later. In the view of the one editor, Candace Waid, she had drafted about two thirds of the novel at the time of her death. In 1938 a decision was taken to publish it accompanied by Wharton's original synopsis. In spite of attempts to 'finish' it by others, this remains the most satisfactory way to read it. Before she died, Edith Wharton wrote to her friend Bernard Berenson quoting a passage from her diary two years earlier. She had asked herself: 'What is writing a novel like?' and told her diary that the answer was: 'The beginning: A ride through a spring wood. The middle: The Gobi desert. The end: Going down the Cresta run.' As far as *The Buccaneers* was concerned, she told her diary, 'I am now in the middle of the Gobi desert.'[78]

Shortly after the annulment was granted, Consuelo enjoyed a short Cresta run in relation to the Balsan family. She married Jacques in a Catholic ceremony and was then taken to meet members of his extended family at Châteauroux for the first time. About twenty brothers, sisters and cousins were assembled in a room dominated by Madame Charles Balsan, Jacques' aunt and head of the family. 'Greeting me with affection, she then called her children by name and presented them, with a word of kindly appraisal for each.'[79] She presented Consuelo with a family heirloom, and later, when Consuelo wrote to thank her both for the gift and to say that she understood Madame Balsan's reasons for excluding her from the family circle before her Catholic marriage, Madame Charles read the letter aloud to the assembled family. It would perhaps have been easy, given Jacques Balsan's warmth and worldly charm, to underestimate his attachment to his French roots, but Consuelo never made this mistake and after his death she commissioned a memorial to him for the park at Châteauroux.

The annulment proceedings, however, left Alva stranded in a 'Gobi desert' of her own. Her reputation never recovered from the highly coloured version of the story filtered through witness statements, ecclesiastical summaries and layers of re-translation that appeared in 1926. The tribunal had been allowed to think that Alva's insistence on marrying Consuelo to the Duke had been 'driven by her desire for a title of nobility',[80] for no ecclesiastical judge would have understood her desire to liberate Consuelo from the 'great gilt cage' of life as the wife of an American plutocrat, and it might have undermined the case. Now, any attempt to argue that her motives were something other than tyrannical ambition also undermined the argument that there had been coercion. 'There is nothing so stimulating as opposition',[81] she wrote in 1922. But the peculiar circumstances of the annulment proceedings meant that this time opposition was impossible. Although Consuelo wrote that Alva, 'with her usual courage, remained undaunted', it seems likely that the whole episode left them both feeling deeply miserable. 'I suffered to see her in so unfavourable a light, knowing that she had hoped to ensure my happiness with the marriage she had forced upon me.'[82] But if Consuelo was appalled at the damage she had inflicted on her mother, Alva would have found being muzzled almost unbearable. She would have been the first to realise that without the mitigating factors – her own experience, her desire to empower Consuelo – she looked cruel, shallow and foolish, both to her fellow feminists and to the many people ranged against her and now enjoying her discomfiture.

It would have been little consolation that this trashing of her reputation was partly her own fault. The annulment would never have caused the storm it did, had Alva not assiduously courted publicity for the best part of five decades, sweeping up both Consuelo and the Duke in her wake whether they liked it or not. She was also the victim of her own taste for belligerence for without Bishop Manning's reaction and the subsequent row in the press, the Vatican would not have felt compelled to release the text of its judgement with all its damaging detail. It is highly unlikely that Alva agreed to give evidence in a spirit of expiation, for she was never on record regretting her decision to manoeuvre Consuelo into marriage with the Duke of Marlborough. But it must have felt harsh that in trying to help her daughter, she ended up being punished as if she had indeed wished to expiate a crime.

The punishment, in the end, was not that the world would now regard her as a wicked woman and insanely ambitious mother. She had ridden out this kind of criticism before. To Alva, it was arguably much worse that all her years of activity in the 'theatre of life' since Oliver's death were now to be swept away by a tide of detritus from an earlier life, pushing her back – in the perception of the world – inside the very cage from which she had tried to free Consuelo. From now on she would go down in history not as an important feminist and influential public figure, but as a tyrannical society woman from the Gilded Age, operating in that enclosed space where there was only sunshine by proxy. For a woman who suffered from claustrophobia it was an unhappy thought.

One of the most extraordinary things about Mrs Belmont, Sara Bard Field once said, was her power of analysis, her 'strange grasp'.[83] So it is perhaps not surprising that as the furore raged on both sides of the Atlantic, Alva was spotted alone on the SS *Berengaria* sailing from America to Europe; and that even though it was Thanksgiving she did not join Queen Marie of Rumania and the captain for luncheon. 'Mrs Belmont avoids all the other passengers and spends considerable time alone in the salon reading,' reported a correspondent. And when Alva finally returned to Paris, all she could say to the press she had courted for so long was 'Please go away.'[84]

12

French lives

SHORTLY AFTER SHE DECIDED to support the American suffrage campaign in 1909, Alva started a collection of newspaper articles which were divided into two sets of bound volumes. The first concerned the progress of the American suffrage movement as a whole. The second set of clippings, painstakingly glued in and dated by a secretary even when they were downright rude, related to Alva's own suffrage activities. Before she died, Alva became worried about the fate of these volumes. 'Writers on the Suffrage Campaign in the future will be able to find every important item in them, and I trust the headquarters will appreciate the importance of taking great care of the clippings,'[1] she wrote, having decided to leave them all to the National Woman's Party. Her concern was also driven by another impulse. The volumes relating to Alva herself represented the way she wanted her story to be told, and the historian Peter Geidel suggests convincingly that both her attempts to describe her life before suffrage, and her account of meeting the Pankhursts, were designed to supplement them. When she thanked the National Woman's Party for its gift of a volume of clippings about the opening of Alva Belmont House in Washington in 1931, Alva wrote: 'I think my family in years to come will appreciate what it all means.'[2]

Understandably perhaps, there is not one word in her clippings books about the controversy surrounding the annulment, though one exchange of letters in the New York press was thought sufficiently relevant to the position of women to be collected into a pamphlet which has surfaced in a collection in New York Public

Library relating to women and religion.[3] Significantly there is a long gap in Alva's clippings books after the annulment controversy died down, lasting nearly a year. Alva spent most of 1927 in France completing a realignment of her life so that she would be close to Consuelo, to whom she seems to have borne no animus over the affair. This rearrangement required some organisation for by early 1926 the Balsans had felt that they would like a retreat from the demands of their house in Paris in the summer months. They had found what they were looking for in the chateau at St Georges-Motel near Dreux in the Eure, the house that Consuelo would later say she loved most. 'It was a tall and elegant house built of pink bricks and was capped by a high roof of blue slates. In its narrow centre were evenly spaced windows through which one saw running water and green parterres beyond. At the two ends were towers bathed in a wide moat whose waters were deep and clear. These waters also enclosed the forecourt, which one entered across a bridge through an iron grille that bore the stamp of Louis XII,' she wrote. It was said that Henry of Navarre had stayed at the chateau the night before the battle of Ivry. The woods that surrounded it had been 'planted for pleasure in days when life had an unhurried ease' and where 'one saw down an alley a stone boar on its pedestal, recalling scenes of past hunts' and where even now 'a stag from the forest beyond would be driven to bay in our river.'[4]

Alva was so taken by the Balsans' chateau at St Georges-Motel that in September 1926 she bought her own, the Domaine d'Auger-ville-la-Riviere, near Malesherbes in the Loiret, about two hours away by car. This was another enchanting moated house, dating from the fifteenth century, built from white stone with grey slate towers standing in a park which Alva furnished with small spotted deer who were fed chocolate at dusk. Consuelo thought the legend that this house had once belonged to the French merchant Jacques Coeur, and that he had willed it to his daughter, particularly appealed to her mother. 'From this,' wrote Consuelo, 'she derived the vicarious pleasure any tribute to the female hierarchy gave her.'[5] Jacques Coeur's great house in Bourges had been one source of inspiration for 660 Fifth Avenue, and according to another legend, he had known and supported Joan of Arc which pleased Alva even more.

From 1927 the lives of both Alva and Consuelo were split between winters in Eze and summers in the countryside of northern France with sojourns in their Paris houses throughout the rest of the year. For Alva life had now come full circle. She would spend her final years in the country she had loved as a girl and which had inspired so much in New York. Consuelo was under no illusion that she made the move just to be closer to her daughter: 'Always an inveterate builder, she welcomed, I knew, the opportunity to build a new home in a new country, being really happy only when thus employed.'[6] Alva's compulsion to rearrange everything around her showed no sign of abating. Once she had restored the chateau at Augerville to its former glory there was still scope for improving the French landscape. 'She was for ever critically surveying her demesne. Walking in the garden with Jacques and me, she would suddenly stop us and, pointing to the river which flowed past the house, would say, "This river is not wide enough; it should be twice as large"; and when we next came an army of workmen would have enlarged it. A great forecourt separated the village from the house. It was sanded instead of being paved. "This is all wrong, it should be paved," my mother commented severely; and the year of her death old paving stones brought from Versailles covered the court.'[7]

The other reason for Alva's self-exile from America was that she had become isolated from her former society friends most of whom she now despised. She had returned to Newport for what turned out to be her last summer season in 1922 while she was still convinced that the National Woman's Party could become a serious third force in American politics, only to be disgusted yet again by the apathy of society women. In spite of the fact that she had 'tried over and over again to explain to them how they might contribute through influence and interest to the removal of injustice imposed'[8] they simply would not be roused. At seventy-one and feeling that she had no wish to grow old in Newport society, Alva sold her house at 477 Madison Avenue in 1924 and Oliver's extensive book collection the following year. In 1925 she closed Beacon Towers permanently and put it on the market. This long disengagement left only Marble House, which she was reluctant to part with after the demolition of 660 Fifth Avenue by developers in the early 1920s. Detaching herself from America was one thing; seeing her great American architectural creations razed to the ground was quite another.

Consuelo later wrote that she and her mother grew very close after Alva's move to France 'each sharing the other's interests'.[9] However, Consuelo was probably fortunate that her mother's enthusiasm for the National Woman's Party – of which she was still president – gradually returned after she settled in France. 'Europe's Embattled Feminists Make Headquarters at Belmont Chateau', said the *New York Sun*; 'All kinds of feminist irons are heated in the fires at Augerville.'[10] Not all the irons were political. In 1927 Doris Stevens, who had eventually married Dudley Field Malone, decided to divorce him. Alva strongly encouraged this in her most declamatory style. 'Believe me, though lonely, oh! Very lonely at times, to walk alone is divinely great and the very man who would take this from you, later *will* leave you.'[11] Having filed her divorce papers, Doris went to stay with Alva at Augerville, and talked to her frankly of her concerns about money and her plans to start a law degree so that she could support herself. At this point Alva appears to have said she would definitely leave Doris a considerable sum in her will. 'We have made agreement. She has promised to put me in her will because of all I have done for her. I am greatly relieved – said I could do law later – after she was gone – very sad.'[12] However it was made, the agreement was not clear enough to prevent bitter confusion in the years ahead.

Alice Paul came to stay at Augerville in the summer of 1927 too, for almost two months. She was anxious to keep Alva's interest in the NWP going, concerned by her diminishing contributions, and keen to secure her assistance with new headquarters, for the building that Alva had bought for them on Capitol Hill was in the process of being compulsorily purchased by the federal government to make way for the Supreme Court building. Alva was delighted by

Alice Paul's visit in 1927, for she was one of the very few people for whom she had unqualified respect. 'Do you know that in former times some of my friends thought me a little "tocquee" (ie crazy) over my fancy for Alice Paul until they had to admit that I was right in all I had told them about Alice's wonderful capacities,' she said to feminist Lies Van de Lehaln the year before her death. 'I wish she would realize that I am her greatest friend.'[13] By the end of the visit Alice Paul had secured Alva's agreement that the money paid by the government for the old building could be used for another.

The visit also seems to have given Alice Paul some insight into the president's troubled soul, for she encouraged her to revive the idea of writing her memoirs. (The process began again when Mary Young arrived from America to take up the position of Alva's secretary in the end of 1928. Like the first attempt these memoirs were never completed or extensively revised, but contain more detail and fewer literary flourishes, suggesting that the method was closer to straightforward dictation than the memoir ghostwritten by Sara Bard Field.) At the end of 1927, in November, Alva went back to America to approve the building chosen for the new headquarters and to arrange an auction of the contents of Beacon Towers which she had sold to the Hearsts. Although she did not know it at the time, this would be her last visit to America.

Content for the moment that matters were progressing as they should be in America, Alva turned her attention to her own back yard. The association between her house and Jacques Coeur may have been tenuous; the connection between Jacques Coeur and Joan of Arc was probably non-existent; but Alva became so annoyed that the church at Augerville did not have its own tribute to Joan of Arc that she decided to put matters right. Because she was a Protestant she prevailed upon her friend, Elizabeth Lehr – who was not only a frequent visitor but a Catholic – to pay for it. 'Fully aware of Mrs Lehr's parsimonious habits, my mother did not entrust her with the choice of the statue, but herself selected a fine life-sized example,'[14] wrote Consuelo. The ceremony to induct Joan of Arc on 9 July 1928 bore Alva's stamp from start to finish. The procession from chateau to church was led by a well-drilled row of little girls from the village dressed in matching white tunics designed by Alva, in time to a march specially composed in Joan of Arc's honour on trumpet, fife and drum. Bearers of banners in gold, blue and scarlet walked in front of the new statue which was carried beneath a canopy by four strong men.[15] As the procession neared the church, the choir chanted religious anthems and Consuelo remembered that the villagers knelt. The clergy came last, resplendent in gold embroidery, with the bishop most resplendent of all – even Alva forbore from telling him what to wear.

The procession moved into the church so that Joan could be ensconced in a candle-lit niche that had been specially prepared for her. The only problem was that Elizabeth Lehr would not stop talking. 'My mother, who could always be relied upon to dominate a situation, furiously and loudly observed: "Bessie, will you shut up!"' wrote Consuelo. 'Thankfully I observed that, obedient as we all were to such admonishments, Mrs Lehr thereafter maintained the dignified deportment the occasion decreed.'[16] Afterwards, there was lunch for Alva's guests at the chateau, where Consuelo discovered that the French visitors were at a loss to understand why an American Protestant should wish to exalt a French Catholic saint. They were vastly amused to be told that they had just participated in a ceremony honouring Joan of Arc's militancy (and, since French women did not get the vote until after the Second World War, they were possibly none the wiser).

In spite of Alice Paul's best efforts, the different perspectives of Augerville and Washington soon caused the relationship between Alva and the NWP to fray, with the excitable personality of Doris Stevens at the centre of much of the friction. In August 1928, Doris led delegates from a newly formed equal rights commission of women from North and South America to Paris, in the hope of presenting an equal rights treaty to governments gathered together to sign the Kellogg-Briand Pact. When the American ambassador refused to help them gain access, Doris Stevens and her colleagues forced their way in to a post-signing reception at Rambouillet, promptly getting themselves arrested for a few hours and generating much heated publicity. Instead of being pleased, Alva thought this was a grave miscalculation. She regarded herself as a much better judge of the outlook of the governments present and was furiously angry with Doris for setting back the cause of equal rights with 'hoodlum methods'. Doris refused to listen to Alva and 'seemed very excitable, very nervous and living in a state of exaltation', whereupon Alva wrote to Washington insisting that mechanisms were put in place to stop such unilateral action in future and that she would never work with Doris Stevens again.[17]

Although she eventually changed her mind, Alva found it difficult to forgive her. 'I am not placed in a position to forgive people. I am working for the welfare of the Woman's Party,'[18] she wrote to its chairman. Doris Stevens eventually wrote her an apologetic letter that helped to mollify. 'Many times I would be happy to see you even if you scolded me – The more I see of people . . . I realize . . . how easy it is to work with you. You are never petty, *never*. You are always forthright and direct . . . both in approval and disapproval.'

Alva, who was now seventy-five, responded with a long and emotional letter in which she told Doris how much she hated ageing and dreaded death. 'During the long hours of the night, when I do not sleep, I realize what I am unable to do out here, it is dreadful to grow old, to know that the body stops the will . . . Perhaps witnessing so many funerals of noted men, this last few days may have made me blue, forgive.' And with this Alva sent Doris a cheque 'for yourself alone'. Doris responded equally emotionally on 29 April. 'Your note made me weep, to think of you sleepless in the long nights.'[19] But in July 1929 there was more friction with the NWP itself. There was a mix-up over the date of the opening of the new headquarters for which Alva had paid, and she nearly travelled to the US from France for the dedication ceremony for no reason. She refused to relent when they offered another date to suit her, and broke off contact with the NWP for several months.

Instead, she decided to go to Egypt for what she called 'a last look round'. The country had fascinated her since the days she had worked with Richard Morris Hunt and this time she was accompanied by Consuelo, Jacques, her grandson William K. Vanderbilt III (son of Willie K. Jr), the Marquis Sommi Picardi and Elsa

Maxwell. Millicent Hearst was so taken with the idea that she joined the expedition with a party of seven guests of her own. The two parties travelled up the Nile in *dahabiyas* to Luxor, making a visit to the Valley of the Kings led by Howard Carter who had discovered the tomb of Tutankhamen seven years earlier. Although she was short of money and was taken along as a treat by Alva, Elsa Maxwell refused to be upstaged by the civilisation of ancient Egypt and threw a surprise party in the desert, arranging for sixty guests to be transported to a village ten miles from Luxor by camel and donkey. When it was explained to her that there was no wood for bonfires in the desert, she ordered in dead palm trees by camel. 'It was an expensive proposition but well worth the dramatic effect. At midnight a huge, red moon came up as though ordered by Cecil B. De Mille and outlined what appeared to be the silhouettes of enormous, fantastic birds floating slowly towards us. The bizarre illusion was created by camels carrying more palm trees across the desert.'[20] The high point was breakfast at sunrise on the site of the ancient temples of Karnak. The high point for Elsa Maxwell came when Prince Duoud, 'the son of the former Khedive', assured her that it was the first time since the dawn of history that anyone had succeeded in making the desert look attractive.

The expedition to Egypt gave Alva the fillip she needed. She did not break off relations with the National Woman's Party and turned her attention to one final campaign. A 'World Conference for Codification of International Law' was planned for the Hague in 1930. Alva had become much exercised by the disparities in nationality rights of women on marriage – an issue which related directly to Consuelo who had ceased to be an American as soon as she married the Duke in 1895. At NWP headquarters in Washington there was considerable disagreement about the importance of this issue in view of the depression gripping America. Peter Geidel points out, however, that Alva was right to pursue it for 'in the first Hague conference, the only legal subject that applied to women was nationality'.[21] Ironically, Doris Stevens thought Alva was principally spurred on by irritation with Consuelo who refused to take any interest in this matter because she was delighted to have become French by virtue of marriage to Jacques. Though mother-daughter relations were generally warm, there was always room for tension. 'Mrs Belmont wanted the law changed, she said, so that at least no woman could lose her nationality unless she wanted to by choice and then she could really show up her daughter's supineness if she chose the latter course,'[22] wrote Doris Stevens later. On bad days, according to Doris, Alva could really rant about Consuelo's 'unloving attitude, her spinelessness about men, her indifference to losing her nationality'[23] – in other words, Consuelo no longer automatically fell into line with Alva's opinions. And as a result, Alva reserved her moods for Doris.

When the Hague conference took place, Alva's persistence resulted in a success for the NWP. Highly inequitable principles affecting women on marriage were almost built into the new convention until Doris Stevens, acting properly this time, alerted President Hoover in a telegram whereupon he instructed the American delegation not to vote for the convention at all. The consequence was that the following year the League of Nations created a Woman's Consultative Committee to listen to representations by international women's groups.

By 1931, however, Alva was almost unable to enjoy this triumph for she was seventy-eight, her blood pressure problems had worsened and she was finding it increasingly difficult to move around independently. She was too unwell to travel to Washington for the re-scheduled opening of Alva Belmont House. She continued to make her views known on a number of issues, suggesting, for instance, that the NWP change its name to 'the International Woman's Party'; but there was little appetite for this in Washington. In the course of the year she found it so difficult to walk that she took to being pushed around the grounds of Augerville by her secretary Mary Young in a Bath chair. Consistent to the end, this was a very elegant and rather grand Bath chair once used by Queen Victoria which Jacques had modified so that it could be pulled by a (specially imported Sicilian) donkey.

Alva battled on through 1931, agitating at Doris Stevens who was in the middle of law exams, to come to Europe and join others based in Geneva working towards the new League of Nations Consultative Committee on the Nationality of Women, established partly in response to NWP action the year before. Doris Stevens felt unable to leave the States until her exams were over. Alva became angry with her because – and this was typical – she had sent her another cheque to 'go on a spree' a few months before and seemed to feel that this bought her the right to summon Doris whenever she wished. Alice Paul, meanwhile, did her best to carry out Alva's wishes, crisscrossing Europe and supplying facts for an article Alva was writing about the nationality issue. By the time Doris turned up at the chateau, Alva was so angry about her apparent lack of commitment to the work in Geneva that Alice Paul had to intervene on Doris's behalf.

There was a dreadful evening when Alva, as if unhinged by her growing powerlessness, insisted that Doris and Alice Paul should play cards and she would watch. 'Throughout this play she constantly upbraided Paul and almost slapped her hands in anger if she did not play fast enough or did not drop a card instantly,'[24] claimed Doris. There was misunderstanding and confusion about tickets to America on the *Aquitania*. When challenged by Alva, Doris Stevens became hysterical and cried, making Mary Young angry. 'There are some things you just do not say to an elderly woman,'[25] she said later. Alva seems to have felt that Doris Stevens was lying about money, though she did not say so to her face. When she had gone, Alva responded by revoking a legacy of $50,000 she had left Doris Stevens in her will, asking Elsa Maxwell and Mary Young to witness the new codicil.

In January 1932, Alva was still well enough to travel to Eze-sur-Mer where her health briefly improved. She bought back Belcourt to save it from a tax auction, giving Perry Belmont the use of it during the Newport season, and in March 1932 she sent from France her last official communication to the NWP, advising the organisation to hold on to its bonds despite economic depression, saying that they would recover in time.

On 12 May, however, she suffered her first major stroke at 9 Rue Monsieur which left her paralysed down one side. She was moved to Augerville in August, accompanied by Consuelo and Jacques, Dr Edmund Gros (William K.'s doctor and founder of the Lafayette Escadrille), four nurses, Mary Young and her sister, Matilda, who was visiting Europe that summer to keep Mary company. 'Mrs O. H. P. Belmont's recovery from her recent critical illness is just another example of her indomitable spirit which refuses to be conquered by anything, even death itself. All her life she has battled for one thing or another,'[26] said one newspaper. While she still could, she finalised the sale of Marble House to the Prince family for just over $100,000 after she learned that they planned to keep the house largely as she had designed it. There was an occasional flash of her old fire and bullishness: one morning she took Dr Gros down to the village school to talk to the children about the importance of cleaning their teeth, and handed out fifty toothbrushes and tubes of toothpaste, to the great excitement of all concerned.[27]

Like the Commodore, her final illness was protracted. 'Poor thing, she really is so valiant ,' wrote Matilda Young to her mother. 'I took hold of the paralysed hand, which has feeling in it now and really looks as well as the other. It had quite a grasp in it, but when I started to go, it was difficult to free my hand. She said: "Now that's a funny hand. It can take hold of things but it won't let go." And she laughed a little – which brought tears to my eyes.'[28]

Harold came to stay – much to the excitement of the entourage for he was one of America's most eligible bachelors, an international bridge player and outstanding sailor who had become a celebrity after winning the America's Cup in 1930. Matilda Young thought he understood Alva's paralysis much better than his siblings and noticed that 'he seems to be feeling very tenderly towards his mother'.[29] He bought her a film projector and tried to design a game she could play with one hand. After dinner, when Alva was in bed, he and Dr Gros amused themselves by pushing each other round the corridors in Alva's Bath chair. Consuelo and Jacques visited frequently. When Elsa Maxwell came to see Alva she found her in a state of irritation with her nurses who had allowed her dyed red hair to turn grey. 'Don't let her change nurses too often,'[30] wrote Dr Gros to Matilda Young – a private joke for Alva was frequently dissatisfied with her attendants. She rallied sufficiently for plans to be made to go to Lou Sueil with Consuelo but in November 1932 she had a second stroke and the plans were abandoned. On 26 January 1933, Alva took a turn for the worse and died with Consuelo and a doctor beside her at 9 Rue Monsieur just before seven in the morning.

In command to the last, Alva left full and precise instructions for her funeral. She had handed them to Alice Paul during one of her visits to Augerville saying: 'Any man who's been in the public eye and been of service to the country, when he dies there's a public ceremony ... nobody ever pays any attention to the death of a woman who has made a great gift to her country in the way of working for some reform.'[31] Alva told Alice Paul to give the instructions to her lawyer in New York and to make certain they were carried out. Her funeral was to be a last defiant gesture in the direction of anyone who ever thought that Alva Belmont, also Alva Vanderbilt, nee Alva Erskine Smith, was merely a spectator in the theatre of life rather than one of its leading players.

Alva's sons worked alongside Doris Stevens and Florence Bayard Hilles of the National Woman's Party to give Alva the funeral she wanted. After a short service in Paris, Consuelo and Jacques (and the Marquess and Marchioness of Blandford) escorted her body back to America on board the SS *Berengaria*. Eight sailors carried the coffin down the gangplank where it was met on the New York dockside by Willie K. Jr and Harold and an escort of honour from the NWP, headed by Alice Paul and Doris Stevens. Consuelo was taken aback by the change of pace between the old world and the new. 'Leaving the old world, with its marked respect for the departed, where men have time to doff their hats and women to cross themselves in greeting, I found it strange to be met in New York by policemen on motor-bicycles who preceded the hearse while we raced to St Thomas Church.'[32] Here, a further group of forty women from the National Woman's Party stood sentinel to the coffin until the funeral service.

Around fifteen hundred people came to St Thomas Church on Sunday 12 February at 2 p.m. Observers were as struck by the incongruity of Alva's last great event as they had been by the Marble House rallies and the motley crew who ate in her PEA lunchrooms. The *New York Times* reporter was fascinated by the contrast between the 200 NWP delegates in gold, white and purple and the ancient limousines drawing up at the curb of the church. 'Passers-by on Fifth Avenue stopped and turned to gaze with curiosity at elderly men and women whose dress and bearing would have been more familiar a generation ago,'[33] he wrote. A press release from the NWP drew special attention to the presence of representatives of the shirt-waist strikers, and listed representatives of nineteen women's organisations including two arch rivals, the National League of Women Voters and the Suffragette Fellowship of English Militants.[34]

Alva had specified that she wished her funeral service to be conducted by a woman. This did not happen. Her children thought the funeral should be held at St Thomas Church because it was large enough for the spectacular public farewell that Alva so desired. This caused no friction with Alice Paul, probably because the suffragist spirit of Alva's instructions was faithfully observed. The coffin was preceded up the aisle of St Thomas Church by twenty feminist honorary pall-bearers including Doris Stevens and Alice Paul herself. It was followed by Consuelo and Jacques, Mr and Mrs William K. Vanderbilt Jr, Harold Vanderbilt, Mrs Cornelius Vanderbilt Snr, the Marquess and Marchioness of Blandford and Mr and Mrs Perry Belmont. The family was followed in turn by an escort of representatives from women's organisations while women in robes bearing white, purple and gold banners processed down the side aisles and stood to attention throughout the short service.

Having corralled New York society into a large public funeral in association with America's leading feminists, Alva now subjected unreconstructed mourners to one last broadside on the theme of militancy. Before the service began, her coffin was draped with an old picket banner embroidered with the slogan 'Failure is Impossible', the last public words of suffrage campaigner Susan B. Anthony. The banner was then carried up the aisle at the head of the procession of women. The hymns Alva selected included 'Still, Still With Thee' by Harriet Beecher Stowe and 'The March of the Women', the battle song composed for the Pankhurst's WSPU in 1911 by Dame Ethel Smythe, a militant suffragette herself.

Comrades – ye who have dared
First in the battle to strive and sorrow!
Scorned, spurned – nought have ye cared,
Raising your eyes to a wider morrow
Ways that are weary, days that are dreary,
Toil and pain by faith ye have borne;
Hail, hail – victors ye stand,
Wearing the wreath that the brave have won![35]

Even this, however, was rather eclipsed by the startling words of the hymn that Alva wrote in honour of herself. Here, the middle verse in particular suggested that she was as confident of success in the Afterlife as she was about almost everything else:

No waiting at the gates of Paradise.
No tribunal of men to judge.
The watchers of the tower proclaim
A daughter of the King.
For a soul has risen today.
Halleluiah! Halleluiah!
Hosanna to our Lord.[36]

After the service three motor-buses took delegates from the NWP and other women's organisations to Woodlawn Cemetery, while members of the family made their way by motor car. A crowd of curious onlookers gathered to see Alva finally laid to rest alongside Oliver in St Hubert's Chapel which she had designed in a last obeisance to French Gothic architecture. The robed delegates preceded the hearse up the steps of the chapel, and formed a guard of honour, banners aloft. A mixed quartet, accompanied by four women buglers, sang 'Softly E'er the Light of Day' and 'Abide with Me' as Dr Brooks, who had conducted the service, led the family, Alice Paul and Doris Stevens into the chapel. The final moment came with the sounding of taps – the haunting melody played by a lone bugler that is often heard at American military funerals and played over the graves of soldiers. 'The relatives returned to their automobiles and drove away, leaving small groups of mourners silent on the hillside in the lengthening shadows, while the entrance to the tomb was covered with banked flowers and with multi-hued banners standing upright in the snow,'[37] wrote *The New York Times*. 'Failure is Impossible' was buried with Alva in St Hubert's Chapel.

All of this would have pleased Alva, but it was probably not quite enough. She had a secret fantasy which she confided to Doris Stevens at their final tea together in 1931, the 'first and only time I ever saw Mrs Belmont weep'. 'She said that . . . she would like me to present to her son WK at an appropriate time, the erecting to her of a monument. She described exactly what she wanted; a heroic figure of herself in the open air in Washington, the space to be set aside by the government, the base of the monument to contain a bas relief depicting various scenes which occurred in Washington – riots by the police and by the mob, women being loaded into petrol wagons, women arrested for petitioning President Wilson – in short, she wanted cut in stone the sacrifices which so many women had made in going to prison for this idea.'[38]

It is not clear whether Doris Stevens ever put this proposal to Willie K. Jr but the longed-for monument never materialised. Instead, Alva had to be content with encomiums that appeared at the time of her death such as: 'There is not a woman living today who is not nearer the benefits and beauties of freedom because of Mrs Belmont.'[39] This was, in many ways, a fair assessment. She was not a suffrage leader of the importance of Alice Paul, Anna Howard Shaw or Carrie Chapman Catt, but she played a key role in bringing votes for women nearer, and her contribution has been underestimated.

Alva was critical to reviving the fortunes of the National American when she joined in 1909, paying for the removal of headquarters from Ohio to New York and making generous subventions to keep it going. She was highly effective in translating society experience into a political campaign for which there was no precedent in America, impressing on the suffrage movement the importance of positive action, public impact, strategically placed buildings and expert manipulation of the press. She was well ahead of her time in perceiving the importance of alliances with other groups of women, regardless of whether they were shirt-waist workers, black women, society women or agricultural workers and regardless too of contradictions such as her simultaneous support for black women and the racist suffrage campaign of the southern states. In becoming the main financial backer of the National Woman's Party, she helped to introduce a new, confrontational element to the suffrage campaign, based in Washington, adapted from the English militant campaign, and led by Alice Paul. This was highly effective in holding male politicians from the President downwards accountable for their failure to deliver the vote, and in pioneering the idea of a suffrage amendment to the constitution alongside the state-by-state campaign, an approach which eventually won the day and also became the basis of Carrie Chapman Catt's 'Winning Plan'. Histories of the American suffrage movement are frequently partisan; but a balanced view suggests that though the conservative National American represented the views of the majority of American women, female suffrage in America would have taken much longer without the National Woman's Party's success in putting President Wilson on the defensive by picketing the White House and forcing him to declare in its favour.

Alva's heroine-worship of Joan of Arc is important too, for it reflected a judgement about men and women that is much closer to modern feminism than that of 'majority' suffragism led by Carrie Chapman Catt and supported by Consuelo. Alva's view – which chimed with that of her colleagues in the NWP – was that men and women were equal in every respect other than biology, that once the chains of male domination had been cast aside women were capable of anything including leading armies to war.

Alva learnt her feminist radicalism in part from the Pankhursts, but at her boldest she drew on the imagery of slavery that she had experienced as a child, which resonated with the Marxist language of exploitation used by radical twentieth-century feminists. In an unpublished article she wrote: 'There is no more subtle and pernicious form of slavery in the world than the subjection of women to men. It is buttressed by tradition and custom; by science and religion; by the present economic and political system; by the mercantile view of love; by the romance of literature . . . It has been the more degrading because women have not realised the extent of their slavery.'[40] Expressed in its starkest form, this was quite at odds with the view of conservative suffragism that men and women inhabited 'separate spheres' and that what was needed was a balance. This difference of view lay at the heart of the controversy about the equal rights amendment which would rage for decades after her death. As one of its earliest supporters, she would have been delighted to hear that the amendment was eventually passed in Congress and incensed that it was never ratified. It would also have infuriated her to learn that when a new generation of feminists emerged in the late sixties and early seventies, and realised that their 'new' movement had a remarkable history of its own in the National Woman's Party, the more radical were turned away in alarm by an organisation already more concerned with past glory than future action.

Alva's strengths were wit, charm, intelligence and energy. The adjectives most often applied to her by her admirers are 'fervent', 'vehement', 'determined'; the nouns 'clarity' and 'fearlessness'. She was, said Alice Paul, 'a born, born fighter',[41] which was very useful to the suffrage campaign. But the frightening belligerence and aggression that drew her to militancy also damaged her legacy. 'I wanted to be a sort of female knight rescuing other women,' she told Sara Bard Field.[42] But this was sharply at odds with how others saw her: 'She has all the crushing force of people who have had power all their lives and have never been constrained by hard necessity to be gentle to others,'[43] said one. These characteristics may have made her a natural politician and a great exponent of militant feminism, but a world view conceived in the language of slavery and dominance drove her to control and rearrange the lives of everyone around her, in spite of herself, and she could treat those she regarded as inferior with attitudes ranging from contempt to violent cruelty. It made her an impossible mother for a teenage daughter, though Alva also gave Consuelo just enough confidence to stand her ground when she first came to England and the basis of unexpected strength of character once she escaped Alva's control.

Belligerence and aggression also came to work against Alva's standing as a public figure. Carrie Chapman Catt disliked her so intensely that Alva was virtually written out of histories of the National American, while comparable histories of the NWP were never written. Bishop Manning's feud with Alva unleashed a view of her as a socially ambitious and tyrannical mother – a view that would stand and obliterate the rest of her achievements. Perhaps even Edith Wharton's dislike of her worked against her, for they should have understood each other. Edith Wharton's best novels anatomise precisely the tyranny of men over powerless women in the closed, claustrophobic world of nineteenth-century New York. Perhaps, in the end, only Edith Wharton could have done justice to the price paid by Alva for challenging its rules.

The verdict of a very different commentator, Elsa Maxwell, remains one of the most astute and interesting. 'Probably the most misunderstood woman in America,' she wrote. 'She was twenty years ahead of her generation.'[44] Elsa Maxwell would have been intrigued by the final pages of Alva's clippings books, for in the five years before her death she started to include articles about the achievements of younger women and less about herself. A 'bobbed head girl', Miss G. A. Nairn, wins the chancellor's gold medal for classics at Cambridge; Amelia Earhart makes a solo flight across the Atlantic; a woman becomes senator of the Czechoslovak Legislative Assembly; Fru Betsy Kjelsberg becomes Norway's first woman factory inspector; Marie Curie receives an honorary degree.[45] 'The important thing is knowing how to live,' Alva told Elsa Maxwell just before she died. 'Learn a lesson from my mistakes. I had too much power before I knew how to use it and it defeated me in the end. It drove all sweetness out of my life except the affection of my children. My trouble was that I was born too late for the last generation and too early for the next one. If you want to be happy, live in your own time.'[46]

* * *

It was perhaps inevitable, given that so much of Alva's life was dominated by feuds, fights and confrontation, that part of her legacy was another quarrel. Most of her will was straightforward. Consuelo was the residual legatee – the American estate alone was valued at over $1.4 million. Marble House was left to Harold (although it had been sold by the time she died) and 9 Rue Monsieur and the chapel at Woodlawn Cemetery to Willie K. Jr . There were legacies for her great-grandchildren; for the children of August Belmont who had attended Oliver's funeral; $100,000 for Jacques; jewellery for Mrs William K. Vanderbilt; and legacies for the long-suffering Azar and another servant, Ernest Mangold. The NWP received $100,000 and the statue of Joan of Arc which Oliver had given her as a birthday present in 1902 and stands in Sewall-Belmont House to this day.

The problem, inevitably, was caused by the change Alva made to her will in 1931 revoking the legacy of $50,000 to Doris Stevens. Doris, to whom this came as a complete shock, decided to contest it and prepared a long and embittered chronological account of her relationship with Alva as part of her case – an angry self-pitying description of what it was really like to be Alva's surrogate militant suffragist daughter; the way she used her money as an instrument of power even with her colleagues in the National Woman's Party; her relentless and overbearing demands; and her tyrannical behaviour. Doris Stevens's own diary entry in 1927, however, would seem to point to the root of the problem. 'We have made agreement. She has promised to put me in her will because of all I have done for her. I am greatly relieved – said I could do law later – after she was gone – very sad.'[47] In other words, Doris thought that the money was largely her reward for work done in the past, but Alva made it conditional on Doris postponing her law degree and helping her in future. In 1931, when Alva changed her will, Doris Stevens was unavailable to come to Europe at her behest because she was busy taking law exams. When she did appear, Alva thought she had behaved badly.

Doris Stevens failed to win her case. She never forgave Alice Paul whom she held responsible for alienating Alva's affections, and was still plotting against her in 1946. Elsa Maxwell, who was a witness to the change in the will, later wrote that when Alva offered to leave her a legacy she had refused it, sensing, perhaps, that it could all too easily turn into the snare it subsequently became for Doris Stevens.[48] Alice Paul was certain, however, that Consuelo and her brothers came to a private arrangement with Doris afterwards, and gave her at least part of the disputed sum.[49] She thought they made the gesture to prevent further damage to Alva's reputation; but perhaps Alva's children also understood how difficult it was to cope with Mrs Belmont from a position of dependence. Consuelo especially may have felt a twinge of guilt at the extent to which Doris Stevens drew Alva's fire in her old age and understood, better than most people, that hysteria was an understandable reaction.

One effect of Alva's death on Consuelo was that she suddenly realised how much she had missed the company of her brothers during her decades in Europe. Alva's child-rearing techniques may have been idiosyncratic but she had produced interesting children – 'the most successful of William Henry's grandchildren',[50] in the view of Louis Auchincloss, noting that they had inherited Alva's intelligence but not her vile temper. Willie K. Jr had by now married Rosamund Warburton, having successfully dodged several of his mother's matrimonial schemes throughout the 1920s. Alva once told Elsa Maxwell rather snappishly that he was her favourite child because: 'He's the only Vanderbilt in captivity who ever got over his accident of birth.'[51] Consuelo thought he had inherited a large part of their father's charm and loved him for his 'joyousness and overwhelming spirits',[52] and he certainly appears to have been the more life-enhancing of her brothers. He combined Alva's love of travel with Vanderbiltian enthusiasm for velocity.

There may not be universal gratitude that it was Willie K. Jr who helped introduce the motor car to America by importing some of the earliest and most elegant racing cars from Europe and encouraging American manufacturers by starting the Vanderbilt Cup Race on Long Island in 1904. But these were the days when motoring was still an adventure, when one took one's chances with the new touring hotels, punctures were routine, signposts unreliable and there was much dependence on maps issued by the Royal Automobile Club. Willie K. Jr's most passionate interest, however, was collecting marine specimens for what would become the Vanderbilt Museum at Centerport, Long Island. Holding a navigation licence that entitled him to command any ship, he sailed all over the world in the 1920s in his yacht, *Ara*, returning on one occasion with over fifty specimens of fish, a collection of parrots, and fourteen tiny canaries, two of which were snowy white. As Louis Auchincloss points out, marine biologists should be indebted to him for specimens such as 'the golden flutemouth, the calico razor wrasse, the Moorish idol, the fringed pipefish, and the blueline butterfly',[53] if only for their names.

Harold, known to his friends as 'Mike', was a rather different character, and already a celebrity by the time of Alva's death. He was regarded as having a brilliant analytical mind by almost everyone with whom he came into contact. He was the last Vanderbilt on the board of directors of the New York Central, where he was responsible for managing a difficult change from steam locomotives to diesel. He is also regarded as the father of contract bridge, creating the system known as the club convention. 'Some may find it distasteful to contemplate the enormous role that this game plays on social occasions and in the lives of the elderly, retired and ill, or that it was formerly played in the long days of middle- and upper-income women before work became the prerogative of both sexes, but I would argue that Harold, for better or worse, was a major force in American social history,'[54] writes Louis Auchincloss. Alva always refused to play bridge by Harold's new rules, and she would not have approved of some of his remarks in his book *Contract Bridge* (1929): 'Many mistakes . . . are due to the chit-chat which prevails at too many card tables . . . Ladies, of course, are the worst offenders,'[55] he wrote.

On the other hand, Alva was immensely proud of Harold's yachting achievements, paying him the ultimate compliment by sticking news of his first America's Cup victory in the back of her suffrage clippings book. Harold is credited with helping to design a series of J-class sailing yachts, and a formula known as the Vanderbilt start, captaining victorious crews in the America's Cup no fewer than three times. His first boat the *Enterprise*, beat Sir Thomas Lipton's boat the *Shamrock*, in 1930, and he followed this up with victories in the *Rainbow* in 1934 and the *Ranger* in 1938, until war in Europe put an end to this type of yachting. Harold was not a particularly approachable man. One of his racing crew wrote: 'It was not too easy to become intimate with Mike. He had a brilliant mind, as proven by his prowess in bridge and his business acumen. He was socially prominent and supremely confident to the point of being overbearing. When I first knew him as a teenager I thought of him with a mixture of awe and fear.' If Willie K. Jr inherited a large part of his father's charm, Harold inherited at least some his characteristics from his mother. 'His mind was razor sharp and the power of his conviction so strong that, even if you thought he was wrong, you sure as hell thought twice before disagreeing. And even on the rare occasions when he was wrong, it was tough to convince him.'[56]

Matilda and Mary Young thought that Harold was shy when they met him on his visits to Augerville. But in the weeks following the funeral Consuelo felt that she came to know him better. She and Jacques spent a few weeks at his magnificent new house at Manalapan near Palm Beach in Florida where they also met Gertrude, whom he would shortly marry. 'From a childhood memory Harold now became a very dear brother, whose sensitive nature I learned to appreciate,'[57] she wrote. It seems likely that Consuelo and Jacques also went to stay with Willie K. and Rosamund at their house on Fisher Island, near Miami, which came complete with seaplane hangar and an eleven-hole golf course where each hole was named after one of Willie K. Jr's yachts. This trip to Florida sowed the seeds of another idea that quickly took root.

When, in 1934, the Demarest estate on Hypoluxo Island came up for sale just across the waterway from Harold's new Palm Beach house, the Balsans decided to buy it. The political outlook in Europe was already unsettled, and it seemed wise to keep Alva's legacy in

America. They commissioned Maurice Fatio – a fashionable architect who had built Harold's house – and by 1936 Casa Alva, as the new house was called, was ready. Built in the Spanish style, the house had views overlooking a long lawn to the south fringed by a lagoon, bamboo and swaying palms. Following Alva's example, many of the fireplaces and doors were shipped to Florida by the Paris art dealer Robert de Gallea. Consuelo is rumoured to have spent about $2 million – her mother's legacy and more – in installing the French *boiseries* she had always loved in eleven of its rooms and creating another refined and elegant house that would turn out to be a haven sooner than the Balsans expected.

Alva's death in January 1933 was followed by two more in the family. The first was the death of Willie K. Jr's only son, William K. Vanderbilt III. A big-game hunter and a traveller like his father, he was only twenty-six, and his death was all the more tragic because he was killed in a motoring accident. The second was the death at Blenheim of the Duke of Marlborough on 30 June 1934. He once told Gladys that Consuelo would wreck the existence of anyone with whom she was associated; but it was his own that became wretched. Gladys's hesitation before their engagement soon proved to be amply justified, for within a few years their marriage had disintegrated in an atmosphere of horrifying rancour. Friends of the Duke such as the Churchills did not warm to Gladys and the antipathy was mutual. Oxfordshire neighbours were frosty too, and although Gladys made new friends such as Lady Ottoline Morrell, she soon missed her old Bohemian life in France. There were moments when she was able successfully to integrate the old life with the new – for instance, by inviting the sculptor Jacob Epstein to Blenheim to sculpt a bust of Sunny. Even this was interrupted by a row and although the work was completed in 1926 – along with a bust of Ivor Churchill – Epstein left Blenheim, like other artists long before him, 'out of spirits, and out of pocket'.[58]

'How can you spend *all* your days in Hyperboria? What are your brain-cells working on?'[59] wrote Walter Berry. Increasingly

Gladys had no reply to this question. She tried to throw herself into life at Blenheim, creating a large rockery at the foot of the lake near Capability Brown's 'Grand Cascade' while Sunny supervised the construction of beautiful new water terraces designed by Achille Duchêne, but it was not enough to stop their quarrels becoming obvious. 'There had been an occasion on which he had struck Gladys on the way to a luncheon party, causing her to arrive with a black eye. There were rows about servants, Marlborough claiming that Gladys upset them and they left, and an atmosphere soon reigned in which neither party could do anything right.'

'In her misery, Gladys responded by kicking out,'[60] writes Hugo Vickers. Once, when Sunny was talking about politics to guests at dinner, Gladys suddenly shouted 'Shut up! You know nothing about politics. I've slept with every prime minister in Europe and most kings. You are not qualified to speak.'[61] Gladys's beautiful face was now disfigured by her own hand. The paraffin wax she had insisted on having injected into the bridge of her nose in 1903 had slowly run down to her jaw, destroying its shape and giving her chin an ulcerated appearance. Though her eyes never lost their startling blue, Gladys's looks continued to deteriorate throughout the 1930s so that by 1943, only nine years after the 9th Duke's death, Sir Henry 'Chips' Channon described her in his diaries as a 'terrifying apparition'.[62]

By 1931 the Duke left Gladys at Blenheim and went to live in London. Gladys took to breeding Blenheim spaniels. Clarissa Churchill remembered visiting the palace as a nine-year old to find the Great Hall divided into dog pens and smelling horribly. Gladys's trust fund had been badly affected by the 1929 crash and her appearance became more eccentric too, for she took to wearing old court dresses held up with safety pins. When he came to Blenheim the Duke would stay at the Bear Hotel in Woodstock and Gladys startled Anita Leslie by showing her the revolver she kept in her bedroom in case he ever came through the door. The Duke developed a passion (unrequited) for Lady Lindsay-Hogg and attracted unkind comment by visiting nightclubs. Gladys said of him: 'The keynote of the Duke of Marlborough's character I found in Brodie in "Hatter's Castle" and his mood toward me is – "I'll wipe them out as I destroy all who offend me. I'll smash everybody that interferes with me. Let them try to do it. Whatever comes I am still myself." A black vicious personal pride like a disease that gets worse and worse.'[63]

The final straw seems to have come in 1932 when the Duke yelled at Gladys at lunch (rather as he had done at Consuelo): 'Bah! I have no consideration for you and never have had.'[64] Gladys, according to Hugo Vickers, found herself deserted on all sides. In the end, the Duke gave orders to close down Blenheim and evict her. She took photographs as her luggage was loaded into a van. 'Good-bye to all that!' she wrote beside them later. As the Duke moved back into Blenheim (where a guest claimed almost to have fainted because of the smell), Gladys moved into their London house at 7 Carlton House Terrace. The Duke cut off the gas and telephone, and the newspapers began to scent a story as she cooked by candlelight and friends smuggled in food. Gladys was evicted from here too, and cut out of the Duke's will; but when he died a few sensitive relations and friends such as Lily de Clermont-Tonnere wrote to Gladys, realising that in spite of everything, the Duke's death would upset her.

In the end, however, even funeral eulogies and obituaries could not entirely disguise the difficulties caused by the Duke's temperament throughout his life. Lord Castlerosse wrote in the *Sunday Express* that he would miss him but that: 'To me he was a pathetic figure like a lonely peacock struggling through deserted gardens . . . The Duke of Marlborough was the last duke who firmly believed that strawberry leaves could effectively cover a multitude of sins.'[65] Winston Churchill was, of course, more sensitive: 'He was always conscious that he belonged to a system which had passed away, and he foresaw with not ill-founded apprehension that the world tides which were flowing would remorselessly wash away all that was left.' But even he observed that his cousin 'sacrificed much – too much – for Blenheim'.[66]

* * *

In France, by contrast, Consuelo's life moved forward happily. The idea of a purely social life had always made her uneasy and by the mid-1920s this was set right when French social workers approached her for help with fundraising for a splendidly French project – a hospital for the bourgeoisie who could not afford to use clinics patronised by aristocrats and were thus forced to resign themselves to beds in public wards alongside the working classes.

Consuelo leapt at the chance of deploying her old professional skills, drawing on her international circle for help. Even President Lebrun was charmed into attending one event, solemnly telling her that his presence added many hundreds of francs in value to the occasion. When the Fondation Foch finally opened, Consuelo was awarded the Légion d'Honneur. 'It was all so very French,' she wrote later. She had refused any kind of public function, so the Minister for Public Health came to the Paris house and presented it to her in front of the assembled household. Since he was not decorated himself he asked Jacques, who was, to pin on the cross. He would not, however, 'be denied the accolade, which he begged my husband's permission to confer'.[67] The Fondation Foch rapidly acquired a reputation as a leading French hospital. (A further – unwelcome – accolade was conferred by the Germans in 1940 when they ejected the French patients and reserved it for themselves.)

If St Georges-Motel was intended as a retreat it was far from lonely, and not always peaceful. The gardens were brought back to life along with the house; parterres restored, boxwood traceries planted, water gardens created and fountains playing, the estate was thrown open to the village for the annual *fête champêtre*, 'to which chatelaines from properties near-by brought their households, and peasants came in all their finery. The village priest and the lady of light virtue were equally welcome.' The children ate at long tables set out under trees decked with bunting, laughing at clowns to the sound of the village band. Shocked by how little French children seemed to play, Consuelo opened up one corner of the estate for a vacation school she started in the village, and one year its children danced a minuet for her benefit, emerging from the chateau 'like little ghosts from an elegant past'.[68] Although

Consuelo and Jacques had no grandchildren of their own, they soon made up for it. In addition to the vacation school, Consuelo set up a sanatorium for eighty children recovering from operations or requiring preventive care. At the request of the Ministry of Health, she then opened an open-air shelter in the woods nearby for children in the early stages of tuberculosis.

Smaller houses on the estate were often lent to friends, notably the Moulin de Montreuil which was occupied by the painter Paul Maze and his family in the summer months (in the view of one authority Paul Maze did some of his best work at St Georges-Motel). His presence there attracted other painters including Dunoyer de Segonzac, Simon Levy and Odette de Garet. The pianist Yvonne Lefebure spent the last summer before the war in one of the cottages. Paul Maze's daughter, Pauline, was also a talented pianist and could be heard practising her scales across the fields. Sometimes, the manifold activities at St Georges-Motel caused problems of their own. One (temporary) lady superintendent of one of the children's houses – for whom the children did not care – had a daughter who was a ballerina with the Paris Opera. 'The ballerina, alas, could not practise her *entrechats* as did Pauline her scales. Jealousy therefore marred what might have become a friendship. In the frustrated idleness of an enforced seclusion the children had got on the ballerina's nerves. Paul's nerves were also taut, and there were sharp encounters in which Mme La Directrice's authority suffered, greatly to the children's joy.'[69] But these were small difficulties caused by a concentration of artistry in one space, and Consuelo later wrote that it forever caused her a pang to think of the careless gaiety of those pre-war summers.

By 1935 a constant undertow of anxiety about the international situation had begun to disrupt the idyll. 'An Englishman told me this summer that he had heard Vansittart tell the Prince of Wales that the situation was far too serious for the Prince to cruise on the M. and as a European war was due within the next 10 years if it came now rather than later it made little difference. If Vansittart and Eden are responsible for England's foreign policy Heaven protect us,'[70] Consuelo wrote to Churchill on 12 October 1935. At this point they were briefly in the same camp. Consuelo, who increasingly saw international politics exclusively from the perspective of French security, was worried that France would be forced into war with Italy over its stance on Abyssinia at a time when Churchill, who was out of office after the defeat of the Tories in 1929, was also arguing for sanctions against Italy, rather than resolute military action. 'Good Luck to you and may you soon be Minister of Defence to protect the peace,'[71] wrote Consuelo.

Winston and Clementine Churchill visited St Georges-Motel several times during the 1930s. Paul Maze's presence in the Moulin de Montreuil was an added draw. Though Clementine Churchill did not greatly care for him, Winston regarded him as a close friend and warmly welcomed his advice on painting. At St Georges-Motel they could meet on neutral ground, surrounded not only by 'paintatious' scenery but by staff available to improve on nature when artists ran into difficulties. On one visit, Churchill decided he wished to paint the moat with ripples in the water. When a lack of wind meant that ripples were not forthcoming he sent to Dreux for a photographer and deployed two of the Balsans' gardeners to row out in boat and make ripples with the oars. 'I can still see the scene with Winston personally directing the manoeuvre,' wrote Consuelo. 'The photographer running around to do the snapshots – the gardeners clumsily belabouring the water. With characteristic thoroughness Winston persisted until all possibilities had been exhausted and the photographer, hot and worried, could be heard muttering, "Mais ces Anglais sont donc tous maniac".'[72]

St Georges-Motel was also the scene of one of Churchill's more curious artistic creations. He was painting a view of the long canal from the lawn on the front of the house when Paul Maze and three other artists whom Consuelo had invited for lunch crowded round him to observe his work. 'Undaunted by such critical observers, he drew four brushes from his stock and handing them round said, "You Paul, shall paint the trees – you, Segonzac, the sky – you, Simon Lévy, the water and you, Marchand, the foreground, and I shall supervise." Thus I later found them busily engaged. Winston, smoking a big cigar, a critical eye on the progress of his picture, now and then intervened – "a little more blue here in your sky, Segonzac – your water more shadowed, Lévy – and, Paul, your foliage a deeper green just there". It was all I could do to drag them away to luncheon.'[73] Ivor and Jacques also pitched in to help, and the painting was signed by them all.

Consuelo's long-standing friendship with Churchill, however, was not enough to prevent their views on how to tackle the threat posed by the rise of Hitler from diverging sharply. By the late 1930s she was seeing European politics exclusively through the eyes of the French and taking on many of Jacques' views. Her support for the Munich Agreement was based partly on her view, shared by many, that France was in no state to confront Nazi aggression. Just before the signing of the agreement in September 1938, she wrote to Churchill to thank him for sending her a copy of the final volume of his biography of the 1st Duke of Marlborough. 'Mr Chamberlain has we hope saved the present situation, but how about the future of the Balkans!' There was also an unmistakeable whiff of anti-Semitism in the sentence which followed, typical of many French conservatives in the late 1930s who preferred to blame 'Jewish disloyalty' than face up to less palatable reasons for France's travails. 'The Jewish press in the USA is very disappointed that there is to be no war – sentiment is hardening against them everywhere.'[74]

A month later, after both France and Britain had signed the Munich Agreement, there was a much brisker exchange. 'My dear Winston, It is reported in the French press that in your broadcast to the United States you say that France's good name is tarnished. If this is the award meted out to an ally for loyally subscribing to Mr Chamberlain's policy it is not one likely to create good feeling and I regret it. Aff. Consuelo.'[75] Churchill replied immediately with a very tart note. 'The actual words used by me were: "We have sustained an immense disaster. The renown of France is dimmed. In spite of her brave, efficient army, her influence is profoundly diminished." It is impossible to imagine any more moderate statement of the painful events that have occurred.'[76]

Consuelo may have supported Chamberlain's policy of appeasement but deep down she sensed that war was inevitable. Her sense of impending doom only increased when she attended the coming-out ball of her granddaughter, Lady Sarah Spencer-Churchill, at Blenheim on 7 July 1939. Jacques and Consuelo had been among the earliest visitors to Blenheim after the 9th Duke's death and the sight of Blenheim occupied by a happy family and her daughter-in-law's keen engagement in local life laid many ghosts to rest. 'How rewarding are my memories of Blenheim in my son's time when his life, with Mary and his children, was all that I wished mine could have been,'[77] Consuelo wrote rather wistfully in her memoirs. On this occasion, however, she went there with 'anxious forebodings'. 'At dinner, sitting next to Monsieur Corbin, the popular French ambassador, I found it difficult to share the diplomatic detachment his conversation maintained ... I suffered the same unease that had afflicted me once in Russia when, surrounded by the glittering splendour of the Czar's Court, I sensed impending disaster.'[78]

Lady Sarah Spencer-Churchill's coming-out ball was the most magnificent of the 1939 season and lingered long in the memories of those who attended as a potent reminder of a lost world, the last private ball where the footman not only wore livery but powdered their hair. Consuelo was not alone in sensing it was the end of an era for least one debutante was already training to be a nurse in the event of war and was so exhausted that she didn't dance at all.[79] 'I supped with Winston and Anthony Eden and wandered out to the lovely terraces Marlborough had built before his death,' wrote Consuelo. That night she almost found it in her heart to forgive him. 'With their formal lines and classic ornaments, they were the right setting for so imposing a monument as Blenheim Palace.'[80]

In August, Winston Churchill invited himself for a few days' rest with Consuelo and Jacques at St Georges-Motel, after making a tour of the Maginot Line. He was now sixty-four and a campaign was afoot in Britain to bring him back into government. Though he was optimistic about French military strength, his description of the gap between the end of the Maginot Line and the sea as dependent on 'pillboxes and wire', was far from reassuring. That weekend in August 1939, St Georges-Motel looked heartbreakingly beautiful.

'Appreciation of those halcyon summer days was heightened by our consciousness that the sands of peace were fast running out,' recalled Mary Soames (then Mary Churchill), who had travelled with her mother to Normandy to join her father. 'There was swimming and tennis (so greatly enjoyed by Clementine) and *fraises des bois*. Winston painted several lovely pictures of the beautiful old rose-brick house and grounds, in company with Paul Maze, who was staying with his family at Le Moulin on the estate. We visited Chartres Cathedral and were drenched in the cool blueness of the windows: "Look thy last on all things lovely every hour".'[81] But it was a 'prickly weekend' where good humour quickly gave way to eruptions of tension.[82]

Churchill himself later remembered that it was 'a pleasant but deeply anxious company . . . even the light of this lovely valley at the confluence of the Eure and the Vesgre seemed robbed of its genial ray. I found painting hard work in this uncertainty.'[83] When Churchill went down to Le Moulin to paint with Paul Maze on 20 August, he turned to him and said: 'This is the last picture we shall paint in peace for a very long time.' 'What amazed me was his concentration over his painting,' wrote Paul Maze in his diary. 'No one but he could have understood more what the war meant, and how ill prepared we were.'[84]

On 21 August 1939, Paul Maze recounted that 'Winston was fuming but with reason as the *assemblée* didn't see any danger ahead.' Paul Maze was even warned off talking to Churchill by a pro-appeasement guest, the debonair Evan Charteris. 'As Charteris was walking up the stairs to go to his bedroom he shouted to me, "don't listen to him. He is a warmonger".'[85] Mary Soames remembers being grateful to Consuelo for taking her off to visit the children in the sanatorium, as well as a gift of a most expensive handbag far beyond her sixteen years.[86]

In the end, however, the visit was cut short on 23 August when Russia and Germany signed a pact of mutual non-aggression, leaving Hitler free to turn on Poland. 'We had to go home: my father left at once for London by air. My mother and I followed the next day, and as we passed through Paris on that golden summer evening, the Gare du Nord teemed with soldiers – the French army was mobilizing,'[87] recalled Mary Soames. Churchill's secretary, Mary Shearburn, travelled with him by car between Dreux and Paris on the way to the airport, taking notes. 'The corn was ripe and, in its heaviness, it looked like the golden waves of a gently undulating sea. Mr Churchill grew graver and graver as he sat wrapped in thought, and then said slowly and sorrowfully: "Before the harvest is gathered in – we shall be at war." '[88]

When Germany invaded Poland on 1 September 1939, it marked the beginning of the *drôle de guerre* or 'phoney war' – a phrase Consuelo later hated. 'Somewhere in the hinterland of my consciousness lies the sadness, the haunting anxiety of that cold and desolate winter . . . In that snow-covered garden everything seemed tense – *waiting*.'[89] Jacques formally rejoined the French army and tried to raise interest in a new Lafayette Escadrille with others, including Dr Gros (still practicing medicine in Paris), in a futile attempt to encourage early American entry into the war.[90] Consuelo spent the long winter evenings huddled up in a fur coat by the fire in the company of Paul Maze and Basil Davidoff who had worked for the Balsans during the 1930s and replaced their land agent after he was called-up. On Friday 10 May 1940, however, everything changed when Consuelo's maid woke her with the news that the Germans had invaded Holland, Belgium and Luxembourg and were marching south.

One difficulty was that with the advent of war, the children in the sanatorium had been joined by several hundred evacuees from Paris. In the summer of 1939 and again in 1940, 'there were children riding in the forest or jumping their ponies over hurdles in the fields, children playing tennis and swimming in the pool, children fishing for trout in the rivers or canoeing on canals, children playing golf or bicycling in the gardens,' leaving Consuelo feeling that, like the old woman who lived in a shoe, 'I had so many I very nearly did not know what to do.'[91] The evacuees from Paris caused much more trouble than the children in the sanatorium. They were a wild bunch who 'displayed a marvellous ingenuity in destruction. On every visit I had to register a new complaint.' Consuelo was heartily relieved when she finally enlisted the daughter of a French cavalry officer who was able to exert such magical authority that she even managed to stop them from singing the 'Internationale' or from balling their fists and yelling 'Heil Hitler, who comes to deliver us!'.[92]

With the German advance the issue now was evacuation: Consuelo told her maid to pack a valise and leave it under the bed. The most pressing problem was the question of what to do with all the children, particularly those in the sanatorium. Parents came from Paris to beg the Balsans to keep the children where they were. Consuelo was told by the authorities that she was not authorised to move the evacuees in any case. In spite of the fact that Consuelo, Paul Maze and Basil Davidoff all sensed the position was hopeless, Jacques – in common with many Frenchmen – continued to view unfolding events through the prism of the Great War and maintained a faith in the French Army and the strength of the Maginot Line. By the end of the day, 10 May, however, it was clear that the French army were having no success at all in turning back the German invasion. 'Returning home from the sanatorium I found Paul Maze, his daughter Pauline, my husband and Davidoff listening to the latest news. The cold impersonal voice of the speaker as hour by hour in measured tones he announced the German advance in all its incredible swiftness was somehow shocking. We studied our maps in sickening apprehension. Everyone of us knew in our hearts that there was no hope, but our lips were sealed.'[93]

Rather than move the children south to safety at this point, the Balsans were overwhelmed by a new problem as the area round St Georges-Motel was overrun with refugees. For some, this was the second or third time they had had to flee in the face of the remorseless German advance. For others, the prospect of a German invasion triggered painful memories of another war and another time. On 11 May, returning from a service at the village church where 'women who had been widowed in the last war [were] now interceding for peace', Consuelo met one of her own gardeners. 'He was an old man and when I spoke to him he seemed obsessed. "Cette fois ils nous auront" – 'This time they'll get us,' he kept repeating. I tried to reassure him and I remember saying, "The Americans will surely liberate you,' but he shook his head. "Trop tard," he said. And for him it was too late, for as the Germans moved in he shot himself.'[94]

On 17 May the *préfet* called Consuelo to a meeting in Evreux where he asked her to help with an emergency order to accommodate an influx of 45,000 refugees from the north. Nothing in Consuelo's life had prepared her for what she saw and heard the following day as they started to arrive. 'Many had been obliged to walk, and on the road they had been raked by machine-guns from low-flying planes. A distraught woman told me that her two children had been killed walking a few yards in front of her. "I saw the aviator's eyes as he aimed at them," she kept repeating in her frenzied grief. They were all in desperate straits, their garments caked with blood, their shoes in shreds . . . In the wards doctors and nurses were removing bandages and we saw the serious wounds bombs and bullets had made. In some cases gangrene had already set in, necessitating amputations.' In the face of this tide of human misery, the Balsans rushed out to buy clothes for the refugees. 'I shall always remember a young woman's joy when we spread a pretty little dress on her bed: "C'est pour moi? Oh, Madame, comme vous êtes bonne!" and for a moment she forgot the bullet in her breast and the loss of all she had.' One woman toiled up the stairs of the makeshift hospital in Evreux with seven children in tow, the eldest boy carrying the youngest child in his arms. She was already in labour as Jacques drove her to the maternity hospital where she gave birth to a boy.[95]

As the days passed, the refugees coming into the village were arriving from zones that were ever closer and the need to evacuate the children from the sanatorium before the Germans got to St Georges-Motel became pressingly urgent. The problem was that, however much Consuelo worried, there was no official plan for evacuation and it was only with great difficulty that she secured the right from the Ministry of Health to evacuate the children at all.

The local authorities were now far too preoccupied with finding beds, housing, medical care and food for the ever-swelling tide of refugees coming into the area to plan any evacuation away from it. Joining the flight of refugees to the south was an individual decision: the *exode* of millions in France in June 1940 was a popular movement, 'a contagion, but not one produced purely by fantasy'.[96] People left their homes in a spirit of great uncertainty and when the enemy was on the doorstep this could lead to irrational decisions: Consuelo met one old man who had fled his home with nothing more than ten little ducklings in a basket.

There was reluctance on the part of the Balsans to succumb to the mood of panic, but it was now clear that moving the children was the more responsible course of action and they began to rehearse drills with them. As 'the wind from the north was bringing the sound of guns ever more clearly to our ears', Consuelo and Jacques, still believing that the Germans would not arrive for a few days, set off south by car to look for a temporary sanatorium, as they had been advised to do by the authorities. As they left it seemed that nothing would ever touch St Georges-Motel. 'The fountains I so loved were throwing their sun-tipped jets into the still air; the children's laughter rang happily as they played near-by . . . I prayed it would be spared,'[97] wrote Consuelo.

Aiming for Pau near the Spanish frontier, where it was thought that the children would be finally be safe, the Balsans made an overnight stop at Châteauroux, where 'German planes dropped bombs close to our cloth factories and we were awakened at dawn by anti-aircraft fire'. When they arrived, they found Pau overcrowded and unpleasant, its position near the Spanish border resulting in an atmosphere of suspicious tension where 'distrust was rampant; arrests were numerous'. They located a villa which would serve their purposes and house the children on a temporary basis. On the second night, the government commandeered their hotel and they went to stay with Jacques' brother Etienne. At 4 a. m. the next morning they set out to make their way back to Normandy.

The side of the road going north was eerily quiet, while the side of the road going south was nose-to-tail with traffic and 'in the stark disillusionment of the faces I saw I realised the stakes were heavily loaded'. When the Balsans stopped at a café in Periguex, they met two friends who told them that the Germans had come to St Georges-Motel, and that the French government had moved to Bordeaux. Jacques refused to believe this rumour, certain that it was an exaggeration. As they continued their journey north, Consuelo scanned the south-bound traffic for signs of the household cars which she knew would be now be heading for Pau if the story were true. In a piece of luck incomprehensible to anyone accustomed to modern roads, she eventually spotted Albert, the Balsans' butler, with their driver. He saw them too but was unable to stop. A man of considerable resource, he then worked out that his employers would spend the night at Châteauroux and telephoned to tell them what had happened. 'We were then told that the Germans were indeed in our village, which had been evacuated. Saint Georges had been bombed, but neither the chateau nor the sanatorium had been hit. The hospital at Dreux, on the contrary, had received a direct hit, and we heard that our *agent*'s wife, who had that day gone to be delivered of her child, had been killed, together with her newly-born child. One more tragedy among so many when nerves were taut and sensibilities flexed is best ignored, and we chose rather to rejoice in the news that the sanatorium children had safely escaped.'

What Consuelo failed to mention in her memoirs was that it was Paul Maze who led the children from the sanatorium to safety, an oversight that can best be accounted for by embarrassment in the face of what was perceived, perhaps unfairly, to be a considerable debacle. With hindsight the children should either have been returned to their parents or evacuated sooner, and better plans should have been laid; but the Balsans were scarcely alone in failing to predict the fall of France. They were also caught out by official insistence that the children should not be moved without permission, by parents begging them to keep the children where they were and new demands on their philanthropy as St Georges-Motel suddenly turned into a refugee reception area. In her memoirs, however, Consuelo does give credit to the daughter of the French cavalry officer who took charge of 120 evacuee children when they were deposited by the French army just beyond the fighting in the Vendôme, struggled to look after them for three weeks, then made her way back with them after the armistice to St Georges-Motel in most difficult circumstances. Here, she informed the German general who had taken up residence in the chateau that it was now up to him to feed and house the children till they could be returned to their parents. 'Which he did,' wrote Consuelo.

Ironically, the unfortunate impression which persisted long afterwards – that the Balsans had panicked and fled – had been caused rather, according to Consuelo's account, by their anxiety for the children: for instead of returning to St Georges-Motel they decided to go south again to see them into the villa in Pau. Here, rumours that an armistice would be agreed grew more insistent every day. Since both the chateau and the Paris house were in the northern part of France under enemy occupation, it appeared that Consuelo's income from America would be frozen, and that further support for the sanatorium was impossible. The villa they had rented in such haste was unsuitable for long-term occupation, in any case, and they were now faced with returning the children who were well enough to their homes and dispersing those who were sick to other institutions.

There is no other account than Consuelo's of what happened next. In an attempt to find out what was happening the Balsans drove to Bordeaux, the temporary seat of the French government after the fall of Paris. In the crowd they met two American friends, one of whom was an important official in the Red Cross who 'looked harassed and insistently begged me to leave the country. He told me that I figured on the Nazi hostage list. Only a few months back, Baron Louis de Rothschild had been imprisoned in Vienna and millions had been extorted from his family before he was returned. We were advised to cross the border at once – for it would soon be closed.' Extremely worried, they found Jacques' chief, the Ministre de l'Air, in the new government offices in Bordeaux. The impending armistice made life easier, for Jacques was demobilised immediately. However, the minister also insisted that Jacques should take

Consuelo to America as quickly as possible and that the necessary visas should be secured immediately. Jacques and Consuelo then went to a café, which they found packed with a silent crowd. 'Suddenly the radio broke into the familiar bars of the "Marseillaise", which by now I had learned to associate with disaster. We all rose to our feet as if impelled. Then came the short and shattering announcement: "The French Government has asked for an armistice, which has been granted." In the ensuing stillness, as men squared their jaws and women wept, we were terrorised by three terrific claps of thunder which rent the air in rhythmic sequence, as if the heavens themselves were moved.'

Convinced that it was now essential to get Consuelo out of the country, the Balsans became mired in the kind of bureaucratic complications which arise when bureaucrats are besieged by too many people in difficult circumstances. They were told by a junior official at the American consulate that they could not have the usual visas to go to Florida but would have to travel as emigrants. To travel as an emigrant Consuelo first had to produce a birth certificate (which she never had), a marriage certificate and divorce certificate (which were both in Paris). They were then taken aside by a friend and told that the best approach was to get visas for Spain and Portugal at Bayonne and to get the American visa in Lisbon. This was a matter of urgency, however, as the border was likely to close at any moment. They made their way to Bayonne, where they were offered a room – and astonishing hospitality under the circumstances – first by a man directing traffic, and later at the home of a garage mechanic, where they slept on two sofas.

Once again the Spanish and Portuguese embassies in Bayonne were in a state of siege. Once again, they were assisted by a letter from a friend to the Portuguese consul. On the way back to Bayonne in the car, the Balsans gave a lift to a man and his wife who were travelling on foot. This small good turn would soon pay off. When the Balsans reached the Portuguese consul at six the following morning, queues of refugees were already stretching in pouring rain far down the street. They waited for two hours with an increasingly desperate crowd when suddenly, a door opened and they were beckoned up the stairs and past the others by the man to whom they had given a lift. Acting on a chance remark of Consuelo's that the Portuguese consul could do probably do with extra help, this enterprising character had found himself a job. He now signalled to the Balsans to jump the queue. Consuelo and Jacques found themselves having to push through the crowd to reach the consulate door at the top of wooden stairs. The necessary visas were obtained in seconds; it was then a matter of having to descend the same rickety wooden stairs past those whom they had queue-barged. But once again, their new friend proved himself equal to the task by shouting: 'Look out-look out – the stairs are giving way – they were never meant to bear so great a weight,' and as the crowd descended the stairs in a hurry, the Balsans were able to make a fast exit. They exercised prerogative once again at the Spanish embassy by waving a letter from another friend who had been a former Spanish ambassador to France in the face of its startled staff. At every turn, Consuelo did the barging since Jacques, who was in his officer's uniform, refused to resort to such methods – 'But neither, I shrewdly calculated, could he refuse to follow me, since alone I might easily have been manhandled,' she wrote later.

From Bayonne they drove to the frontier with Spain. Here, there were no friends to help. They had great trouble getting seats on the train to Lisbon, for which extortionate prices were now being charged. French currency was no longer honoured in Portugal after the fall of France, and for the first time in her life, Consuelo found herself short of money and in the hands of small-time crooks. The Balsans were forced to sell their car for a fraction of its market price in exchange for train tickets which had been unobtainable a few minutes earlier, leaving Consuelo feeling 'outraged by the incredible bargain' won by the haggling dealer. At the Portuguese border there was a further nightmare when their passports ('now more precious than jewels') were removed. In the event, they only managed to retrieve the passports in Lisbon and secure the necessary visas twenty-four hours before they left Portugal on a Clipper seaplane, tickets having been arranged by Consuelo's brothers in America.

The evening before their departure, they received an invitation to dine with the Duke of Kent who was on an official visit to Portugal. He and Consuelo had met less than a year before at the coming-out ball at Blenheim. The Balsans initially declined because they had nothing to wear – the only clothes they had with them were the ones they had taken for a brief visit to Pau. The Duke of Kent, whose solicitude Consuelo found most touching, assured them it did not matter. When they presented themselves at dinner, however, it turned out that it did, for his entourage displayed no understanding of the ordeals of occupation, and every symptom of the British insularity Consuelo had always so disliked. 'How odd it seemed to sit at a formal dinner again free of anxiety and care – how little these people knew of the storm and stress of a country overrun by a ruthless enemy. That we had no evening clothes seemed strange to them.' But the ordeal – slight by many people's standards due to the Balsans' network of contacts and influence – was very nearly over. It was Consuelo's first flight by seaplane. 'As we moved through the waters and rose to our flight, I looked at the blue sky above and the slowly fading coast beneath and felt I had embarked on a celestial passage to a promised land.'

13
Harvest on home ground

ALVA MAY HAVE CONSIDERED her daughter's attitude towards losing her nationality on marriage 'supine', but Consuelo finally came round to her mother's point of view by the time she wrote her memoirs in the early 1950s. She was now back in her native land after a long life under three flags 'having regained a citizenship I would never have resigned had the law of my day permitted me to retain it'.[1] On arrival in the United States in 1940 she had to change herself back into an American by way of a naturalisation certificate. Some of Consuelo's characteristics listed on the certificate had changed remarkably little in the forty-five years she had lived outside America though her hair was now white, and she had gained half an inch in height on the *World*'s long-distance estimate before the wedding in 1895. The immigration authorities were either not interested in whether her nose was 'rather slightly retroussé' or did not feel it was appropriate to mention it:

Age: Sixty-three
Sex: Female
Color: White
Complexion: Fair
Color of Eyes: Hazel
Color of Hair: Gray
Height: 5 feet 8½ inches
Weight: One hundred and thirty pounds
Visible distinctive marks: None
Marital status: Married
Former nationality: French[2]

The naturalisation process signalled that the Balsans intended to live in America regardless of the outcome of the war, for certificates were only issued to those who intended to take up residence permanently. Consuelo and Jacques now settled themselves into a new life. One of their first purchases was another magnificent house, also designed by Maurice Fatio (the architect of Casa Alva) and his partner William A. Treanor – Old Fields, near Oyster Bay on Long Island, New York. The house and its 135 acres of landscaped gardens and golf course had been designed by Fatio and Treanor in 1934 and had been sold to the Balsans by Dorothy Schiff, the Kuhn, Loeb heiress and publisher of the *New York Post*.[3] Its exterior was designed in American neo-Georgian style, but its interior quickly became unmistakably Consuelian, its main rooms filled with French *boiseries*, French eighteenth-century wall panels and mirrors, and rooms leading elegantly from one to another through Louis Quinze double doors. From the time they bought Old Fields, the Balsans followed a pattern typical of some rich Americans, spending the winter in Florida and the summer months on Long Island, an arrangement supplemented by the lease of an apartment in New York, first at 825 Fifth Avenue, and later (by way of the Carlyle) at 1 Sutton Place South.

By January 1942 they were entertaining again. On her first visit to Palm Beach in over twenty years, Elsa Maxwell stayed with the Balsans. Always impeccable hosts, they excelled themselves on 10 January by producing for lunch the one thing Elsa Maxwell liked more than anything else – a celebrity guest in the form of (Prime Minister) Winston Churchill. America had entered the Second World War the day after the attack on Pearl Harbor on 7 December 1941, and Churchill had come to America for talks with President Roosevelt. He travelled to Florida to give his hosts at the White House a few days' rest, accompanied by his doctor, Lord Moran. They stayed in a secluded bungalow at Pompano, near Miami, closely guarded by Secret Service men and kept at a distance from the press who were told, somewhat unconvincingly, that Churchill was a 'Mr Lobb', an English invalid requiring peace and quiet, and that his Principal Private Secretary, John Martin, was the butler.

Lord Moran wrote in his diary: 'Oranges and pineapples grow here. And the blue ocean is so warm that Winston basks half-submerged in the water like a hippopotamus in a swamp.'[4]

Elsa Maxwell 'strained slightly an acquaintanceship of thirty years' standing'[5] by asking Churchill his opinion of her idol General de Gaulle. According to her account of the conversation at lunch, Churchill replied: 'My greatest cross is the Croix de Lorraine'[6] – an estimate of de Gaulle that Elsa Maxwell later discovered to be correct. During this lunch she attempted to lift Churchill's mood by asking him about Rudolf Hess: 'The Prime Minister chuckled when he described the pink polish Hess wore on his toenails. He clenched his famous bulldog jaw and murmured grimly, "The Nazis will lose." '[7]

After lunch at Casa Alva, Churchill and his entourage returned to Washington by train. (Warned to be extremely careful about what he said on the telephone, Churchill told Roosevelt: 'I mustn't tell you on the open line how we shall be travelling, but we shall be coming by puff puff.'[8]) Lord Moran reported in his diaries that after they left the Balsans, 'Winston mused a little, and then said half to himself: "Wealth, taste and leisure can do these things, but they do not bring happiness," '[9] indicating that he may have thought that Consuelo and Jacques were far from content.

If so, Churchill's judgement was shrewd, for though the Balsans adored each other and led a halcyon existence compared to anyone in Europe, there were tensions because Jacques was deeply uneasy at being in America while France was at war. His anxiety was compounded by his faith in Marshal Henri-Philippe Pétain, who had emerged as the leader of defeated France in June 1940 and who then negotiated the armistice with Germany that divided France into two zones: a northern zone (which included Paris and St Georges-Motel) under German occupation, and a southern unoccupied zone administered by the French government, under Pétain, from Vichy. Jacques' confidence in Pétain's actions was characteristic of many who had fought in the Great War and remembered him as the hero of Verdun. Moreover, support inside France for both Pétain and the armistice of June 1940 was almost universal. 'Survivalism had its own irrefutable logic and those who said "no" to the Armistice and rejected Pétain seemed a greater danger to society and appeared to be acting more blindly and absurdly than those who accepted him,' wrote one commentator.[10]

This did not mean that Jacques – or Consuelo – in any way supported fascism, unlike some of her old friends in England such as Lord Castlereagh, (now Lord Londonderry). Jacques' attitude towards Pétain and the 1940 armistice could be best described as *nationalisme germanophobe*. He already had a gallant war record in fighting Germany from 1914–18. His attempts to re-form the Lafayette Escadrille had been designed to bring America into the war early in order to defeat Nazism. But in June 1940, he saw a negotiated settlement with Germany that allowed millions of French refugees to return home as the only realistic solution. Jacques also appears to have shared the view of many that Pétain was playing an elaborate double-game with the Nazis, and would ultimately outwit them. Such views led to a serious quarrel with Elsa Maxwell who saw Pétain as the ringleader of French defeatists. 'When the French capitulated and Vichy became no more than a political *maison de passé* under the cunning manoeuvres of Laval, I did not spare my pen to heap abuse on Marshal Pétain,' she wrote. 'As a result I received a note from Colonel Jacques Balsan, "I shall never forgive you. Marshal Pétain will prove to be the saviour of France." I replied, "History will record who is right."'[11] If Elsa Maxwell is to be believed, this quarrel was never mended and their friendship ended.

As the war progressed, Jacques had to go through the painful process of facing up to his misplaced faith in Pétain, as the Vichy government entered into full collaboration with the Nazi regime. This made him even more frustrated at being exiled from the theatre of war, as a letter written to Myron Taylor on 17 May 1942 indicated. After Pearl Harbor, Jacques seems to have had hopes of setting up an American version of the Foreign Legion. 'Dear Friend,' his letter read, 'I want to tell you how much I want to serve the United States. I have tried in vain to find the means of succeeding. Please read my letter to the President which I attach hereto, and if you do not find it useless, will you post it?' In his letter to President Roosevelt Jacques wrote: 'We are in the United States a great number of foreigners who are attached to the American nation with all our hearts. It is impossible for us, we confirmed it, to be incorporated in the American Army but you might form a Foreign Legion in the American Army so that it would be possible for us to devote ourselves to the American nation as we would ardently like to.'[12]

Jacques' age was against him. At seventy-four he was unable to find a satisfactory role and had to content himself with helping Consuelo chair a fundraising exhibition of French and English art treasures in New York in aid of the American Women's Voluntary Services. The exhibition was held at the end of 1942 in the Parke-Bernet Galleries, run by a society committee which was chaired by Consuelo and included her cousin Ruth Twombly. The lenders included many names from Consuelo's past – Sloan, Wilson, Auchincloss, Crowinshield, Goelet, Dupont, Webb and Lady Ribblesdale, formerly Ava Astor. The exhibition included over a hundred paintings from private American homes by Boucher, Chardin, David, Greuze, Fragonard, Lancret and Hubert Robert, but it was not, said its catalogue, merely an assemblage of the past. It was 'a manifestation, a declaration of faith in the principles of the eighteenth century. For in the age of reason man knew that democracy was a political impossibility unless it was accompanied by an aristocracy of thought.'[13]

Preferring action to representing an aristocracy of thought, Jacques finally found a solution to his problem in 1943 when General Cochet of the French air force escaped from France via Spain and made his way to London. Cochet had chosen to remain in Vichy France after the armistice, calling for resistance against the Nazi regime as a moral act based on French values which crystallised as a series of tracts entitled *Tour d'Horizon*. Each of these started by expressing loyalty to Pétain while calling for opposition to Nazism. Cochet's position eventually became so untenable that he was arrested, but then escaped. It is likely that Jacques' reaction to the Nazi occupation of France followed the same trajectory as his old air force colleague. Soon after General Cochet went to London to join the Free French in the spring of 1943, Jacques crossed the Atlantic to join him, installed himself in the Ritz and assisted Cochet with the coordination of the numerous resistance groups now converging on London, working from one of the many Free French administrative offices scattered around Mayfair. He also helped General Sice in planning the feeding of French children after liberation. Ivor too had a relationship with the Free French, as honorary secretary of the Association of Friends of French Volunteers, a welfare organisation that looked after members of the Free French forces while they were in London.

However relieved Jacques may have felt to be in England making a contribution to the eventual liberation of France, Consuelo was wracked with anxiety about him. In September 1943 her nerve finally cracked. She wrote to Winston Churchill on 23 September and congratulated him on his 'masterly conduct' of the war before coming to the crux of the matter. 'I am writing to you because I am anxious about Jacques. He has never stood the cold and damp of northern winters and at his age I fear he might easily get pneumonia. I know he is bent on following the French General Staff but if he could (without his guessing outside intervention) be sent here on a mission and return to England in the spring it would be a very great relief to my anxiety. He is 75 and although young and well for his years – imprudent. Don't answer this letter and if you feel unable to interfere I shall understand and please let it remain a secret between us.'[14]

Churchill's capacity to attend to such requests from relations in the face of his commitments as Prime Minister of a country at war was one of the most extraordinary things about him (or the most regrettable, according to some). November 1943 saw the start of the Tehran Conference in which the 'Big Three' met to discuss the direction of the war, and during which Churchill was deeply involved in negotiations with Roosevelt and Stalin over Operation Overlord – the Allied cross-Channel landing in France planned for the summer of 1944. However, he still found time to pass Consuelo's letter on to an attaché, Major Morton, and entered into the spirit of the plot by scribbling: 'You might like to know how this stands, and where this officer is likely to be sent. Seventy-five is certainly old for the war. On no account let it be known that I am making any inquiries.'[15]

As it happened, Jacques called on Major Morton soon afterwards in an attempt to see Churchill with 'a Frenchman just out of France'.[16] (Though Major Morton did not mention this in his minute to Churchill, Jacques' anonymous companion was likely to have been a member of the French resistance pleading the cause of Leon Noel, French ambassador to Poland and one of the signatories of armistice.[17]) Major Morton reported back to Winston Churchill: 'I took the opportunity of enquiring tactfully on the lines of your Minute without disclosing your interest . . . Though working in London at present, he is going to America in about a fortnight's time in connection with the children's relief work. I submit that it would be up to his wife to try to keep him there when he gets there. Incidentally he seems very well at present, surprisingly so for his age.'[18]

Churchill immediately dictated a telegram to Consuelo which read: 'I understand Jacques will be with you soon and that he is surprisingly well.'[19] History does not relate whether Jacques ever heard about Consuelo's attempts to get him back to America, or if she ever discovered just how little Churchill had to do with it, but the Prime Minister received a delighted telegram expressing her heartfelt thanks.

Jacques' arrival in the United States at the end of 1943 was timely since he was able to be at Consuelo's side when her brother Willie K. Jr died on 8 January 1944, after several months of illness from a heart condition. On his return Jacques began to argue the case for stopping the American blockade on food relief to France that was being applied as a way of shortening the war. Consuelo contributed a long article to the *Herald Tribune* articulating French frustration at American failure to grasp the realities of daily life in France, where, in the towns and less fertile areas of the French countryside, the civilian population was close to starvation. She suggested that during the winters of 1940–41 and 1941–42, the Germans had allowed emergency supplies to reach their destinations unimpeded and that if the Allies did not do something to help France soon they would find themselves 'liberating a cemetery'.[20]

In October 1944, following the success of Operation Overlord, Jacques (and possibly Consuelo) went back to France and arranged for the Paris house to be used by British service women in need of somewhere to rest. In Normandy, according to a letter from Consuelo's son the 10th Duke of Marlborough to Churchill, 'several hundred refugees were now billeted on St Georges-Motel'.[21] Otherwise the house and its contents had an escape which seemed almost miraculous. Although it had been used as a base for airmen from the Luftwaffe, it had been treated with respect, and had even survived visits from the notoriously light-fingered Hermann Goering. The Luxembourgeois butler Louis Hoffman, who stayed on there, endured one hair-raising moment when he was arrested as a spy and threatened with a firing squad, but he was eventually released unharmed.

The explanation for the safe passage of St Georges-Motel and its furniture and paintings through the war lay with Basil Davidoff, who had played a central role in the lives of the Balsans from the late 1930s. A Russian who had fled to Europe at the time of the revolution in 1917, he was highly educated, was a polished conversationalist and could speak five languages. Consuelo 'found him in a bank in Monte Carlo', according to her granddaughter Lady Rosemary Muir, and gave him the role of a superior major domo. Davidoff mingled with the guests at Lou Sueil and St Georges-Motel and made sure that everything ran smoothly during house parties – matching tennis partners, setting up bridge games, finding golf caddies and ensuring that everyone was content. As a White Russian, Basil Davidoff was less troubled than most by the German occupation of France. Indeed, he longed for the Nazis to invade Russia too and dispose of the Bolsheviks. When officers of the Luftwaffe moved into St Georges-Motel, Davidoff dined with them in the evenings and, unknown to the Balsans, continued to act as major domo in an arrangement so cosy that the officers even arranged for the return of a beautiful stallion belonging to Jacques that had been sent to Germany. Relations became distinctly less friendly when a copy of a telegram from Consuelo to Winston Churchill congratulating him on becoming Prime Minister in May 1940 was discovered in the local post office, but regardless of the *froideur* that then ensued there is no doubt that the Balsans' collection at St Georges-Motel was largely preserved as a result of Basil Davidoff's unfortunate act of loyalty in cultivating the Luftwaffe.

After the war the great paintings and best furniture were shipped to America, accompanied soon afterwards by Louis Hoffman, who married Consuelo's maid Annie. But the circumstances surrounding the survival of the contents of St Georges-Motel were something that Consuelo – who would show signs of anti-German prejudice for the rest of her life – found difficult to face and which she never discussed. She was always generous by instinct, but it is likely that deep unease may have made it easier to sell the great Renoir 'La Baigneuse' for $115,000 as a contribution towards food relief for French children after the war; and she also gave the Boldini portrait to the Metropolitan Museum of Art in New York. Basil Davidoff, meanwhile, took up residence in one of the outbuildings at Lou Sueil where there was no running water or heating. Jacques had rented out the house to an American during one of his wartime trips to Europe, in a gentleman's agreement so half-baked that it proved impossible to evict this unwelcome tenant until 1953. The house was eventually made over by Consuelo to her grandchildren who had the greatest difficulty persuading Basil Davidoff that he would be better off in a comfortable flat in Monte Carlo, provided by the family. Lady Rosemary Muir supplemented Basil Davidoff's Balsan pension out of her own pocket. When she helped him to pack, she discovered that all he owned were some old clothes belonging to Jacques, snapshots of his house in Kiev and a photograph of the Tsar. He continued to yearn for the return of the Tsar until he died.

Consuelo refused to return to France to see her houses before they were sold after the end of the war, feeling, perhaps, that happy memories had become tainted by the complexities of survival in wartime, and that it was better to look forward than back.

Nonetheless, her strong identification with France persisted. She was rewarded by the French government on 24 July 1947 for her support when she was appointed *officier* of the Légion d'Honneur for all her efforts 'in the field of Public Relief'.[22] Insults directed at the French could rouse her to formidable anger: Elsa Maxwell recalls that soon after the war, the Earl of Carnarvon (known as 'Porchy') incurred Consuelo's wrath by making rude remarks about France when he was staying at Casa Alva. 'In the middle of dinner one night the conversation had turned with mounting acrimony onto the international situation and France's role in post-war Europe ... Whereupon Porchy, falling back on the traditional schoolboy expression, suddenly delivered himself of the opinion that "the French were a lot of frogs, anyway". As Madame Balsan is married to a Frenchman and devoted to France the fat was in the fire. Icily, firmly and irrevocably the ultimatum was delivered to poor Porchy across the plates and glass: "Will you kindly leave my table and my house this instant," Mme Balsan demanded. Whereupon, his dinner half eaten, he left the room, went upstairs and had his bags packed and left the house.'[23]

Alva's love of France as a girl, and her attempts to bring French style to America, finally came full circle in the stunning aristocratic French interiors created by Consuelo and Jacques in Old Fields and Casa Alva. 'Paintings by Fragonard, Oudry, Hubert Robert, Cézanne, Pissarro, Renoir, Utrillo, Segonzac; lacquer cabinets on gilded stands of the seventeenth and eighteenth century; Aubussons, Savonneries, and Chinese rugs; Regence secretaries, Louis Quinze sofas and bergères; Clodion *terres cuites*, chinoiserie figurines, Meissen and Chi'en Lung birds and fish with mounts of eighteenth-century French gilt bronze ... The mere bald recital is liable to impress,'[24] wrote Valentine Lawford for *Vogue*. But the resemblance of Consuelo's American houses to her mother's Gilded Age palaces was only superficial for the priceless collection that surrounded the Balsans had nothing to do with ostentation nor was it even simply the legacy of the Balsans' shared past and exquisitely refined taste.

'The more one became acclimatized to the atmosphere of these extraordinary houses, the firmer was one's conviction that to impress was not their owner's need or intention,'[25] wrote Lawford. Flowers were arranged, 'as a painter will use colours', beneath the treasures on the wall, the pictures were hung – and re-hung – at a height where Consuelo could enjoy them and he noticed that she often sat in a spot where the view of the room was most perfect, or would move her chair so that the light fell on one of her paintings in the way she liked best. When they travelled north or south each year, the Balsans took their favourite works of art with them, sending them a few days ahead by road and rail. When two of Consuelo's walnut bergère chairs were sold much later at auction in New York, labels explaining their precise location in each house could still to be found on their undersides.[26]

The impression of a small perfect French world set down unexpectedly in the middle of America was reinforced by the fact that Consuelo and Jacques spoke French to each other. After the arrival of Louis Hoffman and his wife from France, so did most of the servants. 'Everything was perfect,' says Margarette Blouin, French governess to the children of Consuelo's granddaughter Lady Sarah Russell, who had been found for them by Consuelo who insisted that they had to speak French properly. 'The flowers were perfect, the food was perfect, she was beautiful and everyone always looked marvellous.'[27] 'The *éclat* of it all was probably more stupendous in Florida,' Lawford thought. 'Greeted each time one came through the front door by Cézanne's "Bridge" on the far side of the hall, and turning towards a room of painted walls brought from Hamilton Palace, one found it frankly inconceivable that such a treasure house should actually have been superimposed on Hypoluxo Island, still in process of growing, every year a little longer out of the unspeakable bed of the lagoon.'[28]

It was almost too much for the Churchills. They stayed with the Balsans at Casa Alva after the end of the war, in January 1946, where it is thought that Churchill may have worked on the famous speech delivered at Fulton Missouri on 5 March, where he spoke of an 'iron curtain' and called for a Western alliance that later emerged as NATO. 'There is a Norwegian Butler, flocks of pale blue parlour-maids, delicious food, ancient French panelling and Aubusson carpets, with a faint feeling of "ennui" pervading the whole,' wrote Clementine Churchill to her daughter Mary Soames on 27 January 1946.[29] Though it brought back memories of the 1930s in France, the war 'made this elegant life and its setting seem remote and out of touch',[30] Mary Soames wrote of one of her parents' stays there, an understandable reaction when Europe was still gripped by hardship. Another guest, Laura Canfield, found Churchill difficult and moody during this visit but thought that this was caused by the shock of his election defeat in 1945 rather than by the enervating effect of Casa Alva.

As far as some of the Balsans' American friends were concerned, however, Jacques was not always sufficiently remote for members of the opposite sex who caught his eye. Those who remember him consistently testify in the same breath to his warmth and kindness and to the fact that it was most unwise for any attractive woman to find herself alone in a room with him, a tendency which became worse as he grew older. Some of his reputation can be attributed to the difference between prim New England society and the style of a lady's man of the *belle époque*. It was also attributed by some of his younger relations to his fondness for rejuvenation treatments involving monkey glands, pioneered by their old neighbour at Lou Sueil, Serge Voronoff. One sensitive French nephew thought that sometimes Jacques felt restless and confined in his post-war gilded exile, and noticed him using up surplus energy by hacking away at the jungle undergrowth surrounding Casa Alva with a machete. Consuelo – though not her granddaughters – remained wholly unruffled by his pouncing tendencies, expecting little else from a Frenchman of Jacques' age and class. It was clear to all that however Jacques behaved when his wife's back was turned, the Balsans loved one another. He was, says one relation, a perfect foil for her and continued to buzz round her with *petits soins*. Another relative describes his affection for his wife as so overwhelming that she would 'bat him away like a slightly annoying insect'.

In the final phase of her life, Consuelo appeared an intimidating figure to some. Her powerful character, her deafness, her height, her straight back and elegant beauty could make her seem inaccessible to those who did not know her well. She controlled her hearing aid by a small dial on her chest, and had a disconcerting habit of switching it off sharply when she was bored, unaware that the click of the switch could be heard by everyone present. Her hospitality was flawless but formal, even by the standards of the 1950s. In late 1945, the 10th Duke's eldest daughter, Sarah, who had married American publisher Edwin Russell in 1940, went to live in the United States permanently, becoming the grandchild on whom Consuelo relied as though she were a daughter. The Russells' four daughters (one of whom was named Consuelo; another, Jacqueline, was named after Jacques) thought of Consuelo as 'Granny' – though she was, of course, their great-grandmother – and spent many holidays with her on Long Island and Palm Beach. Much as they loved Consuelo and Jacques (who would start every holiday by taking them all off to buy piles of boxes of holiday shoes), some of Alva's drive for control undoubtedly resurfaced in Consuelo in old age. Perfect manners were expected. For lunch, the main meal of the day, the girls changed into tickly white organza dresses, had their own small chairs and were expected to make polite conversation regardless of the grandeur of the guests. Mademoiselle Blouin clearly remembers the thunderstruck expressions on the faces of passers-by on annual visits paid by Consuelo and her great-granddaughters to the Radio City Christmas Show in New York as the chauffeur-driven limousine pulled up on Sixth Avenue and Consuelo stepped out in her famous pearls.

Those able to find a way past her intimidating elegance – and there were many – knew that there was a very different person behind this *grande dame*: kind-hearted, affectionate, interested and cultured. There was a constant stream of visitors to both houses: young and old, distinguished and unimportant, 'authors and career women, lonely or just tired' according to Valentine Lawford. The staff stayed. Mademoiselle Blouin once found a new black cook from Georgia crying in the garden because no white employer had ever spoken to her so kindly before. 'Beautiful' and 'riddled with charm and really rather clever',[31] was the verdict of the young

Louis Auchincloss, who married Adele Lawrence, and thus into the Vanderbilts. Consuelo's granddaughter, Rosemary, (a daughter of the 10th Duke) often travelled to Casa Alva with her parents after the Second World War. She thought that her grandmother's upbringing at the hands of Alva made her particularly sensitive to the aspirations of the young, especially young women who wished to get out and see something of the world before they married. On one occasion, when she was seventeen, Lady Rosemary arrived at Casa Alva with her father, hoping that she would be allowed to travel on to Arizona where she had been invited to stay by an American friend. Her father refused to countenance this proposal and was adamant that she would do nothing of the kind. Consuelo stepped into the fray, told her son firmly that he was being wholly unreasonable and insisted that Rosemary should be encouraged to see a different side of America while she could. She then undermined paternal authority entirely by giving her granddaughter the money to make the trip.

Consuelo did not, however, win all her battles with the 10th Duke. In his fifties, in the late 1940s, he decided not to hand over Blenheim Palace to the National Trust after the war, and worked extremely hard to restore it after the depredations of wartime, opening it to the public in 1950. In every other respect he appeared to the world as a splendidly ducal anachronism ('antediluvian' is the word used by one commentator). The Balsans' American friends could be forgiven for finding him baffling on visits to the States since he was famously incomprehensible even to the English. Much taller than his father (he inherited the Commodore's height) 'Bert' barely moved his lips when he spoke, rarely removed his pipe from his mouth, and thus reduced all his sentences, most of which ended in 'What?', to an inaudible rumble. His deep dislike of appearing sentimental masked kindness and humour, but it took time to get the hang of its surreal streak, and some people never quite managed. His opening sally to Blenheim's historian David Green, who later became a friend, was: 'My main burden at the moment is this bloody roof, what? When we had that heavy fall every man jack in the place had to help sweep the snow off it. One of them disappeared in a drift. I'm told they've found him, but they know I can't go up there because of vertigo.'[32] When the idea of putting on son et lumière at Blenheim was under discussion his chief concern was that the public would 'copulate and poach the pheasants, what?'[33] When Richard Dimbleby turned to the 10th Duke in the Long Library during a television interview and said: 'They tell me that you have the biggest private organ in England,' he smiled and said, 'I would like to think so.'[34] He was deeply attached to his mother who loved him dearly but even she could do nothing about his dress sense. During American visits – in a desperate attempt to make him presentable – she 'would order for him a range of matching outfits, only to be faced next morning with a heartbreaking assortment of oddments, some new, some old, topped with an ancient pullover and a favourite jacket which she hoped had been given away'.[35]

There were some who preferred to think of Consuelo as a Proustian figure whatever she was like: the designer Christian Bérard, for instance, almost 'passed out with excitement'[36] when he discovered that she was sitting in front of him at the Paris Opera. In reality, Consuelo was far more adaptable than most Proustian figures, an advantage in a world that was changing fast. The New York Central, source of so much Vanderbilt wealth, found itself challenged by new competitors. At the end of the Second World War, Commodore Vanderbilt would probably have divested himself of trains and started an airline. Harold Vanderbilt, though effective in seeing through the change to diesel, was unable to fend off a challenge from corporate raider Robert Young which ended Vanderbilt involvement in the New York Central. It was a Pyrrhic victory for Young who was eventually forced to merge with the New York Central's great rival, the Pennsylvania Railroad. The merger took place in such an atmosphere of fraud and scandal that Young felt that the New York Central was finished and shot himself in the library of his house in Palm Beach in 1958.

The association between the Vanderbilts and society was

sustained by Grace Vanderbilt, now *the* Mrs Vanderbilt and society queen, wife of Consuelo's cousin Cornelius who once said despairingly that his wife had become a 'waltzing mouse'. Leading society was 'a full-time profession which taxed all her resources every waking moment',[37] wrote her son, who rarely saw his mother while she partied with her friends in Europe. Consuelo had known Grace Vanderbilt since she was a girl, but she kept her distance from a cousin-in-law who is said to have entertained 30,000 guests in one year at William Henry's house at 640 Fifth Avenue. The Vanderbilts started to leave mid-town Fifth Avenue during the Depression in the 1930s. Grace was the last to go in 1944, moving north to 1048 Fifth Avenue, a fine house by Carrière and Hastings that is now the Neue Galerie. Grace called it 'the gardener's cottage' but even here she often had thirty guests for lunch and a hundred for dinner. Her final years echoed those of Mrs Astor, for she was bedridden for two years before she died in 1953, entertained mainly by a companion who would retell and tell again stories of *the* Mrs Vanderbilt's great social triumphs of the past. After Grace Vanderbilt left, 640 Fifth Avenue was bought by Lord Astor of Hever and knocked down to make way for a bank. 'In the greatest of American society rivalries,' writes Jerry Patterson, 'the Astors in a sense had the last word.'[38]

Consuelo's only surviving aunt, William K.'s sister Florence Twombly, was also fading and finally died in 1952 aged ninety-eight. William Henry Vanderbilt's favourite daughter, she made her own contribution to *The Glitter and the Gold* by supplying her niece with details of where Consuelo was born, important because there was no other record on account of the vexatious missing birth certificate. Florence Twombly's passing marked the end of an era, for she was the only surviving Vanderbilt to have started life on Staten Island, not that one would ever have imagined this to look at her. From the age of ten, when the family moved to Manhattan on the Commodore's instructions, she never lived anywhere but Fifth Avenue and her country estate, Florham, where there were still thirty gardeners, four footmen, and eight housemaids at the time of her death. She never gave interviews, writes

Jerry Patterson, but 'her arrival at the opening night of the Metropolitan Opera in her maroon limousine with maroon-liveried attendants was watched with closer attention than the proceedings on the stage'.[39]

Between 1950 and 1951, feeling that the life of her youth was changing irrevocably, Consuelo started jotting down some notes about the past. One person with whom she first discussed the idea of a memoir was the young lawyer, writer, friend and relation by marriage, Louis Auchincloss, then at the beginning of a literary career that would produce more than fifty novels, short stories and other works. 'I suggested that she open her book with a chapter on a picturesque and opulent American childhood in the eighteen-eighties and nineties, with sylvan scenes in Newport and a Prendergast picture of Central Park and its baby carriages and starchily uniformed nurses; then move swiftly on to dancing classes and balls in gilded ballrooms, and last – bang – hit her reader with a chapter called "The High Price of Dukes".' This did not appeal, however:

> She didn't like the idea at all. When I pointed out that there had been many books about American heiresses marrying European nobles, but that hers would be the most dramatic of all, involving as it would the greatest heiress and the greatest duke, she protested that, as her title would imply, Blenheim Palace was just the glitter and her real life, the one devoted to social work and to her second husband, Jacques Balsan, the gold. Well, that of course was her prerogative, and it would have been ungracious of me to insist that Blenheim Palace was what the public really wanted to read about. 'And of course I can't put in a book what a beast Marlborough was,' she added pensively.[40]

Consuelo persevered and sketched out a book plan of her own which was accepted for publication by Harper & Brothers. Since it first appeared in the US in 1952, critics of *The Glitter and the Gold* have described it 'ghostwritten'[41] but the correspondence between Consuelo and editor Cass Canfield makes it clear that this was not the case. From the outset Consuelo made it clear that she would not accept a 'ghost'. 'I would not like it to become a book written by a professional such as the Duke of Windsor's,' she wrote to him, 'for with all its faults it is a personal story and memoirs to me lose their interest if the writer's way of seeing and saying things is not adhered to.'[42] Cass Canfield agreed and assigned an editor to the book, Marguerite Hoyle, with whom Consuelo formed a warm relationship and who provided outstandingly tactful editorial service. Miss Hoyle's impact can particularly be seen in the later chapters for when Consuelo was trying to finish the book in the autumn of 1951 she fell ill, was admitted to hospital and ran out of energy. Miss Hoyle stepped in to help – one of her contributions was to restructure Consuelo's account of her life in France round each of her houses. But when the book was ready for publication Cass Canfield suggested that it should be described as 'autobiography' rather than 'memoirs' because 'the former clearly indicates that you wrote the book, as you did, instead of a ghost.'[43]

The Glitter and the Gold changed focus as she wrote it. Consuelo had originally envisaged a series of pen portraits of a vanished age, but was encouraged to write a more personal story by her publishers who sensed, correctly, that this is what the public would want to read. She always felt uneasy about this; and it was not long before art began to imitate life as she found herself having to negotiate a path between competing forces while insisting that her story remain her own. Some suggestions were straightforward and helpful. The art critic Stuart Preston pointed out that many people would be interested in her impressions of Edith Wharton. Others were more difficult to accommodate for even comments from her publishers were contradictory. Cass Canfield was taken aback at the contrast between the warmth and charm of the elderly lady in his office and the embittered tone of an early draft of her manuscript and encouraged her to emphasise 'the pleasant and gay times you enjoyed',[44] while Marguerite Hoyle wanted to learn more about her disagreements with the Duke. Consuelo protested at the idea of writing about this at all, but then went much further than she originally intended.

Consuelo wrote to Cass Canfield that she wished the main theme of her memoir to be the 'drive to democracy' which 'as far as I can remember always possessed me together with a restless desire to work – so much more satisfying than the perpetual round of a mundane existence'.[45] This inevitably meant suggesting that the Edwardian aristocratic world was a feudal anachronism; and that meant upsetting her son and daughter-in-law, the 10th Duke and Duchess of Marlborough. The Duchess was shocked by what she felt was the anti-English tenor of *The Glitter and the Gold* and criticism of his father would have been far worse, the Duke told David Green, if he had not 'blue-pencilled freely'.[46] The person to whom the difficult task of reading the manuscript for errors in good taste, spelling and grammar was finally entrusted was Lord Ivor Churchill who also saw the book through further changes required for the English edition. Consuelo herself remained extremely sensitive to criticism that in publishing a memoir that described her life at Blenheim she was somehow being vulgar. One symptom of this was a furore over the issue of the cover design with Consuelo sending the telegram: 'Am distressed and surprised to hear from Adelaide Leonard that jacket has been decided without consulting me stop combination of jacket photograph in coronation dress and name glitter and gold strike me as most unpleasantly vulgar. stop. excuse my frankness stop know that in England effect would be deplorable. stop. refuse to have my photograph on jacket which I would like simple and plain.'[47]

Just before the memoirs were published in the States, Cass Canfield wrote to Winston Churchill at Consuelo's suggestion asking him if he might provide a foreword, or even a favourable endorsement which could be printed on the book jacket. Churchill never responded directly to either suggestion, although he did write to Consuelo that *The Glitter and the Gold* 'was a graceful and readable account of a vanished age' which he was sure would 'command wide attention and interest'. He suggested that the chronology needed more attention, felt that 'the pack could be shuffled with advantage' in places, and remarked that 'chronology is the secret of narrative'. Though he thought that the account of the Balsans' escape from France made a moving end to the story, and that the title (which was suggested by Harper & Brothers) was 'brilliant', he was greatly concerned about references to the Asquiths which he felt would cause their children pain and anger. 'I would not put in the story of your dining with him [Asquith] so soon after Raymond was killed in action, nor do I think you will want to leave in the account of Margot's behaviour when she was your guest at Eze.'[48]

Consuelo immediately removed any mention of dining with Asquith after Raymond's death, but the story of Margot's behaviour when she was newly widowed remained, with a note of explanation about the idiosyncratic nature of grief. Churchill wrote a separate letter to Cass Canfield, returning the proofs where he had marked a few points that had occurred to him. 'I do not, however, wish to be quoted,'[49] he said firmly. The reason for this, though he did not say so, was that Churchill was deeply unhappy about the way Consuelo had 'blackened' Sunny's name – a reaction that cast a shadow over the friendship between Churchill and Consuelo for quite some time.

The American edition of *The Glitter and the Gold* attracted favourable reviews which commended a fascinating story, well told, by an author with a keen eye. It stayed for over twenty weeks on the US bestseller list, though it never rose higher than third place in the face of competition from the actress Tallulah Bankhead's autobiography – which contained the startling revelation that she only lost her virginity when she was thirty-five – and the Bible in the Revised Standard Version. Meanwhile, some changes for the British edition were insisted upon by English lawyers. Consuelo's description of Brown's Hotel as 'frowsty' and miserable compared to the great hotels of Paris was considered libellous. Gladys Deacon was changed to a 'young American girl' in the story of the visit of the Crown Prince to Blenheim and alterations were made to avoid any suggestion that she had pursued the Crown Prince or pressured him into giving her his ring. 'You will be amused to know that all references to Gladys Deacon as regards the Crown Prince story will have to be omitted in England because the lady in question sued a paper before the war as it contained a cartoon in which a vicar was discussing his rose gardens with his gardener and suggested putting the Duchess of Marlborough in the same bed as the Reverend Wilkes,' wrote Consuelo to Marguerite Hoyle. 'As she is now rather mad one must be careful.'[50]

Consuelo's great-granddaughters were always struck by the fact that she never had any photographs of Alva on display, and that the only record of their great-great grandmother was a portrait that hung on one staircase at Casa Alva. Perhaps Consuelo did not need mementos of her mother to remember her, for the only criticism of *The Glitter and the Gold* that truly stung was directed at her treatment of Alva. 'I am being very much criticised for what I have said of my mother which is causing me pain,' she wrote to Miss Hoyle. 'People evidently do not recognise that she was a "personality" and to be judged as such.'[51] Marguerite Hoyle wrote a long and sensitive letter in response:

> I am distressed that you are being hurt by criticism for what you said of your mother. It seemed to me that you showed great understanding of her point of view, and sympathy for her as a person (what the psychiatrists call 'sceptical empathy' and a difficult thing to achieve). Certainly your affection for her is apparent throughout, and she emerges a very vivid and remarkable character. Friends with whom I have talked, and who are reading the story . . . have invariably spoken of the good taste and dignity with which you handle some very difficult personal relationships. I think the important thing is that you yourself know you were objective and that in treating her as a personality in her own right you were paying her a compliment that she herself would have appreciated. Also, I think, you find reassurance in the fact that your own family – Lord Ivor, for instance, of whose sensitive taste and good judgement you can have no doubt – saw no cause for criticism.[52]

This did not reassure. Consuelo combed through what she had written in an attempt to head off such criticisms for the English version. She removed all references to the installation of a bowling alley at Augerville clearly feeling that an affectionate story sounded like callous mockery of the infirmity of old age – Alva had commissioned it when she could no longer walk and the only people who used it were her nurses. To ensure that no-one thought that she was poking fun at her mother's feminism, Consuelo amended her description of Alva's funeral, adding: 'It was only fitting that such tribute should be paid to the courage she had shown in braving popular prejudice and established custom to secure better conditions for women the world over.'[53] However, she made no changes at all to the early part of her memoir in which she recounted her childhood, her romance with Winthrop Rutherfurd and the circumstances leading up to her marriage to the Duke, which, she repeated, was 'forced upon me'.

When the book was published in Britain in 1953, comment was generally favourable though opinion was more divided than it had been in America. One camp of criticism was represented by Katherine Whitehorn, writing in *John O'London's Weekly*: 'It seems harsh to remark that the book would have been better if she had given us rather less discreet portraits of her husband and late Victorian personalities.'[54] Another faction was represented by those close to the Churchills. Lord Birkenhead (son of F. E. Smith) had often spent Christmas at Blenheim as a child after the Marlboroughs separated and while he commented favourably on Consuelo's depiction of 'An extraordinary world . . . which proves its author to be a woman of wit, observation and strong character', he thought that her portrait of her first husband was a serious breach of good taste. The point was put even more forcefully by Randolph Churchill in *Punch*. 'Madame Balsan indulges herself in many criticisms of a man who has been dead for twenty years, and who was the father of her two sons; and it is painful to record that after a lapse of nearly sixty years, and in the sunset of her life, she finds it decent to criticise even his table manners.'[55] Randolph Churchill then went on to query just how 'democratic' life had ever been in Newport, and commented that 'it is scarcely to be wondered at that she found fault with the plumbing at Blenheim, or complained that there were "only six housemaids"' – entirely missing the fact that Consuelo never claimed that life in Newport was democratic, and that low staffing levels at Blenheim were hard on the housemaids, not her.

Indeed, some of Randolph Churchill's sallies demonstrate with startling clarity the complacent hauteur of the English – and particularly the English male – that Consuelo had encountered as an eighteen-year-old and to which she had so frequently objected. Castigating her accounts of 'elegant slumming', Randolph Churchill remarked with lofty disdain that it was a bad American habit to present oneself as an innocent victim of a corrupt Old Europe; asked British readers to bear in mind the (vulgar) American public for which it was written; and finished by damning the memoir with faint praise, saying that it was well-informed and lively 'as one would expect from a beautiful and cultivated American woman who had the advantage of living for nearly half a century in Europe'.[56]

Though *The Glitter and the Gold* has been reprinted since the 1970s, it has fared less well since its initial reception. It is undoubtedly uneven. Its early chapters are animated by descriptions of Alva, life at Blenheim, her bitterness towards the Duke and Consuelo's dislike of the inwardness of the English aristocracy, all conveyed with a strength of feeling that may have surprised even its author. The book's energy runs out towards the end though it flashes back into life with her account of the Balsans' departure from France in June 1940. Even the early chapters go in and out of focus, at some moments remarkably outspoken, and at others frustratingly reticent. It suffers from being passed through too many hands, as if too many competing voices have – at least in places – drawn its sting. Like all memoirs, its view is partial and there are places where friends and supporters – like Paul Maze in 1940 – should probably have been given more credit. Its historical accuracy is sometimes questionable, as Consuelo herself candidly admitted: 'I would not have been likely to forget my conversation with the Czar,' she said to one interviewer. 'And then I found I remembered the *trend* of things, if not the exact thing.'[57] Although Consuelo did have a streak of a vanity that surfaces from time to time throughout the memoir, Daphne Fielding points out that the inclusion of observations like J. M. Barrie's 'I would stand all day in the street to see Consuelo Marlborough get into her carriage' are in the Edwardian tradition of 'dew drops', exquisite compliments collected in the manner of Fabergé eggs,[58] rather than rampant narcissism; and it is unreasonable to expect such a memoir to be frank about love affairs.

The most serious accusation directed at *The Glitter and the Gold*, however, is that in presenting herself as the victim of her mother's wiles, Consuelo simply recycled lies about coercion told at the time of the annulment. Leaving aside the judgement that the allegation of coercion was not based on a lie, this is a most peculiar allegation for it is difficult to understand why she should have taken such a step in the first place. If Consuelo knew that the annulment proceedings had been 'cooked up' she would have insisted on keeping to her original plan for a series of vignettes, or never embarked on a memoir at all. It is more likely that in writing *The Glitter and the Gold* she decided it was time to tell a story that had become part of her persona in her own words, and in less colourful language than the version that appeared courtesy of the press in 1926.

In one important respect, however, Consuelo exacerbated the effect of the annulment proceedings on Alva's reputation – by failing to offer any serious analysis of her reasons for overriding her daughter's feelings in marrying her off to an English duke. Though she mentions more than once that it all seemed quite rational to Alva who was only thinking of her daughter's happiness, Consuelo never explains why Alva took such a step, perhaps because she never quite believed her mother's reasons in the first place. Consuelo spends more time – understandably – outlining why she felt compelled to obey Alva. Readers are therefore left with the strong impression that Alva's behaviour over Consuelo's first marriage was the result of nothing more complex than overweening social ambition. When *The Glitter and the Gold* was published this instantly became the version of the story that reviewers repeated and one that remains in circulation to this day – even the current jacket cover of *The Glitter and the Gold* describes Alva as 'almost certifiably socially ambitious'.[59]

Provided *The Glitter and the Gold* is read as a period piece, 'blue-pencilled' by many, written in the 1950s before the revival of feminism, it remains a rewarding account. Consuelo's observations about late-Victorian and Edwardian life in England are a treasure trove of detail for social historians, by whom she is often quoted. Her depiction of the hauteur and insularity of English aristocracy, and her boredom and frustration with the protocol of what was, in many ways, still an eighteenth-century way of life, is all the better for the fact that she was never persuaded by it. Her account of her powerlessness to prevent her own marriage to a man whom she did not love, remains harrowing. Indeed, the strength of feeling with which she tells this part of her life story is the most convincing argument against those who have suggested it was all made up.

It largely accounts for continued interest in the memoir, and for attempts to tell her story by others, which began almost immediately the book was published. David O. Selznick expressed interest in the film rights seeing it as a star vehicle for his wife Jennifer Jones. 'I gather that you were highly doubtful about this,' wrote Cass Canfield to Consuelo, 'and must say that I agree as to the difficulty of your being able to supervise the making of a motion picture so that you would be assured it was in good taste.'[60] When *The Glitter and the Gold* was finally published in England, one person decided to forgive Consuelo after all. On 8 June 1953, she received a cablegram at Old Fields in Oyster Bay, Long Island. 'Blenheim Sunday night,' it read. 'A wonderful day here. Many memories and thoughts of you. Love Winston.'[61]

In 1953, Consuelo watched a British coronation from afar. Her granddaughter Rosemary took her place as a maid of honour to Elizabeth II, and Winston Churchill attended as Prime Minister. Meanwhile, Ivor had married Elizabeth Cunningham quietly in 1947 and they had a son, Robert, in 1954. Although Ivor had worked hard to make *The Glitter and the Gold* a book of which his mother could be proud, they only met intermittently when she returned to

Europe. He hated flying and always found reasons for refusing to travel to the US to see her. The arrival of his son Robert in 1954 meant a further delay, and then it was too late. She was already saddened by his failure to appear when 1956 was darkened by the worst possible news. Often in fragile health, Ivor now had an inoperable brain tumour.

The autumn of 1956 was a black time. Ivor died in September, aged fifty-seven, and Consuelo was devastated. They had always been very close. After she married Jacques, he had often stayed with them in France. Mother and son developed a very similar taste in painting and painters, particularly the work of André Dunoyer de Segonzac, a friend of Paul Maze and an habitué of St Georges-Motel. Obituaries particularly stressed Lord Ivor Churchill's role as a collector and connoisseur of French painting. 'From his casual manner at exhibitions nobody would have supposed that he was keenly interested in and well informed about what he saw, but it was only necessary to exchange a few words with him in front of a picture to find that he missed nothing and had very definite tastes.'[62] Others testified to his intelligence and wide range of interests which latterly focused on building up a prize-winning Guernsey herd.

While he had been close to Consuelo, Ivor shared his father's drive for perfection. Pale and fragile, he was much more similar in appearance to his father than his elder brother, the 10th Duke, and shared many of the 9th Duke's mannerisms. 'It was a puzzle to his admirers why Ivor Churchill should so obstinately have remained a private individual and have never applied to public affairs his exceptional talents and his special gifts of understanding and concentration,' wrote Harold Nicolson. 'Was it ill health that debarred him from more overt activity? As a boy, he had been delicate and even frail, but in adult age he was in no sense lacking in vitality. Was it indolence? His languor was apparent only: He was capable of intense application.' Harold Nicolson went on to list his erudite interests – building up a collection of first editions while still at school, his horses, his pictures, his rare plants and fruit trees at his garden in Steep in Hampshire. 'How then are we to explain his disinclination for public activity? He was himself worried by his own diffidence and would seek to analyse his self-distrust. Yet his hesitations were due surely to his standard of perfectibility, to his refusal to compete with others in struggles where the best can be but seldom chosen and but seldom attained.'[63]

Consuelo was not able to travel to England to see Ivor before he died because Jacques was also in decline. She therefore had the grim task of merely reading the testimonials to her son, which were later pasted into a scrap book. Jacques had a stroke that came on top of the gradual onset of Alzheimer's. Though he became very quiet, his reckless streak did not diminish. Consuelo's anxiety about this was shared by Louis Hoffmann, who found him collapsed on a golf course at the age of eighty-eight. He died within weeks of Ivor on 4 November. A burial requiem mass was celebrated in New York in the Roman Catholic Chapel of the Helpers of the Holy Soul at 118 East Eighty Sixth Street for close members of the family: his favourite nephew François Balsan; Harold and his wife Gertrude; Lady Sarah Russell; and Count Jean de Lagarde, the French Consul General. Afterwards, his body was flown to Paris where he was buried in the Balsan family tomb in Paris in the Cimetière Montmartre. 'Society is deeply mourning the passing of one of its most colourful figures – the kindly and once-swashbuckling Col. Louis Jacques Balsan. It is to the credit of Col. Balsan that he was not only a World War I hero, and one of France's most daring pilots . . . but he was also a good husband, which is quite a tribute in our age,' read one obituary. 'For there is no question that Consuelo Vanderbilt, who suffered unhappiness at the hands of the Duke of Marlborough, found joy and a happy life with her French husband. And today all her friends feel deeply for her in her bereavement.'[64]

Shortly before Jacques' death, the Balsans had sold Old Fields to the committee of the Pine Hollow Country Club, and bought a smaller house in Southampton, Long Island. Consuelo sold Casa Alva soon after Jacques' death and bought a smaller house in Palm Beach itself, on El Vedado Drive. 'One thing I learnt from Granny,' said one of her great-granddaughters, 'is that you don't sit around moping, you get on and change the things that are bothering you'[65] – an admirable principle, but less easy at seventy-nine. The courage that had been one of Consuelo's most striking characteristics throughout her life emerged once again. Valentine Lawford was much impressed by the decisive way in which she moved onwards in the face of a terrible double blow. 'The manner in which she took these, her deepest losses, like the manner in which she took leave of the two houses which had formed such an intrinsic, living part of her recent life, and set about at once creating new rooms and new gardens, should be enough to shame one out of any incipient, vicarious nostalgia about mere periods or possessions, however irreplaceable,' he wrote. It was, he thought 'further proof that real creativity is often most intimately related to self-discipline, and that a clear eye to the future can be a sign that its owner knows just as clearly that the past's wounds, unsalved by courage, will never even halfway heal.'[66]

As Consuelo created two new homes in Palm Beach and Southampton, the past continued to call. On his very first visit to the United States in 1895, Churchill had met a newly married Sunny and dined with the Vanderbilts in New York. In 1959, Winston Churchill, now retired from politics and very frail, decided he wished to make a visit to America in the interests of Anglo-American friendship, a cause to which he had for a long while been passionately committed. In spite of his doctor's warning that he was mad to take the risk, Churchill flew to the States, in a BOAC Comet, on 4 May 1959. He was given a presidential welcome by Eisenhower, who took three days out of his schedule to entertain one of America's great heroes, organised two stag dinners at the White House with Second World War generals, and flew him by helicopter to his farm at Gettysburg, giving Churchill a chance to inspect the American Civil War battlefield from the air. After the Washington visit, Churchill travelled back to New York, where it was obvious to his old friend Bernard Baruch that he was exhausted. Baruch told reporters that Sir Winston would not be seeing anyone, but a few hours before his departure for England, he confounded everyone by making an exception, and went off to have tea with Consuelo at 1 Sutton Place South. After 1959, Churchill only passed through New York briefly one more time, and they probably both sensed that this was the last time they would meet in a lifelong friendship that had survived many tests.[67]

By 1960, at the age of eighty-two, Consuelo had become a star of Southampton society, her silvery elegance as celebrated as ever. That summer, Southampton's society magazine the *Diplomat* called her 'the acknowledged beauty of the Southampton season',[68] saying that her social appearances were now rare and made any party she attended an occasion. She also contributed an article on entertaining to the *New York Journal*. 'To be a good hostess, one's dominant preoccupation should . . . be the pleasure of one's guests, not the vanity of one's person. The woman who entertains because of her Lowestoft service, her lace tablecloth, her old English silver, her hothouse flowers, her epicurean meal and the vision of herself in her newest gown and jewels is not likely to produce happy reactions,' she wrote. The quality of the food was more important than the table décor, for 'hunger cannot be assuaged by a display of china'. A party of six or eight was the ideal number because conversation could be kept general. Music should not be an adjunct to conversation – this was an insult to the musicians as well as the conversational powers of the guests. Asked by the magazine to provide a perfect luncheon menu for a warm June day she suggested 'Cold *madrilène* soup, *coulibac de saumon* made with salmon, rice, eggs and pastry – a *chaudfroid* of chicken with a salad, and an orange dessert'. One could reduce the likelihood of problems by never accepting invitations one did not wish to return, and 'eliminating bores and bounders'.[69]

In 1961 there was more sadness when Consuelo's daughter-in-law, Mary, Duchess of Marlborough, died of cancer. By now, Consuelo was dogged by ill-health herself and was unable to attend the funeral, though she made suggestions for the music at the funeral service which was later used at her own. Increasingly frail, she was visited frequently by her granddaughter, Lady Sarah Russell, with an ever more watchful eye. Her engagement with life around her continued, nonetheless. She topped up a scholarship for French students at Harvard Business School that she and Jacques had started some years before; she leapt to defend Louis Auchincloss when a rude review of *Portrait in Brownstone* appeared in the *Herald Tribune*; but she was too unwell to travel as far as Newport to look round Marble House in 1963 after Harold Vanderbilt bought it back from the Prince family and handed over Alva's greatest architectural achievement to the Preservation Society of Newport County (though she offered to pay for the restoration of the parquet floor in the place where she had so often felt that her person was dedicated to whatever final disposal Alva had in mind).[70]

In 1962, when she was eighty-five, Consuelo became the subject of one more story – her own. It was told at the behest of a different kind of storyteller – Diana Vreeland, editor-in-chief of *Vogue*, then at the beginning of a legendary association with the magazine. Consuelo's trousseau had provided an early editor of *Vogue* with the scoop of her career in 1895. Now Consuelo gave the career of the latest editor of *Vogue* another helpful boost. Joining the magazine in 1962, Diana Vreeland selected Consuelo as the prototype for one of her first commissions – a series of profiles of the 'lifestyle' of the elegant and wealthy, an approach that seems commonplace now but was pioneering in its time. Vreeland immediately decided to use photographs of Consuelo taken by the photographer Horst Bohrman (always known as 'Horst'), and asked him to take some more. She also asked Horst's partner, Valentine Lawford, to write a profile of Consuelo that was to be ground-breaking in style. 'Horst was told to photograph not only Madame Balsan's paintings, furniture, and objects, but her door handles, her flowers, her plates and knives and forks, and the bathroom details – even the soap. I myself was to write down everything I knew about Consuelo, and to describe in depth the extraordinary refinement of her taste.'[71]

Horst and Valentine Lawford were well placed to do this. They had both known Consuelo for years. A personable and charming British diplomat before he became a writer, Valentine Lawford knew many of Consuelo's younger relatives and friends on both sides of the Atlantic and was invited to stay at Casa Alva in 1948, where he asked permission to take Horst after a trip they made together round Mexico, Guatemala and Cuba. In spite of her prejudice against Germans, Consuelo took to Horst, though she complained his voice was pitched so low that she could not understand what he said. While they were staying she allowed him to take as many photographs as he wanted of Casa Alva and thereafter he photographed Consuelo and her houses many times. This was particularly easy to arrange on Long Island for Horst owned a house in Oyster Bay himself after 1947, and they became neighbours. His photographs included a black-and-white portrait photograph for *Vogue* which accompanied publication of *The Glitter and the Gold* in 1952.

In order to write the *Vogue* profile, Valentine Lawford caught up with Consuelo at Garden Side in Southampton in 1963, the year before she died. This was a story of 'the return of the native, the completion of a circle, the harvest on home ground',[72] he wrote. Even her closest friends had wondered whether she could recreate the atmosphere of her other American houses in this one, aged eighty, but they need not have feared, he said. Though Garden Side had no garden to speak of when Consuelo bought it, there were now vistas, lawns, begonias, roses the size of dahlias and dahlias the size of cabbages. Windows had been enlarged and a long drawing room had been added to the side, lined once again with *boiseries* from a French chateau, and hung with yellow silk curtains. It was somehow typical of her that this was the house that her long-serving domestic staff liked best, he thought, but both houses – Garden Side and the smaller house at Palm Beach 'enclose and encourage the same way of life, even if on a smaller scale, as it were in a lower key, than their predecessors'.[73]

And Consuelo herself? Her patina was not one of old age, nor even of rarity, wrote Lawford. The keynotes of her personality remained self-discipline and youthfulness. Entertaining was on a lighter scale and she was grateful for 'the stair rail on a downward journey and the elevator on her way up',[74] but there were times when she struck those who watched her with her great-grandchildren 'as being at heart as untouched and unprejudiced as the smallest of them',[75] and family photographs suggest that children were always quite at ease with her. There were many grace notes of the younger Consuelo. At eighty-six she had just launched an appeal for the Southampton Hospital Building Fund and he noticed that she hurried to the front door when the mail arrived, feeling each envelope for contributions. Valentine Lawford found her modernity astonishing. 'How hard it is sometimes to remember that the woman who has just been commenting, let us say, on *Lolita*, was twice commanded to "dine and sleep" with Queen Victoria.'[76]

She still had a circle of admirers, especially bachelors who sat at her feet and worshipped her. 'Her incisive opinions on world affairs, and her warmth draw men, young and old, to her side, as was the case in her Edwardian heyday and after the briefest of encounters, even a college boy will concede her lovely appearance,'[77] wrote the *Diplomat* in 1960. One of Consuelo's favourite elderly bachelors (belonging to the group whom she liked to describe as 'safe') was a Southampton neighbour, Edward Crandall, who came to call, played bridge and kept her *au courant* with everything that was happening – New York exhibitions, new writers, new plays, new books. Lawford noticed that she still read three or four books each week. Her great-granddaughter Serena, who lived with her for six months when she was seventeen, also found Granny's engagement with contemporary life remarkable. 'She was very, very modern – she had always read the latest book, seen the latest movie.' Serena felt slightly envious when two young men – Horst and Lawford – called for Granny in an elegant sports car and whisked her off to the cinema in the afternoons. 'If she were alive today, she would have a word-processor. And she would use it. She would have loved e-mail.' She had the appearance of a *grande dame*, but 'if you sat down in front of her and talked to her directly, you could soon find yourself talking about all kinds of things you wouldn't dream of discussing with another adult. She was a huge influence.'[78]

Consuelo died at Garden Side on Sunday 6 December 1964, aged eighty-seven, with her granddaughter Sarah at her side – the 10th Duke was already on his way but arrived too late to see her before she died. Her American funeral was held at St Thomas Episcopal

Church on Fifth Avenue – the same church in which she had married the Duke and said farewell to Alva, though it had been rebuilt after a fire in 1906. There were flowers from Harold who was too unwell to attend, and from Consuelo's only surviving bridesmaid, Daisy Post, now Mrs Louis Brugiere of Newport. As Harold Wilson flew into Washington for talks on nuclear defence with President Lyndon Johnson and Martin Luther King called for a grand alliance of American intellectuals in the fight for civil rights, hundreds of mourners came to the church and sat in pews that had only just ceased to be reserved for Morgans, Goelets, Belmonts and Vanderbilts three years beforehand, in 1961. The list of mourners in *The New York Times* read like a roll call from an American social history book: Jays, Whitneys, Huttons, Fords, Aldriches, Astors, Szechenyi, Szapary. This time there were Vanderbilts a-plenty: Alfred Gwynne Vanderbilt II and his half-brother, William H. Vanderbilt, were ushers and honorary pall-bearers. So was Colonel Serge Obolensky, guest at Blenheim, guest at Lou Sueil and now a neighbour in Southampton. There was a sixty-voice choir which sang hymns of Consuelo's choosing. As was customary in American Episcopalian services, there was no eulogy.

As far as *The New York Times* was concerned, however, there was one other guest at the funeral – the story of her first marriage, retold for the benefit of its readers on Thursday 10 December.[79] But it was Consuelo's own decision to let the story follow her to her grave. In her will she stated that she wished to be interred at Bladon 'next to my son Ivor', amazing her English family who believed that they been left in no doubt as to her feelings about Edwardian England after she published *The Glitter and the Gold*. And so, in the last journey in a long international life, her coffin was taken from St Thomas Church and flown to England, where there was more astonishment when English customs officials insisted that her coffin should be opened and searched for drugs (or so it is said).

In its simple way, her interment in the churchyard of St Martin at Bladon was as stylish as everything else about her. Only a few people gathered in the chill, foggy morning to walk behind the coffin, 'trailing white chrysanthemum petals that fell from it, as it was carried into the churchyard'. The 10th Duke of Marlborough followed behind with his children, the Marquess of Blandford, Lady Sarah Russell, Lady Caroline Waterhouse, and Lady Rosemary Muir. Clementine Churchill was accompanied by Randolph, Sarah and Mary, though Winston Churchill, who had just celebrated his ninetieth birthday, was unable to attend for he was no longer mobile. Two of Jacques' nephews came from France. The flowers came from Blenheim, the wreaths made up by Tompkins, the head gardener. Estate workers and villagers who remembered her stood at a respectful distance. Some of them had been school children when she first arrived at Blenheim in April 1896 and remembered lining up on the platform at Woodstock station to wave and cheer. 'She was very kind and amicable,' said one old man watching from the lichgate.[80]

Because of Ivor, the decision to be buried in Bladon appeared simple, but it was also, perhaps, the result of Consuelo's final discussion with herself about who she really was and what mattered most. Most of the alternatives for burial defined her too narrowly, or resulted in definition of her identity by others. She was not a French Roman Catholic. Her embrace of Roman Catholicism had been a matter of form and though under other circumstances she might have wished to be buried with Jacques, the Balsan family tomb in the Cimetière Montmartre in Paris held no appeal. She might have started life as a Vanderbilt, but she had spun a much wider web and there was little attraction in the idea of burial with her Vanderbilt cousins in the Vanderbilt Mausoleum on Staten Island. And there was no question, after fifty-six years of negotiating independence, of settling down beside Alva in the chapel she had designed as a monument to herself and Oliver in Woodlawn Cemetery.

'My years in England, from eighteen to forty-one, were my longest period in one country. Those are the maturing years, and I still feel very English in many things,'[81] Consuelo once said to an interviewer. At Bladon, she was a safe distance from the 9th Duke himself, interred at Blenheim in the family chapel she always thought vainglorious. At Bladon, she was buried just across from

Ivor, close to him in death even if she had been parted from him in the final years of his life; and she would be near others she had also loved, close to Winston Churchill and later to Clementine. It was also an act of kindness for the decision to be buried in Oxfordshire can only have come as balm to those, especially the 10th Duke, who had been beneficiaries of her marriage into the Marlboroughs, but who hated to think of her unhappiness and who had been so sharply reminded by *The Glitter and the Gold*.

The wording on her tombstone told a story of its own. Alva, who had so often felt forced into the role of spectator in the theatre of life, designed a hugely public theatrical farewell and dreamt of a public monument. Consuelo, who had been pitched into a public life by her mother and been a great success, sought the opposite kind of *coup de théâtre* in which it was not the public life that was celebrated but the private. In the end, Sunny had a small victory. She chose to be commemorated as 'a link in a chain' in an English country churchyard. It was not the absent Duke she wished to reassure, however, but surviving members of her family, a signal to her own flesh and blood that the unhappiest but most creative period of her life had been worth it after all: 'In loving memory of Consuelo Vanderbilt Balsan' read the tombstone, 'Mother of the tenth Duke of Marlborough – born 2nd March 1877 – died 6th December 1964.'

Afterword

IN 1973, AFTER SHE LEFT *VOGUE*, Diana Vreeland took on a new role as special adviser to the Costume Institute at the Metropolitan Museum of Art in New York. In 1975 she organised an exhibition called 'American Women of Style'. She defined the ingredients of 'style' in the catalogue in her usual Vreelandese: 'The energy of imagination, deliberation, and invention, which fall into a natural rhythm totally one's own, maintained by innate discipline and a keen sense of pleasure. All who have it share one thing – originality. This is not a dress show.'[1]

For an exhibition that was not a dress show it included a strikingly large number of frocks. They were associated with ten women: Consuelo, her cousin Gertrude Vanderbilt Whitney, who had so objected to the way heiresses were watched in 1895, Mrs Charles Dana Gibson, Elsie de Wolf (Lady Mendl), Mrs John W. Garrett, Isadora Duncan, Rita de Acosta Lydig, Irene Castle, Millicent Rogers and Josephine Baker. Several of these women knew each other well. Some were connected by family ties, and nearly all of them were linked in one way or another by friendship and acquaintance.

Consuelo might not have cared for the way she was displayed by Diana Vreeland in this exhibition. It was, at best, an elusive suggestion of her personality and focused exclusively on one part of her life. Three shop window mannequins were used to display dresses merely 'attributed' to her. These were set against the background of the Boldini portrait that Consuelo gave to the Metropolitan Museum in 1946, and an enlarged photograph of Consuelo as the 9th Duchess of Marlborough in her coronation robes, an image about which she was always acutely sensitive. Insofar as the costumes belonged to Consuelo at all, they were preserved from her time as an English duchess, during the years she lived at Blenheim: an early-twentieth century evening dress of black net and lace over pale blue chiffon; a walking dress of navy wool with ivory lace from 1901; and an afternoon dress of pin-tucked ice blue silk, with bands of turquoise velvet accented with mink, from 1900. One of Diana Vreeland's favourite aphorisms was: 'Never worry about the facts, just project an image to the public.'[2] The 'American Woman of Style' presented here was in fact an English duchess, with images from a time in Consuelo's life when she often felt that she was, indeed, little more than a shop window mannequin.

'What sells is hope,'[3] Vreeland once said, echoing the sentiments of *Mumsey's Magazine*'s article about the Vanderbilts in 1901. She and Alva had much in common. They both embraced the power of imagery and a stylish public persona. Like Alva, Vreeland was drawn to a dream world of aristocracy that had roots in the French eighteenth century, imperial China, the *belle époque* and the style of the English country gentleman, themes of four of her costume exhibitions during the 1980s. Diana Vreeland borrowed images of Consuelo for her exhibition of the *belle époque* at the Costume Institute too, one of several exhibitions heavily criticised for exalting unbridled wealth and ostentation without reference to the social conditions that sustained it, and for appearing to celebrate – at least by association – the 'Dynasty' culture of the Reagan era, that bore so many similarities to the cruelties of the Gilded Age.[4]

It could be said, therefore, that many years after her death, Consuelo-the-duchess was co-opted once again by a strong-minded female visionary. It was even possible to stare at her by making one's way to Fifth Avenue, as in 1895, though on this occasion she could only be examined through glass in the Metropolitan Museum. Even here, however, she was placed at the service of a world view with which she would profoundly have disagreed. 'What do I think about the way most people dress? Most people are not something one thinks about,'[5] was another Vreeland bon mot. In these exhibitions, images of Consuelo were used to sanctify a morality-free world of ostentation, greed and stylishness that had turned its back on the problems of 'most people' outside. It was the 'glitter' of Consuelo's life with the Duke that was being celebrated here, rather than the 'gold' of her life with Jacques. There was no mention of her work as a serious philanthropist. She might have smiled a wry smile at the thought that, yet again, this was happening without her consent.

Or would she? A closer look at the catalogue for 'American Women of Style' written by the exhibition's curator, Stella Blum, suggests that Consuelo might not have minded as much as all that. This was, after all, an exhibition in the Costume Institute, so clothes obviously counted for something. Otherwise, the criteria for inclusion were remarkably wide-ranging. The women did not have to be good-looking. Only three of them – including Consuelo – could be described as beautiful. They certainly did not have to be rich, or professionally accomplished, though at least five of them were. They did not even have to be fashionable. According to Stella Blum, the common denominators that united these women were something quite different. They had to have 'a strong creative drive that looked for a perfect expression'; they had to be 'daring in a positive way'; they all 'insisted on living usefully'. Most were 'involved in social movements, humane causes and even politics, some actually before women were doing such things'. All had an 'iron-willed self-discipline' and 'left an imprint on everything they did'. Set against these criteria, their clothes were a 'manifestation of their individuality'.[6] Leaving aside the objection that the coronation robes of an English duchess were not a manifestation of individuality and never would be, Consuelo was an outstanding example of such stylishness.

In spite of all the privileges conferred by great riches, starting out life as a Gilded Age heiress brought disadvantages. Wealth was a 'great gilt cage' of its own, a solipsistic life led within the gilded aviary of Gilded-Age society. Many of the difficulties in Consuelo's life stemmed from her strong desire to step outside the cage, without wishing to leave completely. Alva ultimately rejected the caged life herself, though she tried ruling it first before she reached for an escape route. Long before Alva's life was dominated by the suffrage campaign, however, she arranged a way out for Consuelo – or thought she did. Eleven years of 'an endlessly spread red carpet' with the 9th Duke of Marlborough opened up Consuelo's world, but failed to solve the problem that she wanted more. Yet one of Consuelo's most striking characteristics was her ability to break through barriers, in spite of the deafness that sometimes threatened to cut her off entirely. Consuelo, said Viscount Churchill, was the only person who 'passed completely through the barrier separating child from grown-up'.[7] Many others noticed this too, from those who spoke up for her as a Progressive in 1917, to the old lady in the Oxfordshire village of Long Hanborough who still had dreams about her years later.

Stepping out of the great gilt cage was never as easy as it looked. It would have been easy for Consuelo to follow the example of Alva in her society phase and withdraw into a world of opulent fantasy, but she never did. Throughout her life she demonstrated an extremely practical streak of kindness which was still evident in her will. It was clear that she thought long and hard about how best to leave her estate to her grandchildren, according to their various circumstances. This posthumous thoughtfulness was extended to her friends as well: one admirer remembered in her will – Edward Crandall – was not well-off and Consuelo's instructions to her executors with regard to Edward could not have been more down-to-earth. There were two Pissarros and a Segonzac in her house at El Vedado Drive in Palm Beach. But rather than give these to Mr Crandall as a sentimental keepsake which he would never have sold himself, she told her executors to sell the paintings and give him the cash which she clearly realised he needed.[8]

Those who knew her well were always struck by her modernity. In a recent exhibition of Horst's photographs at the National Portrait Gallery in London, images of Consuelo from the 1950s did not seem out of place among photographs of Jacqueline Kennedy, Steve McQueen or even the model Verushka. Though Consuelo was much older than almost every other subject in the exhibition, she looked just as glamorous and thoroughly at home. There was no sense of withdrawal here. Refusal to become detached required great self-discipline and self-reliance, however, and in Consuelo's case it also meant resisting self-pity. Her creative vision involved retaining what was best about her life and discarding the detritus of the rest without sentimentality.

Alva and Consuelo reached for different solutions to the problems caused for talented and energetic women by the sharp division between public and private life in the late-nineteenth and early-twentieth centuries. Consuelo's life was much happier because, unlike Alva, she never drove away affection and love. Alva achieved more for the history books and contemporary women have several reasons to thank her. In one important respect, however, Consuelo's was the more modern persona, as Diana Vreeland instinctively realised when she made her the prototype for a new way of writing about the truly stylish for *Vogue*. In spite of the many voices throughout her life who told her what to be, she negotiated her way through and found an authentic – and extremely stylish – way of telling her own story. This was not the story told in *The Glitter and the Gold*. It was the story of the way she lived, the causes she pursued and the ambience she created – the manner in which she created her own reality. To this extent, Diana Vreeland was right. Consuelo, was indeed an American Woman of Style; and an American aristocrat.

ACKNOWLEDGEMENTS

I am deeply indebted to many people for their help and encouragement as I wrote this book. It would not have been written at all without Candia McWilliam's kindly insistence that I contact Clare Alexander at Gillon Aitken to discuss the idea; nor without Clare's consummate professional guidance and support thereafter. Hugo Vickers was generosity itself from the earliest stages. Faced with someone who was patently a novice, he provided invaluable and consistently good advice on how to set about telling the story of another person's life. I have drawn heavily on his excellent biography of Gladys Deacon and am most grateful to him for allowing me access to her papers. I would particularly like to thank Serena Balfour for her support from the outset, for talking about her great-grandmother Consuelo on various occasions and for lending me Consuelo's scrapbook and photograph album; the Duke of Marlborough, Lady Rosemary Muir, Lady Soames and other members of the Spencer-Churchill family who kindly agreed to be interviewed at length about their memories; Jacqueline Williams (another great-granddaughter) who allowed generous access to L.H. Prost's catalogue of the Balsan collection; and William Lee, who was exceptionally magnanimous in making available references from the diaries of Grand Duke Dmitry at breathtakingly short notice.

Others who knew Consuelo provided many valuable insights. Stuart Preston was a fund of information; Mademoiselle Blouin, governess to the daughters of Lady Sarah Russell provided a unique perspective; Aimée Balsan kindly introduced me to members of the Balsan family who remembered Consuelo and Jacques before 1940; and Louis Auchincloss gave me the benefit of his astute judgement over a memorable lunch at the Colony Club in New York. Owners and occupiers of houses belonging to Consuelo and Alva have been both welcoming and knowledgeable. I owe special thanks to the owners of Crowhurst, William E. Benjamin of Casa Alva, Catherine Hamilton of St Georges-Motel, His Excellency Dilip Lahiri, Mark Del Priore of the Pine Hollow Country Club and the staff of the Chateau Golf at Augerville-la-Rivière. Marble House is run by the Preservation Society of Newport County. I am most grateful to its curator, Paul Miller, and to Andrea Carneiro for all their help, but above all to associate curator Charles J. Burns, who arranged access to the Society's archives and several visits to Marble House over a period of three years, and whose capacious knowledge of the Vanderbilts and interest in the project has enhanced each visit.

Many other people have been magnanimous with their time and knowledge. My special thanks go to Michael Harvey for introducing me to the Balsan family; Jane Lady Abdy for advice about the Souls and the work of Helleu; Eric Homberger for his expert knowledge of New York society in the 1870s; David Gilmour for advice about Curzon; Alastair Gray for bringing the presence of Keir Hardie aboard the *Campania* to my attention; Eleni Bide for her work on Edwardian jewellery; H.R. Kedward, John Forster (education officer at Blenheim Palace), Jack Renton, Professor Kathleen Burk, Harle Tinney at Belcourt Castle, Newport, Rhode Island, Charlotte Alston, Narayan Naik, Katie Mackenzie Stuart, Virginia Murray of John Murray Publishers, Robert de Balkany, Daisy Hay and James Dunkerley for their help on specific questions; Charlotte Ward-Perkins for her help with selecting illustrations; Candia McWilliam, Martine Stewart, Dr Elisabeth Kehoe, and Erik Tarloff for reading the manuscript with such care and attention and making many valuable suggestions; and Betsy Newell and Peter and Virginia Carry for hospitality in New York over a period of four years. I would also like to express my gratitude to Dr Peter Geidel. Writing about Alva's life would have been much more difficult without his work but my efforts to track him down and thank him in person have so far proved fruitless.

While writing this book I consistently underestimated how much research there was to do, usually after I'd packed up and arrived home again. I am extremely grateful to all those who came to my aid: Jessie Carry Saunders and Elizabeth Sodel in New York; Stephane Porion in France; Daisy Hay in Oxford; Matthew Brown and Andrea

Cox during the final stages in London; and Flora Joll and Anthony Cummins who scrutinised the proofs. I would particularly like to thank Dr Chloe Campbell for her invaluable help, which drew on her expertise as a historian of Edwardian England and involved many trips to the British Library Newspaper Library at Colindale on my behalf; Bryn Harris for assistance with Latin translation; Aoife Ní Luanaigh for her help with translation from French, particularly the short stories of Paul Morand; and Gareth Prosser.

Staff of libraries and archives in both the U.S. and the U.K. have been unfailingly helpful. I would like to express my gratitude to Jennifer Spencer, collections manager at Sewall-Belmont House in Washington, for much assistance including access to Alva's books of suffrage clippings; Florence Ogg, director of archives and collections at the Suffolk County Vanderbilt Museum, Centerport, New York; Bertram Lippincott at Newport Historical Society, Newport, Rhode Island; John Sledge, architectural historian at Mobile Historic Development Commission for his help and advice about Alva's early years in Mobile, Alabama; Mari Nakahara for assistance with the Richard Morris Hunt Archives at the Octagon Museum in Washington; Cate Lynch at the Surrogate's Court of Suffolk County, Riverhead, New York; Peter J. Blodgett, H. Russell Smith Foundation Curator of Western Historical Manuscripts at the Huntington Library, San Marino, California; Elizabeth J. Dunn at the Rare Book, Manuscript and Collections Library at Duke University, North Carolina; John Pinfold and his staff at the Vere Harmsworth Library in the Rothermere American Institute, Oxford; Mike Bott at Reading University Library; Dr Kate Fielden, Curator, Bowood House; the staff of the Bodleian Library, Oxford; the staff of New York Public Library; the staff of the Manuscripts Division of the Library of Congress; the staff of the New-York Historical Society's reading room; the Historical Society of Palm Beach County; Alan Packwood and his staff at the Churchill Archives; and the staff of the British Library and the London Library.

I am deeply grateful to Michael Fishwick at HarperCollins for his outstanding editorial guidance, and to Terry Karten for all her help from New York. It has been a pleasure to work with Kate Johnson, Kate Hyde, Annabel Wright and Cathie Arrington during the production of this book. The support of my family throughout has been unflagging and magnificent. Daisy helped with research, read the book in draft and was indisputably my sternest critic. Marianna accompanied me cheerfully on several trips to New York and Newport where she made enthusiastic attempts to emulate a Vanderbilt lifestyle. As always, Janet Fenton held things together and saved everyone from neglect. My husband Michael Hay combined a demanding job with executive-producing trips to America and France on my behalf and driving hundreds of miles on arrival. His evenings and weekends were interrupted for months by endless demands that he critique yet another draft. He could not have given me more help and sage advice with more good humour and I cannot thank him enough.

I would also like to thank the following for allowing me to quote material:

George Mann Books for permission to quote from *The Glitter and the Gold*; the Huntington Library, San Marino, California, for permission to quote from the papers of Charles Erskine Scott Wood; the Rare Book, Manuscript and Special Collections Library at Duke University for permission to quote from the papers of Matilda Young; the Schlesinger Library, Harvard University for permission to quote from the papers of Doris Stevens; *Country Life* for permission to quote from 'Blenheim Fifty Years Ago: Memoirs of Gentleman's Service'; Hugo Vickers for permission to quote from the papers of Gladys Deacon; the Regional Oral History Office, the Bancroft Library, Berkeley, California for permission to quote from oral histories of Alice Paul and Sara Bard Field; Curtis Brown for permission to quote from the letters and works of Sir Winston Churchill, Clementine Churchill and Lady Soames; New-York Historical Society for permission to quote from the Vanderbilt family papers.

NOTES

The following abbreviations are used in the notes.

SOURCES

CA	Churchill Archives
CAP	Church Army Papers
CESWP	Charles Erskine Scott Wood Papers
CVBS	Consuelo Vanderbilt Balsan Scrapbook
DSUP	Doris Stevens's Unprocessed Papers
GDP	Gladys Deacon Papers
GVWP	Gertrude Vanderbilt Whitney Papers
HRP	Harper and Row Papers
LCW	Library of Congress, Washington
NWPP	National Woman's Party Papers
VA	Vanderbilt Archives
VP	Vanderbilt Papers

PERSONAL ABBREVIATIONS

9th Duke	9th Duke of Marlborough
C	Consuelo
CESW	Charles Erskine Scott Wood
CSC	Clementine Spencer-Churchill
GD	Gladys Deacon
SBF	Sara Bard Field
WSC	Winston Spencer-Churchill

PROLOGUE

1 Quoted in B. H. Friedman, *Gertrude Vanderbilt Whitney* (Garden City, New York: Doubleday & Company, Inc., 1978), pp. 96–7.
2 *New York World*, 29 September 1895. Other characteristics in the list were: Face – somewhat oval, Complexion – Clearest olive, with rosy cheeks, Mouth – Small and without character, Teeth – White, regular and well kept, Lips – full and describing a Cupid's bow, Accomplishments – Music, painting, languages, Chief accomplishment – none, Ears – small and close to the head. Head – well-rounded and well poised, Special Fad – None, Favorite Color – Pink, Favorite Sport – Tennis, Favorite Exercise – Bicycling, Favorite Flower – American Beauty rose.
3 'No-one knows what it all costs – or cares,' added the report. 'A Fortune in Flowers', *Town Topics*, 31 October 1895, p. 21. Also see *The New York Times*, 7 November 1895; 'Saunterings,' *Town Topics*, 31 October 1895.
4 *The New York Times*, 7 November 1895.
5 *World*, 7 November 1895.
6 *World*, 7 November 1895.
7 *World*, 29 September 1895.
8 *Munsey's Magazine*, Vol XXII, No. 4, January 1900, pp. 467–8.
9 *The New York Times*, 7 November 1895.
10 Interview with an anonymous guest, *The New York Times*, 7 November 1895.
11 Interview with an anonymous guest, *The New York Times*, 7 November 1895.

1 THE FAMILY OF THE BRIDE

1 Letter to Alva Belmont from genealogist Lily van der Schalk, 4 May 1931, CVBS.
2 W. J. Lane, *Commodore Vanderbilt: Epic of the Steam Age* (New York: Alfred A. Knopf, 1942), pp. 5–8. Also see J. E. Patterson, *The Vanderbilts* (New York: Harry N. Abrams, Inc., 1989), pp. 12–15. Both draw on the work of the Commodore's first biographer, William A. Croffut.
3 See D. A. Silver, *The Entrepreneurial Life* (New York: John Wiley and Sons, 1986), pp. 35–7.
4 Quoted in Patterson, *Vanderbilts*, p. 59.
5 Quoted in W. Andrews, *The Vanderbilt Legend: The Story of the Vanderbilt Family, 1794–1940* (New York: Harcourt, Brace and Company, 1941), p. 19.
6 Quoted in Andrews, *Vanderbilt Legend*, p. 19.
7 Only one New York railroad, the Erie, ever eluded the Commodore in a much publicised battle with his arch rival, Jay Gould. During the 1870s, the Vanderbilt railroad empire continued to expand westwards as the business acquired the Michigan Central Railroad and the Lake Shore and Michigan Southern Railroad. After these acquisitions Vanderbilt trains ran from New York as far as Chicago, Cincinnati and St Louis.
8 M. Josephson, *The Robber Barons: The Great American Capitalists, 1861–1901* (New York: Harcourt, Brace and Company, 1934), p. 72.
9 K. C. Schlichting, *Grand Central Terminal: Railroads, Engineering and Architecture in New York City* (Baltimore: John Hopkins University Press, 2001), p. 40.

10 M. Twain, 'Open Letter to Commodore Vanderbilt', *Packard's Monthly*, March 1869.

11 In the memoirs dictated to Mary Young after 1928, Alva described the Commodore as 'one of the handsomest, most intelligent, and most interesting men I had ever met', though she conceded that his manner was overbearing: *Belmont Memoirs* (*Young*), Matilda Young Papers, Rare Book, Manuscript, and Special Collections Library, Duke University, North Carolina, p. 89. This second set of memoirs was dictated to Alva's secretary, Mary Young, some time after she took up the post in 1928. They then passed into the care of her sister, Matilda Young. Since Mary Young was not a writer like Sara Bard Field, it is assumed that the greater part of this memoir was dictated by Alva.

12 Cornelius Vanderbilt to Frank Armstrong Crawford, 24 October 1868, Vanderbilt family folder, VP, New York Historical Society.

13 F. Crowinshield, 'The House of Vanderbilt', *Vogue*, 15 November 1941, p. 38.

14 A. T. Vanderbilt, *Fortune's Children: The Fall of the House of Vanderbilt* (New York: William Morrow and Company, 1989), p. 26.

15 Quoted in Vanderbilt, *Fortune's Children*, p. 15.

16 Vanderbilt, *Fortune's Children*, p. 40.

17 Quoted in Vanderbilt, *Fortune's Children*, p. 55.

18 F. L. Ford, 'The Vanderbilts and The Vanderbilt Millions', *Munsey's Magazine*, Vol XXII (No. 4, January 1900).

19 J. Foreman and R. P. Stimson, *The Vanderbilts and The Gilded Age: Architectural Aspirations 1879–1901* (New York: St Martin's Press, 1991), p. 28.

20 C. V. Balsan, *The Glitter and the Gold* (Maidstone: George Mann, 1973; first published 1953), p. 4.

21 Quoted in *The New York Times*, 'The Vanderbilt Will Case', 6 February 1879.

22 Quoted in *The New York Times*, 'Vanderbilt Will Case'.

23 Quoted in Vanderbilt, *Fortune's Children*. p. 87.

24 P. Geidel, PhD thesis: 'Alva E. Belmont: A Forgotten Feminist' (Columbia University, 1993), p. vi. He points out that even scholars have difficulty spelling Alva's name correctly.

25 Balsan, *Glitter*, p. 4: 'Her father, who owned plantations near Mobile, was ruined by the liberation of the slaves.'

26 *Belmont Memoirs* (*Field*): CESWP, Huntington Library, San Marino, California, p. 1. This first set of memoirs was ghostwritten by Sara Bard Field after conversations with Alva in Newport, August 1917. Sara Bard Field's papers include notes of conversations with Alva and a first draft. The project was never completed.

27 SBF to CESW, 29 July 1917, CESWP.

28 E. S. Martin, *The Unrest of Women* (New York and London: D. A. Appleton and Company, 1913), pp. 52–3.

29 Balsan, *Glitter*, p. 5.

30 *Belmont Memoirs* (*Young*), p. 38a.

31 *Belmont Memoirs* (*Young*), p. 9.

32 H. Fuller, *Belle Brittan on a Tour, at Newport, and Here and There* (New York: Derby & Jackson, 1858), p. 112.

33 T. C. De Leon, *Belles, Beaux and Brains of the 60s* (London: T. Fisher Unwin, 1909), p. 182.

34 Geidel, 'Forgotten Feminist', p. 48.

35 *Belmont Memoirs* (*Field*), p. 10.

36 *Belmont Memoirs* (*Young*), pp. 9 and 11.

37 *Belmont Memoirs* (*Field*), p. 9.
38 *Belmont Memoirs* (*Young*), p. 30.
39 Though this 'letter' may have been the basis for a newspaper or magazine article. See National Woman's Party Papers (NWPP): 1913–72, Series 1, Correspondence, 1913–72, Alva Belmont Correspondence Scrapbook, Glen Rock, New Jersey, Microfilming Corporation of America, p. 16.
40 *Belmont Memoirs* (*Field*), p. 25.
41 *Belmont Memoirs* (*Field*), p. 7.
42 *Belmont Memoirs* (*Field*), p. 10.
43 *Belmont Memoirs* (*Field*), p. 11.
44 *Belmont Memoirs* (*Field*), pp. 7 and 9.
45 Geidel, 'Forgotten Feminist', p. 51.
46 Geidel, 'Forgotten Feminist', pp. 12–13.
47 Eric Homberger, *Mrs Astor's New York: Money and Social Power in a Gilded Age* (New Haven and London: Yale University Press, 2002), p. 10.
48 Quoted in Homberger, *Mrs Astor's New York*, p. 10.
49 R. Townsend, *Mother of Clubs, Being the History of the First Hundred Years of the Union Club 1836–1936* (New York: privately printed, 1936), p. 70.
50 'Every little southerner I met at dancing classes was a "wicked rebel", to be pinched if possible', Mrs George Cornwallis-West, *The Reminiscences of Lady Randolph Churchill* (Bath: Cedric Chivers Ltd, 1973; first published London: Edward Arnold, 1908), p. 2.
51 *Belmont Memoirs* (*Young*), p. 47.
52 *Belmont Memoirs* (*Field*), p. 15.
53 A. Horne, *The Fall of Paris* (London: Pan Macmillan Ltd, 2002; first published Macmillan, 1965), p. 16.
54 Horne, *Fall of Paris*, p. 16.
55 Quoted in Horne, *Fall of Paris*, p. 7.
56 *Belmont Memoirs* (*Young*), p. 72.
57 I am indebted to Peter Geidel for his work on tracing the Smiths' movements between New York and Paris and their history of house ownership in New York between 1865 and 1874.
58 *Belmont Memoirs* (*Field*), p. 21.
59 M. Twain and C. D. Warner, *The Gilded Age* (New York and Oxford: Oxford University Press, 1996; first published Hartford: American Publishing Company, 1873), p. v.
60 Homberger, *Mrs Astor's New York*, p. 8.
61 Homberger, *Mrs Astor's New York*, p. 181.
62 W. McAllister, *Society As I Have Found It*, in series: *The Leisure Class of America* (New York: Arno Press, 1975; first published New York: Cassell Publishing Co., 1890) pp. 349–50.
63 Quoted in Homberger, *Mrs Astor's New York*, p. 181.
64 J. E. Patterson, *The First Four Hundred: Mrs Astor's New York in The Gilded Age* (New York: Rizzoli Publications, 2000) p. 59.
65 L. Auchincloss, *The Vanderbilt Era: Profiles of A Gilded Age* (New York: Collier Books, Macmillan Publishing Company, 1989), p. 189.
66 Quoted in Homberger, *Mrs Astor's New York*, p. 193.
67 *Belmont Memoirs* (*Field*), p. 32.
68 C. Tomalin, *Samuel Pepys: The Unequalled Self* (London: Viking Penguin, 2002), p. 9.
69 *Belmont Memoirs* (*Young*), pp. 75–6.
70 Handwritten notes on early recollections, *Belmont Memoirs* (*Field*), p. 3.
71 *Belmont Memoirs* (*Young*), p. 79.
72 Quoted in E. Eliot, *Heiresses and Coronets: The Story of Lovely Ladies and Noble Men* (New York: McDowell, Obolensky 1959) p. 72.

73 Homberger, *Mrs Astor's New York*, p. 199.
74 I am most grateful to Eric Homberger for his advice on this circle.
75 SBF to CESW from Newport, 31 July 1917, CESWP.
76 Quoted in D. Silverman, *Selling Culture: Bloomingdale's, Diana Vreeland and The New Aristocracy of Taste in Reagan's America* (New York: Pantheon Books, 1986), p. 14.
77 Gustav Lening quoted in Homberger, *Mrs Astor's New York*, p. 29.
78 *Belmont Memoirs (Field)*, p. 7.
79 E. Wharton, *The House of Mirth* (London: FilmFour Books with Macmillan & Co., 2000), pp. 46 and 38.
80 *Town Topics*, 12 June 1913.

2 BIRTH OF AN HEIRESS

1 *Belmont Memoirs (Field)*, pp. 34–5.
2 SBF to CESW, 29 July 1917, CESWP.
3 *The New York Times* quoted in Geidel, 'Forgotten Feminist', p. 16.
4 Quoted in Geidel, 'Forgotten Feminist', p. 16.
5 Quoted in Homberger, *Mrs Astor's New York*, p. 11.
6 Homberger, *Mrs Astor's New York*, p. 12.
7 *Belmont Memoirs (Field)*, p. 35.
8 *Belmont Memoirs (Young)*, pp. 89–90.
9 A. Churchill, *The Upper Crust: An Informal History of New York's Highest Society* (Englewood Cliffs, New Jersey: Prentice Hall Inc., 1970), p. 120. This Bouncer's ball took place on 27 December 1874.
10 Quoted in Eliot, *Heiresses*, p. 81.
11 *World*, 19 August 1892.
12 Entry for 28 June 1876, Frank Armstrong Crawford Vanderbilt (Mrs Cornelius Vanderbilt), *Diaries 1876–1878*, BV Vanderbilt, VP.
13 Quoted in Patterson, *Vanderbilts*, p. 58.
14 *The New York Times*, 'Vanderbilt Will Case', 6 February 1879.
15 Homberger, *Mrs Astor's New York*, p. 6.
16 Foreman and Stimson, *Vanderbilts and Gilded Age*, p. 175.
17 Foreman and Stimson, *Vanderbilts and Gilded Age*, p. 173.
18 Foreman and Stimson, *Vanderbilts and Gilded Age*, p. 177.
19 *Belmont Memoirs (Young)*, p. 92.
20 *Belmont Memoirs (Field)*, p. 38.
21 *Belmont Memoirs (Field)*, p. 37.
22 D. Lowe, *Beaux Arts New York* (New York: Whitney Library of Design, Watson-Gupthill Publications, 1998), p. 13.
23 Foreman and Stimson, *Vanderbilts and Gilded Age*, p. 35.
24 Quoted in Lowe, *Beaux Arts*, p. 17.
25 *Belmont Memoirs (Young)*, p. 111.
26 Quoted in Geidel, 'Forgotten Feminist', p. 19.
27 See 'Supper Room, W. K. Vanderbilt', No. 81.7267, Prints and Drawings Collection, The Octagon Museum, The American Architectural Foundation, Washington DC.
28 *Belmont Memoirs (Field)*, p. 39.
29 *Belmont Memoirs (Field)*, p. 40.
30 Homberger, *Mrs Astor's New York*, p. 271.
31 'We wanted the money power,' Ward McAllister wrote, 'but not in any way to be controlled by it,' McAllister, *Society*, p. 214.
32 Quoted in Homberger, *Mrs Astor's New York*, p. 272.
33 McAllister, *Society*, p. 351.
34 *Belmont Memoirs (Young)*, p. 109.
35 McAllister, *Society*, pp. 352–3.
36 *Belmont Memoirs (Young)*, p. 108.
37 Patterson, *Vanderbilts*, p. 130.
38 McAllister, *Society*, pp. 353–4.
39 *World*, 28 March 1883.
40 Balsan, *Glitter*, p. 8.

41 Balsan, *Glitter*, p. 8.
42 *Belmont Memoirs* (*Field*), p. 49.
43 *Belmont Memoirs* (*Field*), p. 49.
44 *New York City Journal*, 21 October 1909.
45 *Belmont Memoirs* (*Field*), p. 49.
46 *Belmont Memoirs* (*Field*), p. 50.
47 *Belmont Memoirs* (*Field*), p. 50.
48 *Belmont Memoirs* (*Field*), p. 11.
49 *Belmont Memoirs* (*Field*), p. 20.
50 Auchincloss, *Vanderbilt Era*, p. 83.
51 'Early Training of Children' in Anon., *The Fireside Miscellany and Young People's Encyclopedia* (New York: 1854; see www.merrycoz.org/training.htm), p. 60.
52 Balsan, *Glitter*, p. 6.
53 Balsan, *Glitter*, p. 7.
54 Balsan, *Glitter*, p. 5.
55 Balsan, *Glitter*, p. 9.
56 Balsan, *Glitter*, pp. 10–11.
57 *Belmont Memoirs (Young)*, p. 97.
58 Balsan, *Glitter*, p. 16.
59 *Belmont Memoirs* (*Field*), p. 55.
60 *New York City Journal*, 21 October 1909.
61 *Belmont Memoirs* (*Field*), p. 54.
62 *Belmont Memoirs (Young)*, pp. 94–5.
63 Balsan, *Glitter*, p. 11.
64 Balsan, *Glitter*, p. 11.
65 Balsan, *Glitter*, p. 12.
66 *Belmont Memoirs (Young)*, p. 96.
67 Balsan, *Glitter*, p. 11.
68 Quoted in Auchincloss, *Vanderbilt Era*, p. 153. Wharton wrote to Ogden Codman in 1897: 'I wish the Vanderbilts didn't retard culture so very thoroughly. They are entrenched in a sort of Thermopylae of bad taste, from which no force on earth can dislodge them.'
69 *Belmont Memoirs* (*Field*), pp. 35–6.
70 Quoted in Patterson, *Vanderbilts*, pp. 115–16.
71 Quoted in Patterson, *Vanderbilts*, p. 116.
72 W. H. Vanderbilt's will is included as Appendix G in W. A. Croffut, *The Vanderbilts and the Story of Their Fortune* (Chicago: Bedford, Clarke & Co.,1886), p. 298.
73 D. Turbeville and L. Auchincloss, *Newport Remembered: A Photographic Portrait of A Gilded Past*, (New York: Harry N. Abrams Inc., 1994), pp. 16–17.
74 C. P. B. Jefferys, *Newport: A Short History* (Newport, Rhode Island: Newport Historical Society, 1992), p. 49.
75 H. James, 'The Sense of Newport', *Harper's Monthly Magazine* Vol. CXIII, No. 675, p. 354, reprinted in *A Selection of Historical and Social Articles pertaining to Newport, Rhode Island*, compiled by Florence Archambault (Newport, Rhode Island: Historic Newport Publishers, 1997). James went on to say: 'while their averted owners, roused from a witless dream, wonder what in the world is to be done with them. The answer to which, I think, can only be that there is absolutely nothing to be done; nothing but to let them stand there always, vast and blank, for reminder to those concerned of the prohibited degrees of witlessness, and of the peculiarly awkward vengeances of affronted proportion and discretion.'
76 E. Warburton, *In Living Memory: A Chronicle of Newport 1888–1988* (Newport, Rhode Island: Newport Savings and Loan Assoc./Island Trust Co., 1988), p. 4.
77 G. MacColl and C. Wallace, *To Marry an English Lord* (New York: Workman Publishing 1989), pp. 160–61.
78 M. K. Van Rensselaer, *The Social Ladder* (London: Eveleigh Nash & Co., 1925), pp. 243–44.
79 'W. K. Vanderbilt mansion, Marble House', No. 79.3682, Prints and Drawings Collections, The Octagon Museum.

80 *New York Times* 1st October 1886.
81 'Log of the *Alva*', privately printed, in the Collection of the Preservation Society of Newport County.
82 *Belmont Memoirs (Young)*, p. 131.
83 Balsan, *Glitter*, p. 12.
84 'Log of the *Alva*', 22 March 1887.
85 Balsan, *Glitter*, p. 12.
86 Balsan, *Glitter and the Gold*, p. 15.
87 M. Strange, (B. M. L. Tweed/B. Oelrichs), *Who Tells Me True* (New York: C. Scribner and Sons, 1940), p. 62.
88 *Belmont Memoirs* (*Field*), p. 55.
89 'People' in Gertrude Vanderbilt's private papers, entries for 16 September 1894 (Mo Taylor); 5 September 1894 (Dick Wilson); and 9 September 1894, GVWP.
90 F. A. Sloane, *Maverick in Mauve*, with a commentary by Louis Auchincloss (Garden City, New York: Doubleday & Company, Inc., 1983), p. 126.
91 Balsan, *Glitter*, p. 21.
92 Balsan, *Glitter*, p. 5.
93 2 March 1893, William Gilmour, *Notebooks*, in the Collection of Preservation Society of Newport County.

3 SUNLIGHT BY PROXY

1 SBF to CESW, 29 July 1917, CESWP.
2 J. D'Emilio and E. B. Freedman, *Intimate Matters: A History of Sexuality in America* (New York: Harper and Row, 1988), p. 182.
3 D'Emilio and Freedman, *Intimate Matters*, pp. 182–3.
4 *Belmont Memoirs (Young)*, p. 154.
5 *Belmont Memoirs (Young)*, p. 153.
6 *Belmont Memoirs (Young)*, pp. 153–4.
7 *Belmont Memoirs (Young)*, p. 153.
8 *Belmont Memoirs (Young)*, pp. 152–3.
9 *Belmont Memoirs* (*Field*), pp. 59–60.
10 *Belmont Memoirs* (*Field*), p. 10.
11 *World*, 6 March 1895.
12 Balsan, *Glitter*, p. 23.
13 J. Cherol, 'Historic Architecture: Richard Morris Hunt, Mr and Mrs William K. Vanderbilt's Marble House in Newport', *Architectural Digest* (October 1985), p. 134.
14 See P. F. Miller, 'The Gothic Room in Marble House', *Antiques*, August 1994.
15 Balsan, *Glitter*, p. 19.
16 Balsan, *Glitter*, p. 20.
17 Balsan, *Glitter*, p. 20.
18 *Belmont Memoirs* (*Field*), p. 60.
19 Strange, (B. M. L. Tweed/B. Oelrichs), *Who Tells Me True*, p. 24.
20 D. Black, *The King of Fifth Avenue* (New York: Dial Press, 1981), p. 707. Black suggests that the relationship 'was becoming a talking point'.
21 'Log of the *Alva*', entries for 1 Feb 1889 and 4 Feb 1890.
22 Auchincloss, *Vanderbilt Era*, p. 49.
23 Strange (B. M. L. Tweed/B. Oelrichs), *Who Tells Me True*, p. 23.
24 E. Drexel Lehr (Lady Decies), *King Lehr and The Gilded Age* (London: Constable and Co., 1935), p. 14.
25 E. G. Slocum, 'Memories of Bellevue Avenue: The Story of A Newport Family', *Newport History*, Vol. 67, Part 1 (Newport Historical Society: Number 230, Summer 1995), p. 48.
26 Balsan, *Glitter*, p. 23. The Log of the *Valiant* does not list Winthrop Rutherfurd as a guest on the cruise, though he was apparently on the pier to wave farewell.
27 *World*, 24 November 1893.
28 *Belmont Memoirs (Young)*, p. 136.
29 Balsan, *Glitter*, p. 24.
30 N. Nicholson, *Mary Curzon: A Biography* (London: Phoenix Orion, 1998; first published

London: Weidenfeld and Nicolson, 1977), p. 111.
31 Quoted in Lord Newton, *Lord Lansdowne: A Biography* (London: Macmillan and Co. Ltd, 1929), pp. 58–9.
32 D. Cannadine, *Aspects of Aristocracy* (New Haven: Yale University Press, 1994), p. 84.
33 *Belmont Memoirs (Young)*, p. 138.
34 Newton, *Lord Lansdowne*, p. 126.
35 Balsan, *Glitter*, p. 24.
36 Balsan, *Glitter*, p. 24.
37 'Log of the *Valiant*', 19 January 1894, VP.
38 'Log of the *Valiant*', 31 January 1894, (William Kissam Vanderbilt) BV Vanderbilt, VP.
39 Balsan, *Glitter*, p. 29.
40 Balsan *Glitter*, p. 25.
41 Balsan *Glitter*, pp. 1–2.
42 Balsan *Glitter*, pp. 26–7.
43 E. de Gramont, *Pomp and Circumstance* (London: Jonathan Cape, 1929), p. 222.
44 de Gramont, *Pomp*, p. 213.
45 de Gramont, *Pomp*, p. 213.
46 Balsan, *Glitter*, p. 27.
47 *Town Topics*, 9 August 1894.
48 de Gramont, *Pomp*, p. 222.
49 Balsan, *Glitter*, p. 29.
50 Balsan, *Glitter*, pp. 29–30.
51 Balsan, *Glitter*, p. 7.
52 Balsan, *Glitter*, p. 30.
53 *Town Topics*, 17 March 1910.
54 Balsan, *Glitter*, p. 30.
55 Frances, Duchess of Marlborough to Lady Randolph Churchill, 19 July 1894, CHAR 28/42/23, CA.
56 R. Brandon, *The Dollar Princesses* (New York: Alfred A. Knopf Inc.,1980), p. 20.
57 A. Logan, *The Man Who Robbed The Robber Barons*, (London: Victor Gollancz, 1966), p. 136.
58 *Town Topics*, 19 July 1894.
59 *Town Topics*, 26 July 1894.
60 *Town Topics*, 13 December 1894.
61 *Town Topics*, 4 October 1894.
62 Balsan, *Glitter*, p. 32.
63 *Belmont Memoirs (Field)*, p. 63.
64 *Belmont Memoirs (Field)*, p. 62.
65 *Belmont Memoirs (Young)*, p. 152.
66 *Belmont Memoirs (Field)*, pp. 65–6.
67 *Town Topics*, 13 December 1894.
68 22 December 1894, Gilmour, *Notebooks*.
69 1 January 1895, Gilmour, *Notebooks*.
70 *World*, 16 January 1895.
71 *World*, 6 March 1895.
72 *World*, 6 March 1895.
73 Interview with Sara Bard Field by Amerlia R. Fry, recorded between 1 October 1959 and 31 October 1963, available from the Online Archive of California (http://ark.cdlib.org./ark:/13031/htisp300ini), 26, Adjustment. Digression: Mrs Belmont's Divorce, p. 389.
74 Though she did occasionally come to stay at Blenheim, signing the visitor's book twice, in June and November 1897.
75 *World*, 6 March 1895.
76 *Town Topics*, 14 March 1895.
77 'Log of the *Valiant*', 16 March 1895.
78 *Town Topics*, 11 April 1895.
79 Quoted in Patterson, *Vanderbilts*, p. 152.
80 *World*, 6 March 1895.
81 William Kissam Vanderbilt, Vanderbilt Probate Papers, File No. 24398, Surrogate's Court of Suffolk County, Riverhead, Long Island, New York.
82 *Belmont Memoirs (Field)*, p. 64.
83 *Belmont Memoirs (Field)*, p. 64.
84 *World*, 14 February 1895.
85 *World*, 16 February 1895.
86 G. Myers, *History of The Great American Fortunes, Vol 11* (Chicago: Charles H. Kerr and Co., 1909–1910), p. 276.
87 Gustav Myers calculates that as much as $220 million drained from the United States to Europe as a result of international marriages up to 1909: Myers, *Great American Fortunes*, p. 274.
88 Quoted in Geidel, 'Forgotten Feminist', p. 30.

89 Balsan, *Glitter*, p. 32.
90 Balsan, *Glitter*, p. 32.
91 Balsan, *Glitter*, p. 32.
92 Balsan, *Glitter*, p. 33.
93 *Town Topics*, 20 June 1895.
94 *Town Topics*, 20 June 1895.
95 Balsan, *Glitter*, p. 35.
96 Balsan, *Glitter*, p. 35.
97 Balsan, *Glitter*, p. 35.
98 *Town Topics*, 4 July 1895.
99 Balsan, *Glitter*, p. 33.
100 Balsan, *Glitter*, p. 36.
101 Balsan, *Glitter*, p. 36.
102 See M. H. Rector, *Alva, That Vanderbilt-Belmont Woman: Her Story As She Might Have Told It* (Wickford, RI: Dutch Island Press, 1992).
103 Balsan, *Glitter*, p. 36.
104 Balsan, *Glitter*, p. 37.
105 See, for example, draft chapter, 'Consuela's [sic] marriage' in *Belmont Memoirs* (*Field*), no page numbers: 'I had been brought up for a great part of my life abroad where parents chose the mates for their children . . . It is the parents' solemn duty to form that circle and to know to what sort of alliance her son or daughter is exposed'.
106 Balsan, *Glitter*, p. 37.
107 Sloane, *Maverick*, p. 40.
108 Balsan, *Glitter*, p. 37.
109 Balsan, *Glitter*, p. 9.
110 Quoted in *The New York Times*, 23 July 1920. According to the newspaper, this interview was originally given in 1905.
111 Balsan, *Glitter*, p. 37.
112 Balsan, *Glitter*, p. 38.

4 THE WEDDING

1 *Labour Leader*, 7 September 1895.
2 *Labour Leader*, 14 September 1985.
3 *New York Herald*, 25 August 1895.
4 *Town Topics*, 29 August 1895.
5 *New York Herald*, 26 August 1895.
6 *Newport Mercury*, 31 August 1895.
7 *Newport Journal*, 29 August 1895.
8 'Log of the *Valiant*', entry for 28 August 1895.
9 *Town Topics*, 5 September 1895.
10 *New York Herald*, 29 August 1895.
11 *New York Herald*, 29 August 1895.
12 *Newport Journal*, 31 August 1895.
13 *Town Topics*, 5 September 1895.
14 *New York Herald*, 29 August 1895.
15 *Newport Mercury*, 31 August 1895.
16 *Belmont Memoirs (Young)*, p. 146.
17 28 August 1895, Gilmour, *Notebooks*.
18 *Town Topics*, 5 September 1895.
19 Quoted in M. H. Elliott, *This Was My Newport* (Cambridge, Mass.: The Mythology Co., 1944), pp. 202–3.
20 Balsan, *Glitter*, p. 39.
21 Balsan, *Glitter*, pp. 39–40.
22 *Newport Journal*, 31 August 1895.
23 Gilmour, *Notebooks*, 1 September 1895.
24 *New York Herald*, 2 September 1895.
25 *Town Topics*, 12 September 1895.
26 See D. Riggs, *Keelhauled: Unsportsmanlike Conduct and The America's Cup* (London: Stanford Maritime, 1986), p. 46.
27 *Town Topics*, 19 September 1895.
28 *Town Topics*, 19 September 1895.
29 *Town Topics*, 19 September 1895.
30 *World*, 13 October 1895.
31 Balsan, *Glitter*, p. 38.
32 19 and 20 September 1895, Gilmour, *Notebooks*.
33 Balsan, *Glitter*, p. 40.
34 Balsan, *Glitter*, p. 40.
35 *Town Topics*, 26 September 1895.
36 *World*, 22 September 1895.
37 *Newport Journal*, 19 October 1895.
38 *New York Herald*, 21 September 1895.
39 Balsan, *Glitter*, p. 41.
40 William Kissam Vanderbilt Probate Papers, File No. 24398.
41 Balsan, *Glitter*, p. 41.
42 Strange (B. M. L. Tweed/B. Oelrichs), *Who Tells Me True*, p. 26.
43 *The New York Times*, 7 November 1895.

44 Quoted in G. Juergens, *Joseph Pulitzer and The New York World* (Princeton, New Jersey: Princeton University Press, 1966), p. 192.
45 *World*, 20 October 1895.
46 In an interview with the newspaper, the Duke said that reports of misbehaviour with Tottie Coughdrops were 'unjust and unkind. I did not go on the stage and was not put off. I knew nothing about that until I saw it in the papers. The report was false and had nothing to do with my return.' *New York Tribune*, 16 October 1895.
47 *World*, 15 October 1895.
48 *World*, 16 October 1895.
49 *New York Herald*, 19 October 1895.
50 *World*, 6 October 1895.
51 *World*, 20 October 1895. Alva bought the house, which stood at the southeast corner of Madison Avenue and 72nd Street, on 8 March from Ruth A. Brown for $250,000.
52 *World*, 22 October 1895.
53 Balsan, *Glitter*, p. 40.
54 *New York Herald*, 26 October 1895.
55 *Town Topics*, 31 October 1895.
56 E. W. Chase, and I. Chase, *Always in Vogue* (New York: Doubleday & Company, Inc., 1954), p. 38.
57 *Vogue*, Vol. VI, No.20 (14 November 1895).
58 Balsan, *Glitter*, p. 41.
59 *New York Herald*, 31 October 1895.
60 Balsan, *Glitter*, p. 41.
61 *New York Times*, 5 November 1895.
62 Balsan, *Glitter*, p. 41.
63 *World*, 3 November 1895.
64 Balsan, *Glitter*, pp. 41–2.
65 *The New York Times*, 7 November 1895.
66 *New York Herald*, 7 November 1895.
67 *New York Times*, 7 November 1895.
68 *World*, 7 November 1895.
69 Balsan, *Glitter*, p. 42.
70 *World*, 7 November 1895.
71 Balsan, *Glitter*, p. 42.
72 *The New York Times*, 25 November 1926.
73 *The New York Times*, 7 November 1895.
74 Balsan, *Glitter*, p. 42.
75 *World*, 7 November 1895.
76 *The New York Times*, 7 November 1895.
77 *World*, 7 November 1895.
78 *The New York Times*, 7 November 1895.
79 *New York Herald*, 7 November 1895; Balsan, *Glitter*, p. 42.
80 Balsan, *Glitter*, p. 44.
81 *The New York Herald*, 7 November 1895.
82 Balsan, *Glitter*, p. 46.
83 Balsan, *Glitter*, p. 42–3.
84 *New York Herald*, 7 November 1895.
85 *The New York Times*, 7 November 1895.
86 'Consuela's [sic] marriage', *Belmont Memoirs* (*Field*).
87 SBF to CESW, 15 August 1917, CESWP.
88 Interview with Sara Bard Field by Amerlia R. Fry, 12 July 1962, 29, 'The Woman's Party, Peace and Religion: "Animus" versus "Anima"', p. 543.
89 'Consuela's [sic] marriage', *Belmont Memoirs* (*Field*).
90 'Consuela's [sic] marriage', *Belmont Memoirs* (*Field*).
91 Wharton, *House of Mirth*, p. 53.
92 *Belmont Memoirs* (*Field*), pp. 51–2.
93 E. Wharton, *French Ways and Their Meaning* (Lenox, Mass., Lee, Mass.: Edith Wharton Restoration at the Mount and Berkshire House Publishers, 1997; first published London: 1919), pp. 115 and 116.
94 'Consuela's [sic] marriage' in *Belmont Memoirs* (*Field*).
95 *Town and Country*, 18 January 1902.
96 *The New York Times*, 21 March 1944.
97 Wharton, *House of Mirth*, p. 53.

98 'Consuela's [sic] marriage', *Belmont Memoirs* (*Field*).
99 *Town Topics*, 'The Widow on the Marlborough Case', 8 November 1906.
100 *Belmont Memoirs* (*Field*), p. 7.
101 Eliot, *Heiresses*, pp. 176–7.
102 'Consuela's [sic] marriage', *Belmont Memoirs* (*Field*).
103 Quoted in Turbeville and Auchincloss, *Newport Remembered*, p. 117.

5 BECOMING A DUCHESS

1 M. Ashley, *Churchill as Historian* (London: Secker and Warburg, 1968), pp. 4 and 5.
2 Patterson, *Vanderbilts*, p. 149.
3 A. L. Rowse, *The Early and The Later Churchills* (London: Macmillan and Company Ltd, 1958), p. 272.
4 Rt Hon. W. Spencer-Churchill and C. C. Martindale, *Charles IXth Duke of Marlborough, K.G.* (London: Burns Oates & Washbourne Ltd, 1934), pp. 6–8 passim.
5 M. Soames, *Clementine Churchill* (London: Doubleday, Transworld; revised and updated version, 2002), pp. 47–8.
6 Quoted in H. Montgomery-Massingberd, *Blenheim Revisited: The Spencer-Churchills and Their Palace* (London: The Bodley Head, 1985), p. 45.
7 Rowse, *Churchills*, p. 215.
8 Montgomery-Massingberd, *Blenheim Revisited*, p. 104.
9 Cannadine, *Aspects of Aristocracy*, p. 133.
10 Rowse, *Churchills*, p. 253.
11 Maud Lansdowne to GD, undated but probably *c*.December 1921, GDP.
12 Rowse, *Churchills*, p. 263.
13 P. Magnus, *King Edward VII* (London: John Murray, 1964), pp. 90–1 passim.
14 D. Cannadine, *The Decline and Fall of the British Aristocracy* (London: Papermac, revised edition, 1996), p. 17.
15 'Consuela's [sic] marriage', *Belmont Memoirs* (*Field*).
16 Balsan, *Glitter*, p. 46.
17 *New York Herald*, 11 November 1895.
18 *New York Herald*, 11 November 1895.
19 *Town Topics*, 21 November 1895. The magazine frequently satirised Alva's fondness for publicity. On 7 November 1895, for example, Colonel Mann wrote: 'Queens must not hide their doings from their subjects. Mrs Vanderbilt has not tried to do so . . . It is not too much to say that the future of female underclothing will be momentously affected by the light which the public has received . . . To have furnished the press, with no ungrudging hand, authentic information as to the smallest details of a thrilling and historical event, to have tossed away that pretence of secrecy with which humbler folk affect to enwrap themselves, surely merits commendation from a thoughtful and intelligent public.'
20 WSC to Lady Randolph Churchill, 10 November 1895 in R. S. Churchill, *Winston S. Churchill Volume 1, Companion Part 1, 1874–1896* (London: Heinemann, 1967), p. 597.
21 WSC to Jack Churchill, 15 November 1895 in Churchill, *Churchill Volume 1*, pp. 599–600; CHAR 28/21/85–89, CA.
22 Balsan, *Glitter*, p. 40.
23 *New York Herald*, 18 November 1895.
24 *New York Herald*, 18 November 1895.
25 *Belmont Memoirs* (*Young*), p. 157.
26 *Town Topics*, 9 January 1896.

27 *Town Topics*, 23 January 1896.
28 William Kissam Vanderbilt Probate Papers, File No. 24398.
29 *Belmont Memoirs* (*Field*), chapter: 'My Second Marriage', quoted in Geidel, 'Forgotten Feminist', p. 42.
30 The Marlboroughs booked a passage on the *Fulda* rather than a luxurious transatlantic liner, like the *Campania*, because the *Fulda* sailed straight to Genoa, a plan that soon changed.
31 Balsan, *Glitter*, p. 46.
32 Balsan, *Glitter*, p. 48.
33 Balsan, *Glitter*, p. 49.
34 Balsan, *Glitter*, pp. 49–50.
35 Balsan, *Glitter*, p. 51.
36 Rowse, *Churchills*, p. 248.
37 D. Green, *The Churchills of Blenheim* (London: Constable and Company Ltd, 1984), p. 137.
38 H. Vickers, *Gladys, Duchess of Marlborough* (New York: Holt, Rinehart and Winston, 1979), p. 172.
39 *Town Topics*, 2 January 1896. The Duke was trying to insure Consuelo's half of the marriage settlement against her death.
40 Balsan, *Glitter*, p. 53.
41 *Town Topics*, 23 January 1896.
42 Balsan, *Glitter*, p. 53.
43 G. Horne, 'Blenheim Fifty Years Ago: Memoirs of Gentleman's Service', *Country Life* (23 February 1945).
44 Balsan, *Glitter*, p. 53. Frances, Countess of Warwick, also remarked that she had little control over dresses designed by Jean Worth, though she minded less than Consuelo because she considered him to be an artist. 'He would study his subject as a painter would study a woman sitting for her portrait. In silent attendance, his satellites would wait on him, watchful that not a movement of theirs should disturb his concentration. "This," he would murmur, "is the colour". With each "this" the gown would grow under the magic of his fingers. He would make us pose, just as Sargent and Carolus-Duran and other great painters did when I sat for them': Frances, Countess of Warwick, *Afterthoughts* (London: Cassel and Company Ltd, 1931) pp. 160–1.
45 Quoted in P. Proddow and M. Fasel, *Diamonds: A Century of Spectacular Jewels* (New York: Harry N. Abrams, 1996), p. 224.
46 Quoted in P. Hinks, *Nineteenth Century Jewellery* (London: Faber, 1975), p. 38.
47 See S. Bury, *Jewellery 1789–1910, The International Era, Vol. 11 1862–1910* (Woodbridge: Antique Collectors Club, 1991), p. 711.
48 G. F. Kunz and C. H. Stevenson, *The Book of the Pearl: The History, Art, Science and Industry of the Queen of Gems* (New York: Dover, 1993; first published London: Macmillan, 1908), p. 440.
49 See Hinks, *Nineteenth Century Jewellery* and D. J. Jackson, *From Slave to Siren – The Victorian Woman and Her Jewellery from Neoclassic to Art Nouveau*, Exhibition Catalogue, Duke University Museum of Art, 1971.
50 M. Fowler, *Blenheim: Biography of a Palace* (London: Viking, 1989), p. 182.
51 Balsan, *Glitter*, p. 54.
52 Lord Dufferin and Ava to 9th Duke, 18 March 1896, Social Letters, Marlborough Papers, LCW.
53 Balsan, *Glitter*, p. 54.
54 Spencer-Churchill and Martindale, *Charles IXth Duke*, p. 5.
55 Rowse, *Churchills*, p. 250.
56 Cannadine, *Decline and Fall*, pp. 26–7 passim.
57 Balsan, *Glitter*, p. 54.
58 Balsan, *Glitter*, p. 55.

59 Balsan, *Glitter*, p. 55.
60 Cornwallis-West, *Reminiscences*, p. 47.
61 E. Wharton, *The Buccaneers* (London: J. M. Dent and Everyman, 1993), p. 157.
62 Balsan, *Glitter*, p. 56.
63 Balsan, *Glitter*, p. 55.
64 Balsan, *Glitter*, p. 55.
65 Balsan, *Glitter*, p. 58.
66 Balsan, *Glitter*, pp. 56–7.
67 Balsan, *Glitter*, p. 57.
68 Balsan, *Glitter*, p. 59.
69 *Oxford Times*, Saturday 4 April 1896. The homecoming took place on Tuesday 31 March but the newspaper, like *Jackson's Oxford Journal*, was published weekly (and still is).
70 *Oxford Times*, Saturday 4 April 1896.
71 Balsan, *Glitter*, p. 59.
72 The mayor's speech as reported by the *Oxford Times*, Saturday 4 April 1896.
73 The Duke's speech as reported by the *Oxford Times*, Saturday 4 April 1896.
74 *Oxford Times*, Saturday 4 April 1896.
75 *Oxford Times*, Saturday 4 April 1896.
76 *Oxford Times*, Saturday 4 April 1896.
77 *Jackson's Oxford Journal*, Saturday 4 April 1896.
78 Horne, 'Blenheim Fifty Years Ago', p. 327.
79 Balsan, *Glitter*, p. 60.
80 Balsan, *Glitter*, p. 60.
81 *Oxford Times*, Saturday 4 April 1896.
82 Cannadine, *Decline and Fall*, p. 24.
83 Balsan, *Glitter*, p. 70.

6 SUCCESS

1 Balsan, *Glitter*, p. 80.
2 Balsan, *Glitter*, p. 60.
3 Balsan, *Glitter*, p. 45.
4 Balsan, *Glitter*, p. 63.
5 Horne, 'Blenheim Fifty Years Ago', p. 326.
6 J. Lees-Milne, *Ancestral Voices* (London: Chatto & Windus, 1975) p. 188.
7 Horne, 'Blenheim Fifty Years Ago', p. 326.
8 Horne, 'Blenheim Fifty Years Ago', p. 326.
9 Balsan, *Glitter*, p. 67.
10 Balsan, *Glitter*, p. 69.
11 Balsan, *Glitter*, p. 68.
12 Horne, 'Blenheim Fifty Years Ago', p. 327.
13 Green, *Churchills of Blenheim*, p. 137.
14 *Glitter and Gold* reviews, VA.
15 Earl of Carnarvon, *No Regrets* (London: Weidenfeld and Nicolson, 1976), p. 140.
16 Balsan, *Glitter*, p. 68.
17 Balsan, *Glitter*, p. 65.
18 Balsan, *Glitter*, p. 69.
19 Balsan, *Glitter*, p. 70.
20 Balsan, *Glitter*, p. 76.
21 Balsan, *Glitter*, p. 71.
22 Balsan, *Glitter*, p. 76.
23 Quoted in Nicholson, *Mary Curzon*, p. 97.
24 Quoted in Proddow and Fasel, *Diamonds*, p. 335.
25 Viscount Churchill, *All My Sins Remembered* (London: William Heinemann Ltd, 1964), pp. 33–4.
26 Lady Fortescue, *There's Rosemary, There's Rue* (London: Black Swan, Transworld, 1993; first published London: William Blackwood, 1939), p. 34. Seventeen-year-old Winifred Fortescue was equally enchanted when she first met Consuelo, probably around 1905–6: 'Never before had I seen anyone so lovely, so exquisitely dressed, so perfectly simple and sweet despite her great possessions as this wonderful Duchess,' (p. 35). Consuelo befriended Winifred Fortescue and was consistently kind to her. Consuelo suggested that Winifred took up a career on the

stage to assist her family's finances, looked after her while she was at drama school in London and paid for her wedding trousseau. Winifred Fortescue was married from Sunderland House.

27 N. Armstrong, *Victorian Jewellery* (London: Studio Vista, 1976), p. 135.
28 Balsan, *Glitter*, p. 77.
29 Quoted in Nicholson, *Mary Curzon*, p. 98.
30 Balsan, *Glitter*, p. 85. Frances, Countess of Warwick, also thought that Ascot was vulgar, calling it a 'stilted, expensive, extensive and over-elaborate' garden party. She greatly preferred Newmarket (see *Afterthoughts*, p. 168).
31 Balsan, *Glitter*, pp. 82–3.
32 Though there may have been earlier meetings in London, this is the first on record. Harold Vanderbilt, who was educated at boarding school like his older brother Willie K. Jr, lived with Alva and Oliver Belmont until he came of age.
33 Balsan, *Glitter*, p. 84.
34 G. Cornwallis-West to WSC, 5 October 1904, CHAR 28/37/32, CA.
35 P. Lytton to WSC, 14 September 1907, CHAR 1/66/27, CA.
36 Balsan, *Glitter*, p. 84.
37 Balsan, *Glitter*, pp. 86–7 passim.
38 J. Mordaunt Crook, *The Rise of the Nouveaux Riches: Style and Status in Victorian and Edwardian Architecture* (London: John Murray, 1999), p. 67.
39 Quoted in Rowse, *Churchills*, p. 280.
40 Rowse, *Churchills*, pp. 247–8.
41 Lord Knollys to 9th Duke, 10 August 1896, Social Letters, LCW.
42 Lord Knollys to 9th Duke, 21 August 1896, Social Letters, LCW.
43 Lady Randolph Churchill to WSC, 13 November 1896, CHAR 1/8/72, CA.
44 Quoted in Balsan, *Glitter*, p. 92.
45 *Jackson's Oxford Journal*, 28 November 1896.
46 Balsan, *Glitter*, p. 93.
47 *Jackson's Oxford Journal*, 28 November 1896.
48 *Jackson's Oxford Journal*, 28 November 1896.
49 Fowler, *Blenheim*, pp. 201–2.
50 *Jackson's Oxford Journal*, 5 December 1896.
51 *Jackson's Oxford Journal*, 5 December 1896.
52 Horne, 'Blenheim Fifty Years Ago', p. 327.
53 *Town Topics*, 3 December 1896.
54 See letter from Sir Arthur Ellis (equerry) to 9th Duke, 30 November 1896, Social Letters, LCW.
55 Balsan, *Glitter*, p. 92.
56 Balsan, *Glitter*, p. 94.
57 Lady Randolph Churchill to WSC, 27 November 1896, CHAR 1/8/75, CA.
58 WSC to Lady Randolph Churchill from Calcutta, 23 December 1896, CHAR 1/22/37, CA
59 Horne, 'Blenheim Fifty Years Ago', p. 327.
60 Balsan, *Glitter*, p. 94.
61 Lady Randolph Churchill at Blenheim to WSC in India, 24 December 1896, CHAR 1/8/81–82, CA.
62 Balsan, *Glitter*, p. 95. According to Gerald Horne, the Duke and Duchess took fewer staff to Melton Mowbray: the groom of the chambers, two footmen, three housemaids, three kitchen staff, an 'oddman', himself, the stud groom, a second horse-man and stable helpers. Everyone, including the horses, travelled by special train; Horne, 'Fifty Years Ago: Memoirs of Gentleman's Service', in *Country Life*, p. 328.
63 Balsan, *Glitter*, p. 95.

64 From the autobiography of Jean Worth, quoted in S. Murphy, *The Duchess of Devonshire's Ball* (London: Sidgwick & Jackson, 1984), p. 63.
65 *Town Topics*, 22 April 1897.
66 Lady Randolph Churchill to WSC, 21 September 1897, CHAR 1/8/107, CA.
67 Horne, 'Blenheim Fifty Years Ago', p. 328.
68 Balsan, *Glitter*, p. 98.
69 Balsan, *Glitter*, p. 98.
70 Balsan, *Glitter*, p. 98.
71 Balsan, *Glitter*, p. 66.
72 G. Cornwallis-West, *Edwardian Hey-days* (London and New York: Putnam, 1930), p. 128.
73 Balsan, *Glitter*, p. 104.
74 Balsan, *Glitter*, p. 102.
75 Balsan, *Glitter*, pp. 103–4 passim.
76 Lord Lansdowne to 9th Duke, *c.*15 August 1895, LCW.
77 WSC to 9th Duke, 26 January 1899, CHAR 28/26/23, CA.
78 Princess Daisy of Pless, *From My Private Diary* (London: John Murray, 1931), p. 47.
79 Balsan, *Glitter*, pp. 88–9.
80 Balsan, *Glitter*, p. 133.
81 Cannadine, *Aspects*, p. 77.
82 Balsan, *Glitter*, p. 107.
83 Balsan, *Glitter*, p. 122.
84 Quoted in D. Stuart, *Dear Duchess: Millicent Duchess of Sutherland 1867–1955* (London: Victor Gollancz Ltd, 1982), p. 97.
85 Balsan, *Glitter*, p. 124.
86 Quoted in Stuart, *Dear Duchess*, p. 98.
87 Balsan, *Glitter*, pp. 125–26 passim.
88 Balsan, *Glitter*, p. 132.
89 Balsan, *Glitter*, p. 132.
90 Cannadine, *Aspects*, p. 82.
91 Balsan, *Glitter*, pp. 138–9.
92 Cannadine, *Aspects*, p. 86.
93 Balsan, *Glitter*, p. 139.
94 Mary Curzon to Lady Randolph Churchill, 18 May 1903, quoted in Cornwallis-West, *Reminiscences*, pp. 277–8.
95 Quoted in Cannadine, *Aspects*, p. 87.
96 Balsan, *Glitter*, p. 140.
97 Quoted in Cannadine, *Aspects*, p. 88.
98 Quoted in Cannadine, *Aspects*, p. 89.
99 Quoted in Cannadine, *Aspects*, p. 90.

7 DIFFICULTIES

1 C to Cass Canfield at Harper & Brothers, 27 July 1952, HRP.
2 'The Widow on the Marlborough Case', *Town Topics*, 8 November 1906.
3 Balsan, *Glitter*, p. 50.
4 Balsan, *Glitter*, p. 60.
5 Cornwallis-West, *Reminiscences*, p. 47.
6 Cornwallis-West, *Reminiscences*, p. 47.
7 Balsan, *Glitter*, p. 148.
8 9th Duke to WSC, CHAR 1/57/25, CA.
9 Balsan, *Glitter*, p. 119.
10 Balsan, *Glitter*, p. 109.
11 Balsan, *Glitter*, p. 100.
12 Balsan, *Glitter*, pp. 104–5 passim.
13 Balsan, *Glitter*, p. 105.
14 Balsan, *Glitter*, p. 106.
15 Balsan, *Glitter*, p. 113.
16 *Town Topics*, 4 January 1900.
17 *Town Topics*, 1 November 1906.
18 W. S. Churchill, *My Early Life: A Roving Commission* (London: Mandarin, Octopus, 1990; first published London: Thornton Butterworth, 1930), pp. 365–6.
19 Balsan, *Glitter*, p. 108.
20 *Town Topics*, 3 January 1901.
21 Balsan, *Glitter*, p. 107.
22 *Town Topics*, 3 January 1901.
23 *The New York Times*, 22 April 1901.
24 *The New York Times*, 22 April 1901.
25 Balsan, *Glitter*, p. 98.
26 Vickers, *Gladys*, p. 41. Hugo Vickers is clear that Consuelo and Gladys never met in

Newport for Consuelo spent much time in Europe and Gladys was away at boarding school.

27 Quoted in Vickers, *Gladys*, p. 54.
28 Quoted in Vickers, *Gladys*, p. 61.
29 Quoted in Vickers, *Gladys*, p. 66.
30 9th Duke to GD from Harrogate, August 1901, GDP.
31 Balsan, *Glitter*, p. 116.
32 C to GD, probably January 1902, GDP.
33 C to GD, October 1901, GDP.
34 C to GD, probably January 1902, GDP.
35 C to GD, October 1901, GDP.
36 C to GD, October 1901, GDP.
37 9th Duke to GD, undated letter from Warwick House, GDP.
38 Quoted in Vickers, *Gladys*, p. 74.
39 Quoted in Vickers, *Gladys*, p. 111.
40 Quoted in Vickers, *Gladys*, p. 75.
41 Balsan, *Glitter*, p. 142.
42 *Town Topics*, 25 February 1904.
43 Balsan, *Glitter*, p. 144.
44 Balsan, *Glitter*, p. 144.
45 Daisy of Pless, *My Private Diary*, p. 69.
46 Quoted in Vickers, *Gladys*, p. 81.
47 Vickers, *Gladys*, p. 83.
48 Vickers, *Gladys*, p. 83.
49 Quoted in Vickers, *Gladys*, pp. 85–6.
50 Alfred Lyttleton to 9th Duke, 17 December 1905, Marlborough Papers, LCW.
51 Daisy of Pless, *My Private Diary*, p. 129.
52 Balsan, *Glitter*, p. 135.
53 Quoted in E. Kehoe, *Fortune's Daughters* (London: Atlantic Books, 2004), p. 211.
54 *Town Topics*, 2 June 1904.
55 Quoted in Vickers, *Gladys*, p. 92.
56 C to GD from Sunderland House, 13 July 1904, GDP.
57 9th Duke to GD from Sunderland House, 13 July 1904, GDP.
58 9th Duke to GD, second letter, from Sunderland House, 13 July, GDP.
59 *Town Topics*, 5 October 1905.
60 *Town Topics*, 1 November 1906.
61 Balsan, *Glitter*, p. 146.
62 Balsan, *Glitter*, p. 146.
63 Balsan, *Glitter*, p. 109.
64 Cannadine, *Decline and Fall*, p. 189.
65 Balsan, *Glitter*, p. 136.
66 Quoted in Anne de Courcy, *Circe: The Life of Edith, Marchioness of Londonderry* (London: Sinclair-Stevenson, 1992), pp. 73–4.
67 de Courcy, *Circe*, p. 74.
68 P. Craigie to Rev. W. Brown, 6 May 1906, Pearl Craigie Papers.
69 WSC to Lady Randolph Churchill, 13 October 1906, CHAR 28/27/63, CA.
70 Lady Randolph Churchill to WSC, 16 October 1906, CHAR 28/78/44, CA.
71 George Cornwallis-West to WSC, 20 October 1906, CHAR 1/57/7, CA.
72 George Cornwallis-West to WSC, 20 October 1906, CHAR 1/57/7, CA.
73 Quoted in Friedman, *Gertrude Vanderbilt*, p. 229. William Kissam Vanderbilt remarried in 1903. His second wife was Anne Harriman Sands Rutherfurd, daughter of Oliver Harriman and twice widowed. Her first husband was Samuel S. Sands and her second was Lewis Morris Rutherfurd, Winthrop's brother. Consuelo was very fond of her stepmother with whom she shared an interest in philanthropy. 'Visits to my father were particularly pleasant, for I rejoiced in the happiness he had found in his second marriage. My stepfather had a gay and gentle nature,' she wrote. From the time of their marriage, William K. and his second wife spent much time in France, where he became an important figure in French horse racing, acquiring a racing stable at Poissy.
74 Quoted in Friedman, *Gertrude Vanderbilt*, p. 230.

75 George Cornwallis-West to WSC, 21 October 1906, CHAR 1/57/9, CA.
76 Lady Randolph Churchill to C, 2 November 1906, CHAR 28/78/45, CA.
77 9th Duke to WSC, CHAR 1/57/27, CA.
78 Quoted in Friedman, *Gertrude Vanderbilt*, p. 230.
79 Hugh Cecil to WSC, undated, October 1906, CHAR 1/57/18, CA.
80 Hugh Cecil to WSC, 31 October 1906, CHAR 1/57/19, CA.
81 9th Duke to WSC, CHAR 1/57/22, CA.
82 Ivor Guest to WSC, CHAR 1/57/51, CA.
83 WSC to 9th Duke, 4 January 1907, CHAR 1/65/1, CA.
84 WSC to 9th Duke, 4 January 1907, CHAR 1/65/1, CA.
85 9th Duke to WSC, 31 January 1907, CHAR 1/65/8, CA.
86 *Town Topics*, 1 November 1906.
87 Daisy of Pless, *My Private Diary*, p. 207.

8 PHILANTHROPY, POLITICS AND POWER

1 Theodore Roosevelt to Whitelaw Reid, 27 November 1906, quoted in Vickers, *Gladys*, p. 108.
2 Lady Fortescue, *There's Rosemary* (London: Black Swan, Transworld, 1993; first published London: William Blackwood & Sons, 1939), pp. 69–70.
3 *The Times*, Friday 29 June 1906.
4 *John Bull*, 5 December 1908, CAP.
5 *Leicester Mercury*, 3 December, year unknown, CAP.
6 Press cutting, no source, CAP.
7 Balsan, *Glitter*, p. 149.
8 Balsan, *Glitter*, pp. 149–50.
9 K. D. McCarthy (ed.), *Lady Bountiful Revisited: Women, Philanthropy and Power* (New Brunswick: Rutgers University Press, 1990), p. 1.
10 See, for example, F. Prochaska's *Women and Philanthropy in Nineteenth-Century England* (Oxford: Oxford University Press, 1980) and P. Thane, 'The Social, Economic and Political Status of Women', in P. Johnson (ed.), *Twentieth Century Britain: Economic, Social and Cultural Change* (London: Longman, 1994).
11 Prochaska, *Women and Philanthropy*, p. 2.
12 The Duchess of Marlborough, 'The Position of Women – III', *The North American Review*, p. 352.
13 *The New York Times*, 8 March 1908.
14 Speech made by Winston Churchill, 11 October 1906.
15 *The New York Times*, 19 July 1908; *New York Times*, 15 November 1908.
16 Balsan, *Glitter*, p. 154.
17 Daisy of Pless, *My Private Diary*, p. 241.
18 Unidentified magazine clipping, one in a series by Dame Anna Neagle, CVBS.
19 *The New York Times*, 1 April 1908. Jacob Riis, author of the groundbreaking study of poverty in New York, *How The Other Half Lives*, was another guest speaker at this dinner.
20 *New York Evening Journal*, 1 April 1908.
21 Duchess of Marlborough, 'Position of Women – I', *North American Review*, Vol. 189, No. 1 (January 1909), p. 11.
22 Duchess of Marlborough, 'Position of Women – I', p. 15.
23 Duchess of Marlborough, 'Position of Women – III', *North American Review*, Vol. 189, No. 3, (March 1909), p. 3.
24 Duchess of Marlborough, 'Position of Women – II', *North American Review*, Vol. 189, No. 2 (February 1909), p. 189.
25 Duchess of Marlborough, 'Position of Women – III', p. 4.

26 Duchess of Marlborough, 'Position of Women – III', p. 8.
27 Duchess of Marlborough, 'Position of Women – III', p. 9.
28 *The New York Times*, 5 November 1909.
29 'Consuela's [sic] marriage', *Belmont Memoirs* (*Field*).
30 Willie K. Vanderbilt Jr spent the early years of his marriage working in his father's office at Grand Central Station, but he was equally interested in sailing and won the Lipton Cup in 1900 before turning his attention to motor racing and automobile development.
31 Turbeville and Auchincloss, *Newport Remembered*, p. 82.
32 E. Drexel Lehr, *King Lehr and the Gilded Age* (London: Constable & Co., 1935), pp. 134–7 passim.
33 Drexel Lehr, *King Lehr*, p. 41.
34 Turbeville and Auchincloss, *Newport Remembered*, p. 36.
35 Drexel Lehr, *King Lehr*, pp. 210–12 passim.
36 *Town Topics*, 10 June 1897.
37 Logan, *Robber Barons*, p. 164.
38 *Town Topics*, 19 March 1908.
39 *Town Topics*, 29 September 1904.
40 *Belmont Memoirs* (*Young*), p. 162.
41 *The New York Times*, 8 January 1944.
42 *Town Topics*, 18 July 1907.
43 Drexel Lehr, *King Lehr*, pp. 108–9.
44 See Geidel, 'Forgotten Feminist', pp. 43–4.
45 Drexel Lehr, *King Lehr*, p. 133.
46 Drexel Lehr, *King Lehr*, p. 134.
47 *Town Topics*, 3 September 1908.
48 *Belmont Memoirs* (*Field*), 'My Second Marriage', quoted in Geidel, 'Forgotten Feminist', p. 66.
49 Letter from Alva Belmont to August Belmont, 8 July 1908, Belmont Papers.
50 18 June 1908, Gilmour, *Notebooks*.
51 *Belmont Memoirs* (*Young*), p. 167.
52 *Town Topics*, 9 July 1908.
53 *Town Topics*, 3 September 1908.
54 *Town Topics*, 3 September 1908.
55 Quoted in Geidel, 'Forgotten Feminist', p. 78.
56 Mrs O. H. P. Belmont, 'Why I am a Suffragist', *World Today*, Vol. XXI (October 1911), p. 1172.
57 Quoted in Drexel Lehr, *King Lehr*, p. 159.
58 Alva to SBF after Field's son had been killed in a car accident, 5 January 1919, CESWP.
59 Belmont, 'Why I am a Suffragist', p. 1172.
60 Belmont, 'Why I am a Suffragist', p. 1172.
61 Belmont, 'Why I am a Suffragist', p. 1172.
62 *Town Topics*, 26 August 1909.
63 *New York Mail and Express*, 26 August 1909.
64 Drexel Lehr, *King Lehr*, p. 208.
65 Drexel Lehr, *King Lehr*, p. 206.
66 Alva Belmont, 'Autobiographical Note', DSUP, pp. 2–3.
67 Belmont, 'Autobiographical Note', pp. 1–4 passim.
68 Figures quoted in Geidel, 'Forgotten Feminist', p. 84.
69 Belmont, 'Autobiographical Note', p. 8.
70 *New York City Evening World*, 25 August 1909.
71 *New York Evening Sun*, 24 August 1909.
72 *Town Topics*, 5 August 1909.
73 *World*, 24 August 1909.
74 *New York Evening Sun*, 24 August 1909.
75 *Spectator*, 3 September 1909.
76 *Town Topics*, 26 August 1909.
77 *World*, 25 August 1909.
78 Drexel Lehr, *King Lehr*, p. 208.
79 Drexel Lehr, *King Lehr*, pp. 209–10.
80 *New York Herald*, 25 August 1909.
81 Drexel Lehr, *King Lehr*, p. 210.

9 OLD TRICKS

1 Quoted in Geidel, 'Forgotten Feminist', p. 140.

2 See Geidel, 'Forgotten Feminist', p. 155.
3 *New York Evening World*, 24 January 1911.
4 Quoted in Geidel, 'Forgotten Feminist', p. 129.
5 *Washington Times*, 30 September 1909.
6 Quoted in Geidel, 'Forgotten Feminist', p. 181.
7 *New York Sun*, 17 September 1910.
8 Quoted in Geidel, 'Forgotten Feminist', p. 181.
9 See Geidel, 'Forgotten Feminist', p. 182.
10 'Autobiographical Note', DSUP, p. 13.
11 *New York Journal*, 24 December 1910.
12 *Washington Post*, 19 June 1909; *World*, 16 June 1909.
13 *New York Evening Journal*, 25 June 1909.
14 'Autobiographical Note', DSUP, p.10.
15 *New York Sun*, 18 September 1909.
16 'Autobiographical Note', DSUP, p. 2.
17 'Autobiographical Note', DSUP, pp. 12–3.
18 *The Woman's Journal*, 24 August 1910.
19 *Daily Mirror*, 12 June 1910; *New York Evening Post*, 16 July 1910.
20 *New York Mail*, 13 May 1911; *Los Angeles Record*, 19 May 1911.
21 *New York Tribune*, 22 June 1911.
22 *New York American*, 23 June 1911.
23 Quoted in Geidel, 'Forgotten Feminist', p. 207.
24 *New York Herald*, 5 November 1911.
25 *New York Press*, 7 January 1912.
26 *Jacksonville Times*, 18 February 1912.
27 *New York Press*, 7 January 1912.
28 Memo on Elsa Maxwell, DSUP, p. 14.
29 *The New York Times*, 9 July 1911.
30 *Chicago Tribune*, 9 June 1912.
31 Balsan, *Glitter*, p. 170.
32 Balsan, *Glitter*, p. 156.
33 M. J. Tuke, *A History of Bedford College for Women, 1849–1937* (Oxford: Oxford University Press, 1939), p. 213.
34 Tuke, *Bedford College*, p. 213.
35 Duchess of Marlborough, 'Hostels for Women', in *The Nineteenth Century and After*, Vol IX (January-June 1911), p. 866.
36 Duchess of Marlborough, 'Hostels for Women', p. 864.
37 E. Ross, 'Good and Bad Mothers: Lady Philanthropists and London Housewives before the First World War', in McCarthy (ed.), *Lady Bountiful Revisited*, p. 276.
38 *The New York Times*, 27 May 1913.
39 Editorial in the *Daily Citizen* quoted in *The New York Times*, 29 May 1913.
40 Mrs O'Sullivan, *Woman's Municipal Party: Formation and Work – A Statement*, November 1913.
41 C to Marie Stopes, 7 October 1913, Stopes Papers, British Library, Add. 58682.
42 'Duchess Launches a Party', *The New York Times*, 14 March 1914.
43 Quoted in Balsan, *Glitter*, p. 169.
44 Quoted in Balsan, *Glitter*, p. 168.
45 Quoted in *Pittsburgh Times*, 18 November 1913.
46 Quoted in *New York Tribune*, 22 April 1913.
47 *Kansas City Journal*, 1 May 1913.
48 *St Louis Republic*, 1 May 1913.
49 An English correspondent writing in the *Houston Chronicle*, 21 April 1913.
50 *New York Journal*, 30 April 1913.
51 Quoted in Geidel, 'Forgotten Feminist', p. 412.
52 'Autobiographical Note', DSUP, p. 22.
53 Balsan, *Glitter*, p. 170.
54 Balsan, *Glitter*, p. 172.
55 'Autobiographical Note', DSUP, p. 23.
56 NWPP: 1913–1972, Series 1, Correspondence,1913–1972, Alva

Belmont Correspondence Scrapbook, pp. 12–13.
57 'Autobiographical Note', DSUP, p. 24.
58 *The New York Times*, 13 September 1913.
59 Quoted in Geidel, 'Forgotten Feminist', pp. 439–40.
60 Quoted in Geidel, 'Forgotten Feminist', p. 439.
61 'Autobiographical Note', DSUP, p. 25.
62 Quoted in Geidel, 'Forgotten Feminist', p. 456.
63 Alice Paul to Mary Hutchinson Page, 3 July 1914, NWPP, LCW.
64 Quoted in Geidel, 'Forgotten Feminist', p. 498.
65 *New York City Press*, 28 June 1914.
66 *The New York Times*, 6 July 1914.
67 *The New York Times*, 27 June 1914.
68 *The New York Times*, 6 July 1914.
69 *Providence RI Journal*, 7 July 1914.
70 *New York Herald*, 9 July 1914.
71 Quoted in Geidel, 'Forgotten Feminist', p. 507.
72 *The New York Times*, 9 July 1914.
73 *New York City American*, 18 July 1914.
74 Doris Stevens to Alice Paul, 8 July 1914, NWPP, LCW.
75 *New York Call*, 12 July 1914.
76 Unidentified press cutting, 12 July 1914, Alva Belmont Clippings Book, 1914. There were reports of Consuelo's speech in the *New York Evening Sun, Brooklyn Eagle, England Mercury, New York City American, Boston Post, Washington Post, Oakland Tribune, Syracuse Herald, Cleveland Plain Dealer, Baltimore Sun, Atlanta Constitution* and *Louisville Post*, among others.
77 *Town Topics*, 16 July 1914.
78 Doris Stevens to Alice Paul, 8 July 1914, NWPP, LCW.
79 Alice Paul to Alva Belmont, 11 July 1914, NWPP National Woman's Party Papers LCW.
80 Alice Paul to Alva Belmont, 11 July 1914, NWPP, LCW.
81 Alice Paul to Alva Belmont, 16 July 1914, NWPP, LCW.
82 Doris Stevens to Alice Paul, undated July 1914, NWPP, LCW.
83 *Chicago Record*, 21 July 1914.
84 *Town Topics*, 23 July 1914.
85 *Belmont Memoirs (Young)*, p. 163.
86 *Town Topics*, 30 July 1914.
87 *Holyoke Transcript*, 24 July 1914, quoted in Geidel, 'Forgotten Feminist', p. 513.
88 *Dunkirk New York Observer*, 16 July 1914.

10 LOVE, PHILANTHROPY AND SUFFRAGE

1 *Town Topics*, 6 August 1914.
2 Lord Lansdowne to Duke of Marlborough, 9 May 1911 Marlborough Papers, LCW.
3 Balsan, *Glitter*, p. 158.
4 Balsan, *Glitter*, p. 189.
5 Balsan, *Glitter*, p. 162.
6 M. G. Hoare to C., 14 August 1953, VA.
7 Quoted in J. Musson, *The English Manor House: from the Archives of Country Life* (London: Aurum Press, 1999), p. 65.
8 C. Aslet, *The Last Country Houses* (New Haven and London: Yale University Press, 1982), p. 162.
9 P. Morand, *Nouvelles Complètes*, Vol. 1., edited by Michel Collomb and translated by Aoife Ní Luanaigh (Paris: Éditions Gallimard, 1992), pp. xl, 750 and 748.
10 Morand, *Nouvelles Complètes*, p. 758.
11 Morand, *Nouvelles Complètes*, pp. 764 and 748.
12 Consuelo to Lord D'Abernon, 14 September 1912 in Add. MS 48939, D'Abernon Papers, British Library.
13 Duke of Marlborough to GD from Hotel Bristol, Beaulieu-sur-Mer, 18 March 1913, GDP.
14 Duke of Marlborough to GD

from Hotel Bristol, 18 March 1913, GDP.

15 Duke of Marlborough to GD, Easter Sunday 1913, GDP.
16 Duke of Marlborough to GD, 13 September 1912, GDP.
17 Clipping from *Town Topics*, 15 March 1913, GDP.
18 *The Times*, 22 November 1910.
19 Duke of Marlborough to GD, 17 July 1915, GDP.
20 Lord Lansdowne to WSC, 7 July 1915, CHAR 2/67/26, CA.
21 Duke of Marlborough to GD, 8 July 1915, GDP.
22 Balsan, *Glitter*, p. 178.
23 Balsan, *Glitter*, p. 177.
24 Balsan, *Glitter*, p. 183.
25 Duke of Marlborough to GD, 27 December 1917, GDP.
26 Duke of Marlborough to GD, 19 January, probably 1916, GDP.
27 Duke of Marlborough to GD, undated, GDP.
28 D. Stevens, *Jailed For Freedom: American Women Win the Vote*, edited by C. O'Hare from the original edition (Troutdale: New Sage Press, 1995; first published in 1920), p. 44.
29 R. B. Mackay, A. K. Baker and C. Traynor (eds) *Long Island Country Houses and Their Architects 1860–1940* (New York: Society for the Preservation of Long Island Antiquities in association with W. W. Norton & Co., 1997), p. 231.
30 Quoted in Geidel, 'Forgotten Feminist', p. 566.
31 E. Maxwell, *The Celebrity Circus* (London: W. H. Allen, 1964), p. 12.
32 Maxwell, *Celebrity Circus*, p. 13.
33 Memo on Elsa Maxwell, DSUP, p. 13.
34 *The New York Times*, 28 December 1915.
35 Mrs O. H. P. Belmont and Elsa Maxwell, (music and lyrics by Elsa Maxwell), *Melinda and Her Sisters* (New York: Robert J. Shores, 1916), pp. 3–4.
36 *Theater Magazine*, February 1916.
37 Belmont and Maxwell, *Melinda and Her Sisters*, pp. 30, 32. Alva is also said to have secretly contributed $10,000 secretly to the Southern Woman Suffrage Conference which was explicitly against black women having the vote saying: 'I plead guilty to so strong a desire for the political emancipation of women that I am not at all particular as to how it shall be granted,' (see C. Stasz, *The Vanderbilt Women: dynasty of wealth, glamour and tragedy*; this edition San Jose, New York: to Exel 1999, first published New York: St Martin's Press, 1991, p. 215).
38 *New York Telegraph*, 19 February 1916.
39 *New York Sun*, 19 February 1916.
40 Memo on relationship with Mrs Belmont, DSUP, p. 6.
41 *Alice Paul*, Regional Oral History Office, University of California, Berkeley, interview by Amerlia R. Fry, November 1972 and May 1973, 'Congressional Union Becomes the Woman's Party', pp. 347–48.
42 Memo on relationship with Mrs Belmont, DSUP, p. 6.
43 SBF to CESW, 2 August 1917, CESWP.
44 SBF to CESW, 16 August 1917, CESWP.
45 Alva E. Belmont, *Log of the Seminole* (New York: privately printed, 1916), p. 67.
46 Belmont, *Log of the Seminole*, pp. 67–8.
47 Belmont, *Log of the Seminole*, p. 64.
48 Belmont, *Log of the Seminole*, p. 64.
49 Belmont, *Log of the Seminole*, p. 28.
50 Belmont, *Log of the Seminole*, p. 23.
51 Belmont, *Log of the Seminole*, p. 33.

52 Belmont, *Log of the Seminole*, pp. 48–9.
53 Stevens, *Jailed For Freedom*, p. 58.
54 Quoted in Geidel, 'Forgotten Feminist', p. 630.
55 *Sara Bard Field*, Online Archive of California, (http://ark.cdlib.org./ark:/13031/htisp300ini), interview by Amerlia R. Fry between 1 October 1959 and 31 October 1963, XXIV, 'Politics, Newport and Tragedy, Mrs Belmont's "Autobiography" ', p. 369.
56 SBF to CESW, 16 August 1917, CESWP.
57 SBF to CESW, 27 July 1917, p. 3, CESWP.
58 SBF to CESW, 16 August 1917, CESWP.
59 SBF to CESW, 24 August 1917, p. 1, CESWP.
60 SBF to CESW, 27 July 1917, p. 3, CESWP.
61 SBF to CESW, 27 July 1917, p. 3, CESWP.
62 SBF to CESW, 29 July 1917, p. 3, CESWP.
63 *Sara Bard Field*, Online Archive of California, interview by Amerlia R. Fry between 1 October 1959 and 31 October 1963, XXVI, 'Adjustment, Digression: Mrs Belmont's Divorce', p. 391.
64 SBF to CESW, 29 July 1917, p. 7, CESWP.
65 SBF to CESW, 31 July 1917, p. 3, CESWP.
66 *Sara Bard Field*, Online Archive of California, 'Mrs Belmont's "Autobiography" ', p. 371.
67 *Sara Bard Field*, Online Archive of California, 'Mrs Belmont's "Autobiography" ', pp. 371–72.
68 Notes on conversation with Elsie Powers, Collection of the Preservation Society of Newport County.
69 *Sara Bard Field*, Online Archive of California, interview by Amerlia R. Fry between 1 October 1959 and 31 October 1963, XXIX, 'The Woman's Party, Peace and Religion', 12 July 1962 (no page nos).
70 'Note de service pour le feuillet du personnel du Commandant Balsan', probably written in 1919 and supporting the award of the *Croix de Guerre avec Palme*. Jacques had already been made Chevalier of the Légion d'Honneur in January 1913 for his Moroccan expeditions.
71 Quoted in P. M. Flammer, *The Vivid Air: the Lafayette Escadrille* (Athens, Georgia: The University of Georgia Press, 1981), p .11.
72 Montgomery-Massingberd, *Blenheim Revisited*, p. 172.
73 Montgomery-Massingberd, *Blenheim Revisited*, p. 172.
74 Balsan, *Glitter*, p. 179.
75 Balsan, *Glitter*, p. 179.
76 Balsan, Glitter, p. 144.
77 *World*, 13 November 1917.
78 Quoted in Geidel, 'Forgotten Feminist', p. 646.
79 Quoted in Geidel, 'Forgotten Feminist', p. 648.
80 Quoted in Geidel, Forgotten Feminist, p. 650.
81 Balsan, *Glitter*, p. 158.
82 W. Lee to author, April 2005.
83 W. Lee to author, April 2005.
84 Quoted in *The Times*, 24 February 1922.
85 Balsan, *Glitter*, p. 212.
86 Balsan, *Glitter*, p. 189.
87 Balsan, *Glitter*, p. 187.
88 Balsan, *Glitter*, p. 189.
89 *The New York Times*, 24 March 1920.
90 The Duke of Marlborough to GD, 10 November 1920, GDP.
91 Quoted in Vickers, *Gladys*, p. 173.
92 Quoted in Vickers, *Gladys*, p. 173.
93 Quoted in Vickers, *Gladys*, p. 175.
94 Balsan, *Glitter*, p. 190.

11 A STORY RE-TOLD

1 Balsan, *Glitter*, p. 113.
2 Balsan, *Glitter*, pp. 193–94.

3 William K. Vanderbilt Probate Papers, File No. 24398. It was also reported in the press that he transferred $15 million to Consuelo shortly before his death but there is no record of this in his probate papers.
4 Balsan, *Glitter*, p. 194.
5 L. H. Prost, *Collection de Madame et du Colonel Balsan* (Paris: privately printed *c*. 1930).
6 Balsan, *Glitter*, p. 197.
7 Balsan, *Glitter*, p. 203.
8 R. Cameron, *The Golden Riviera* (London: Weidenfeld and Nicolson, 1975), p. 46.
9 G. Vanderbilt and Lady Furness, *Double Exposure: A Twin Autobiography* (London: Frederic Muller Ltd, 1959), p. 185.
10 CSC to WSC *c*. 8 March 1925, CHAR 1/179/5–7, CA.
11 Soames, *Clementine*, p. 288. Winston Churchill's description of the Riviera scenery as 'paintatious' appears on p. 287.
12 CSC to WSC, 24 February 1924, quoted in M. Soames (ed.), *Speaking for Themselves: The Personal Letters of Winston and Clementine Churchill* (London: Black Swan, 1999; first published London: Doubleday, 1998), p. 279.
13 CSC to WSC, c. 8 March 1925, CHAR 1/179/5–7, CA.
14 Balsan, *Glitter*, p. 190.
15 S. Obolensky, *One Man in His Time: The Memoirs of Serge Obolensky* (New York: McDowell, Obolensky, 1958), p. 303.
16 Balsan, *Glitter*, p. 222.
17 Balsan, *Glitter*, p. 219.
18 CSC to WSC, *c*. 8 March 1925, CHAR 1/179/5–7, CA.
19 Balsan, *Glitter*, p. 218.
20 Cameron, *Golden Riviera*, p. 46.
21 CSC to WSC, *c*. 8 March 1925, CHAR 1/179/5–7, CA.
22 Balsan, *Glitter*, p. 204.
23 'Chez Madame Balsan' in Edith Wharton Papers.
24 Quoted in Adeline R. Tintner, 'Consuelo Vanderbilt, John Esquemeling and the Buccaneers', in *Edith Wharton in Context: Essays on Intertextuality* (Tuscaloosa and London: The University of Alabama Press, 1999), p. 146.
25 Balsan, *Glitter*, p. 215.
26 *Washington Star*, 2 April 1922.
27 Memorandum: General Statement, p. 7, DSUP.
28 S. J. Morris, *Rage for Fame: The Ascent of Clare Boothe Luce* (New York: Random House, 1997), p. 114. Clare Boothe Luce is quoted as saying in her diary: 'I don't like older women who get a crush on girls, and several of them around here make entirely too great a fuss over me for it to be comfortable.'
29 Memo: General Statement, p. 8, DSUP.
30 Maxwell, *I Married the World*, pp. 81–2.
31 Memo: General Statement, p. 10, DSUP.
32 Memo: General Statement, p. 4, DSUP.
33 Memo: General Statement, p. 6, DSUP.
34 E. Drexel Lehr (Lady Decies), *The Turn of the World*, (Philadelphia: J. B. Lippincott Co., 1937), p. 44.
35 'Woman as Dictators', *Ladies Home Journal*, 39, September 1922.
36 'Woman as Dictators'.
37 'Woman as Dictators'.
38 'Woman as Dictators'.
39 'Woman as Dictators'.
40 *New York City Journal*, 25 October 1922.
41 *The New York Times*, 1 October 1922.
42 Quoted in Geidel, 'Forgotten Feminist', p. 692.
43 Geidel, 'Forgotten Feminist', p. 693.
44 *Chicago Tribune*, 1 June 1926.
45 Quoted in Geidel, 'Forgotten Feminist', p. 705.

46 Quoted in Geidel, 'Forgotten Feminist', p. 706.
47 CSC to WSC from Grantully Castle, Perthshire, 6 September 1926, quoted in Soames (ed.), *Speaking for Themselves*, p. 299.
48 CSC to WSC, Grantully, 6 September 1926, in Soames (ed.), *Speaking for Themselves*, p. 299.
49 CSC to WSC, Grantully, 6 September 1926, Soames (ed.), *Speaking for Themselves*, p. 299.
50 See Vickers, *Gladys*, pp. 208–09.
51 Balsan, *Glitter*, p. 192.
52 Balsan, *Glitter*, p. 193.
53 Balsan, *Glitter*, p. 192.
54 'Marlborough Gets Marriage Annulled by Catholic Court', *The New York Times*, 13 November 1926.
55 'Marlborough Gets Marriage . . .', *The New York Times*, 13 November 1926.
56 'Duchess Obtained Annulment Decree in Marlborough Suit', *The New York Times*, 14 November 1926.
57 'Women have given their time, their energy and their money to support the church,' she wrote. 'We are allowed to sit in the pews, but not to stand in the pulpit. The men of the church accept our support, but are not willing to share their exalted position with us. We are required to acknowledge man as our spiritual superior. We do not acknowledge him as such, and we know that Christ did not acknowledge him.' Women should be allowed to be priests, argued Alva. 'If you are worthy to be a priest be a priest, but if I am worthy to be a priest I shall be a priest too. We will both be priests.' 'Woman as Dictators', *Ladies Home Journal*.
58 Memorandum: General Statement, p. 12, DSUP.
59 'Mrs Belmont Twits Bishop About Gift', *The New York Times*, 28 April 1926.
60 *Chicago Tribune*, 1 June 1926.
61 Elsa Maxwell, *I Married the World* (London: William Heinemann Ltd, 1955), p. 80.
62 *The New York Times*, 15 November 1926.
63 *The New York Times*, 18 November 1926.
64 *The New York Times*, 22 November 1926.
65 *The New York Times*, 23 November 1926.
66 *The New York Times*, 19 November 1926.
67 *The New York Times*, 20 November 1926.
68 *The New York Times*, 26 November 1926.
69 *The New York Times*, 26 November 1926.
70 *The New York Times*, 25 November 1926. These extracts were translated from an advance copy of the summary that appeared in *Acta Apostolicae Sedis* in December 1926. The translation that appeared seems to have been commissioned by *The New York Times*. Most of it was from the witness statements in French rather than the summary in Latin.
71 *The New York Times*, 18 November 1926.
72 *World*, 17 November 1926.
73 'Former Duchess Was Not Driven Into Marriage', *People*, 21 November 1926.
74 *The New York Times*, 21 November 1926.
75 *The New York Times*, 27 November 1926.
76 *The New York Times*, 28 November 1926.
77 *The New York Times*, 30 November 1926.
78 Quoted in Candace Waid's introduction to *The Buccaneers* (London: J. M. Dent, 1993; Everyman 1996), p. xix.

79 Balsan, *Glitter*, p. 193.
80 *The New York Times*, 25 November 1926 and *Acta Apostolicae Sedis*, December 1926.
81 'Woman as Dictators', *Ladies Home Journal*.
82 Balsan, *Glitter*, p. 192.
83 *Sara Bard Field*, Online Archive of California (http://ark.cdlib.org/ark:/13030/htip300ini), interview by Amerlia R. Fry recorded between 1 October 1959 and 31 October 1963, XIX, 'The Demands of the Suffrage Movement; The Politics of Women's Suffrage', p. 301.
84 *The New York Times*, 1 December 1926.

12 FRENCH LIVES

1 Quoted in Geidel, 'Forgotten Feminist', p. 758.
2 Quoted in Geidel, 'Forgotten Feminist', p. 758.
3 'The Marlborough-Vanderbilt marriage annulment': letters of the Right Reverend John J. Dunn V.G. Auxiliary Bishop of New York, and of Charles C. Marshall to *The New York Times*, November-December 1926.
4 Balsan, *Glitter*, p. 224.
5 Balsan, *Glitter*, p. 228.
6 Balsan, *Glitter*, p. 189.
7 Balsan, *Glitter*, p. 228. She also medievalised her estate by building a long crenellated wall round its boundary.
8 Quoted in Geidel, 'Forgotten Feminist', p. 694.
9 Balsan, *Glitter*, p. 185.
10 *New York Sun*, 28 September 1928.
11 Quoted in Geidel, 'Forgotten Feminist', p. 708.
12 Quoted in Geidel, 'Forgotten Feminist', p. 748.
13 Quoted in Geidel, 'Forgotten Feminist', p. 748.
14 Balsan, *Glitter*, pp. 228–29. In fact the statue, which is still in the church at Augerville, is not life-sized.
15 *The New York Times*, 7 September 1928.
16 Balsan, *Glitter*, p. 229.
17 Geidel, 'Forgotten Feminist', p. 714.
18 Quoted in Geidel, 'Forgotten Feminist', p. 751.
19 This exchange is quoted in Geidel, 'Forgotten Feminist', pp. 753–54.
20 Maxwell, *I Married the World*, p. 174.
21 Geidel, 'Forgotten Feminist', p. 719.
22 Memo: General Statement, p. 20, DSUP.
23 Additional memorandum concerning events at Augerville, July 1931, p. 13, DSUP.
24 Additional memorandum Concerning Mary Young, p. 10, DSUP.
25 Matter of Alva E. Belmont Deceased, Stenographer's Minutes, New York, 19 May 1933.
26 Unidentified newspaper cutting, Alva Belmont's Clippings Book, 1932.
27 Matilda Young to her mother, 25 August 1932. Matilda Young was the sister of Mary Young, Alva's secretary from 1928. According to Doris Stevens, Mary Young was often very unhappy on account of Alva's rages and Matilda seems to have joined her in Augerville out of solidarity, sometimes running errands for Alva and helping to push her round the grounds of the chateau. Matilda Young Papers.
28 Young to her mother, 13 August 1932, Matilda Young Papers.
29 Young to her mother, 12 September 1932, Matilda Young Papers.
30 Dr Edmund Gros to Matilda Young, 13 September 1932. Dr Gros sounds more fun than his

colleague Dr Fuller who was described by Matilda as 'a nice little man but he doesn't row a boat as well as Dr Gros'.

31 *Alice Paul*, Regional Oral History Office, University of California, Berkeley, interview by Amerlia R. Fry, November 1972 and May 1973, in IV: 'The Equal Rights Amendment: Coup D'Etat Within the Women's Party 1946–1947', p. 564.
32 Balsan, *Glitter*, p. 229.
33 *The New York Times*, Monday 13 February 1933.
34 Press release for Alva's funeral on Sunday 12 February 1933, prepared by National Woman's Party, Jane Norman Smith Papers.
35 Quoted by the Women's Resource Project at www.ibiblio.org.
36 'Hymn to be sung at Mrs Belmont's funeral', Jane Norman Smith Papers.
37 *The New York Times*, 13 Monday 1933.
38 Memo: General Statement, p. 28, DSUP.
39 Eulogy given by Doris Stevens at Alva's funeral, Sunday 12 February 1933, DSUP.
40 'Why Women Went to Jail', draft of article by Alva Belmont, DSUP.
41 *Alice Paul*, Regional Oral History Office, University of California, Berkeley, interview by Amerlia R. Fry, November 1972 and May 1973, in III: 'The Suffrage Campaign Reviewed', p. 331.
42 *Belmont Memoirs (Field)*, p. 61.
43 *Chicago Tribune* (French edition), 1 June 1926.
44 Maxwell, *I Married the World*, pp. 70, 80.
45 Alva Belmont's Clippings Book, 1932.
46 Maxwell, *I Married the World*, pp. 84–5.
47 Quoted in Geidel, 'Forgotten Feminist', p. 748.
48 Maxwell, *I Married the World*, p. 84.
49 *Alice Paul*, Regional Oral History Office, University of California, Berkeley, interview by Amerlia R. Fry, November 1972 and May 1973, in IV: 'The Equal Rights Amendment', p. 564.
50 Auchincloss, *Vanderbilt Era*, p. 110.
51 Maxwell, *I Married the World*, p. 83.
52 Balsan, *Glitter*, p. 219.
53 Auchincloss, *Vanderbilt Era*, p. 110.
54 Auchincloss, *Vanderbilt Era*, p. 110.
55 Harold S. Vanderbilt, *Contract Bridge: Bidding and The Club Convention* (New York and London: Charles Scribner's Sons, 1929), p. 11.
56 Robert N. Bavier, *The America's Cup: an Insider's View: 1930 to the Present* (New York: Bob Bauier Dodd Mead and Co. Inc., 1986) p. 12.
57 Balsan, *Glitter*, p. 230.
58 Quoted in Vickers, *Gladys*, p. 202.
59 Quoted in Vickers, *Gladys*, p. 205.
60 Vickers, *Gladys*, p. 223.
61 Vickers, *Gladys*, p. 223.
62 Quoted in Vickers, *Gladys*, p. 1.
63 Quoted in Vickers, *Gladys*, pp. 227–28.
64 Quoted in Vickers, *Gladys*, p. 228.
65 Quoted in Vickers, *Gladys*, p. 237.
66 Spencer-Churchill and Martindale, *Charles IXth Duke of Marlborough*, p. 7.
67 Balsan, *Glitter*, p. 202.
68 Balsan, *Glitter*, p. 225.
69 Balsan, *Glitter*, p. 233.
70 C to WSC, 12 October 1935, CHAR 2/237/111, CA.
71 C to WSC, 12 October 1935, CHAR 2/237/111, CA.
72 Balsan, *Glitter*, p. 227.
73 Balsan, *Glitter*, p. 227.
74 C to WSC, 20 September 1938, CHAR 8/596, CA, quoted in M. Gilbert, *Winston S. Churchill, Vol*

V, Companion Part 3: Documents, The Coming of War 1936–1939 (London: Heinemann, 1982), pp. 1169–70.
75 C to WSC, 18 October 1938, CHAR 2/332, CA, quoted in Gilbert, *Churchill, Vol V, Companion Part 3*, p. 1232.
76 WSC to C, 20 October 1938, CHAR 2/332, CA, quoted in Gilbert, *Churchill, Vol V, Companion Part 3*, p. 1232.
77 Balsan, *Glitter*, p. 236.
78 Balsan, *Glitter*, pp. 235–36.
79 A. Lambert, *1939: The Last Season of Peace* (London and New York: Weidenfeld and Nicolson, 1989), pp. 170–71.
80 Balsan, *Glitter*, p. 236.
81 Soames, *Clementine*, pp. 313–4.
82 Mary Soames to author, June 2002.
83 WSC, 'Recollections', 20 August 1939, 4/114, CA, quoted in Gilbert, *Churchill, Vol V, Companion Part 3*, p. 1591.
84 Paul Maze diary, 20 August 1939, from Maze Papers quoted in Gilbert, *Churchill, Vol V, Companion Part 3*, p. 1591.
85 Maze diary, 20 August 1939, from Maze Papers quoted in Gilbert, *Churchill, Vol V, Companion Part 3*, p. 1592.
86 Mary Soames to author, June 2002.
87 Soames, *Clementine*, p. 314.
88 W. H. Thompson, 'Recollections (Sixty Minutes with Winston Churchill)', 22 August 1939, quoted in Gilbert, *Churchill, Vol V, Companion Part 3*, p. 1592.
89 Balsan, *Glitter*, p. 238.
90 H. Molloy, *The Lafayette Escadrille* (New York: Random House, 1964), pp. 288–89.
91 Balsan, *Glitter*, pp. 225 and 230.
92 Balsan, *Glitter*, pp. 230–32, passim.
93 Balsan, *Glitter*, pp. 240–41.
94 Balsan, *Glitter*, p. 242.
95 Balsan, *Glitter*, pp. 244–45 passim.
96 H. R. Kedward, *Resistance in Vichy France* (Oxford: OUP, 1978), p. 8.
97 Balsan, *Glitter*, p. 247. All direct quotations to the end of the chapter now come from Balsan, *Glitter*, pp. 247–59 passim.

13 HARVEST ON HOME GROUND

1 Balsan, *Glitter*, p. ix.
2 US Naturalisation Certificate, 8 November 1940, CVBS.
3 Mackay et al (eds), *Long Island Country Houses*, p. 401. Old Fields was originally built by Treanor and Fatio for George Backer and was adapted from the design of a mid-eighteenth century Tidewater plantation house.
4 Diary entry for 5 January 1942 in Lord Moran, *Winston Churchill: The Struggle for Survival 1940–1965* (London: Heron Books, 1966), p. 21.
5 Maxwell, *I Married the World*, p. 18.
6 Maxwell, *I Married the World*, p. 18.
7 Maxwell, *I Married the World*, p. 18.
8 Quoted in M. Gilbert, *Winston S. Churchill 1941–45, Vol. VII: Road to Victory* (London: Heinemann, 1986), p. 38.
9 Diary entry for 9 January 1942 in Lord Moran, *Winston Churchill*, p. 22.
10 H. R. Kedward, *Resistance in Vichy France* (Oxford: Oxford University Press, 1978), p. 16.
11 Maxwell, *I Married the World*, p. 292.
12 Jacques Balsan to Myron Taylor and Franklin D. Roosevelt, 17–19 May 1942, Franklin D. Roosevelt Digital Archives (www.fdrlibrary.marist.edu/psf/box 51).
13 Catalogue: 'French and English Art Treasures of the Eighteenth

Century', Parke-Bernet Galleries, 20–30 December 1942, Foreword by Francis Henry Taylor. The exhibition was in aid of the American Women's Voluntary Services and the catalogue was presented by Consuelo to the Library of Congress on 16 June 1943.

14 C to WSC, 23 September 1943, CHAR 20/95A/74, CA.

15 Prime Minister's Personal Minute to Desmond Morton (the Prime Minister's Personal Assistant), Serial no DM9/3, 3 November 1943, CHAR 20/95A/77, CA.

16 Major Morton to WSC, 8 November 1943, CHAR 20/95A/79, CA.

17 See Foreign Office Minutes on visit by Colonel Balsan, 21 September 1943, FO/954/8B, National Archives.

18 Major Morton to WSC, 8 November 1943, CHAR 20/95A/79, CA.

19 WSC to C, 16 November 1943, CHAR 20/95A/82, CA.

20 *The Herald Tribune*, 19 December 1943.

21 10th Duke of Marlborough to WSC, October 1944, saying that his mother has had '800 refugees billeted on her' and asking whether he could accompany WSC on his next visit to France so that he could 'motor from Paris to see her'. This suggests that Consuelo did return briefly to St Georges-Motel in October 1944 with Jacques. CHAR 1/380/35–36, CA.

22 French Consul General to C, New York, 30 July 1947, CVBS.

23 Maxwell, *I Married the World*, pp. 291–92.

24 V. Lawford, 'Madame Jacques Balsan: Portrait of a Unique American' was first published in *Vogue*, 1 February 1963 and then republished in Valentine Lawford, *Vogue's Book of Houses, Gardens, People* – with photographs by Horst and an introduction by Diana Vreeland – (London: Bodley Head, 1968), p. 8.

25 Lawford, 'Madame Jacques Balsan', in *Vogue's Book of Houses*, p. 8.

26 Sale catalogue: 'Estate of Lady Sarah Consuelo Spencer-Churchill', Tuesday 15 May 2001, Doyle New York, p. 49.

27 Margarette Blouin to author, 3 April 2001.

28 Lawford, 'Madame Jacques Balsan', p. 8.

29 Quoted in M. Soames, *Winston Churchill: His Life as a Painter* (London: Collins, 1990), p. 151.

30 Quoted in Soames, *Churchill: His Life as a Painter*, p. 151.

31 Louis Auchincloss to author, Wednesday 25 July 2001.

32 Green, *Churchills of Blenheim*, p. 164.

33 Green, *Churchills of Blenheim*, p. 185.

34 Green, *Churchills of Blenheim*, p. 180.

35 Green, *Churchills of Blenheim*, p. 163.

36 V. Lawford, *Horst: His Work and His World* (London: Viking, 1985) p. 353.

37 C. Vanderbilt Jr, *Queen of the Golden Age: The Fabulous Story of Grace Wilson Vanderbilt* (Maidstone, Kent: George Mann, 1999; first published New York: McGraw-Hill, 1956), p. 268.

38 Patterson, *Vanderbilts*, p. 281.

39 Patterson, *Vanderbilts*, p. 248.

40 Auchincloss, *Vanderbilt Era*, p. 105.

41 See, for example, Stasz, *The Vanderbilt Women* p. 364: 'a highly sanitized, ghostwritten view of her life'.

42 C to Cass Canfield, 10 July 1951, HRP.

43 Cass Canfield to C, 4 January 1952, HRP.

44 Cass Canfield to C, 17 July 1951, HRP.
45 C to Cass Canfield, 27 July 1951, HRP.
46 Green, *Churchills of Blenheim*, p. 176.
47 Telegram from C to Cass Canfield, 28 December, 1951.
48 WSC to C, 27 June 1952, HRP.
49 WSC to Cass Canfield, 30 June 1952, HRP.
50 C to Marguerite S. Hoyle, 23 October 1952, HRP.
51 C to Marguerite S. Hoyle, 5 August 1952, HRP.
52 M. S. Hoyle to C, 11 August 1952, HRP.
53 Balsan, *Glitter*, pp. 229–30.
54 *John O'London's Weekly*, 1 May 1953.
55 *Punch*, 13 May 1953.
56 *Punch*, 13 May 1953.
57 *New York Herald Tribune*, interview with Consuelo by John K. Hutchens, 28 September 1952.
58 D. Fielding, *The Face on The Sphinx: A Portrait of Gladys Deacon, Duchess of Marlborough* (London: Hamish Hamilton, 1978), p. 47.
59 See jacket cover of Balsan, *The Glitter and the Gold* (1973 edition).
60 Cass Canfield to C, 7 October 1952.
61 WSC to C, London, 8 June 1953, CVBS.
62 *The Times*, Obituary, Tuesday 18 September 1956.
63 *The Times*, Harold Nicholson, Wednesday 19 September 1956.
64 Press cutting, November 1956, CVBS.
65 Serena Balfour to author, 23 April 2001.
66 V. Lawford, 'Madame Jacques Balsan', in *Vogue's Book of Houses*, p. 9.
67 R. H. Pilpel, *Churchill in America 1895–1961: An Affectionate Portrait* (London: New English Library, 1977; first published Harcourt Brace Jovanovich, 1976), p. 270.
68 *Diplomat*, October 1960. According to the article, she outclassed 'Mrs "Chessy" Lewis Amory, Mrs Donald Leas, Mrs Byrnes McDonald' and 'any of the deb contingent'.
69 *The New York Journal-American*, undated cutting, CVBS.
70 C to Katherine Warren, Preservation Society of Newport County, 2 October 1964.
71 Lawford, *Horst*, p. 354.
72 V. Lawford, 'Madame Jacques Balsan', p. 7.
73 V. Lawford, 'Madame Jacques Balsan', p. 10.
74 V. Lawford, 'Madame Jacques Balsan', p. 11.
75 V. Lawford, 'Madame Jacques Balsan', p. 10.
76 V. Lawford, 'Madame Jacques Balsan', p. 10.
77 *Diplomat*, October 1960.
78 Serena Balfour to author, 23 April 2001.
79 *The New York Times*, 10 December 1964.
80 *Oxford Times*, 18 December 1964.
81 *New York Herald Tribune*, interview with Consuelo by J. Hutchens, 28 September 1952.

AFTERWORD

1 D. Vreeland, 'American Women of Style', exhibition catalogue (New York: Costume Institute, Metropolitan Museum of Art, 1975), with an introduction by curator, S. Blum (no page nos).
2 See www.canadianinteriordesign.com/diana_vreeland.htm: Diana Vreeland 1906–1989.
3 See www.canadianinteriordesign.com/diana_vreeland.htm: Diana Vreeland 1906–1989.
4 See D. Silverman, *Selling Culture: Bloomingdale's, Diana Vreeland and the New Aristocracy of Taste*

in Reagan's America (New York: Pantheon Books, 1986), passim.

5 See www.Bartleby.com/63/89/6089.html: 'Simpson's Contemporary Quotations', compiled by James B. Simpson, 1988.

6 S. Blum introduction to Vreeland, 'American Women of Style', exhibition catalogue.

7 Viscount Churchill, *All My Sins Remembered* (London: William Heinemann Ltd, 1964), pp. 33–4.

8 Consuelo Vanderbilt Balsan: 'Last Will and Testament'.

BIBLIOGRAPHY

I. Works by Consuelo (in chronological order)

Marlborough, the Duchess of, 'The Position of Women', I, II and III, *North American Review*: Vol. 189: no. 1 (January 1909); Vol. 189: no. 2 (February 1909); Vol. 189: no. 3 (March 1909)

—— 'Hostels for Women', *The Nineteenth Century and After*, May 1911, Vol. IX

—— 'Saving The Children,' Lady Priestley Memorial Lecture, *National Health Society*, 29 June 1916

Balsan, Consuelo Vanderbilt, *The Glitter and the Gold* (Maidstone: George Mann, 1973; first published New York: Harper & Brothers, 1952; London: William Heinemann, 1953)

II Works by Alva (in chronological order)

Printed works

Belmont, Alva Erskine, *Log of the Seminole* (New York: privately printed, 1916)

—— *Melinda and Her Sisters*, with Elsa Maxwell, music and lyrics by Elsa Maxwell (New York: Robert J. Shores, 1916)

Selected Articles

Belmont, Mrs O. H. P., 'Woman's Right to Govern Herself', *North American Review*, November 1909, Vol. 190

—— 'Belief in Women is Belief in Women's Suffrage,' *Women's Magazine*, December 1909

—— 'Woman and the Suffrage,' *Harper's Bazaar*, March 1910

—— 'Woman Suffrage as it Looks Today,' *Forum*, March 1910

—— 'How Can Women Get the Suffrage?' *Independent*, 31 March 1910

—— 'Why I Am a Suffragist,' *World Today*, October 1911, Vol. XXI

—— *Chicago Sunday Tribune* series, April-November 1912:

'How Suffrage Will Protect Women From Men Who "Sow Wild Oats"', 28 April 1912; 'Why Women Need the Ballot', 12 May 1912; 'Votes For Women Will Improve Existing Conditions', 26

May 1912; 'A Son Loses Respect For His Mother the Day He Votes', 2 June 1912; 'A Girl? What a Pity It Was Not a Boy!', 9 June 1912; 'In What Respect Do Women Differ From Slaves or Serfs?', 16 June 1912; 'Woman's Suffrage Raises the Quality of the Electorate', 30 June 1912; 'Woman Suffragists Ask For Progressive Constitution', 7 July 1912; 'Do Not Let the Women Vote – Slogan of Political Bosses', 14 July 1912; 'In Nonvoting States Women are Classed with Lunatics', 21 July 1912; 'Man Has Failed to Care For Women and Children', 28 July 1912; 'Are Politicians Seeking a Flirtation with Suffragists?', 4 August 1912; 'What Place Will Women Take in Our Political Life?', 15 September 1912; 'We Must Not Be Made the Laughing Stock of the Voters', 27 October 1912

—— 'The Liberation of a Sex', *Hearst's Magazine*, April 1913, Vol. 23

—— 'Jewish Women in Public Affairs', *American Citizen*, May 1913, Vol. 232

—— Foreword to Christabel Pankhurst, 'Story of the Woman's War,' *Good Housekeeping*, November 1913, Vol. 57

—— 'New Standards for Business Women', *Business Woman's Magazine*, January 1915

—— 'Women as Dictators', *Ladies Home Journal*, September 1922, Vol. 39

—— 'Are Women Really Citizens?', *Good Housekeeping*, September 1931, Vol. 93

III Manuscript Sources Consulted

Alva Belmont Clippings Books, Sewall-Belmont House and Museum, Washington DC

Alva Belmont Correspondence Scrapbook, National Woman's Party Papers: 1913–1972, Microfilming Corporation of America *c*.1981, Women's Library, London Metropolitan University

Asquith Papers, Bodleian Library, University of Oxford

Bedford College Archives, Royal Holloway, University of London

Belmont Family Papers, Rare Book and Manuscript Library, Columbia University

Belmont Memoirs (Field), Papers of Sara Bard Field, in Charles Erskine Scott Wood Papers, Huntington Library, San Marino, California

Belmont Memoirs (Young), Papers of Matilda Young, Rare Book, Manuscript and Special Collections Library, Duke University, Durham, North Carolina

Church Army Papers, Cambridge University Library

Churchill Papers, Churchill Archives, Churchill College, University of Cambridge

Consuelo Vanderbilt Balsan Scrapbook and Photograph Album, courtesy of Serena Balfour
Curzon Papers, British Library
D'Abernon Papers, British Library
Desborough Papers, Hertfordshire County Council Record Office
Doris Stevens Unprocessed Papers, Schlesinger Library, Radcliffe Institute, Harvard University, Box 9, Folders 290–292
Edith Wharton Papers, Beinecke Library, Yale University
Gertrude Vanderbilt Whitney Papers, Archives of American Art, Smithsonian Institution, Washington, donated by the Whitney Museum of American Art
Gladys Deacon Papers, courtesy of Hugo Vickers
Harper and Row Archives, Rare Book and Manuscript Library, Columbia University
Helen Sioussat Papers, University of Maryland
Jane Norman Smith Papers, Schlesinger Library, Radcliffe Institute, Harvard University
London Metropolitan Archives
Marie Stopes Papers, British Library
Marlborough Papers, Manuscripts Division, Library of Congress
National Archives, Kew
National Woman's Party Papers, the Suffrage Years, 1913–1920, Group 1, Reels 10–12, Microfilming Corporation of America *c.* 1981, Library of Congress
Pearl Craigie Papers, Reading University Library
Richard Morris Hunt Archives, Prints and Drawings Collections, The Octagon Museum. The American Architectural Foundation, Washington D.C.
Royal Free Hospital Archives, Royal Free Hospital Archives Centre
Shane Leslie Papers, Georgetown University, Washington D.C.
Vanderbilt Archives, Suffolk County Vanderbilt Museum, Centreport, Long Island, New York
Vanderbilt Papers, New-York Historical Society
West Ham Hospital (Queen Mary's Hospital) Archives, Stratford Library Local Archives
William Gilmour Notebooks, Collection of the Preservation Society of Newport County, Newport, Rhode Island
William K. Vanderbilt Probate Papers, file no. 24398, Surrogate's Court of Suffolk County, Riverhead, Long Island, New York

IV Oral History

Fry, Amerlia R., 'Conversations with Sara Bard Field', 1959–1963, Online Archive of California (http://ark.cdlib.org./ark:/ 13031/ htisp300ini), 1 October 1959–31 October 1963, XXIV, Berkeley, Suffragists Oral History Project
—— 'Conversations with Alice Paul', 1972–73, Suffragists Oral History Project, Regional Oral History Office, University of California, Berkeley

V Unpublished and Privately Printed Sources

French and English Art Treasures of the Eighteenth Century (Parke-Bernet Galleries, 1942)
Geidel, Peter, *'Alva E. Belmont: A Forgotten Feminist'*, PhD thesis: Columbia University, 1993
'In the Matter of Alva E. Belmont Deceased', Stenographer's Minutes, New York, 1933
Log of the Alva, Collection of the Preservation Society of Newport County
O'Sullivan, Mrs Fitzstephen, *Woman's Municipal Party: Formation and Work – A Statement* (London: WMP, 1913)
Prost, L. H., *Collection de Madame et du Colonel Balsan* (Paris: c.1930)
Sale Catalogue of the Estate of Lady Sarah Consuelo Spencer-Churchill (Doyle New York: 2001)
Townsend, Reginald, *Mother of Clubs, Being the History of the First Hundred Years of the Union Club, 1836–1936* (New York: The Union Club, 1936)
Vanderbilt Jr, William K., *Log of My Motor* (New York: 1912)
—— *Across the Atlantic with Ara* (New York: 1925)
—— *To Galapagos on the Ara* (New York: 1926)
—— *Fifteen Thousand Miles Cruise with Ara* (New York: c.1928)
—— *Taking One's Own Ship Around the World* (New York: 1929)
—— *Flying Lanes* (New York: 1937)
—— *West Made East with the Loss of a Day* (New York: 1933)

VI Published Sources

Acta Apostolicae Sedis, December 1926
Allen, Frederick Lewis, *The Lords of Creation* (New York: Harper and Brothers, 1935)
Amory, Cleveland, *The Last Resorts* (New York: Harper and Brothers, 1952)

Andrews, Wayne, *The Vanderbilt Legend; the Story of the Vanderbilt Family, 1794–1940* (New York: Harcourt, Brace and Company, 1941)
Anon., *The Fireside Miscellany and Young People's Encyclopaedia* (New York: 1854; see www.merrycoz.org/training.htm)
Archambault, Florence (ed.) *A Selection of Historical and Social Articles pertaining to Newport, Rhode Island*, (Newport, Rhode Island: Historic Newport Publishers, 1997)
Armstrong, Nancy, *Victorian Jewellery* (London: Studio Vista, 1976)
Ashley, Maurice, *Churchill as Historian* (London: Secker and Warburg, 1968)
Aslet, Clive, *The Last Country Houses* (New Haven and London: Yale University Press, 1982)
Auchincloss, Louis, *The Vanderbilt Era* (New York: Collier Books, Macmillan Publishing Co., 1989)
Baker, Paul R., *Richard Morris Hunt* (Cambridge, Mass.: MIT Press, 1980)
Balfour, Frances, *Ne Obliviscaris: dinna forget*, Vols 1 and 2 (London: Hodder and Stoughton, 1930)
Bavier, Robert N., *The America's Cup: An Insider's View, 1930 to the Present* (New York: Dodd Mead and Co., 1986)
Birkenhead, Earl of, *The Life of F.E. Smith: First Earl of Birkenhead, By His Son The Second Earl of Birkenhead* (London: Eyre and Spottiswoode, 1959)
Black, David, *The King of Fifth Avenue* (New York: Dial Press, 1981)
Blume, Mary, *Côte d'Azur* (London: Thames and Hudson, 1992)
Bourget, Paul, *Outre-Mer* (New York: Charles Scribner and Sons, 1895)
Boyer, Paul S. (editor in chief) et al., *Oxford Companion to American History* (Oxford and New York: Oxford University Press, 2001)
Brandon, Ruth, *The Dollar Princesses* (New York: Alfred A. Knopf, Inc., 1980)
Brough, James, *Consuelo: Portrait of an American Heiress* (New York: Coward, McCann & Geoghan, Inc., 1979)
Burnett, Robert N., 'Captains of Industry Part XXIII: William Kissam Vanderbilt', *Cosmopolitan*, 1904, 36, pp. 553–54
Bury, Shirley, *Jewellery 1789–1910, The International Era* (Woodbridge: Antique Collector's Club, 1991)
Cameron, Roderick, *The Golden Riviera* (London: Weidenfeld and Nicolson, 1975)
Cannadine, David, *Aspects of Aristocracy* (New Haven: Yale University Press, 1994)
—— *The Decline and Fall of the British Aristocracy* (London: revised edition, Papermac, 1996; this edition first published by Macmillan General Books, Picador, 1992; first published New Haven and London: Yale University Press, 1990)

Carnarvon, Earl of, *No Regrets* (London: Weidenfeld and Nicolson, 1976)

Catt, Carrie Chapman, 'Too Many Rights: Women Leaders Who Would Go Beyond All Decent Limits', *Ladies Home Journal*, November 1922

Chase, Edna Woolman and Chase, Ilka, *Always in Vogue* (New York: Doubleday & Company, Inc., 1954)

Cherol, John, 'Historic Architecture: Richard Morris Hunt, Mr and Mrs William K. Vanderbilt's Marble House in Newport', *Architectural Digest*, October 1985

Churchill, Allen, *The Upper Crust* (Englewood Cliffs, New Jersey: Prentice Hall, 1970)

—— *The Splendor Seekers* (New York: Grosset and Dunlap, 1974)

Churchill, Randolph S., *Winston S. Churchill, Volume I, Companion Part I, 1874–1896* (London: Heinemann, 1967)

Churchill, Winston S., *My Early Life* (London: Mandarin, Octopus, 1990; first published London: Thornton Butterworth, 1930)

Churchill, Winston S. and Martindale, C. C., *Charles IXth Duke of Marlborough K. G.: Tributes* (London: Burns Oates & Washbourne Ltd, 1934)

Churchill, Viscount, *All My Sins Remembered* (London: William Heinemann Ltd, 1964)

Clifford, Colin, *The Asquiths* (London: John Murray, 2002)

Clews, Henry, *Fifty Years in Wall Street* (New York: Irving Publishing Co., 1908)

Cornwallis-West, George, *Edwardian Hey-Days* (London and New York: Putnam, 1930)

Cornwallis-West, Mrs George (Lady Randolph Churchill), *The Reminiscences of Lady Randolph Churchill* (Bath: Cedric Chivers Ltd, 1973; first published London: Edward Arnold, 1908)

Croffut, W. A., *The Vanderbilts and the Story of Their Fortune* (Chicago: Bedford, Clarke and Co., 1886)

Crook, J. Mordaunt, *The Rise of the Nouveaux Riches: Style and Status in Victorian and Edwardian Architecture* (London: John Murray, 1999)

Crowinshield, Frank, 'The House of Vanderbilt', *Vogue*, 15 November 1941

D'Emilio, John and Freedman, Estelle B., *Intimate Matters: A History of Sexuality in America* (New York: Harper and Row Publishers, 1988)

de Courcy, Anne, *Circe: The Life of Edith, Marchioness of Londonderry* (London: Sinclair-Stevenson, 1992)

de Gramont, E. (Clermont-Tonnerre, E. de), *Pomp and Circumstance* (London: Jonathan Cape, 1929)

De Leon, Thomas Cooper, *Belles, Beaux and Brains of the 60s* (London: T. Fisher Unwin, 1909)

Downing, Antoinette, F. and Scully Jr., Vincent J., *The Architectural Heritage of Newport, Rhode Island 1640–1915* (New York: Clarkson N. Potter Inc., 1967)

Drexel Lehr, Elizabeth, *King Lehr and The Gilded Age* (London: Constable and Co., 1935)

—— (Lady Decies), *The Turn of the World* (Philadelphia: J. B. Lippincott Co., 1937)

Dwight, Eleanor, *Edith Wharton: An Extraordinary Life* (New York: Harry N. Abrams Inc., 1994)

Eliot, Elizabeth, *Heiresses and Coronets: The Story of Lovely Ladies and Noble Men* (New York: McDowell, Obolensky, 1959)

Elliott, Maud Howe, *This Was My Newport* (Cambridge, Mass.: The Mythology Co., University Press Inc., 1944)

Fielding, Daphne, *The Face on the Sphinx: A Portrait of Gladys Deacon, Duchess of Marlborough* (London: Hamish Hamilton, 1978)

Flammer, Philip, *The Vivid Air: The Lafayette Escadrille* (Athens: University of Georgia Press, 1981)

Flexner, Eleanor, *Century of Struggle: The Woman's Rights Movement in the United States* (Cambridge, Mass.: Harvard University Press, 1973; first published 1959)

Ford, Frank Lewis, 'The Vanderbilts and the Vanderbilt Millions', in *Munsey's Magazine*, Vol. XXII, January 1900

Foreman, John and Stimson, Robbe Pierce, *The Vanderbilts and the Gilded Age: Architectural Aspirations 1879–1901* (New York: St Martin's Press, 1991)

Fortescue, Winifred, *There's Rosemary, There's Rue* (London: Black Swan, Transworld, 1993; first published London: William Blackwood, 1939)

Fowler, Marian, *Blenheim: Biography of a Palace* (London: Viking, 1989)

—— *In a Gilded Cage: From Heiress to Duchess* (Toronto: Random House, 1993)

Friedman, B. H., *Gertrude Vanderbilt Whitney* (Garden City, New York: Doubleday & Company, Inc., 1978)

Fuller, Hiram, *Belle Brittan on a Tour, at Newport, and Here and There* (New York: Derby & Jackson, 1858)

Gilbert, Sir Martin, *Winston S. Churchill: Volume V, Companion Part 3: Documents, The Coming of War 1936–1939* (London: Heinemann, 1982)

—— *Winston S. Churchill, 1941–45, Vol. VII: Road to Victory* (London: Heinemann, 1986)

Gilmour, David, *Curzon* (London: John Murray, 1994)

Gray, Christopher, 'Au Revoir to the French Style', *Avenue*, June-August 1987

Green, David, *The Churchills of Blenheim* (London: Constable and Company Ltd, 1984)

Hammack, David C., *Power and Society: Greater New York at the Turn of the Century* (New York: Russell Sage Foundation, 1982)
Hattersley, Roy, *The Edwardians* (London: Little, Brown, 2004)
Hinks, Peter, *Nineteenth Century Jewellery* (London: Faber, 1975)
Hollis, Patricia, *Ladies Elect: Women in English Local Government 1865–1914* (Oxford: The Clarendon Press, 1987)
Homberger, Eric, *Mrs Astor's New York: Money and Social Power in a Gilded Age* (New Haven: Yale University Press, 2002)
Horne, Alistair, *The Fall of Paris* (London: Pan Macmillan Ltd, 2002; first published London: Macmillan, 1965)
Horne, Gerald, 'Blenheim Fifty Years Ago: Memoirs of Gentleman's Service', in *Country Life*, February 1945
Howarth, Patrick, *When the Riviera Was Ours* (London: Routledge & Kegan Paul, 1977)
Hoyt, Edwin Palmer, *The Vanderbilts and Their Fortunes* (New York: Doubleday and Co., 1962)
Irwin, Inez Haynes, *The Story of Alice Paul and the National Woman's Party* (Fairfax, Virginia: Denlinger's Publishers Ltd, 1977; first published 1964)
Jackson, Dora Jane, *From Slave to Siren: the Victorian Woman and Her Jewellery from Neoclassic to Art Nouveau* (Durham, North Carolina: Duke University Museum of Art, 1971)
James, Henry, 'The Sense of Newport', *Harper's Monthly Magazine*, CXIII, no. 675 (43), 1880
Jefferys, C. P. B., *Newport: A Short History* (Newport, Rhode Island: Newport Historical Society, 1992)
Jenkins, Roy, *Churchill* (Basingstoke: Macmillan, 2001)
Johnson, Paul (ed.), *Twentieth-Century Britain: Economic, Social and Cultural Change* (London: Longman, 1994)
Josephson, Matthew, *The Robber Barons* (New York: Harcourt, Brace and Co., 1934)
Juergens, George, *Joseph Pulitzer and The New York World* (Princeton, New Jersey: Princeton University Press, 1966)
Kedward, H. R., *Resistance in Vichy France* (Oxford: Oxford University Press, 1978)
Kehoe, Elisabeth, *Fortune's Daughters: The Extravagant Lives of the Jerome Sisters: Jennie Churchill, Clara Frewen, and Leonie Leslie* (London: Atlantic Books, 2004)
Kunz, George Frederick and Stevenson, Charles Hugh, *The Book of the Pearl: The History, Art, Science and Industry of the Queen of Gems* (New York: Dover, 1993; first published London: Macmillan, 1908)
Lambert, Angela, *Unquiet Souls: The Indian Summer of the British Aristocracy, 1880–1918* (London: Macmillan, 1984)
—— *1939: The Last Season of Peace* (London: Weidenfeld and Nicholson, 1989)

Lane, Wheaton J., *Commodore Vanderbilt: Epic of the Steam Age* (New York: Alfred A. Knopf, 1942)
Lawford, Valentine, *Vogue's Book of Houses, Gardens, People* (London: Bodley Head, 1968)
—— *Horst: His Work and His World* (New York: Viking, 1985)
Lees-Milne, James, *Ancestral Voices* (London: Chatto and Windus, 1975)
Leslie, Anita, *Edwardians in Love* (London: Hutchinson, 1972)
—— *The Gilt and The Gingerbread: an Autobiography* (London: Hutchinson, 1981)
Lespargot, Alain, *Histoire du Château d'Augerville-la-Rivière* (Bulletin de la Société Archéologique de Puiseaux, No. 17, 1993)
Logan, Andy, *The Man who Robbed the Robber Barons* (London: Victor Gollancz, 1966)
Lowe, David, *Beaux Arts New York* (New York: Whitney Library of Design, Watson-Gupthill Publications, 1998)
Lundberg, Ferdinand, *America's 60 Families* (New York: The Vanguard Press, 1937)
MacColl, Gail and Wallace, Carol McD., *To Marry an English Lord* (New York: Workman Publishing, 1989)
Mackay, Robert B., Baker, Anthony K., and Traynor, Carol A., (eds) *Long Island Country Houses and their Architects* (New York: Society for the Preservation of Long Island Antiquities in association with W. W. Norton and Co., 1997)
Magnus, Philip, *King Edward VII* (London: John Murray Ltd, 1964)
Mandler, Peter, *The Fall and Rise of the Stately Home* (New Haven and London: Yale University Press, 1997)
Marlborough, Laura, Duchess of, *Laughter from a Cloud* (London: Weidenfeld and Nicolson, 1980)
Martin, Edward Sandford, *The Unrest of Women* (New York and London: D. A. Appleton and Co., 1913)
Martin, Ralph, *Lady Randolph Churchill: A Biography*. Vol. II, 1969–70 (London: Cassell)
Mason, Herbert Malloy, *The Lafayette Escadrille* (New York: Random House, 1964)
Maxwell, Elsa, *I Married The World* (London: William Heinemann, 1955)
—— *The Celebrity Circus* (London: W. H. Allen, 1964)
Mayhall, Laura E. Nym, *The Militant Suffrage Movement: Citizenship and Resistance in Britain, 1860–1930* (Oxford: Oxford University Press, 2003)
McAllister, Ward, *Society as I Have Found It* (first published New York: Arno Press, 1975; first published Cassell Publishing Co., 1890)
McCarthy, Kathleen D., (ed.) *Lady Bountiful Revisited: Women,*

Philanthropy and Power (New Brunswick: Rutgers University Press, 1990)

Miller, Paul F., 'The Gothic Room in Marble House', *Antiques*, August 1994

Montgomery, Maureen E., *'Gilded Prostitution': Status, Money and Transatlantic Marriages, 1870–1914* (London and New York: Routledge, 1989)

Montgomery-Massingberd, Hugh, *Blenheim Revisited: the Spencer-Churchills and Their Palace* (London: The Bodley Head, 1985)

Moran, Lord, *Winston Churchill: The Struggle for Survival* (London: Heron Books, 1966)

Morand, Paul, *Nouvelles Complètes, Vol. I*, edited by Michel Collomb (Paris: Éditions Gallimard, 1992)

Morris, Sylvia Jukes, *Rage For Fame: The Ascent of Clare Boothe Luce* (New York: Random House, 1997)

Mott, Frank Luther, *American Journalism: A History of Newspapers in the United States through 250 years, 1690–1940* (New York: The Macmillan Company, 1941)

Muir, Robin and Pepper, Terence, *Horst: Portraits* 2001 (London: National Portrait Gallery, published in 2001 to accompany the exhibition 'Horst Portraits: 60 Years of Style', 1 March-3 June)

Murphy, Sophia, *The Duchess of Devonshire's Ball* (London: Sidgwick & Jackson, 1984)

Musson, Jeremy, *The English Manor House: from the Archives of Country Life* (London: Aurum Press, 1999)

Myers, Gustavus, *History of the Great American Fortunes*, Vol II (Chicago: Charles H. Kerr & Company, 1909–1910)

Newton, Lord, *Lord Lansdowne: A Biography* (London: Macmillan and Co. Ltd, 1929)

Nicholson, Nigel, *Mary Curzon* (London: Phoenix Orion, 1998; first published London: Weidenfeld and Nicolson, 1977)

Obolensky, Serge, *One Man in His Time: the Memoirs of Serge Obolensky* (New York: McDowell, Obolensky, 1958)

Patterson, Jerry E., *The Vanderbilts* (New York: Harry N. Abrams Inc., 1989)

—— *The First Four Hundred: Mrs Astor's New York in the Gilded Age* (New York: Rizzoli Publications, 2000)

Pilpel, Robert H., *Churchill in America, 1895–1961: An Affectionate Portrait* (London: New English Library, 1977; first published New York: Harcourt Brace Jovanovich 1976)

Princess Daisy of Pless, *From My Private Diary* (London: John Murray, 1931)

Prochaska, F. K., *Women and Philanthropy in Nineteenth-Century England* (Oxford: Clarendon Press, 1980)

Proddow, Penny and Fasel, Marion, *Diamonds: A Century of Spectacular Jewels* (New York: Harry N. Abrams Inc., 1996)
Pryce-Jones, Alan, 'The Golden Age of Newport', *House and Garden*, June 1987
Pugh, Martin, *The March of the Women: A Revisionist Analysis of the Campaign for Women's Suffrage, 1866–1914* (Oxford: Oxford University Press, 2000)
—— *The Pankhursts* (London: Allen Lane, The Penguin Press, 2001)
Rector, Margaret Hayden, *Alva, That Vanderbilt-Belmont Woman: Her Story as She Might Have Told It* (Wickford, Rhode Island: Dutch Island Press, 1992)
Riggs, Douglas, *Keelhauled: Unsportsmanlike Conduct and the America's Cup* (London: Stanford Maritime, 1986)
Rosen, Andrew, *Rise Up Women! The Militant Campaign of the Women's Social and Political Union, 1903–1914* (London: Routledge and Kegan Paul, 1974)
Rousset-Charny, Gerard, *Les Palais Parisiens de la Belle Époque* (Paris: Délégation à l'action artistique de la Ville de Paris, *c.*1990)
Rowse, A. L., *The Early and The Later Churchills* (London: Macmillan and Co., 1958)
Russell, Vivian, *Gardens of the Riviera* (Boston, New York: Little, Brown and Co., 1993)
Sackville-West, Vita, *The Edwardians* (London: The Hogarth Press, 1930)
Schlichting, Kurt C., *Grand Central Terminal: Railroads, Engineering and Architecture in New York City* (Baltimore and London: The John Hopkins University Press, 2001)
Searle, G. R., *The Quest for National Efficiency: A Study in British Politics and Political Thought 1899–1914* (Oxford: Basil Blackwell, 1971)
—— *The Liberal Party: Triumph and Disintegration 1886–1929* (Basingstoke: Macmillan, 1992)
Silver, David A., *The Entrepreneurial Life* (New York: John Wiley and Sons, 1986)
Silverman, Debora, *Selling Culture: Bloomingdale's, Diana Vreeland and the New Aristocracy of Taste in Reagan's America* (New York: Pantheon Books, 1986)
Singer, Anne, *Paul Maze, the Lost Impressionist* (London: Aurum Press, 1983)
Sirkis, Nancy, *Newport: Pleasures and Palaces*, with an introduction by Louis Auchincloss (New York: Viking Press Inc., 1963)
Sledge, John, 'Alabama's Bengal Tiger: Alva Smith Vanderbilt Belmont', *Alabama Heritage*, Spring 1997
Sloane, Florence Adele (with a commentary by Louis Auchincloss), *Maverick In Mauve* (Garden City, New York: Doubleday & Co., 1983)
Slocum, Eileen G., 'Memories of Bellevue Avenue: The Story of a

Newport Family', *Newport History, Bulletin of the Newport Historical Society*, 1995, 67, Part I

Soames, Mary, *Winston Churchill: His Life as a Painter* (London: Collins, 1990)

—— *Clementine Churchill* (London: Doubleday, Transworld, revised and updated 2002)

—— (ed.), *Speaking For Themselves* (London: Black Swan, 1999; first published London: Doubleday, 1998)

Stasz, Clarice, *The Vanderbilt Women: Dynasty of Wealth, Glamour, and Tragedy* (San Jose, New York: to Excel, 1999; first published New York: St Martin's Press, 1991)

Stevens, Doris, *Jailed For Freedom: American Women Win the Vote* (Troutdale: New Sage Press, 1995; first published New York: Boni and Liveright, 1920)

Stone, Herbert L., *The America's Cup Races* (New York: The Macmillan Co., 1930)

Strange, Michael, *Who Tells Me True* (New York: C. Scribner and Sons, 1940)

Stuart, Dennis, *Dear Duchess: Millicent Duchess of Sutherland 1867–1955* (London: Victor Gollancz, 1982)

Tintner, Adeline R., *Edith Wharton in Context: Essays on Intertextuality* (Tuscaloosa and London: The University of Alabama Press, 1999)

Tomalin, Claire, *Samuel Pepys: The Unequalled Self* (London: Viking, 2002)

Tuchman, Barbara W., *The Proud Tower* (New York: Ballentine Books, 1996; first published: The Macmillan Co., 1966)

Tuke, Margaret, *A History of Bedford College For Women, 1849–1937* (Oxford: Oxford University Press, 1939)

Turbeville, Deborah and Auchincloss, Louis, *Newport Remembered: A Photographic Portrait of a Gilded Past* (New York: Harry N. Abrams Inc., 1994)

Twain, Mark, 'Open Letter to Commodore Vanderbilt', *Packard's Monthly*, March 1869

Twain, Mark and Warner, Charles Dudley, *The Gilded Age* (New York and Oxford: Oxford University Press, 1996; first published Hartford: American Publishing Co., 1873)

Van Rensselaer, May King, *The Social Ladder* (London: Eveleigh Nash & Co., 1925)

Vanderbilt, Arthur T., *Fortune's Children: The Fall of the House of Vanderbilt* (New York: William Morrow and Co., 1989)

Vanderbilt Jr, Cornelius, *Queen of the Golden Age* (Maidstone: George Mann, 1999; first published New York: McGraw-Hill, 1956)

Vanderbilt, Gloria and Furness, Lady Thelma, *Double Exposure* (London: Frederic Muller Ltd, 1959)

Vanderbilt, Harold S., *Contract Bridge: Bidding and the Club Convention* (New York: Charles Scribner's Sons, 1929)

—— *On the Wind's Highway: Ranger, Rainbow and Racing* (New York: Charles Scribner's Sons, 1939)

Van Pelt, John Vredenburgh, *A Monography of the William K. Vanderbilt House, Richard Morris Hunt Architect* (New York: J. V. Van Pelt, 1925)

Vickers, Hugo, *Gladys, Duchess of Marlborough* (New York: Holt, Rinehart and Winston, 1979; first published London: Weidenfeld and Nicolson, 1979)

Vreeland, Diana, 'American Women of Style', exhibition catalogue (New York: Costume Institute, Metropolitan Museum of Art, 1975)

Warburton, Eileen, *In Living Memory: A Chronicle of Newport 1888–1988* (Newport, Rhode Island: Newport Savings and Loan Assoc. Island Trust Co., 1988)

Warwick, Frances, Countess of, *Afterthoughts* (London: Cassell and Co. Ltd, 1931)

Wecter, Dixon, *The Saga of American Society: A Record of Social Aspiration 1607–1937* (New York: C. Scribner's Sons, 1937)

Wharton, Edith, *The House of Mirth* (London: FilmFour Books, 2000; first published London: Macmillan & Co., 1905)

—— *The Age of Innocence* (London: Penguin, 1996; first published New York: D. Appleton and Co., 1920)

—— *The Custom of the Country* (London: Penguin, 1987; first published New York: C. Scribner and Sons, 1913)

—— *The Buccaneers*, with an introduction by Candace Waid (London: Everyman, 1993)

—— Margerie, Diane de and Caws, Mary Ann (eds), *French Ways and their Meaning* (Lenox, Mass., Lee, Mass.: Edith Wharton Restoration at the Mount, Berkshire House Publishers, 1997; first published London: Macmillan, 1919)

—— *A Backward Glance* (New York, London: D. Appleton-Century Co., 1934)

INDEX